AF509460

Health Promoting Effects of Phytochemicals

Health Promoting Effects of Phytochemicals

Editor

Baojun Xu

Basel • Beijing • Wuhan • Barcelona • Belgrade • Novi Sad • Cluj • Manchester

Editor
Baojun Xu
Department of Life Sciences
Beijing Normal University-
Hong Kong Baptist University
United International College
Zhuhai
China

Editorial Office
MDPI
St. Alban-Anlage 66
4052 Basel, Switzerland

This is a reprint of articles from the Special Issue published online in the open access journal *International Journal of Molecular Sciences* (ISSN 1422-0067) (available at: www.mdpi.com/journal/ijms/special_issues/health_effects_phytochemicals).

For citation purposes, cite each article independently as indicated on the article page online and as indicated below:

Lastname, A.A.; Lastname, B.B. Article Title. *Journal Name* **Year**, *Volume Number*, Page Range.

ISBN 978-3-0365-8777-6 (Hbk)
ISBN 978-3-0365-8776-9 (PDF)
doi.org/10.3390/books978-3-0365-8776-9

Contents

About the Editor

Baojun Xu

Dr. Xu is a Chair Professor in Beijing Normal University-Hong Kong Baptist University United International College (UIC, a full English-teaching college in China), Fellow of the Royal Society of Chemistry, Zhuhai Scholar Distinguished Professor, Head of the Department of Life Sciences, Program Director of Food Science and Technology Program, and an author of over 300 peer-reviewed papers. Dr. Xu received his Ph.D. in Food Science from Chungnam National University, South Korea. He conducted postdoctoral research work at North Dakota State University (NDSU), Purdue University, and the Gerald P. Murphy Cancer Foundation in the USA during the years 2005–2009. He completed short-term visiting research at NDSU in 2012 and the University of Georgia in 2014, followed by additional visiting research during his sabbatical leave (7 months) at Pennsylvania State University in the USA in 2016. Dr. Xu is serving as the Associate Editor-in-Chief of *Food Science and Human Wellness*, Associate Editor of *Food Research International*, Associate Editor of *Food Frontiers*, and is an Editorial Board Member of around 10 international journals. He received the inaugural President's Award for Outstanding Research of UIC in 2016 and President's Award for Outstanding Service of UIC in 2020. Dr. Xu was listed within the world's top 2% of scientists by Stanford University in 2020, 2021, and 2022, and he was listed as the best scientist in the world in the field of Biology and Biochemistry by Research.com in 2023.

International Journal of
Molecular Sciences

Correction

Correction: Ou-Yang, F. et al. Antiproliferation for Breast Cancer Cells by Ethyl Acetate Extract of *Nepenthes thorellii* x (*ventricosa* x *maxima*). *Int. J. Mol. Sci.* 2019, 20, 3238

Fu Ou-Yang [1,2], I-Hsuan Tsai [3], Jen-Yang Tang [4,5], Ching-Yu Yen [6,7], Yuan-Bin Cheng [8], Ammad Ahmad Farooqi [9], Shu-Rong Chen [8], Szu-Yin Yu [8], Jun-Kai Kao [10,11,12,*] and Hsueh-Wei Chang [2,3,13,14,15,*]

1   Division of Breast Surgery and Department of Surgery, Kaohsiung Medical University Hospital, Kaohsiung 80708, Taiwan; kmufrank@gmail.com
2   Cancer Center, Kaohsiung Medical University Hospital, Kaohsiung 80708, Taiwan
3   Department of Biomedical Science and Environmental Biology, Kaohsiung Medical University, Kaohsiung 80708, Taiwan; s0932961465@gmail.com
4   Department of Radiation Oncology, Faculty of Medicine, College of Medicine, Kaohsiung Medical University, Kaohsiung 80708, Taiwan; reyata@kmu.edu.tw
5   Department of Radiation Oncology, Kaohsiung Medical University Hospital, Kaohsiung 80708, Taiwan
6   Department of Oral and Maxillofacial Surgery Chi-Mei Medical Center, Tainan 71004, Taiwan; ycysmc@gmail.com
7   School of Dentistry, Taipei Medical University, Taipei 11050, Taiwan
8   Graduate Institute of Natural Products, Kaohsiung Medical University, Kaohsiung 80708, Taiwan; jmb@kmu.edu.tw (Y.-B.C.); highshorter@hotmail.com (S.-R.C.); s91412232@gmail.com (S.-Y.Y.)
9   Institute of Biomedical and Genetic Engineering (IBGE), Islamabad 44000, Pakistan; ammadfarooqi@rlmclahore.com
10   Institute of Biomedical Sciences, National Chung Hsing University, Taichung 40227, Taiwan
11   Pediatric Department, Children's Hospital, Changhua Christian Hospital, Changhua 50006, Taiwan
12   School of Medicine, Kaohsiung Medical University, Kaohsiung 80708, Taiwan
13   Drug Development and Value Creation Research Center, Kaohsiung Medical University, Kaohsiung 80708, Taiwan
14   Institute of Medical Science and Technology, National Sun Yat-sen University, Kaohsiung 80424, Taiwan
15   Department of Medical Research, Kaohsiung Medical University Hospital, Kaohsiung 80708, Taiwan
*   Correspondence: 96777@cch.org.tw (J.-K.K.); changhw@kmu.edu.tw (H.-W.C.); Tel.: +886-4-723-8595 (ext. 1905) (J.-K.K.); +886-7-312-1101 (ext. 2691) (H.-W.C.); Fax: +886-4-723-8847 (J.-K.K.); +886-7-312-5339 (H.-W.C.)

Citation: Ou-Yang, F.; Tsai, I.-H.; Tang, J.-Y.; Yen, C.-Y.; Cheng, Y.-B.; Farooqi, A.A.; Chen, S.-R.; Yu, S.-Y.; Kao, J.-K.; Chang, H.-W. et al. Correction: Ou-Yang, F. et al. Antiproliferation for Breast Cancer Cells by Ethyl Acetate Extract of *Nepenthes thorellii* x (*ventricosa* x *maxima*). *Int. J. Mol. Sci.* 2019, 20, 3238. *Int. J. Mol. Sci.* **2021**, 22, 668. https://doi.org/10.3390/ijms22020668

Received: 16 November 2020
Accepted: 17 November 2020
Published: 12 January 2021

Publisher's Note: MDPI stays neutral with regard to jurisdictional claims in published maps and institutional affiliations.

The authors would like to make corrections to their published paper [1].

There were mistakes in some usages of the chemical name "isoplumbagin" in the original version in Sections 2.1, 3.1, and 3.2. These "isoplumbagin" words should be changed to "plumbagin".

Literature reported that the difference between plumbagin and isoplumbagin is the coupling constant of methyl group (1.5 Hz for plumbagin and 1.2 Hz for isoplumbagin) [2]. We found that the ^{1}H and ^{13}C spectra data (Figures 1 and 2) of the main compound of EANT is plumbagin because our compound shows a J value of 1.5. Therefore, we confirmed the major compound of EANT to be "plumbagin" based on spectra data rather than isoplumbagin.

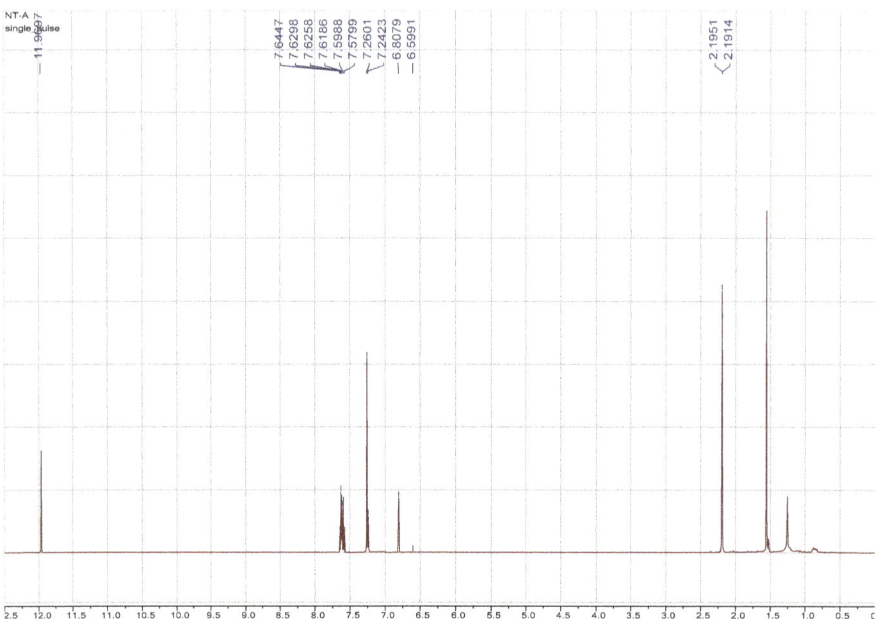

Figure 1. Spectrum data of ^{1}H of the major compound isolated from EANT.

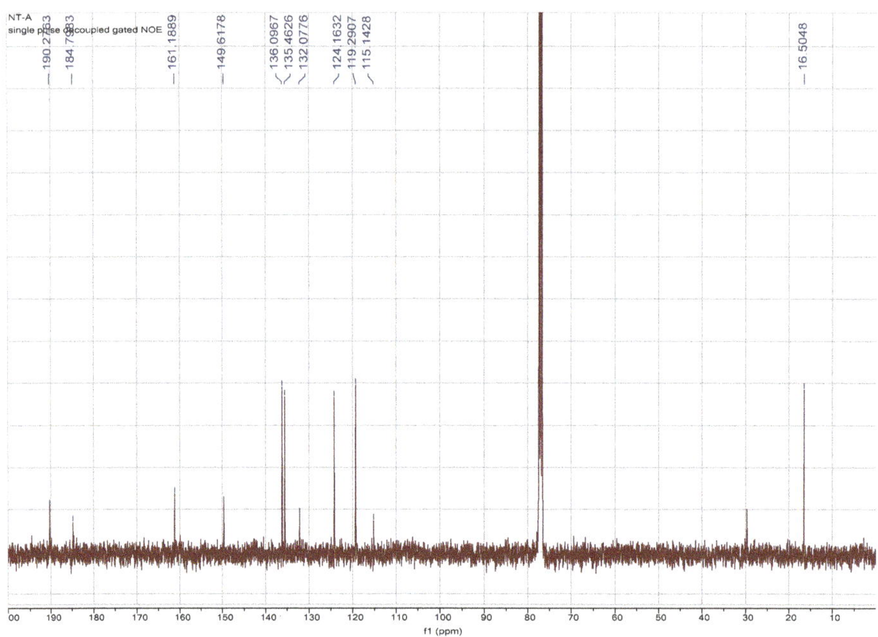

Figure 2. Spectrum data of ^{13}C of the major compound isolated from EANT.

We also further performed the detailed physical characters such as melting temperature. The melting points of these two compounds are different (74–75 °C for plumbagin
and 158–159 °C for isoplumbagin) [2]. After checking the melting point (77–78 °C) of the

major compound isolated from EANT, we realized it should be plumbagin rather than isoplumbagin.

Additionally, the paragraph for the Supplementary Materials also needs to be corrected due to missing words. The authors have corrected the error as shown below. The change does not affect the scientific results. The authors would like to apologize for any inconvenience that may have been caused to readers of the journal. The manuscript will be updated, and the original will remain online on the article webpage.

Please find the correct sentences below (only isoplumbagin in the original paper has been corrected to plumbagin):

2.1. The Identified Components from Fingerprint Profiles of EANT (Page 2: Line 2 of the First Paragraph)

According to HPLC fingerprinting assay (Supplementary Figure S1), the major bioactive components of EANT are plumbagin, *cis*-isoshinanolone, quercetin 3-*O*-(6″-*n*-butyl β-D-glucuronide), and fatty acids.

3.1. EANT Preferentially Inhibits Proliferation of Breast Cancer Cells (page 8: line 6 of the Second Paragraph of 3.1)

In the current study, we found that the major bioactive components of EANT identified by HPLC fingerprinting method were plumbagin [21], *cis*-isoshinanolone [22], and quercetin 3-*O*-(6″-*n*-butyl β-D-glucuronide) [23] (Supplementary Figure S1).

3.2. EANT Induces Oxidative Stress on Breast Cancer Cells (Page 8: line 1 of the Third Paragraph of 3.2)

Plumbagin is a common naphthoquinone in Nepenthes. Moreover, plumbagin but not isoplumbagin is identified in EANT.

Supplementary Materials (page 11)

The following are available online at http://www.mdpi.com/1422-0067/20/13/32 38/s1, Table S1: The HPLC method for fingerprint profile of *N. thorellii* x (*ventricosa* x *maxima*), Figure S1: Components of EANT. (A) Fingerprint profile of EANT. It is monitored at 365 nm. (B) Retention time of plumbagin (NT-A). Volume is 50 μL. It is monitored at 400 nm. (C) Retention time of *cis*-isoshinanolone (NT-B). Volume is 10 μL. It is monitored at 254 nm.

References

1. Ou-Yang, F.; Tsai, I.H.; Tang, J.Y.; Yen, C.Y.; Cheng, Y.B.; Farooqi, A.A.; Chen, S.R.; Yu, S.Y.; Kao, J.K.; Chang, H.W. Antiproliferation for breast cancer cells by ethyl acetate extract of *Nepenthes thorellii* x (*ventricosa* x *maxima*). *Int. J. Mol. Sci.* **2019**, *20*, 3238. [CrossRef] [PubMed]
2. Wang, J.L.; Yang, Z.J.; Wang, J.J.; Tang, W.X.; Zhao, M.; Zhang, S.J. Synthesis of juglone and its derivatives. *Appl. Mech. Mater.* **2012**, *138–139*, 1139–1141. [CrossRef]

International Journal of
Molecular Sciences

Article

Antiproliferation for Breast Cancer Cells by Ethyl Acetate Extract of *Nepenthes thorellii* x (*ventricosa* x *maxima*)

Fu Ou-Yang [1,2], I-Hsuan Tsai [3], Jen-Yang Tang [4,5], Ching-Yu Yen [6,7], Yuan-Bin Cheng [8], Ammad Ahmad Farooqi [9], Shu-Rong Chen [8], Szu-Yin Yu [8], Jun-Kai Kao [10,11,12,*] and Hsueh-Wei Chang [2,3,13,14,15,*]

[1] Division of Breast Surgery and Department of Surgery, Kaohsiung Medical University Hospital, Kaohsiung 80708, Taiwan
[2] Cancer Center, Kaohsiung Medical University Hospital, Kaohsiung 80708, Taiwan
[3] Department of Biomedical Science and Environmental Biology, Kaohsiung Medical University, Kaohsiung 80708, Taiwan
[4] Department of Radiation Oncology, Faculty of Medicine, College of Medicine, Kaohsiung Medical University, Kaohsiung 80708, Taiwan
[5] Department of Radiation Oncology, Kaohsiung Medical University Hospital, Kaohsiung 80708, Taiwan
[6] Department of Oral and Maxillofacial Surgery Chi-Mei Medical Center, Tainan 71004, Taiwan
[7] School of Dentistry, Taipei Medical University, Taipei 11050, Taiwan
[8] Graduate Institute of Natural Products, Kaohsiung Medical University, Kaohsiung 80708, Taiwan
[9] Institute of Biomedical and Genetic Engineering (IBGE), Islamabad 44000, Pakistan
[10] Institute of Biomedical Sciences, National Chung Hsing University, Taichung 40227, Taiwan
[11] Pediatric Department, Children's Hospital, Changhua Christian Hospital, Changhua 50006, Taiwan
[12] School of Medicine, Kaohsiung Medical University, Kaohsiung 80708, Taiwan
[13] Drug Development and Value Creation Research Center, Kaohsiung Medical University, Kaohsiung 80708, Taiwan
[14] Institute of Medical Science and Technology, National Sun Yat-sen University, Kaohsiung 80424, Taiwan
[15] Department of Medical Research, Kaohsiung Medical University Hospital, Kaohsiung 80708, Taiwan
* Correspondence: changhw@kmu.edu.tw (H.-W.C.); 96777@cch.org.tw (J.-K.K.);
Tel.: +886-7-312-1101 (ext. 2691) (H.-W.C.); +886-4-723-8595 (ext. 1905) (J.-K.K.);
Fax: +886-7-312-5339 (H.-W.C.); +886-4-723-8847 (J.-K.K.)

Received: 26 April 2019; Accepted: 28 June 2019; Published: 1 July 2019

Abstract: Extracts from the Nepenthes plant have anti-microorganism and anti-inflammation effects. However, the anticancer effect of the Nepenthes plant is rarely reported, especially for breast cancer cells. Here, we evaluate the antitumor effects of the ethyl acetate extract of *Nepenthes thorellii* x (*ventricosa* x *maxima*) (EANT) against breast cancer cells. Cell viability and flow cytometric analyses were used to analyze apoptosis, oxidative stress, and DNA damage. EANT exhibits a higher antiproliferation ability to two breast cancer cell lines (MCF7 and SKBR3) as compared to normal breast cells (M10). A mechanistic study demonstrates that EANT induces apoptosis in breast cancer cells with evidence of subG1 accumulation and annexin V increment. EANT also induces glutathione (GSH) depletion, resulting in dramatic accumulations of reactive oxygen species (ROS) and mitochondrial superoxide (MitoSOX), as well as the depletion of mitochondrial membrane potential (MMP). These oxidative stresses attack DNA, respectively leading to DNA double strand breaks and oxidative DNA damage in γH2AX and 8-oxo-2′deoxyguanosine (8-oxodG) assays. Overall these findings clearly revealed that EANT induced changes were suppressed by the ROS inhibitor. In conclusion, our results have shown that the ROS-modulating natural product (EANT) has antiproliferation activity against breast cancer cells through apoptosis, oxidative stress, and DNA damage.

Keywords: breast cancer; oxidative stress; carnivorous plants; natural product

---

1. Introduction

Breast cancer is the most common cancer in women, comprises 30% of all new female cancer cases, and accounts for 15% of all female cancer-related deaths [1]. Breast cancer is complex and an effective cure remains elusive. Resistance to clinical drugs and radiation reduces the therapeutic effect against breast cancer and is partly attributed to the loss of apoptosis function [2]. Accordingly, apoptosis-inducing drugs may improve the therapeutic effect against breast cancer cells.

Accumulating evidence suggests that natural products contain many bioactive components for cancer prevention and therapy, and sometimes provide advantages over isolated compounds. This is partly because natural products consist of many bioactive components that can suppress the function of multiple targets [3]. Natural products provide valuable resources for drug discovery for breast cancer treatment through apoptosis [2,4–7]. For example, *Phyla nodiflora* L. extracts were found to inhibit cell proliferation by inducing apoptosis in human breast cancer cells [2]. The study of additional natural products for drug discovery against breast cancer cells is thus warranted.

Nepenthes (also known as tropical pitcher plants) are tropical carnivorous plants. *Nepenthes* comprise a number of natural and cultivated hybrids and present diverse species development. Many species of *Nepenthes* are commonly used in herbal medicine in several Southeast Asian countries [8]. Some types of extracts from *Nepenthes* are known to have anti-bacterial and anti-fungal properties. For example, methanolic extract of *N. bicalcarata* inhibited growth of gram-positive bacteria (*Staphylococcus aureus*, *Bacillus subtilis* and *B. spizizenii*) with the minimum inhibitory concentration (MIC) at 256 µg/mL [9]. Hexane extract of *N. ventricosa* x *maxima* inhibited the growth of several species of fungi such as *Alternaria alternata*, *Aspergillus niger*, *Bipolaris oryzae*, *Fusarium oxysporum*, *Phytophthora capsici*, *Rhizoctonia solani*, *Rhizopus stolonifer* var. stolonifera, and *Sclerotinia sclerotiorum* with MIC values ranging from 7.2 to 43.7 µg/mL [10]. *Nepenthes* extracts have been reported to suppress inflammation [11]. Anti-inflammation drugs frequently show anticancer effects [12–14]. Recently, the methanol extract and its sequential partitions of *N. alata* Blanco as well as its bioactive compound plumbagin demonstrated the anti-breast-cancer effect [15]. Therefore, we hypothesize that extracts from other *Nepenthes hybrid* may have an anticancer effect against breast cancer cells.

This study evaluates the antiproliferation effect from an ethyl acetate extract of *Nepenthes thorellii x (ventricosa x maxima)* (EANT) on breast cancer cells. The underlying mechanisms of antiproliferation (e.g., cell viability, apoptosis, oxidative stress, and DNA damage) were determined on breast cancer cells following EANT treatment.

2. Results

2.1. The Identified Components from Fingerprint Profiles of EANT

According to HPLC fingerprinting assay (Supplementary Figure S1), the major bioactive components of EANT are plumbagin, *cis*-isoshinanolone, quercetin 3-*O*-(6″-*n*-butyl β-D-glucuronide), and fatty acids. In the present study, we focused on the antiproliferation effect of breast cancer cells using EANT and therefore the detailed involvement of these bioactive components on anti-breast cancer effect was not addressed.

2.2. EANT Preferentially Inhibits Viability of Breast Cancer Cells

In Figure 1A, EANT dose-dependently more significantly reduces the viability (%) of two breast cancer cells (MCF7 and SKBR3) than normal breast cells (M10). In Figure 1B, the viability reducing effects of EANT are suppressed by the inhibitor pretreatments for ROS (NAC). Accordingly, the role of oxidative stress and apoptosis in EANT-treated breast cancer cells warrants further detailed investigation. For comparison to clinical drugs, the cell viability of breast cancer cells treated with cisplatin was performed (Figure 1C). The drug sensitivity of EANT in breast cancer cells is higher than that of cisplatin at 24 h treatment. The drug sensitivity of cisplatin in breast cancer cells at 48 and 72 h is higher than that of 24 h treatment.

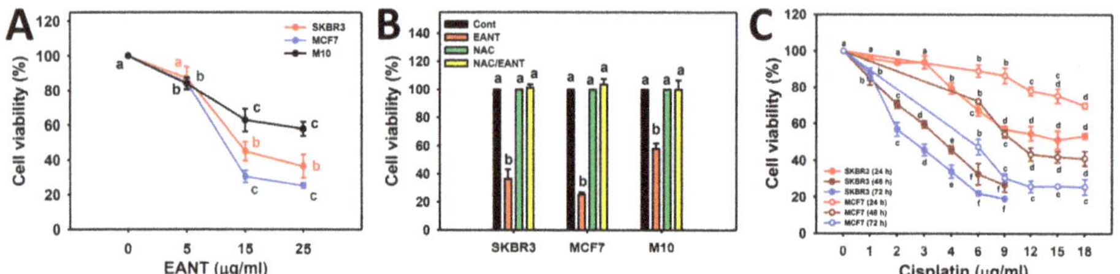

Figure 1. Cell viability following ethyl acetate extract of *Nepenthes thorellii x (ventricosa x maxima)* (EANT) treatment. (**A**) Cell viability of breast cancer cells (MCF7 and SKBR3) and breast normal cells (M10) treated with 0 (control with DMSO only), 5, 15, and 25 µg/mL of EANT for 24 h. (**B**) Cell viability of breast cancer cells after NAC pretreatment (2 mM for 1 h) and EANT post-treatment (25 µg/mL for 24 h), i.e., NAC/EANT. (**C**) Cell viability of breast cancer cells treated with different concentrations of cisplatin for 24, 48, and 72 h. For each cell line, treatments labeled without the same lower-case letters indicate significant difference. $p < 0.05 \sim 0.0001$. Data, mean $\pm$ SD ($n = 3$).

2.3. EANT Changes Cell Cycle Distribution in Breast Cancer Cells

Figure 2A shows the flow cytometry patterns of cell cycle distribution in breast cancer cells (MCF7 and SKBR3) without (up) or with (down) NAC pretreatment. In Figure 2B, the subG1 and G2/M population gradually accumulates and the G1 population gradually decreases in breast cancer cells after EANT treatments. After NAC pretreatments, the subG1 accumulation and cell cycle disturbance recover to the normal distribution as control.

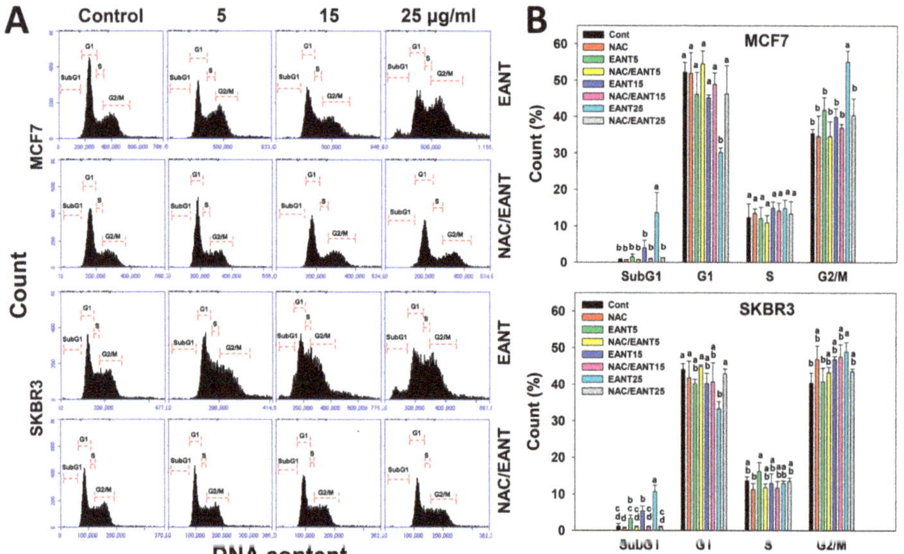

Figure 2. Cell cycle change after EANT treatment. (**A,B**) Cell cycle distribution patterns and statistics. Without or with NAC pretreatment, breast cancer cells (MCF7 and SKBR3) were treated with 0 (control with DMSO only), 5, 15, and 25 µg/mL of EANT for 24 h, i.e., EANT vs. NAC/EANT. For each cycle

phase, treatments labeled without the same lower-case letters indicate significant difference. $p < 0.05{\sim}0.0001$. Data, mean $\pm$ SD ($n = 3$). Positive controls for subG1 accumulation and G2/M arrest were provided in the Supplementary Figure S2A,B.

2.4. EANT Induces Apoptosis in Breast Cancer Cells

The possibility that subG1 accumulation may lead to apoptosis was further examined by flow cytometry. Figure 3A shows the flow cytometry patterns of annexin V/7AAD in breast cancer cells (MCF7 and SKBR3). In Figure 3B (top part), the early apoptosis (%) (annexin V (+)/7AAD (-)) of MCF7 cells is dramatically increased to about 80% in 15 µg/mL of EANT and its late apoptosis (%) (annexin V (+)/7AAD (+)) is increased to 20% compared to the control. In Figure 3B (bottom part), the early and late apoptosis (%) of SKBR3 cells is only mildly increased in 15 µg/mL of EANT compared to the control. In a higher concentration (25 µg/mL), EANT is more likely to induce late apoptosis than early apoptosis in both breast cancer cells.

Figure 3C,E illustrate the roles of ROS and apoptosis effects on flow cytometry patterns of annexin V/7AAD in breast cancer cells (MCF7 and SKBR3). At 12 h of EANT treatment, the early apoptosis population exceeds the late apoptosis population in breast cancer cells. At 24 h of EANT treatment, the early apoptosis population shifts to late apoptosis in breast cancer cells. In Figure 3D,F, the early apoptosis population, that is, annexin V(+)/7AAD (−) (%), increases dramatically for 12 h and then declines at 24 h in breast cancer cells. The late apoptosis population, that is, annexin V(+)/7AAD (+) (%), increases dramatically for 12 and 24 h in breast cancer cells.

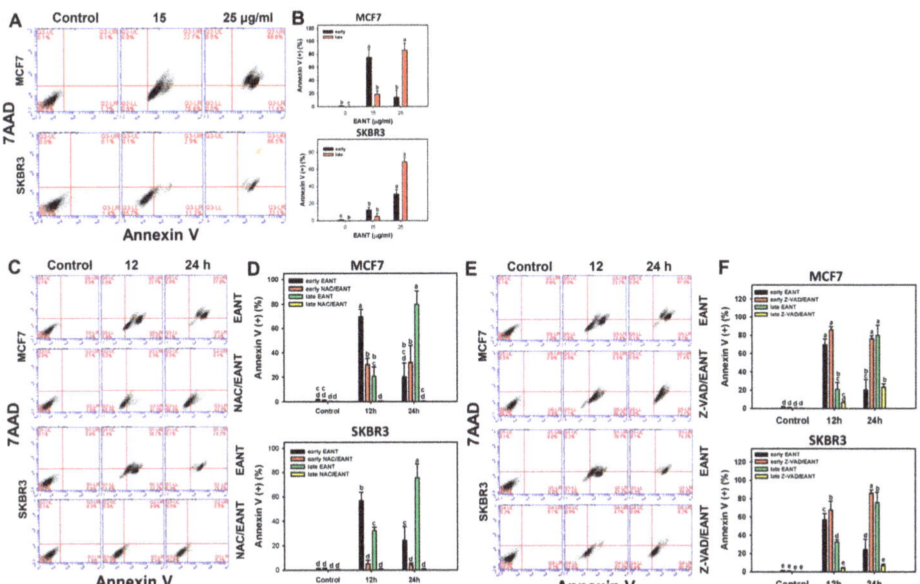

Figure 3. Apoptosis change of annexin V/7AAD after EANT treatment. (**A,B**) Concentration effect of EANT on Annexin V/7AAD patterns and statistics. Breast cancer cells (MCF7 and SKBR3) were treated with control with DMSO only and EANT (15 and 25 µg/mL) for 24 h. Annexin V (+)/7AAD (−) and annexin V (+)/7AAD (+) were respectively regarded as early and later apoptosis. (**C–F**) Time course effect of EANT on Annexin V/7AAD patterns and statistics. Without or with (**C,D**) NAC pretreatment or (**E,F**) Z-VAD pretreatment, breast cancer cells (MCF7 and SKBR3) were treated with 25 µg/mL of EANT for 0 (control with DMSO only), 12, and 24 h. For each cell line, treatments labeled without the same lower-case letters indicate significant difference. $p < 0.01{\sim}0.0001$. Data, mean $\pm$ SD ($n = 3$). Positive controls for apoptosis are provided in Supplementary Figure S2C,D.

For EANT post-treatment for 12 h, NAC or Z-VAD pretreatment inhibits both early and late apoptosis populations. For EANT post-treatment for 24 h, NAC or Z-VAD pretreatment inhibits late apoptosis populations and the late apoptosis population shifts to early apoptosis.

2.5. EANT Induces ROS Production and GSH Depletion in Breast Cancer Cells

Since NAC reverts the EANT induced changes of cell viability, cell cycle distribution, and apoptosis, the status of oxidative stress such as the ROS level in EANT-treated breast cancer cells warrants detailed investigation. Figure 4A,C respectively show flow cytometry patterns of ROS in breast cancer cells (MCF7 and SKBR3) without or with NAC pretreatment. In Figure 4B, the ROS (+) (%) is dramatically increased in both 15 and 25 μg/mL of EANT. In Figure 3D, the ROS (+) (%) increases dramatically within 5 min and is suppressed by NAC pretreatment in breast cancer cells (MCF7 and SKBR3).

Because glutathione (GSH) has a ROS-scavenging function [16], the status of GSH level in EANT-treated breast cancer cells warrants detailed investigation. Figure 4E shows flow cytometry patterns of GSH in breast cancer cells (MCF7 and SKBR3). In Figure 4F, the GSH (+) (%) is dramatically decreased in 25 μg/mL of EANT in breast cancer cells.

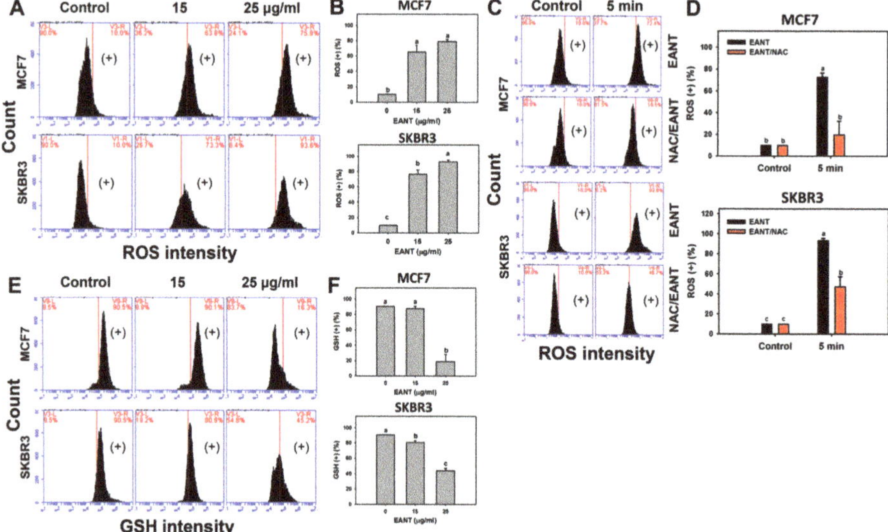

Figure 4. Reactive oxygen species (ROS) and glutathione (GSH) changes after EANT treatment. (**A,B**) Concentration effect of EANT on ROS patterns and statistics. Breast cancer cells (MCF7 and SKBR3) were treated with control with DMSO only and EANT (15 and 25 μg/mL) for 5 min. (**C,D**) Time course effect of EANT on ROS patterns and statistics. Without or with NAC pretreatment, breast cancer cells (MCF7 and SKBR3) were treated with control with DMSO only and 25 μg/mL of EANT for 5 min. (**E,F**) The concentration effect of EANT on GSH patterns and statistics. Breast cancer cells were treated with control with DMSO only and EANT for 24 h. For each cell line, treatments labeled without the same lower-case letters indicate significant difference. $p < 0.05{\sim}0.0001$. Data, mean ± SD ($n = 3$). (+) indicates the ROS or GSH (+) population. Positive controls for ROS and GSH were provided in Supplementary Figure S2E–H.

2.6. EANT Induces MitoSOX Production and MMP Reduction in Breast Cancer Cells

Since NAC reverts the EANT induced changes to cell viability, cell cycle distribution, and apoptosis, the status of oxidative stress such as MitoSOX and MMP levels in EANT-treated breast cancer cells warrants detailed investigation. Figure 5A shows flow cytometry patterns of MitoSOX in breast cancer cells (MCF7 and SKBR3). In Figure 5B, the MitoSOX (+) (%) is dramatically increased in both 15 and 25 μg/mL of EANT. Figure 5C,E respectively show flow cytometry patterns of MitoSOX in breast cancer cells (MCF7 and SKBR3) without or with NAC or MitoTEMPO. In Figure 5D,F, the MitoSOX (+) (%) of breast cancer cells are suppressed by the inhibitor pretreatments for ROS (NAC) or MitoSOX (MitoTEMPO).

Figure 5G shows flow cytometry patterns of MMP in breast cancer cells (MCF7 and SKBR3). In Figure 5H, the MMP (−) (%) is dramatically increased in both 15 and 25 µg/mL of EANT. Figure 5I shows the time course for flow cytometry patterns of MMP in breast cancer cells (MCF7 and SKBR3) without or with NAC. In Figure 5J, the MMP (−) (%) of breast cancer cells is increased at 12 and 24 h and is suppressed by the NAC pretreatment.

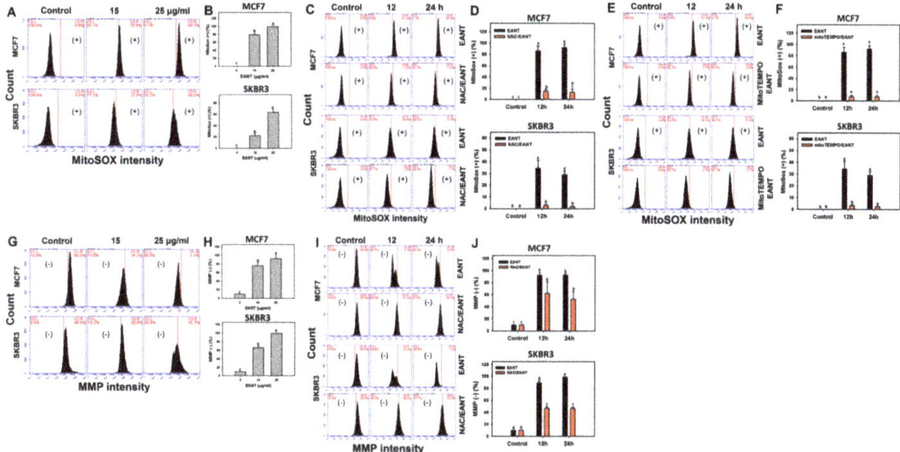

Figure 5. MitoSOX and MMP change after EANT treatment. (**A**,**B**) The concentration effect of EANT on MitoSOX patterns and statistics. Breast cancer cells (MCF7 and SKBR3 were treated with 0 (control with DMSO only), 15, and 25 µg/mL of EANT for 24 h. (**C**–**F**) Time course effect of EANT on MitoSOX patterns and statistics. Without or with (**C** and **D**) NAC pretreatment or (**E**,**F**) MitoTEMPO pretreatment, breast cancer cells (MCF7 and SKBR3) were treated with 25 µg/mL of EANT for 0 (control with DMSO only), 12, and 24 h. Annexin V (+) was counted to apoptosis. (**G**,**H**) The concentration effect of EANT on MMP patterns and statistics. Breast cancer cells (MCF7 and SKBR3) were treated with EANT for 24 h. (**I**,**J**) Time course effect of EANT on MMP patterns and statistics. Without or with NAC pretreatment, breast cancer cells (MCF7 and SKBR3) were treated with 25 µg/mL of EANT for 0 (control with DMSO only), 12, and 24 h. For each cell line, treatments labeled without the same lower-case letters indicate significant difference. $p < 0.05{\sim}0.0001$. Data, mean ± SD ($n = 3$). (+) or (−) respectively indicates the MitoSOX (+) or MMP (−) population. Positive controls for MitoSOX and MMP were provided in Supplementary Figure S2I–L.

2.7. EANT Induces DNA Damage in Breast Cancer Cells

ROS is a DNA damage factor [17] and the DNA damage status in EANT-treated breast cancer cells warrants detailed investigation. Figure 6A shows flow cytometry patterns of γH2AX in breast cancer cells (MCF7 and SKBR3). In Figure 6B, the γH2AX (+) (%) is dramatically increased in both 15 and 25 µg/mL of EANT. Figure 6C shows time course for flow cytometry patterns of γH2AX in breast cancer cells (MCF7 and SKBR3) without or with NAC pretreatment. In Figure 6D, the γH2AX (+) (%) of breast cancer cells are suppressed by NAC pretreatment.

Figure 6E shows flow cytometry patterns of 8-oxodG in breast cancer cells (MCF7 and SKBR3). In Figure 6F, the 8-oxodG (+) (%) is gradually increased in both 15 and 25 µg/mL of EANT. Figure 6G shows time course for flow cytometry patterns of 8-oxodG in breast cancer cells (MCF7 and SKBR3) without or with NAC pretreatment. In Figure 6H, the 8-oxodG (+) (%) of breast cancer cells is increased at 12 and 24 h and is suppressed by NAC pretreatment.

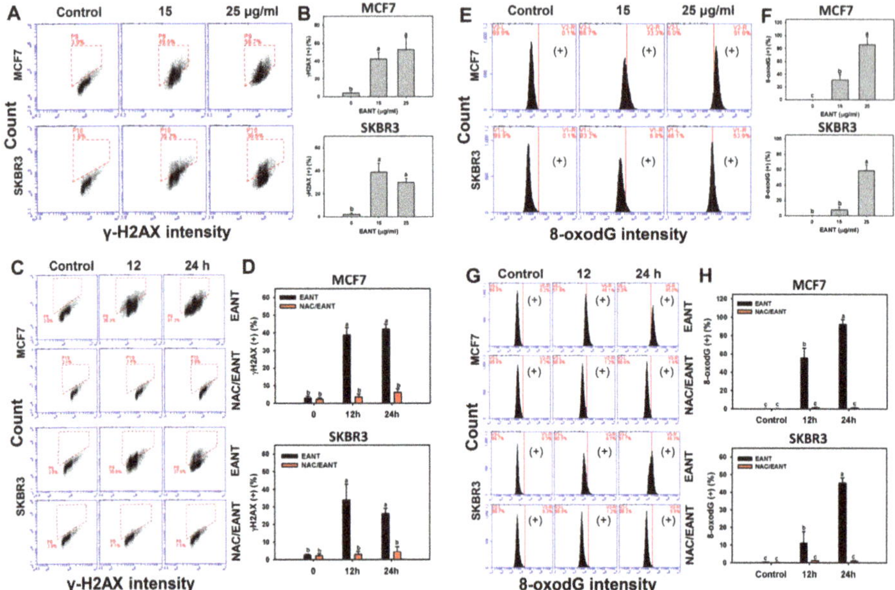

Figure 6. γH2AX and 8-oxodG changes after EANT treatment. (**A**,**B**) The concentration effect of EANT on γH2AX patterns and statistics. Breast cancer cells (MCF7 and SKBR3) were treated with 0 (control with DMSO only), 15, and 25 μg/mL of EANT for 24 h. The dashed box indicates the γH2AX (+) population. (**C**,**D**) Time course effect of EANT on γH2AX patterns and statistics. Without or with NAC pretreatment, breast cancer cells (MCF7 and SKBR3) were treated with 25 μg/mL of EANT for 0 (control with DMSO only), 12 and 24 h. (**E**,**F**) The concentration effect of EANT on 8-oxodG patterns and statistics. Breast cancer cells (MCF7 and SKBR3) were treated with control with DMSO only and EANT for 24 h. (**G**,**H**) Time course effect of EANT on 8-oxodG patterns and statistics. Without or with NAC pretreatment, breast cancer cells (MCF7 and SKBR3) were treated with 25 μg/mL of EANT for 0 (control with DMSO only), 12, and 24 h. For each cell line, treatments labeled without the same lower-case letters indicate significant difference. $p < 0.01$~0.001. Data, mean ± SD (n = 3). (+) indicates the 8-oxodG (+) population. Positive controls for γH2AX and 8-oxodG were provided in Supplementary Figure S2M–P.

3. Discussion

3.1. EANT Preferentially Inhibits Proliferation of Breast Cancer Cells

The antiproliferation effect of cancer cells using Nepenthes has been rarely reported. The current study demonstrates that EANT inhibited the cell proliferation of breast cancer cells more than that of normal breast cells at higher concentrations (15 and 25 μg/mL) (Figure 1A). IC_{50} values of EANT are respectively 10 and 15 μg/mL for MCF7 and SKBR3 cells at 24 h MTS assay. For comparison to extracts from other Nepenthes such as *N. alata* Blanco, IC_{50} values at 24 h MTT assay using MCF7 cells showed 99.78 μg/mL for its methanol extract and showed 36.60, 90.02, 61.13, and > 100 μg/mL for its CH_2CL_2, EtOAc, n-BuOH, and H_2O fractions, respectively [15]. These results suggest that our developed ethyl acetate extract of *N. thorelii* x (*ventricosa* x *maxima*), that is, EANT, has higher sensitivity to breast cancer MCF7 cells than that of *N. alata* Blanco. It is possible that extracts from different Nepenthes may have different components or proportions leading to different drug sensitivity to breast cancer cells.

Antimalarial naphthoquinones such as plumbagin, 2-methylnaphthazarin, octadecyl caffeate, isoshinanolone, and droserone were isolated from *N. thorelii* [18]. Several Nepenthes plants reported that plumbagin is one of the main bioactive components [19,20]. Plumbagin showed IC_{50} values of 1.155 and 0.658 μg/mL to breast cancer MCF7 cells and noncancerous MCF-10A cells was 2.548 and

1.882 µg/mL at 24 and 48 h treatments, respectively [15]. In the current study, we found that the major bioactive components of EANT identified by HPLC fingerprinting method were plumbagin [21], *cis*-isoshinanolone [22], and quercetin 3-*O*-(6″-*n*-butyl β-D-glucuronide) [23] (Supplementary Figure S1). Isoplumbagin also isolated from the common leadwort (*Plumbago europaea*) and showed an anti-Candida effect [21], however, its anticancer effect was rarely reported. *cis*-isoshinanolone also isolated from the plant (*Diospyros shimbaensis*) and showed antioxidant property without testing anticancer effect [22]. Quercetin 3-*O*-(6″-*n*-butyl β-D-glucuronide) also isolated from a slow-growing small leafless desert shrub (*Calligonum polygonoides*) and showed cytotoxic effect against liver cancer HepG2 and breast cancer MCF7 cells [23], that is, 60.46 and 61.4 µg/mL at 72 h sulphorhodamine-B assay. It warrants the detailed investigation for the cytotoxic effects of these components in EANT to breast cancer cells in future.

For comparison to clinical drugs, the IC_{50} value of cisplatin is 15 µg/mL for SKBR3 cells but it is undetectable for MCF7 cells (Figure 1C). IC_{50} value of cisplatin at 48 and 72 h MTS assay is 3.71 and 2.69 µg/mL for SKBR3 cells and 10.45 and 6 µg/mL for MCF7 cells (Figure 1C). Similarly, IC_{50} values of cisplatin at 72 h treatment by MTT assay are 3.36 and 4.44 µg/mL [24] for SKBR3 and MCF7 cells. Although the IC_{50} values of EANT at 24 h treatment are higher than that of cisplatin at 72 h treatment, their treatment times are different. For 24 h treatment, EANT has higher drug sensitivity to MCF7 cells and shows similar sensitivity to SKBR3 cells compared to cisplatin.

3.2. EANT Induces Oxidative Stress on Breast Cancer Cells

Oxidative stress is an imbalance status between ROS production and antioxidant response [25]. When ROS is overproduced, the cellular antioxidant system is unable to remove extra ROS to steady its state, thus reducing the antioxidant level. The current study found that intracellular ROS and mitochondrial superoxide MitoSOX were upregulated and MMP was downregulated following EANT treatment. These results suggest that EANT induces oxidative stress in breast cancer cells (MCF7 and SKBR3). These oxidative stress changes were validated by NAC or MitoTEMPO pretreatments.

GSH is an ROS scavenging antioxidant in cells [16] to maintain oxidative homeostasis. When the ROS level exceeds the ROS cleanness ability of GSH, the intracellular ROS level may increase and the GSH level may decrease. Natural products have been reported to inhibit cellular antioxidant system of cancer cells and lead to cell death for cancer treatments [26]. For example, ethanolic extract of red algae *Gracilaria tenuistipitata* can increase ROS production and decrease the GSH level of oral cancer cells [27]. Consistently, we found that GSH level declined dramatically at higher concentration of EANT (25 µg/mL). It is noted that the GSH level was not completely depleted in breast cancer cells (MCF7 and SKBR3) following EANT treatment. In addition to GSH, this warrants a detailed investigation for the involvement of other non-GSH members in antioxidant systems.

Plumbagin is a common naphthoquinone in Nepenthes, however, plumbagin but not isoplumbagin is identified in EANT. Plumbagin was reported to bind to GSH [28], behave as electrophile against GSH [29], and induce GSH depletion of leukemia Kasumi-1 [30] and U937 [31] cells. However, the GSH effect of isoplumbagin remains unclear. Accordingly, the role of the naphthoquinones isoplumbagin in cytotoxic effects against breast cancer cells needs detailed investigation in future. Moreover, the possible involvement of the other main compounds (*cis*-isoshinanolone and quercetin 3-*O*-(6″-*n*-butyl β-D-glucuronide)) in EANT-induced cytotoxicity of breast cancer cells cannot be excluded.

3.3. EANT Induces Oxidative Stress-Mediated Apoptosis and DNA Damage on Breast Cancer Cells

Oxidative stress is an activator for triggering apoptosis [32]. SubG1 accumulation is one of the apoptosis indicators [33]. In the current study, EANT partly induces subG1 accumulation and increases the intensity of annexin V. These apoptosis events were suppressed by NAC pretreatment, suggesting that EANT induces apoptosis of breast cancer cells depending on oxidative stress.

Moreover, oxidative stress is an activator for triggering DNA damage [34]. In the current study, the γH2AX marker [35] for DNA double strand breaks was induced in breast cancer cells after EANT treatment. The role of oxidative stress in DNA double strand breaks was further confirmed by evidence that NAC pretreatment suppresses the γH2AX intensity in breast cancer cells (MCF7 and SKBR3) following EANT treatment. This finding raises the possibility that oxidative stress may attach DNA to generate other types of DNA damage. In the example of 8-oxodG [36], we further demonstrated that oxidative DNA damage is also induced in breast cancer cells after EANT treatment. Again, this type of oxidative DNA damage in breast cancer cells is suppressed by NAC pretreatment. Therefore, EANT induces apoptosis and DNA damage to breast cancer cells through oxidative stress.

3.4. Conclusion

Extracts from *Nepenthes* plants have been reported to have anti-microorganism effects, but the anticancer effect of *Nepenthes* has rarely been investigated. This study examines the anticancer effect of *Nepenthes* on breast cancer cells in the example of ethyl acetate fraction of *N. thorellii* x *(ventricosa x maxima)* (EANT). EANT shows a greater antiproliferation effect against breast cancer cells than for normal breast cells. Different flow cytometric assays show that EANT induces oxidative stress and apoptosis in breast cancer cells and these changes were suppressed by their inhibitors. In conclusion, EANT exerts an antiproliferation effect on breast cancer cells and the detailed mechanisms as mentioned are explored in a ROS-dependent manner.

4. Materials and Methods

4.1. Nepenthes Extraction and Inhibitors

The aerial parts of *Nepenthes thorellii* x *(ventricosa x maxima)* were collected from Dr. Celica Koo Botanic Conservation Center (KBCC), Pingtung County, Taiwan, in October 2014. This *Nepenthes* hybrid plant was identified by KBCC researcher Ray-Hsuan Kuo, marked with the voucher number K47512 EA, and stored in the Graduate Institute of Natural Products, Kaohsiung Medical University.

The aerial part of plant material (500 g) was sliced and immersed in MeOH (1 L) for three days. The organic solvent was removed by a rotavapor (Heidolph, Schwabach, Germany) to afford a methanol extract (25 g). This extract was partitioned between water and ethyl acetate (EA). The EA-soluble fraction of *N. thorellii* x *(ventricosa x maxima)* (9.5 g) is referred to here as EANT. EANT was stored in −20 °C and dissolved in dimethyl sulfoxide (DMSO) prior to experimentation. All treatments have the same concentration of DMSO, that is, 0.05%.

To examine the role of oxidative stress, cells were pretreated with 2 mM ROS inhibitor *N*-acetylcysteine (NAC) (Sigma-Aldrich; St Louis, MO, USA) [37] for 1 h. The role of mitochondrial oxidative stress was also examined by pretreating cells with 10 μM of the mitochondrial superoxide inhibitor MitoTEMPO (Cayman Chemical, Ann Arbor, Michigan, USA) [38] for 1 h. The role of apoptosis was examined by pretreating cells with 20 μM of an apoptosis inhibitor Z-VAD-FMK (Z-VAD) (Selleckchem. com; Houston, TX, USA) [39] for 2 h.

4.2. HPLC Fingerprint Profile of EANT

In order to understand the abundance and distribution of the components of EANT, the fingerprint profile of EANT was studied. A Shimadzu LC-20AD prominence liquid chromatograph HPLC equipped with SPD-M10A VP diode array detector was used for investigation. Injection of EANT (volume = 100 μL, concentration = 1 mg/mL, in MeOH) was performed by a Shimadzu SIL-10AD VP auto injector. The separation of EANT was executed by a Phenomenex Luna 5 μ C18(2) 100 Å column (250 × 4.6 mm, flow rate = 1.0 mL/min) eluted with two solvents [solvent A = 0.1% trifluoroacetic acid in aqueous solution; solvent B = acetonitrile]. The gradient program was set up as (a) 0 min (20% B), (b) 60 min (100% B), (c) 70 min (100% B), (d) 80 min (20% B). The fingerprint profile and separation method of EANT are shown in Supplementary Figure S1 and Table S1, respectively.

The plant material of *N. thorellii* x *(ventricosa x maxima)* was collected from KBCC. However, it is hard to obtain large amounts of this plant for active component separation and identification. Therefore, we bought another Nepenthes plant (*N. mirabilis*) from a local drug store in China. Because taxonomically related plants usually produce similar secondary metabolites (chemotaxonomic approaches), we used the major compounds isolated from *N. mirabilis* as standard to interpret the components in the fingerprint profile of our research material. We also used repeated column chromatography to isolate those major compounds from *N. mirabilis*. Those major compounds were identified by their NMR and mass spectroscopic data.

4.3. Cell Culture and Viability

Human breast cancer cells (MCF7 and SKBR3) were obtained from the American Type Culture Collection (ATCC, Manassas, VA, USA) and cultured in Dulbecco's Modified Eagle's Medium (DMEM) (Gibco, Grand Island, NY, USA). Human breast normal cells (M10) were obtained from Bioresource Collection and Research Center (BCRC) (Hsinchu, Taiwan) and cultured in Minimum Essential Medium Alpha Medium (αMEM) (Gibco). All cells were supplemented with 10% fetal bovine serum (Gibco) and antibiotics, and maintained in a humidified 5% CO2 incubator at 37 °C. Cell viability was determined by Promega MTS Assay (Madison, WI, USA) as previously described [27].

4.4. Cell Cycle Assay

Cell cycle distribution was determined using 1 µg/mL of 7-aminoactinmycin D (7AAD) (Biotium, Inc., Hayward, CA, USA) for 30 min [40]. Following fixation and washing, cells were suspended in PBS for flow cytometry (Accuri C6, Becton-Dickinson, Mansfield, MA, USA).

4.5. Annexin V/7AAD Assay for Apoptosis

Cells were incubated with annexin V/7AAD reagents (Strong Biotech Corporation, Taipei, Taiwan) for 30 min [7]. Following suspension in PBS, apoptotic cells were counted by flow cytometry (Accuri C6).

4.6. ROS Assay

Cells were incubated with 2 µM of 2′,7′-dichlorodihydrofluorescein diacetate (DCFH-DA) (Sigma-Aldrich, St. Louis, MO, USA) for 30 min at 37 °C [41]. Following suspension in PBS, intracellular ROS levels were measured by flow cytometry (Accuri C6).

4.7. GSH Assay

Cells were incubated with 0.1 µM of CellTracker™Green CMFDA (5-chloromethylfluorescein) (Thermo Fisher Scientific, Carlsbad, CA, USA) for 30 min at 37 °C [42]. Following suspension in PBS, intracellular ROS levels were measured by flow cytometry (Accuri C6).

4.8. Mitochondrial Superoxide Assay

Cells were incubated with 5 µM of MitoSOX™ Red (Molecular Probes, Invitrogen, Eugene, OR, USA) for 30 min at 37 °C [43]. Following suspension in PBS, MitoSOX levels were measured by flow cytometry (Accuri C6).

4.9. Mitochondrial Membrane Potential Assay

Cells were incubated with 50 nM of $DiOC_2$(3) from the MitoProbe™ kit (Invitrogen, San Diego, CA, USA) for 20 min [44]. Following suspension in PBS, MMP levels were measured by flow cytometry (Accuri C6).

4.10. γH2AX Assay

Following fixation, cells were incubated with the primary antibody (1:50 dilution) against p-Histone H2A.X (γH2AX) (Santa Cruz Biotechnology, Santa Cruz, CA, USA) for 1 h at 4 °C [45]. Subsequently, cells were incubated with Alexa Fluor®488-conjugated secondary antibody (Cell Signaling Technology)

(1:10,000 dilution) for 30 min at room temperature. Finally, cells were stained with 1 µg/mL of 7AAD for 30 min. Following suspension in PBS, γH2AX levels were measured by flow cytometry (Accuri C6).

4.11. 8-Oxo-2′deoxyguanosine Assay

Following fixation, cells were incubated with the FITC-conjugated antibody (1:100 dilution) against 8-OxodG (Santa Cruz Biotechnology) for 1 h. Following suspension in PBS, 8-OxodG levels were measured by flow cytometry (Accuri C6).

4.12. Statistical Analysis

Data were analyzed by JMP12 software (SAS Institute, Cary, NC, USA). The significance between multiple comparisons was analyzed by one-way analysis of variance (ANOVA) using the Tukey HSD post-hoc test. Different treatments with the same small letters indicate non-significance, while different treatments without the same small letters indicate significance.

Supplementary Materials: Supplementary materials can be found at http://www.mdpi.com/1422-0067/20/13/3238/s1. Supplementary Table S1. The HPLC method for fingerprint profile of *N. thorellii* x (*ventricosa* x *maxima*). Supplementary Figure S1. Components of EANT. (A) Fingerprint profile of EANT. It is monitored at 365 nm. (B) Retention time of plumbagin (NT-A). Volume is 50 µL. It is monitored at 400 nm. (C) Retention time of *cis*-isoshinanolone (NT-B). Volume is 10 µL. It is monitored at 254 nm. (D) Retention time of quercetin 3-*O*-(6″-*n*-butyl β-D-glucuronide) (NT-E). Volume is 10 µL. It is monitored at 254 nm. UV patterns were inserted in right side for the Supplementary Figure S1B–D. Supplementary Figure S2. Patterns and statistics for positive controls for flow cytometry experiments using breast cancer SKBR3 and MCF7 cells. (A, B) Cell cycle, (C, D) apoptosis, (E, F) ROS, (G, H) GSH, (I, J) MitoSOX, (K, L) MMP, (M, N) γH2AX, and (O, P) 8-oxodG.

Author Contributions: F.O.-Y. and H.-W.C. contributed to original draft preparation. I.-H.T. performed the cytotoxicity assays and flow cytometry. J.-Y.T., C.-Y.Y., and A.A.F. contributed to methodology, data analysis and statistics. Y.-B.C., S.-R.C., and S.-Y.Y., contributed to EANT preparation and fingerprint profile characterization. J.-K.K. and H.-W.C. designed whole experiment and improved the manuscript.

Funding: This work was partly supported by funds of the Ministry of Science and Technology (MOST 107-2320-B-037-016, MOST 107-2314-B-037-048, and MOST 107-2311-B-214-003), the National Sun Yat-sen University-KMU Joint Research Project (#NSYSUKMU 108-P001), the Kaohsiung Medical University Hospital (KMUH107-7R74), Changhua Christian Hospital-KMU Joint Research Project (108-CCH-KMU-008), the Chimei-KMU jointed project (108CM-KMU-11), the Kaohsiung Medical University Research Center (KMU-TC108A03), and the Health and welfare surcharge of tobacco products, the Ministry of Health and Welfare, Taiwan, Republic of China (MOHW 108-TDU-B-212-124016).

Conflicts of Interest: The authors declare no conflict of interest. The funders had no role in the design of the study; in the collection, analyses, or interpretation of data; in the writing of the manuscript, or in the decision to publish the results.

References

1. Siegel, R.L.; Miller, K.D.; Jemal, A. Cancer statistics, 2019. *CA Cancer J. Clin.* **2019**, *69*, 7–34. [CrossRef] [PubMed]

2. Teoh, P.L.; Liau, M.; Cheong, B.E. *Phyla nodiflora* L. Extracts induce apoptosis and cell cycle arrest in human breast cancer cell line, MCF-7. *Nutr. Cancer* **2019**, *71*, 668–675. [CrossRef] [PubMed]

3. Aung, T.N.; Qu, Z.; Kortschak, R.D.; Adelson, D.L. Understanding the effectiveness of natural compound mixtures in cancer through their molecular mode of action. *Int. J. Mol. Sci.* **2017**, *18*, 656. [CrossRef] [PubMed]

4. Asadi-Samani, M.; Rafieian-Kopaei, M.; Lorigooini, Z.; Shirzad, H. The effect of *Euphorbia szovitsii* Fisch. & C.A.Mey extract on the viability and the proliferation of MDA-MB-231 cell line. *Biosci. Rep.* **2019**, *39*, BSR20181538. [PubMed]

5. Jiang, X.; Cao, C.; Sun, W.; Chen, Z.; Li, X.; Nahar, L.; Sarker, S.D.; Georgiev, M.I.; Bai, W. Scandenolone from *Cudrania tricuspidata* fruit extract suppresses the viability of breast cancer cells (MCF-7) in vitro and in vivo. *Food Chem. Toxico.l* **2019**, *126*, 56–66. [CrossRef] [PubMed]

6. Weng, J.R.; Chiu, C.F.; Hu, J.L.; Feng, C.H.; Huang, C.Y.; Bai, L.Y.; Sheu, J.H. A sterol from soft coral induces apoptosis and autophagy in MCF-7 breast cancer cells. *Ma.r Drugs* **2018**, *16*, 238. [CrossRef] [PubMed]

7. Huang, H.W.; Tang, J.Y.; Ou-Yang, F.; Wang, H.R.; Guan, P.Y.; Huang, C.Y.; Chen, C.Y.; Hou, M.F.; Sheu, J.H.; Chang, H.W. Sinularin selectively kills breast cancer cells showing G2/M arrest, apoptosis, and oxidative DNA damage. *Molecules* **2018**, *23*, 849. [CrossRef] [PubMed]

8. Sanusi, S.B.; Bakar, M.F.A.; Mohamed, M.; Sabran, S.F.; Mainasara, M.M. Ethnobotanical, phytochemical, and pharmacological properties of *Nepenthes* species: A review. *Asian. J. Pharm. Clin. Res.* **2017**, *10*, 16–19. [CrossRef]

9. Ismail, N.A.; Kamariah, A.S.; Lim, L.B.; Ahmad, A. Phytochemical and pharmacological evaluation of methanolic extracts of the leaves of *Nepenthes bicalcarata* Hook. F. *Int. J. Pharma. Phyto. Res.* **2015**, *7*, 1127–1138.

10. Shin, K.S.; Lee, S.; Cha, B.J. Suppression of phytopathogenic fungi by hexane extract of *Nepenthes ventricosa* x *maxima* leaf. *Fitoterapia* **2007**, *78*, 585–586. [CrossRef]

11. Thao, N.P.; Luyen, B.T.; Koo, J.E.; Kim, S.; Koh, Y.S.; Thanh, N.V.; Cuong, N.X.; Kiem, P.V.; Minh, C.V.; Kim, Y.H. In vitro anti-inflammatory components isolated from the carnivorous plant *Nepenthes mirabilis* (Lour.) Rafarin. *Pharm. Biol.* **2016**, *54*, 588–594. [CrossRef] [PubMed]

12. Sugata, M.; Lin, C.Y.; Shih, Y.C. Anti-inflammatory and anticancer activities of taiwanese purple-fleshed sweet potatoes (*Ipomoea batatas* L. Lam) extracts. *Biomed. Res. Int.* **2015**, *2015*, 768093. [CrossRef] [PubMed]

13. Orlikova, B.; Legrand, N.; Panning, J.; Dicato, M.; Diederich, M. Anti-inflammatory and anticancer drugs from nature. *Cancer Treat Re.s* **2014**, *159*, 123–143.

14. Akunne, T.C.; Akah, P.A.; Nwabunike, I.A.; Nworu, C.S.; Okereke, E.K.; Okereke, N.C.; Okeke, F.C.; Hsu, T.-C. Anti-inflammatory and anticancer activities of extract and fractions of *Rhipsalis neves-armondii* (Cactaceae) aerial parts. *Cogent Biol.* **2016**, *2*, 1237259. [CrossRef]

15. De, U.; Son, J.Y.; Jeon, Y.; Ha, S.Y.; Park, Y.J.; Yoon, S.; Ha, K.T.; Choi, W.S.; Lee, B.M.; Kim, I.S.; et al. Plumbagin from a tropical pitcher plant (*Nepenthes alata* Blanco) induces apoptotic cell death via a p53-dependent pathway in MCF-7 human breast cancer cells. *Food Chem. Toxicol.* **2019**, *123*, 492–500. [CrossRef] [PubMed]

16. Nunes, S.C.; Serpa, J. Glutathione in ovarian cancer: A double-edged sword. *Int. J. Mol. Sci.* **2018**, *19*, 1882. [CrossRef] [PubMed]

17. Richter, C. Reactive oxygen and DNA damage in mitochondria. *Mutat Res.* **1992**, *275*, 249–255. [CrossRef]

18. Likhitwitayawuid, K.; Kaewamatawong, R.; Ruangrungsi, N.; Krungkrai, J. Antimalarial naphthoquinones from *Nepenthes Thorelii*. *Planta Med.* **1998**, *64*, 237–241. [CrossRef]

19. Schlauer, J.; Nerz, J.; Rischer, H. Carnivorous plant chemistry. *Acta Botanica Gallica* **2005**, *152*, 187–195. [CrossRef]

20. Eilenberg, H.; Pnini-Cohen, S.; Rahamim, Y.; Sionov, E.; Segal, E.; Carmeli, S.; Zilberstein, A. Induced production of antifungal naphthoquinones in the pitchers of the carnivorous plant *Nepenthes khasiana*. *J. Exp. Bot.* **2010**, *61*, 911–922. [CrossRef]

21. Sobhani, M.; Abbas-Mohammadi, M.; Ebrahimi, S.N.; Aliahmadi, A. Tracking leading anti-Candida compounds in plant samples; *Plumbago europaea*. *Iran. J. Microbiol.* **2018**, *10*, 187–193. [PubMed]

22. Aronsson, P.; Munissi, J.J.E.; Gruhonjic, A.; Fitzpatrick, P.A.; Landberg, G.; Nyandoro, S.S.; Erdelyi, M. Phytoconstituents with radical scavenging and cytotoxic activities from *Diospyros Shimbaensis*. *Dis.* **2016**, *4*, 3. [CrossRef] [PubMed]

23. Ahmed, H.; Moawad, A.; Owis, A.; AbouZid, S.; Ahmed, O. Flavonoids of *Calligonum polygonoides* and their cytotoxicity. *Pharm. Biol.* **2016**, *54*, 2119–2126. [CrossRef] [PubMed]

24. Kwon, Y.E.; Park, J.Y.; Kim, W.K. In vitro histoculture drug response assay and in vivo blood chemistry of a novel Pt(IV) compound, K104. *Anticancer Res.* **2007**, *27*, 321–326. [PubMed]

25. Poljsak, B.; Suput, D.; Milisav, I. Achieving the balance between ROS and antioxidants: When to use the synthetic antioxidants. *Oxid. Med. Cell Longev.* **2013**, *2013*, 956792. [CrossRef] [PubMed]

26. Sznarkowska, A.; Kostecka, A.; Meller, K.; Bielawski, K.P. Inhibition of cancer antioxidant defense by natural compounds. *Oncotarget* **2017**, *8*, 15996–16016. [CrossRef] [PubMed]

27. Yeh, C.C.; Tseng, C.N.; Yang, J.I.; Huang, H.W.; Fang, Y.; Tang, J.Y.; Chang, F.R.; Chang, H.W. Antiproliferation and induction of apoptosis in Ca9-22 oral cancer cells by ethanolic extract of *Gracilaria tenuistipitata*. *Molecules* **2012**, *17*, 10916–10927. [CrossRef] [PubMed]

28. Inbaraj, J.J.; Chignell, C.F. Cytotoxic action of juglone and plumbagin: A mechanistic study using HaCaT keratinocytes. *Chem. Res. Toxicol.* **2004**, *17*, 55–62. [CrossRef]

29. Castro, F.A.; Mariani, D.; Panek, A.D.; Eleutherio, E.C.; Pereira, M.D. Cytotoxicity mechanism of two naphthoquinones (menadione and plumbagin) in *Saccharomyces cerevisiae*. *PLoS One* **2008**, *3*, e3999. [CrossRef]
30. Kong, X.; Luo, J.; Xu, T.; Zhou, Y.; Pan, Z.; Xie, Y.; Zhao, L.; Lu, Y.; Han, X.; Li, Z.; et al. Plumbagin enhances TRAIL-induced apoptosis of human leukemic Kasumi1 cells through upregulation of TRAIL death receptor expression, activation of caspase-8 and inhibition of cFLIP. *Oncol. Rep.* **2017**, *37*, 3423–3432. [CrossRef]
31. Gaascht, F.; Teiten, M.H.; Cerella, C.; Dicato, M.; Bagrel, D.; Diederich, M. Plumbagin modulates leukemia cell redox status. *Molecules* **2014**, *19*, 10011–10032. [CrossRef] [PubMed]
32. Kashyap, D.; Sharma, A.; Garg, V.; Tuli, H.S.; Kumar, G.; Kumar, M.; Mukherjee, T. Reactive oxygen species (ROS): An activator of apoptosis and autophagy in cancer. *J. Biol. Chem. Sci.* **2016**, *3*, 256–264.
33. Semaan, J.; Pinon, A.; Rioux, B.; Hassan, L.; Limami, Y.; Pouget, C.; Fagnere, C.; Sol, V.; Diab-Assaf, M.; Simon, A.; et al. Resistance to 3-HTMC-induced apoptosis through activation of PI3K/Akt, MEK/ERK, and p38/COX-2/PGE2 pathways in human HT-29 and HCT116 colorectal cancer cells. *J. Cell. Biochem.* **2016**, *117*, 2875–2885. [CrossRef] [PubMed]
34. Li, Z.; Yang, J.; Huang, H. Oxidative stress induces H2AX phosphorylation in human spermatozoa. *FEBS Lett* **2006**, *580*, 6161–6168. [CrossRef] [PubMed]
35. Banath, J.P.; Klokov, D.; MacPhail, S.H.; Banuelos, C.A.; Olive, P.L. Residual gammaH2AX foci as an indication of lethal DNA lesions. *BMC Cancer* **2010**, *10*, 4. [CrossRef] [PubMed]
36. Cooke, M.S.; Evans, M.D.; Dizdaroglu, M.; Lunec, J. Oxidative DNA damage: Mechanisms, mutation, and disease. *FASEB J* **2003**, *17*, 1195–1214. [CrossRef] [PubMed]
37. Huang, C.H.; Yeh, J.M.; Chan, W.H. Hazardous impacts of silver nanoparticles on mouse oocyte maturation and fertilization and fetal development through induction of apoptotic processes. *Env. Toxicol* **2018**, *33*, 1039–1049. [CrossRef] [PubMed]
38. Wang, T.S.; Lin, C.P.; Chen, Y.P.; Chao, M.R.; Li, C.C.; Liu, K.L. CYP450-mediated mitochondrial ROS production involved in arecoline N-oxide-induced oxidative damage in liver cell lines. *Env. Toxicol* **2018**, *33*, 1029–1038. [CrossRef]
39. Chen, C.Y.; Yen, C.Y.; Wang, H.R.; Yang, H.P.; Tang, J.Y.; Huang, H.W.; Hsu, S.H.; Chang, H.W. Tenuifolide B from *Cinnamomum tenuifolium* stem selectively inhibits proliferation of oral cancer cells via apoptosis, ROS generation, mitochondrial depolarization, and DNA damage. *Toxins* **2016**, *8*, 319. [CrossRef]
40. Vignon, C.; Debeissat, C.; Georget, M.T.; Bouscary, D.; Gyan, E.; Rosset, P.; Herault, O. Flow cytometric quantification of all phases of the cell cycle and apoptosis in a two-color fluorescence plot. *PLoS ONE* **2013**, *8*, e68425. [CrossRef]
41. Yeh, C.C.; Yang, J.I.; Lee, J.C.; Tseng, C.N.; Chan, Y.C.; Hseu, Y.C.; Tang, J.Y.; Chuang, L.Y.; Huang, H.W.; Chang, F.R.; et al. Anti-proliferative effect of methanolic extract of *Gracilaria tenuistipitata* on oral cancer cells involves apoptosis, DNA damage, and oxidative stress. *BMC Complement Altern. Med.* **2012**, *12*, 142. [CrossRef] [PubMed]
42. Lin, K.Y.; Chung, C.H.; Ciou, J.S.; Su, P.F.; Wang, P.W.; Shieh, D.B.; Wang, T.C. Molecular damage and responses of oral keratinocyte to hydrogen peroxide. *BMC Oral Health* **2019**, *19*, 10. [CrossRef] [PubMed]
43. Chang, Y.T.; Huang, C.Y.; Tang, J.Y.; Liaw, C.C.; Li, R.N.; Liu, J.R.; Sheu, J.H.; Chang, H.W. Reactive oxygen species mediate soft corals-derived sinuleptolide-induced antiproliferation and DNA damage in oral cancer cells. *Onco. Targets Ther.* **2017**, *10*, 3289–3297. [CrossRef] [PubMed]
44. Chang, H.S.; Tang, J.Y.; Yen, C.Y.; Huang, H.W.; Wu, C.Y.; Chung, Y.A.; Wang, H.R.; Chen, I.S.; Huang, M.Y.; Chang, H.W. Antiproliferation of *Cryptocarya concinna*-derived cryptocaryone against oral cancer cells involving apoptosis, oxidative stress, and DNA damage. *BMC Complement Altern. Med.* **2016**, *16*, 94. [CrossRef] [PubMed]
45. Chiu, C.C.; Haung, J.W.; Chang, F.R.; Huang, K.J.; Huang, H.M.; Huang, H.W.; Chou, C.K.; Wu, Y.C.; Chang, H.W. Golden berry-derived 4beta-hydroxywithanolide E for selectively killing oral cancer cells by generating ROS, DNA damage, and apoptotic pathways. *PLoS One* **2013**, *8*, e64739. [CrossRef] [PubMed]

International Journal of
Molecular Sciences

Article

Dietary Supplementation with Sea Bass (*Lateolabrax maculatus*) Ameliorates Ulcerative Colitis and Inflammation in Macrophages through Inhibiting Toll-Like Receptor 4-Linked Pathways

Jiali Chen [1,2], Muthukumaran Jayachandran [1], Wenxia Zhang [1], Lingyuqing Chen [1], Bin Du [3], Zhiling Yu [2,*] and Baojun Xu [1,*]

[1] Programme of Food Science and Technology, Division of Science and Technology, Beijing Normal University-Hong Kong Baptist University United International College, Zhuhai 519087, China; kellychan123@126.com (J.C.); jmkbio@uic.edu.hk (M.J.); l630013050@mail.uic.edu.hk (W.Z.); j430013005@mail.uic.edu.hk (L.C.)

[2] Centre for Cancer and Inflammation Research, School of Chinese Medicine, Hong Kong Baptist University, Hong Kong, China

[3] Hebei Key Laboratory of Natural Products Activity Components and Function, Hebei Normal University of Science and Technology, Qinhuangdao 066004, China; bindufood@aliyun.com

* Correspondence: zlyu@hkbu.edu.hk (Z.Y.); baojunxu@uic.edu.hk (B.X.); Tel.: +852-34112465 (Z.Y.); +86-756-3620636 (B.X.)

Received: 25 May 2019; Accepted: 12 June 2019; Published: 14 June 2019

Abstract: Sea bass (*Lateolabrax maculatus*) is a kind of food material commonly consumed in daily life. In traditional Chinese medicinal books, it has been indicated that sea bass can be applied for managing many inflammation-associated conditions. However, the studies on the pharmacological mechanisms of inflammation of sea bass remain scarce. Hence, this study aims to investigate the molecular mechanisms of the anti-inflammatory activity of sea bass. Anti-inflammatory activities of sea bass were assessed using dextran sulfate sodium (DSS)-induced colitis in a mice model and lipopolysaccharide (LPS)-activated macrophages model. Low body weight and short colon length were observed in DSS-fed mice that were significantly recovered upon sea bass treatments. Moreover, the colon histopathology score showed that sea bass-treated mice had decreased crypt damage, focal inflammation infiltration and the extent of inflammation, suggesting that treatment with sea bass could attenuate intestinal inflammation. In addition, the in-vitro study conjointly indicated that sea bass could suppress the inflammatory mediators in LPS-activated macrophage by inhibiting the TLR4-linked pathway. The present findings demonstrated that sea bass has an inhibitory effect on TLR4 signaling; thus, it could be a promising candidate for treating inflammation-associated conditions. A further justification for the clinical application of sea bass in treating inflammation-associated conditions is necessary.

Keywords: inflammation; ulcerative colitis; dietary therapy; TLR4 signaling

---

1. Introduction

Inflammatory bowel disease (IBD) is principally outlined as Crohn's disease and ulcerative colitis (UC). Crohn's disease might have an effect on any part of the digestive tract, from the mouth to the anus with diarrhea and abdominal pain. Ulcerative colitis (UC) mainly presents in the rectal and colonic mucosa and is accompanied by weight loss, diarrhea, abdominal pain, and rectal bleeding. This kind of uncontrolled gut inflammation affects millions of individuals in the world [1,2]. It has been reported that the worldwide incidence and prevalence of UC are increasing, especially in newly industrialized countries. The highest reported prevalence values appeared in Europe and North America [1,2]. In population-based studies, it also indicated that UC patients will have proximal disease extension within 10 years [3]. In the 21st century, systematic research into UC prevention and dietary therapy development seems very significant with an irresistible trend. However, the pathogenesis of IBD remains unclear due to lack of investigation. IBD is a multifactorial disorder induced by the interaction of genetic factors, environment, microbiota, and immune response, which are involved in pathogenesis [4,5]. Recently, many studies have indicated that the breakdown of homeostasis among the immune system, epithelial barrier, and gut microbiome might be the critical underlying mechanism responsible for the development of IBD [4,5]. Previous reports suggested that the products of gut microbiota could positively have an effect on the pathogenesis of inflammation-associated diseases [6,7]. The gastrointestinal tract provides residence to both beneficial and potentially pathogenic microorganisms. The imbalance within the microbiota composition may worsen the dysbiosis in the inflamed gut [6]. Due to the immune modulatory role of gut microbiota, sea bass is hypothesized to have beneficial effects on the host immune response and amelioration of intestinal inflammation. Moreover, the dextran sulfate sodium (DSS) model resembles UC in several pathophysiological and morphological features, including the production of pro-inflammatory cytokines, crypt damage, focal inflammation infiltration, and ulceration [8]. Generally, colitis is induced chemically in this model by adding DSS to the drinking water of mice. It mainly affects the distal colon; some inflammatory responses appeared even in the proximal colon and caecum. The outcome of this model may be affected by the genetic background of the animals, environment, and the DSS concentration [6,8]. Additionally, body weight, feed consumption, and colon length were treated as an indication of the disease severity in the DSS-induced model [9–11]. Therefore, DSS-induced colitis model is employed to study the efficacy of aqueous extract of sea bass (ASB) in managing inflammation-associated conditions (colitis). Toll-like receptor 4 (TLR4) signaling is one of the important mechanisms for inflammation-related studies and it is a key receptor for commensal recognition in gut innate immunity [12]. It was the subject of target inhibition in ulcerative colitis (UC) [12,13]. Moreover, AP-1 and NF-κB were treated as the critical and classical pathways in TLR4 signaling. Many researchers have already elucidated that over-expression of TLR4-linked AP-1 or NF-κB is typical in inflamed colonic tissue [13,14]. Therefore, it is necessary to evaluate the effects of TLR4 signaling in DSS-induced colitis for studying UC in detail.

Sea bass (*Lateolabrax maculatus*) is an economically important cultured fish species and has a long history of managing inflammation-associated conditions. However, the mechanism of action of sea bass needs to be investigated. Therefore, the aim of this study was to study the efficacy of ASB in managing inflammation-associated conditions by in vitro and in vivo experiments. The DSS-induced colitis model was used for discovering ulcerative colitis in vivo. Many studies have reported that the process of colitis is closely linked with neutrophils and macrophages [14–16]. It is well known that macrophages play vital roles in innate immunity for the inhibition of inflammatory cytokines [17,18]. As one of the typical in vitro models for investigating inflammation, lipopolysaccharide (LPS)-activated macrophages were used in this study [17,19,20]. Moreover, many reports have indicated that inflammation in macrophages is closely linked to the activation of TLR4 signaling [21–23]. In addition, NF-kB and AP-1 are the typical pathways in TLR4 signaling which are associated with the inflammation triggered by the innate immune system [24,25]. The activation of NF-kB or AP-1 pathways will lead to the production of a series of inflammatory mediators, including interferon gamma (IFN-γ), tumor necrosis factor alpha (TNF-α), and monocyte chemoattractant protein-1 (MCP-1). The TNF-α is a pleiotropic cytokine, which is an important mediator of inflammation [26]. The inhibition of TNF-α secretion in LPS-induced macrophages results in anti-inflammation [27]. MCP-1 is a highly representative chemokine, critical for the pathogenesis of liver disease and granulomatous inflammation [28,29]. The regulation of TNF-α, MCP-1, and IFN-γ via NF-κB, and AP-1 pathways are important mechanisms in inflammatory responses [28]. The therapeutic potential of sea bass against ulcerative colitis has not yet been discovered. Therefore, a good understanding of the anti-inflammatory activities of sea bass is essential for providing the pharmacological basis for the folk use of sea bass, and further, its application in the medical industry.

2. Results

2.1. Characterization of the Aqueous Extract of Sea Bass (ASB)

To characterize the aqueous extract of sea bass, the crude protein content of ASB, the molecular weight of the composition of protein fractions in ASB, and the composition of amino acids in ASB were evaluated. Results showed that the crude protein value of ASB ranged from 74.91% to 78.87%. As shown in Figure 1, the molecular weight of protein fractions in ASB ranged from 3.3 to 250 kDa. In particular, the molecular weight of ASB protein fractions was distributed around 150, 37, and 10 kDa, respectively. As shown in Supplementary Figures S1 and S2, amino acids in ASB were quantified and identified in the chromatogram.

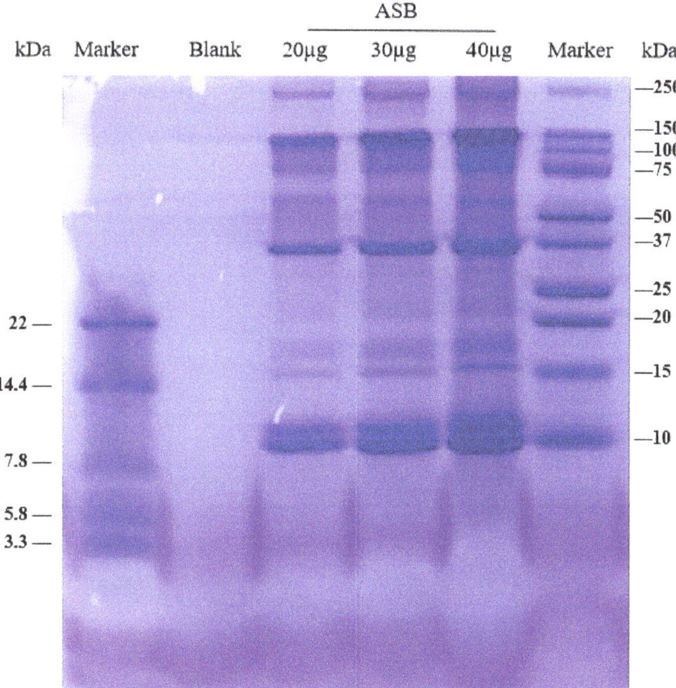

Figure 1. SDS-PAGE images of the aqueous extract of sea bass (ASB).

2.2. ASB Ameliorated DSS-Induced Colitis

DSS-induced colitis model was constructed to explore the role of ASB in UC. After DSS feeding, a significant body weight loss, less fodder consumption, and bloody stools were observed, especially in the DSS group. As shown in Figure 2D, mice treated with ASB or sulfasalazine (SASP) showed body weight recovery compared to the DSS-treated group. Mice body weight did not show any marked changes in the high dosage reference group, which was similar to the control group. In accordance with the results shown in Figure 2D, DAI (Figure 2H) also indicated that mice treated with ASB or SASP could ameliorate the severity of colitis, compared to the DSS-treated group. Meanwhile, the amount of feed consumption also demonstrated a significant improvement upon ASB or SASP treatments in comparison with the DSS-treated group (Figure 2E). As shown in Figure 2A,F, DSS-induced colitis caused a marked decrease in colon length, while it improved upon ASB or SASP treatment. No obvious colon length change was observed in the high dosage reference group, compared to the control group. As is evident from the results, DSS-induced colitis in mice was ameliorated upon ASB treatments at the dosage of 1.125, 2.25, and 4.5 g/kg b.w. (the human equivalent dose (HED): 200 g/60 kg, 400 g/60 kg, 800 g/60 kg, respectively), which was similar to the treatment with SASP.

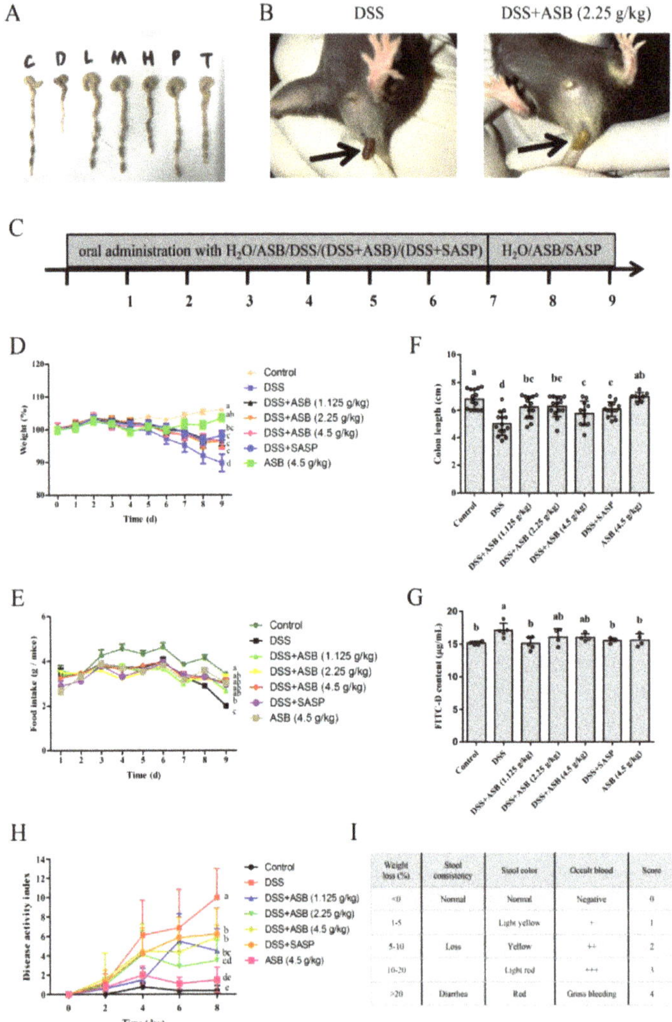

Figure 2. ASB protects against DSS-induced colitis in mice. (**A**) Representative photographs of colon (C: control, $n = 14$; D: DSS model, $n = 14$; L: DSS+ASB low dosage, $n = 14$; M: DSS+ASB medium dosage, n=14; H: DSS+ASB high dosage, $n = 14$; P: SASP, $n = 14$; T: ASB high dosage reference, $n = 8$). (**B**) Effect of ASB on rectal bleeding in DSS-treated mice at day 7. (**C**) Schematic representations of the colitis model. Effect of ASB on body weight (**D**), daily feed content (**E**) and colon length (**F**) at the end of the experiment. (**G**) Effect of intestinal permeability upon ASB treatments at the end of the experiment. (**H**) Disease activity index (DAI) evaluation of mice in each group. (**I**) The scoring criteria for DAI. Results were expressed as mean ± SD. Parameters marked by the same letter are not significantly different. Significance is represented as $p < 0.05$.

2.3. Hematological Parameters

As shown in Table 1, mice with DSS administration showed significant ($p < 0.05$) anemia (lower red blood cell (RBC) levels, hemoglobin (HGB), hematocrit (HCT), mean corpuscular volume (MCV), and platelet distribution width (PDW)) in comparison with the control group. Meanwhile, the results also showed that supplementation of the diet with ASB could ameliorate these symptoms. Significant differences ($p < 0.05$) were found in the RBC levels, HCT, and PDW upon low dosage ASB treatment in comparison with the DSS model group.

Table 1. Hematological parameters in mice.

Hematological Parameters	Control	DSS	DSS+ASB (1.125 g/kg)	DSS+ASB (2.25 g/kg)	DSS+ASB (4.5 g/kg)	DSS+SASP	NVR
RBC (10^{12}/L)	7.39 ± 0.55 a	5.37 ± 1.25 c	6.67 ± 1.15 ab	6.23 ± 0.64 bc	6.01 ± 0.90 bc	6.07 ± 0.56 bc	6.68~8.28
HGB (g/L)	115 ± 8 a	83 ± 20 c	101 ± 17 ab	95 ± 10 bc	93 ± 14 bc	90 ± 8 bc	106~129
HCT (%)	37.2 ± 2.6 a	26.3 ± 6.1 c	32.8 ± 5.7 ab	30.5 ± 3.2 bc	29.3 ± 4.5 bc	29.3 ± 2.5 bc	33.9~41.8
MCV (fL)	50.4 ± 0.8 a	49.0 ± 1.0 b	49.2 ± 0.7 b	48.9 ± 0.5 b	48.8 ± 0.5 b	48.4 ± 1.7 b	49.2~51.3
MCH (pg)	15.5 ± 0.2 a	15.4 ± 0.2 ab	15.2 ± 0.1 b	15.3 ± 0.1 ab	15.5 ± 0.2 a	14.9 ± 0.6 c	15.2~15.9
MCHC (g/L)	308 ± 4 bc	314 ± 8 ab	309 ± 4 bc	314 ± 5 ab	318 ± 5 a	308 ± 4 bc	299~313
PDW	14.8 ± 0.1 a	14.3 ± 0.2 c	14.6 ± 0.2 b	14.5 ± 0.2 b	14.5 ± 0.2 b	14.4 ± 0.1 bc	14.7~14.9
WBC (10^9/L)	3.76 ± 0.65 ab	3.30 ± 0.87 b	5.14 ± 1.10 a	4.07 ± 2.34 ab	2.59 ± 0.95 b	4.02 ± 1.29 ab	2.53~4.62
Neu# (10^9/L)	2.08 ± 0.81 abc	1.52 ± 0.34 bc	2.84 ± 0.61 a	2.29 ± 1.91 ab	1.15 ± 0.52 c	1.86 ± 0.63 abc	0.47~3.01
Lymph# (10^9/L)	1.68 ± 0.35 ab	1.78 ± 0.61 ab	2.29 ± 0.62 a	1.77 ± 0.60 ab	1.43 ± 0.64 b	2.16 ± 0.72 a	1.23~2.31
Mon# (10^9/L)	0 ± 0 b	0 ± 0 b	0.01 ± 0.01 a	0 ± 0 b	0 ± 0 b	0 ± 0 b	0
Eos# (10^9/L)	0 ± 0.01 a	0 ± 0 a	0 ± 0 a	0.01 ± 0.01 a	0 ± 0 a	0 ± 0 a	0~0.02
Bas# (10^9/L)	0 ± 0 b	0 ± 0 b	0 ± 0 b	0 ± 0 b	0.01 ± 0.01 a	0 ± 0 b	0
Neu% (%)	53.3 ± 16.5 a	46.8 ± 6.3 a	55.5 ± 6.7 a	51.6 ± 12.7 a	45.2 ± 13.2 a	46.2 ± 4.9 a	39.6~66.1
Lymph% (%)	46.5 ± 16.4 a	53.1 ± 6.3 a	44.2 ± 6.8 a	48.0 ± 12.4 a	54.1 ± 13.2 a	53.7 ± 5.0 a	33.7~81.3
Mon% (%)	0 ± 0 b	0.04 ± 0.1 ab	0.2 ± 0.2 a	0.1 ± 0.1 ab	0.1 ± 0.2 ab	0.1 ± 0.1 ab	0
Eos% (%)	0.2 ± 0.2 a	0 ± 0 a	0.1 ± 0.1 a	0.2 ± 0.4 a	0.2 ± 0.2 a	0 ± 0 a	0~0.7
Bas% (%)	0 ± 0.1 b	0.1 ± 0.1 b	0 ± 0.1 b	0.2 ± 0.2 b	0.4 ± 0.4 a	0 ± 0 b	0~0.2
RDW-CV (%)	13.3 ± 0.8 b	13.6 ± 2.6 b	13.6 ± 1.3 b	13.7 ± 1.3 b	13.1 ± 0.6 b	17.2 ± 1.6 a	12.6~15.1
PLT (10^9/L)	611 ± 53 a	576 ± 146 a	681 ± 122 a	631 ± 94 a	569 ± 128 a	688 ± 91 a	526~662
MPV (fL)	5.3 ± 0.2 a	5.3 ± 0.2 a	5.3 ± 0.2 a	5.4 ± 0.2 a	5.3 ± 0.1 a	5.3 ± 0.1 a	5.2~5.7
PCT (%)	0.325 ± 0.032 a	0.304 ± 0.072 a	0.359 ± 0.067 a	0.339 ± 0.058 a	0.299 ± 0.065 a	0.363 ± 0.048 a	0.274~0.362

Results were expressed as mean ± SD, $n = 8$ independent experiments. WBC, white blood cell; Neu#, neutrophil values; Lymph#, lymphocyte values; Mon#, monocyte values; Eos#, eosinophil values; Bas#, basophil values; Neu%, Lymph%, Mon%, Eos% and Bas%, percentages of corresponding cell over white blood cell; RBC, red blood cell; HGB, hemoglobin; HCT, hematocrit; MCV, mean corpuscular volume; MCH, mean corpuscular hemoglobin; MCHC, mean corpuscular hemoglobin concentration; RDW-CV, coefficient of variation of erythrocyte distribution width; PLT, platelets; MPV, mean platelet value; PDW, platelet distribution width; PCT, procalcitonin. Hematological inflammatory parameters of each row marked by the same letter are not significantly different. Significance is represented as $p < 0.05$.

2.4. ASB Reduced Intestinal Permeability

Increase in gut permeability was linked with greater susceptibility to colitis. As shown in Figure 2G, the permeability of FITC-Dextran was significantly increased in the DSS-treated group, while such a change was improved upon ASB or SASP treatment. The high dosage reference group did not demonstrate a significant increase in comparison with the control group.

2.5. ASB Reduced Colonic Tissue Damage

As shown in Figure 3A, 1.5% DSS in drinking water resulted in extensive colonic tissue damage, including inflammatory cell infiltration, crypt damage, and focal formation. Results showed that less colonic tissue damage was presented upon ASB treatments, as compared to the DSS-treated group. Diffuse infiltration of inflammation in mucosa and submucosa and crypt damage in colonic tissue was markedly increased in the DSS group, while such changes were significantly suppressed in the ASB-treated groups (Figure 3A,B). Meanwhile, mice in the high dosage reference group presented normal, similar to the control group.

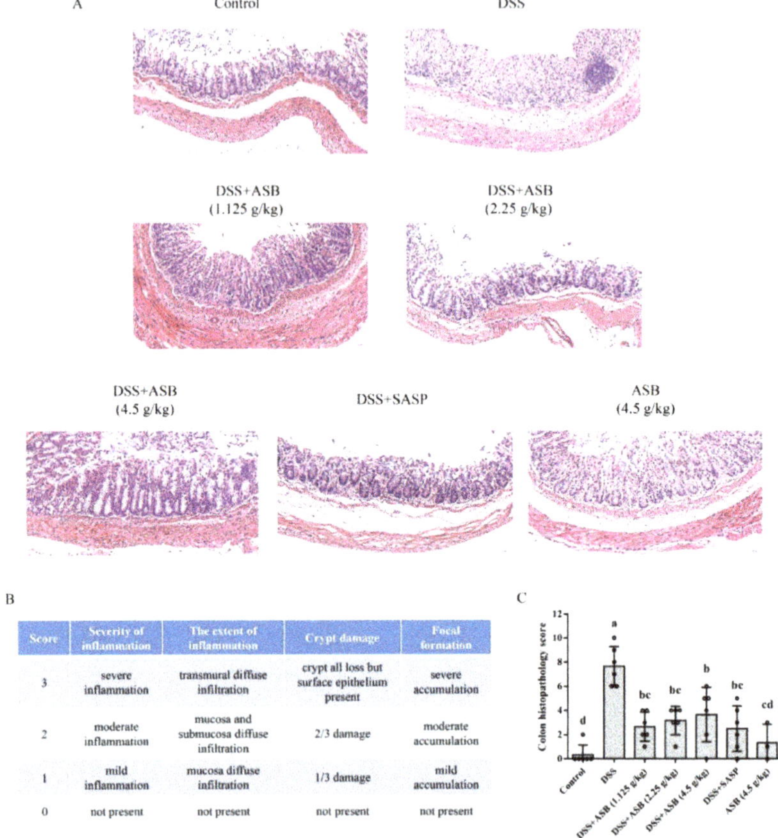

Figure 3. Effect of ASB on histopathological changes of mice in DSS-induced colitis. (**A**) Histological analysis (scale bar: 50 μm); (**B**) chart indicating scoring criteria for the evaluation of intestinal inflammation; (**C**) histological score. Results were expressed as mean ± SD (n = 3~6). Colon histopathology score marked by the same letter is not significantly different. Significance is represented as $p < 0.05$.

2.6. ASB Inhibited the Neutrophil Infiltration in Impaired Colon

Similar to the colon histopathology scores, the expression level of colonic myeloperoxidase (MPO) was greatly up-regulated in the DSS group, while the MPO activities in the ASB-treatment groups were markedly reduced (Figure 4A). As shown in Figure 4B, less MPO-positive cells were detected in the high dosage reference group, which was similar to the control group.

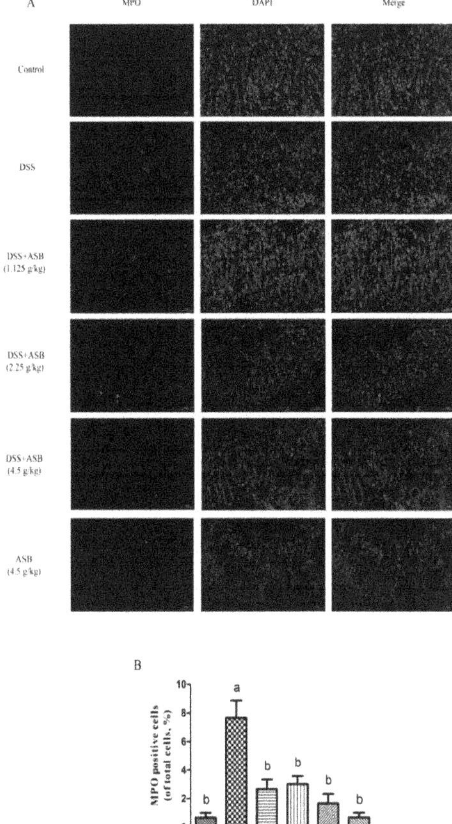

Figure 4. Effect of ASB on the levels of an inflammatory factor in the colon. Expression of MPO in colonic tissue was assessed by immunofluorescence (scale bar: 50 μm) (**A**) and MPO-positive cells quantification (**B**). Results were expressed as mean ± SD (*n* = 3). The expression levels of MPO marked by the same letter are not significantly different. Significance is represented as *p* < 0.05.

2.7. ASB Suppressed the Production of Pro-Inflammatory Mediators in the Impaired Colon

As shown in Figure 5A, treatment with SASP markedly inhibited the secretion of TNF-α (25.06 ± 5.88 pg/mL) in serum as compared to the secretion in the DSS model group (49.77 ± 7.34 pg/mL). ASB treatments with high, medium, and low dosage also inhibited the production of TNF-α (45.84 ± 17.70 pg/mL, 25.45 ± 19.16 pg/mL, 39.96 ± 9.44 pg/mL, respectively) in serum in comparison with the DSS group, similar to the SASP-treated group. Results also indicated that the expression level of TNF-α (19.96 ± 6.48 pg/mL) in serum presents normal in the high dosage reference group, similar to the expression level in the control group (19.57 ± 4.90 pg/mL). Moreover, results in Figure 5B–D showed that ASB treatments and SASP treatment could greatly down-regulate the expression level of pro-inflammatory mediators TNF-α, IFN-γ, and MCP-1 in colonic tissue. TNF-α production (54.22 ± 12.61 pg/mL, 63.51 ± 27.83 pg/mL, 79.43 ± 36.41 pg/mL, respectively) in colonic tissue was significantly suppressed upon ASB treatments with high, medium and low dosage as compared to the production (98.15 ± 33.93 pg/mL) in the DSS group (Figure 5B). As shown in Figure 5C, DSS-induced colitis markedly increased the production of IFN-γ (309.23 ± 122.83 pg/mL) in colonic tissue, while the production was suppressed upon ASB treatments with high, medium and low dosage (235.22 ±

91.03 pg/mL, 219.13 ± 79.14 pg/mL, 219.52 ± 72.98 pg/mL, respectively). Results in Figure 5D also indicated that ASB treatments with high, medium and low dosage could significantly suppress MCP-1 production (201.57 ± 38.34 pg/mL, 191.57 ± 47.17 pg/mL, 229.67 ± 84.42 pg/mL, respectively) in colonic tissue as compared to the production in DSS-induced colitis (278.24 ± 110.97 pg/mL). Meanwhile, results also indicated that the expression level of TNF-α, IFN-γ, and MCP-1 in colonic tissue were normal in the high dosage reference group, similar to the control group.

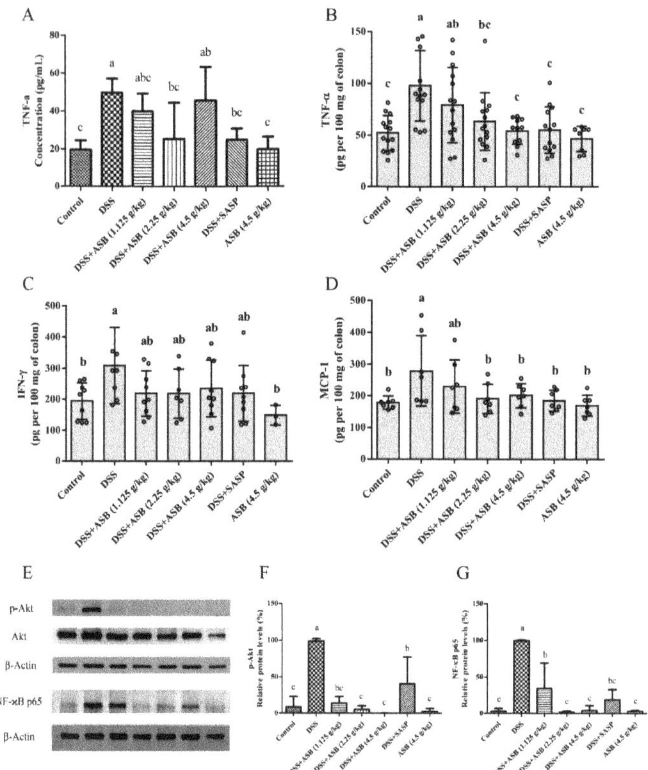

Figure 5. (**A**) Effects of ASB on the production of cytokines TNF-α in the serum of mice with DSS-induced colitis ($n = 3$). Effects of ASB on the production of cytokines (**B**) TNF-α (C: control, $n = 14$; D: DSS model, $n = 14$; L: DSS+ASB low dosage, $n = 14$; M: DSS+ASB medium dosage, $n = 14$; H: DSS+ASB high dosage, $n = 11$; P: SASP, $n = 13$; T: ASB high dosage reference, $n = 8$), (**C**) IFN-γ (C: control, $n = 9$; D: DSS model, $n = 8$; L: DSS+ASB low dosage, $n = 9$; M: DSS+ASB medium dosage, $n = 7$; H: DSS+ASB high dosage, $n = 9$; P: SASP, $n = 9$; T: ASB high dosage reference, $n = 3$), and (**D**) MCP-1 ($n = 7$) in the culture supernatants of colonic tissue of mice with DSS-induced colitis. ASB ameliorates DSS-induced colitis via the TLR4-linked NF-κB signaling pathway. (**E**) Protein levels of p-Akt and NF-κB in the colon were assessed by Western blotting. (**F,G**) Relative protein levels of p-Akt/β-Actin and NF-κB/β-Actin ($n = 3$). Results were expressed as mean ± SD. Parameters marked by the same letter are not significantly different. Significance is represented as $p < 0.05$.

2.8. ASB Improved UC through TLR4 Signaling Inhibition

As shown in Figure 5E–G, the protein levels of NF-κB and p-Akt in colonic tissues were increased in DSS-induced colitis, in comparison with the control group. However, results showed that ASB treatments could significantly down-regulate the protein expression levels of NF-κB and p-Akt in the inflamed colon tissues, similar to the control group.

2.9. ASB Down-Regulated the Expression Levels of Inflammatory Mediators in LPS-Activated Macrophages through TLR4 Signaling Inhibition

As shown in Figure 6C, the production of MCP-1 was markedly increased in the culture of LPS-activated macrophages compared to the control, while the secretion of MCP-1 was significantly suppressed upon ASB treatment in a dose-dependent manner. Moreover, significant ($p < 0.05$) differences were also found in the phosphorylation of TAK1, ERK, JNK, and p38 in TLR4 signaling, except for the protein expression level of ERK at 0.1 mg/mL (Figure 6A,D–G). Due to the phosphorylation of ERK, JNK, and p38, the protein expression levels were down-regulated in the LPS-activated macrophages with dose-dependency upon ASB treatments. Meanwhile, one of the AP-1 components (c-Jun) also significantly reduced the corresponding nuclear localization in LPS-activated macrophages upon ASB treatments (Figure 6C,H,I).

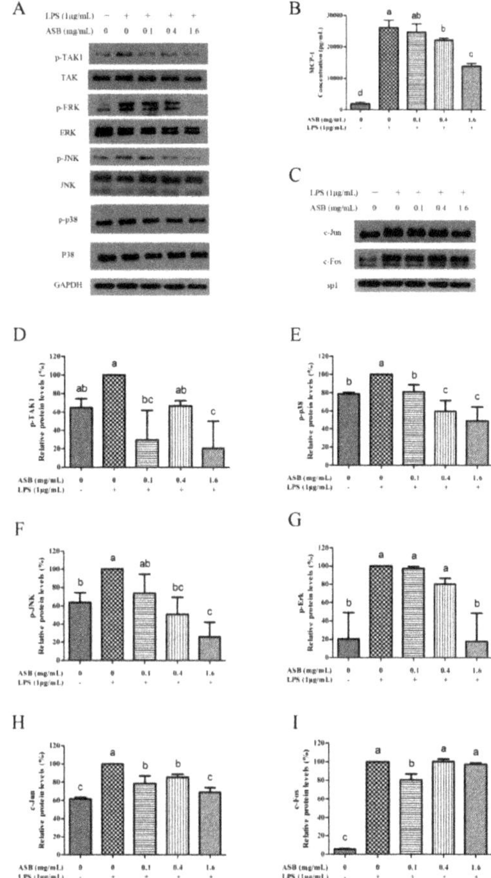

Figure 6. Effects of ASB on the phosphorylation of protein levels on AP-1 pathways in LPS-induced macrophages (**A**), and relative protein levels of p-TAK1 (**D**), p-p38 (**E**), p-JNK (**F**), and p-Erk (**G**). Expression levels of nuclear proteins of transcription factors NF-κB and AP-1 regulated upon ASB treatment (**C**), and relative protein levels of c-Jun (**H**) and c-Fos (**I**). (**B**) Effects of ASB on the secretion of MCP-1 in the culture of LPS-induced macrophages. Results were expressed as mean ± SD. Relative protein levels of each column marked by the same letter are not significantly different. Significance is represented as $p < 0.05$.

3. Discussion

The DSS-induced colitis in C57BL/6 mice showed many similarities in appearance to ulcerative colitis, together with several pathophysiological and morphological features, such as weight loss, shortened colon length, production of inflammatory mediators, crypt damage, and focal inflammation infiltration. Sulfasalazine (SASP), a drug which has been used for treating inflammatory bowel disease (IBD) for decades, is commonly used as a positive control in colitis study [30–32]. The results indicated that ASB treatments showed many similarities to SASP-positive control treatment in treating colitis. In this study, the DSS group demonstrated anemia (lower RBC levels, HGB, HCT, MCV, and PDW), which was significant ($p < 0.05$) as compared to the control group, and which is in agreement with the report by Larrosa et al. (2009) [33]. Meanwhile, results showed that ASB treatments could significantly ameliorate these changes (Table 1). Moreover, body weight, feed consumption, colon length, and DAI were commonly applied as the indicators for evaluating the disease severity of DSS-induced colitis [10,11]. Many researchers have shown that the DSS group could markedly shorten the colon length, lower the body weight, and reduce feed consumption, which is similar to the result in this study [11,34]. Treatments with ASB could significantly prevent colon length shortening, reduce food consumption, and help in losing body weight. The current result (Figure 2B,H) indicated that ASB treatments could reduce rectal bleeding and ameliorate colitis in mice with DSS-induced colitis, similar to the report by Markovic et al. (2016) and Yan et al. (2018) [14,35]. Intestinal permeability was performed to evaluate the barrier function, as it is an important indicator for assessing colitis. Results of various studies indicated that the increase in gut permeability was linked with greater susceptibility to colitis [36,37]. The FITC-Dextran assay was a typical method for the in vivo assessment of intestinal permeability [36,37]. Consistent with the results in colon length and body weight, results showed that ASB could significantly reduce intestinal permeability. Collectively, the results showed that DSS successfully induced colitis, similar to the previous findings. In addition, the results suggest that ASB administration might ameliorate UC.

For further confirmation, histopathology was studied for assessing the degree of colonic tissue damage and neutrophil infiltration. With this context, previous work by Zhu et al. (2017) and Yan et al. (2018) have already indicated that DSS could cause colonic tissue damage, including inflammatory cell infiltration and crypt damage, leading to higher histological score, while such a change could be improved upon suitable treatments [11,14]. The current study showed that ASB could significantly reduce the colonic tissue damage and lead to a lower histological score as compared to the DSS group. Moreover, myeloperoxidase (MPO) is a critical marker for neutrophils, correlating with the extent of neutrophil infiltration [34,38]. Therefore, it is very meaningful to detect the expression level of MPO in colonic tissue to evaluate the degree of colonic damage. Previous studies have already reported that neutrophil infiltration into injured colonic tissue could accelerate the damage of colonic tissue by enzyme MPO [8,34,38]. The current results showed that the expression level of MPO in inflamed tissue could be down-regulated upon ASB treatment (Figure 4A,B). Together, results suggested that ASB treatments with different dosages showed dietary efficacy in ulcerative colitis (UC) amelioration. Based on the evaluation of DAI (Figure 2H), ASB treatment at the dosage of 2.25 g/kg b.w. (the human equivalent dose (HED): 400 g/60 kg) could be treated as the suggested dose for sea bass consumption. Based on the current study, it was clearly indicated that ASB possessed potential therapeutic efficacy against DSS-induced colitis.

Inflammation plays a vital role in DSS-induced colitis. TLR4 signaling, one of the important mechanisms for inflammation, is a key receptor for commensal recognition in gut innate immunity [12,39,40]. TLR4 signaling was the subject of therapeutics (target inhibition) in ulcerative colitis (UC) [12,13,41]. Moreover, AP-1 and NF-κB were treated as the critical and classical pathways in TLR4 signaling [42,43]. Many researchers have already elucidated that over-expression of TLR4-linked AP-1 or NF-κB activation is typical in inflamed colonic tissue [13,14]. Therefore, in vivo and in vitro studies on the efficacy of ASB treatment in inflammation through TLR4 signaling have been studied. The present in vivo and in vitro studies showed that the activation of TLR4 signaling was up-regulated in the DSS-induced colitis and LPS-activated macrophages. However, the up-regulation of TLR4 signaling was markedly inhibited upon ASB treatments. The results indicated that inflammatory mediators in DSS-induced colitis could be inhibited upon ASB treatments, similar to the previous researches [36,44,45]. The in vitro study also indicated that ASB could suppress the inflammatory mediators in LPS-activated macrophage through inhibiting TLR4-linked AP-1 activation. According to the in vivo and in vitro studies, current results demonstrated that ASB treatments ameliorate the intestinal inflammation in the gut and are correlated with the suppression of the activation of TLR4 signaling.

4. Materials and Methods

4.1. Materials and Reagents

Lipopolysaccharide (LPS, purified from *Escherichia coli* O55: B5) and bovine serum albumin were obtained from Sigma Chemical Co. (St. Louis, MO, USA). RAW 264.7 cell line (ATCC no.: TIB-71) derived from murine macrophages was purchased from ATCC (Rockville, MD, USA). TNF-α (D2D4), Akt (11E7), phospho-Akt (Ser473), and NF-κB (p65) were obtained from Cell Signaling Technology (Boston, MA, USA). Myeloperoxidase (MPO) was obtained from Abcam (Cambridge, UK). Other antibodies in NF-κB pathways were purchased from Santa Cruz Biotechnology (Santa Cruz, CA, USA). ELISA kits for the determination of the cytokines TNF-α, MCP-1, and IFN-γ were purchased from Invitrogen (Carlsbad, CA, USA). All other chemicals were of analytical grade.

4.2. Sea Bass Materials and Preparation of Aqueous Extract of Sea Bass

Baijiao sea bass (*Lateolabrax japonicus*) was collected from Estuarine Fisheries Research Institute in Zhuhai, Guangdong Province, China. Aqueous extract of sea bass (ASB) was prepared via steaming, size reduction, sonication, and freeze-drying [46]. Briefly, the edible parts of sea bass were dissected and steamed for 10 min. After steaming, fish bones were removed and the edible parts were homogenized. Then, the homogenized meat was weighed in the beaker, and distilled water (ratio of solid to liquid at 1:5) was added, and then extracted by ultrasonication for 2 h at 90~100 °C. It was followed by freezing at −80 °C and freeze-drying via vacuum freeze—drier (FreeZone Benchtop, Labconco Company, Kansas City, MO, USA). The final products (the aqueous extract of sea bass, ASB) were weighted and stored in −80 °C refrigerator for further analysis.

4.3. Characterization of the Aqueous Extract of Sea Bass (ASB)

Crude protein content was determined using the Kjeldahl method based on the procedure described by Kirk (1950) [47]. The molecular weight of the protein fractions of ASB was analyzed by SDS-PAGE (sodium dodecyl sulphate-polyacrylamide gel electrophoresis) based on the method of Jambrak et al. (2014) with some modifications [48]. The amino acids content of ASB from each batch was analyzed using Biochrom 30 amino acid analyzer (DKSH management Ltd., Shanghai, China).

4.4. Animals

Male C57BL/6 mice (8 weeks, ~22 g) were purchased from SPF (Beijing) Biotechnology Co., Ltd. Animal procedures were approved by the Ethics Committee of Hong Kong Baptist University with an ethical code (REC/18-19/0008, 18 October 2018, committee on the use of human & animal subjects in teaching and research, Hong Kong Baptist University). All animal treatments complied with the "Guide for the Care and Use of Laboratory Animals" published by the National Institutes of Health (NIH).

4.5. Establishment of Ulcerative Colitis (UC) Model

Ulcerative colitis was induced by the oral administration of 1.5% (*w/v*) DSS (relative molecular mass 36–50 kDa; MP Biomedicals) dissolved in drinking water for 7 days according to Heinsbroek et al. (2015) and Zhu et al. (2017) with some modifications [9,11]. Mice were arbitrarily allocated into seven groups: control group (DDI water, $n = 14$ per group), DSS model group ($n = 14$ per group), three ASB treated groups (low-dosage: 1.125 g/kg b.w./day; mid-dosage: 2.25 g/kg b.w./day; high-dosage: 4.5 g/kg b.w./day, $n = 14$ per group), high-dosage reference group (high-dosage: 4.5 g/kg b.w./day, $n = 8$ per group) and sulfasalazine (SASP)-positive control group (SASP, 50 mg/kg b.w./day, $n = 14$ per group). ASB, SASP, and DDI water were given orally by gavage, once daily for 9 days. Meanwhile, drinking water was replaced with 1.5% (*w/v*) DSS solution in the DSS model group, three ASB-treated groups, and SASP-treated group for the first 7 days. Control group and high dosage reference groups received drinking water without DSS throughout the experiment period.

4.6. Intestinal Permeability In Vivo

The measurement of intestinal permeability towards FITC-Dextran (4 kDa, Catalog#4009, Chondrex, Redmond, WA, USA) was performed according to the manufacturer's protocol and Cani et al. (2009) [36]. Briefly, mice that had fasted for 4 h were given FITC-Dextran by oral gavage (500 mg/kg body weight, 25 mg/mL). After maintaining fasting conditions for 3 h, blood was collected from the orbital venous plexus. Then, blood was centrifuged at 14,000× g for 5 min at 4 °C. Plasma was diluted in an equal volume of PBS and read on a fluorescent plate reader at an excitation wavelength of 485 nm and an emission wavelength of 520 nm. A standard curve was prepared by making dilutions of the stock FITC-dextran in normal mouse plasma diluted with PBS.

4.7. Disease Activity Index (DAI)

The characterization of colitis symptoms was monitored by the underlying body weight, stool consistency, stool color, and occult bleeding. The DAI was evaluated based on the scoring methods of Chen et al. (2017) and Yan et al. (2018) with some modifications [10,14]. Weight loss was calculated based on the difference in body weight of mice between day 0 and testing day. Stool consistency and stool color were monitored base on the fecal pellet formation and visible blood by visual identity. Occult blood was analyzed using a fecal occult blood test kit (Nanjing Jiancheng Bioengineering Institute, Nanjing, China). The DAI values were conducted by the sum of the score from weight loss, stool consistency, stool color, and occult bleeding.

4.8. Histology and Immunofluorescence

At the end of the experiment, all mice were euthanized. The colon was dissected and the length between ileo-cecal junction and anal verge was measured. The colonic tissues were fixed in 10% formalin solution overnight and then processed, embedded in paraffin and cut into 4-μm-thick sections. Histopathological examination was conducted by H & E staining according to the manufacturer's instructions. Images were obtained by Nikon Eclipse C1 microscopy (Japan). The histological scoring system (Figure 3B) was described by Xiao et al. (2013) and Winter et al. (2013), with some modifications [34,49]. To quantify the inflammatory responses in colonic tissue, the expression of MPO was detected by immunofluorescence (IF) analysis. The positive expression levels in IF images were processed and quantified using Image-Pro Plus 6.0.

4.9. Cell Culture

RAW 264.7 cells were cultured in DMEM containing 10% FBS and 1% antibiotics (penicillin-streptomycin) at 37 °C under a humidified atmosphere of 95% air and 5% CO_2 by referring to the method described previously [50].

4.10. Western Blotting

For the immunoblot analysis of TLR4, cytoplasmic protein extraction method and analysis were applied, based on the method of Lai et al. (2017) with slight modifications [51]. RAW 264.7 macrophages were grown to confluence in 6-well plates for overnight adhesion and subsequently treated with various concentrations (0.1, 0.4, and 1.6 mg/mL) of ASB for 1 h before LPS (1 µg/mL) stimulation. After 24 h incubation, the cells were collected for Western blotting. The cells were washed with ice-cold PBS twice and then incubated with lysis buffer for 30 min on ice. Supernatants were collected by centrifugation at 13,523× g for 15 min. For nuclear protein extraction, the method of Cheng et al. (2015) was applied, with slight modifications [28]. The collected cells were washed with PBS, and then, hypotonic buffer (15 nM MgCl$_2$, 10 mM KCl, and 20 nM Tris-HCl (pH 7.9)) was added for extraction for 15 min on ice. Then, 12 µL NP-40 (10%, *v/v*) was added for another 10 min. The supernatants were collected as cytoplasmic extracts by centrifugation at 15,777× g at 4 °C for 1 min. The remaining pellets were washed with 100 µL hypotonic buffer and then suspended in a high salt buffer (0.2 mM EDTA, 1.5 mM MgCl$_2$, 0.42 M NaCl, 25% glycerol, and 20 mM Tris-HCl, pH 7.9) on ice for 30 min. The nuclear protein was obtained via centrifugation at 15,777× g for 10 min at 4 °C. For the in vivo experiment, colonic tissues were homogenized in RIPA lysis buffer containing proteinase and phosphatase inhibitor and incubated for 20 min on ice. Supernatants were collected by centrifugation at 13,523× g for 60 min. The concentrations of the extracted proteins were measured using a BSA protein assay. An equal amount of extracted proteins (20–40 µg) was loaded onto the prepared gel (8~12% (*w/v*)) for Western blotting. The membranes were blocked with milk for 1 h and then washed with TBST for 10 min. Primary antibodies were diluted in 3% BSA and overnight cultured with membranes at 4 °C for shaking. Subsequently, membranes were washed with TBST for 20 min three times and then incubated with secondary antibody at room temperature for 1 h. Finally, the membranes were visualized by soaking in a chemiluminescent substrate and then exposed to obtain the signal. Band images were obtained using EPSON scanner, and band densities were analyzed using the Image J software (BioTechniques, New, York, NY, USA).

4.11. Enzyme-Linked Immunosorbent Assay (ELISA) Analysis

The secretion of cytokine (MCP-1) (eBioscience, San Diego, CA, USA) in the culture of LPS-stimulated macrophages was measured using ELISA kits by following the manufacturer's instruction. Additionally, the production of cytokines TNF-α in serum and MCP-1, TNF-α, and IFN-γ in the culture supernatants of colonic tissue of mice upon treatments (eBioscience) were quantified using ELISA kits according to the manufacturer's instructions.

4.12. Statistical Analysis

Analyses were performed in triplicates, and results were expressed as mean ± SD. For multiple group comparisons, one-way analysis of variance (ANOVA) was conducted by Dunnett's post hoc test and applied for determining the significance ($p < 0.05$) differences. Statistical analyses were performed using Microsoft 2016 package and SPSS (SPSS 17.0, SPSS Inc., Chicago, IL, USA).

5. Conclusions

The current results clearly indicate that ASB possesses potential anti-inflammation therapeutic efficacy through inhibiting the activation of TLR4 signaling against DSS-induced colitis and LPS-activated macrophages. According to the in vivo and in vitro studies, the activation of TLR4 signaling was significantly inhibited upon ASB treatments. The production of pro-inflammatory cytokines in inflamed models was markedly reduced upon ASB treatments. Results also indicated that ASB could significantly ameliorate several pathophysiological and morphological features in DSS-induced colitis. The current work illustrated that ASB demonstrated an inhibitory efficacy on TLR4 signaling activation, and thus, could be a promising candidate for treating UC. In addition, it also establishes a pharmacological basis for the folk use of sea bass.

Supplementary Materials: Supplementary materials can be found at http://www.mdpi.com/1422-0067/20/12/2907/s1.

Author Contributions: J.C. performed the in vivo and in vitro experiments, analyzed data and wrote the paper; M.J., W.Z. and L.C. performed part of the in vivo experiments; B.D. performed the determination of amino acids; B.X. and Z.Y. designed the concept, organized the writing and revised the paper.

Funding: This research was funded by grants JCYJ201160229210327924 from STICS, FRG1/16-17/048 and FRG2/17-18/032 from Hong Kong Baptist University, and one research grant from Zhuhai Higher Education Construction Project (Zhuhai Key Laboratory of Agricultural Product Quality and Food Safety).

Conflicts of Interest: The authors declare no conflict of interest.

Abbreviations

ASB	Aqueous extract of sea bass
DSS	Dextran sulphate sodium
ELISA	Enzyme-linked immunosorbent assay
IBD	Inflammatory bowel disease
IFN-γ	Interferon gamma
IF	Immunofluorescence
LPS	Lipopolysaccharide
MPO	Myeloperoxidase
MCP-1	Monocyte chemoattractant protein-1
SASP	Sulfasalazine
TLR	Toll-like receptor
TNF-α	Tumor necrosis factor alpha
UC	Ulcerative colitis

References

1. Fumery, M.; Singh, S.; Dulai, P.S.; Gower-Rousseau, C.; Peyrin-Biroulet, L.; Sandborn, W.J. Natural history of adult ulcerative colitis in population-based cohorts: A systematic review. *Clin. Gastroenterol. Hepatol.* **2018**, *16*, 343–356. [CrossRef] [PubMed]

2. Ng, S.C.; Shi, H.Y.; Hamidi, N.; Underwood, F.E.; Tang, W.; Benchimol, E.I.; Panaccione, R.; Ghosh, S.; Wu, J.C.Y.; Chan, F.K.L.; et al. Worldwide incidence and prevalence of inflammatory bowel disease in the 21st century: A systematic review of population-based studies. *Lancet* **2017**, *390*, 2769–2778. [CrossRef]

3. Qiu, Y.; Chen, B.; Li, Y.; Xiong, S.; Zhang, S.; He, Y.; Zeng, Z.R.; Ben-Horin, S.; Chen, M.H.; Mao, R. Risk factors and long-term outcome of disease extent progression in Asian patients with ulcerative colitis: A retrospective cohort study. *BMC Gastroenterol.* **2019**, *19*, 7. [CrossRef] [PubMed]

4. Loddo, I.; Romano, C. Inflammatory bowel disease: Genetics, epigenetics, and pathogenesis. *Front. Immunol.* **2015**, *6*, 551. [CrossRef] [PubMed]

5. Ananthakrishnan, A.N.; Xavier, R.J.; Podolsky, D.K. Inflammatory Bowel Diseases: Pathogenesis. In *Yamada's Textbook of Gastroenterol*; Wiley-Blackwell: New York, NY, USA, 2015; pp. 1364–1377.

6. Mazmanian, S.K.; Round, J.L.; Kasper, D.L. A microbial symbiosis factor prevents intestinal inflammatory disease. *Nature* **2008**, *453*, 620–625. [CrossRef] [PubMed]

7. Wen, L.; Ley, R.E.; Volchkov, P.Y.; Stranges, P.B.; Avanesyan, L.; Stonebraker, A.C.; Hu, C.Y.; Wong, S.S.; Szot, G.L.; Bluestone, J.A.; et al. Innate immunity and intestinal microbiota in the development of type 1 diabetes. *Nature* **2008**, *455*, 1109–1113. [CrossRef]

8. Hakansson, A.; Badia Tormo, N.; Baridi, A.; Xu, J.; Molin, G.; Hagslatt, M.L.; Karlsson, C.; Jeppsson, B.; Cilio, C.M.; Ahrne, S. Immunological alteration and changes of gut microbiota after dextran sulfate sodium (DSS) administration in mice. *Clin. Exp. Med.* **2015**, *15*, 107–120. [CrossRef]

9. Heinsbroek, S.E.M.; Williams, D.L.; Welting, O.; Meijer, S.L.; Gordon, S.; Jonge, W.J. Orally delivered β-glucans aggravate dextran sulfate sodium (DSS)—Induced intestinal inflammation. *Nutr. Res.* **2015**, *35*, 1106–1112. [CrossRef]

10. Chen, L.; Wilson, J.E.; Koenigsknecht, M.J.; Chou, W.C.; Montgomery, S.A.; Truax, A.D.; Brickey, W.J.; Packey, C.D.; Maharshak, N.; Matsushima, G.K.; et al. NLRP12 attenuates colon inflammation by maintaining colonic mircobial diversity and promoting protective commensal bacterial growth. *Nat. Immunol.* **2017**, *18*, 541–551. [CrossRef]

11. Zhu, W.H.; Winter, M.G.; Byndloss, M.X.; Spiga, L.; Duerkop, B.A.; Hughes, E.R.; Buttner, L.; Romao, E.D.L.; Behrendt, C.L.; Lopez, C.A.; et al. Precision editing of the gut microbiota ameliorates colitis. *Nature* **2018**, *553*, 208–211. [CrossRef]

12. Cui, L.; Feng, L.; Zhang, Z.H.; Jia, X.B. The anti-inflammation effect of baicalin on experimental colitis through inhibiting TLR4/NF-κB pathway activation. *Int. Immunopharmacol.* **2014**, *23*, 294–303. [CrossRef] [PubMed]

13. Kim, K.A.; Lee, I.A.; Gu, W.; Hyam, S.R.; Kim, D.H. β-Sitosterol attenuates high fat diet induced intestinal inflammation in mice by inhibiting the binding of lipopolysaccharide to toll-like receptor 4 in the NF-κB pathway. *Mol. Nutr. Food Res.* **2014**, *58*, 963–972. [CrossRef] [PubMed]

14. Yan, Y.X.; Shao, M.J.; Qi, Q.; Xu, Y.S.; Yang, X.Q.; Zhu, F.H.; He, S.J.; He, P.L.; Feng, C.L.; Wu, Y.W.; et al. Artemisinin analogue SM934 ameliorates DSS-induced mouse ulcerative colitis via suppressing neutrophils and macrophages. *Acta Pharmacol. Sin.* **2018**, *39*, 1633–1644. [CrossRef] [PubMed]

15. Nakatsuji, M.; Minami, M.; Seno, H.; Yasui, M.; Komekado, H.; Higuchi, S.; Fujikawa, R.; Nakanishi, Y.; Fukuda, A.; Kawada, K.; et al. EP4 receptor—Associated protein in macrophages ameliorates colitis and colitis-associated tumorigenesis. *PLoS Genet.* **2015**, *11*, e1005542. [CrossRef] [PubMed]

16. Haribhai, D.; Ziegelbauer, J.; Jia, S.; Upchurch, K.; Yan, K.; Schmitt, E.G.; Salzman, N.H.; Simpson, P.; Hessner, M.J.; Chatila, T.A.; et al. Alternatively activated macrophages boost induced regulatory T and Th17 cell responses during immunotherapy for colitis. *J. Immunol.* **2016**, *196*, 3305–3317. [CrossRef]

17. Liu, Y.T.; Fang, S.L.; Li, X.Y.; Feng, J.; Du, J.; Guo, L.J.; Su, Y.Y.; Zhou, J.; Ding, G.; Bai, Y.X.; et al. Aspirin inhibits LPS-induced macrophage activation via the NF-κB pathway. *Sci. Rep.* **2017**, *7*, 11549. [CrossRef]

18. Pham, T.H.; Kim, M.S.; Le, M.Q.; Song, Y.S.; Bak, Y.; Ryu, H.W.; Oh, S.R.; Yoon, D.Y. Fargesin exerts anti-inflammatory effects in THP-1 monocytes by suppressing PKC-dependent AP-1 and NF-kB signaling. *Phytomedicine* **2017**, *24*, 96–103. [CrossRef] [PubMed]

19. Hsu, C.L.; Lin, Y.J.; Ho, C.T.; Ye, G.C. The inhibitory effect of pterostilbene on inflammatory responses during the interaction of 3T3-L1 adipocytes and RAW264.7 macrophages. *J. Agric. Food Chem.* **2013**, *61*, 602–610. [CrossRef] [PubMed]

20. Jang, K.J.; Choi, S.H.; Yu, G.J.; Hong, S.H.; Chung, Y.H.; Kim, C.H.; Yoon, H.M.; Kim, G.Y.; Kim, B.W.; Choi, Y.H. Anti-inflammatory potential of total saponins derived from the roots of *Panax ginseng* in lipopolysaccharide—Activated RAW 264.7 macrophages. *Exp. Ther. Med.* **2016**, *11*, 1109–1115. [CrossRef]

21. Chen, H.; Sohn, J.; Zhang, L.; Tian, J.; Chen, S.; Bjeldanes, L.F. Anti-inflammatory effects of chicanine on murine macrophage by down-regulating LPS-induced inflammatory cytokines in IκBα/MAPK/ERK signaling pathways. *Eur. J. Pharmacol.* **2014**, *724*, 168–174. [CrossRef]

22. Cheng, B.C.Y.; Ma, X.Q.; Kwan, H.Y.; Tse, K.W.; Cao, H.H.; Su, T.; Shu, X.; Wu, Z.Z.; Yu, Z.L. A herbal formula consisting of Rosae Multiflorae Fructus and Lonicerae Japonicae FLoS inhibits inflammatory mediators in LPS-stimulated RAW 264.7 macrophages. *J. Ethnopharmacol.* **2014**, *153*, 922–927. [CrossRef] [PubMed]

23. Du, B.; Yang, Y.D.; Bian, Z.X.; Xu, B.J. Characterization and anti-inflammatory potential of and exopolysaccharide from submerged mycelial culture of *Schizophyllum commune*. *Front. Pharmacol.* **2017**, *8*, 252. [CrossRef] [PubMed]

24. Delhase, M. Kappa B kinase and NF-kappa B signaling in response to pro-inflammatory cytokines. *Methods Mol. Biol.* **2003**, *225*, 7–17. [PubMed]

25. Ahn, C.B.; Jung, W.K.; Park, S.J.; Kim, Y.T.; Kim, W.S.; Je, J.Y. Gallic acid-g-chitosan modulates inflammatory responses in LPS stimulated RAW264.7 cells via NF-kB, AP-1 and MAPK pathways. *Inflammation* **2015**, *39*, 366–374. [CrossRef] [PubMed]

26. Kumar, S.; Joos, G.; Boon, L.; Tournoy, K.; Provoost, S.; Maes, T. Role of tumor necrosis factor—α and its receptors in diesel exhaust particle-induced pulmonary inflammation. *Sci. Rep.* **2017**, *7*, 711508. [CrossRef] [PubMed]

27. Wu, J.; Liu, Z.; Su, J.; Pan, N.; Song, Q. Anti-inflammatory activity of 3β-hydroxycholest-5-en-7-one isolated from *Hippocampus trimaculatus* leach via inhibiting iNOS, TNF-α, and 1L-1β of LPS induced RAW 264.7 macrophage cells. *Food Funct.* **2017**, *8*, 788–795. [CrossRef] [PubMed]

28. Cheng, B.C.Y.; Yu, H.; Su, T.; Fu, X.Q.; Guo, H.; Li, T.; Cao, H.H.; Tse, A.K.W.; Kwan, H.Y.; Yu, Z.L. A herbal formula comprising Rosae Multiflorae Fructus and Lonicerae Japonicae FLoS inhibits the production of inflammatory mediators and the IRAK-1/TAK1 and TBK1/IRF3 pathways in RAW 264.7 and THP-1 cells. *J. Ethnopharmacol.* **2015**, *174*, 195–199. [CrossRef] [PubMed]

29. Xie, J.X.; Yang, L.; Tian, L.; Li, W.Y.; Yang, L.; Li, L.Y. Macrophage migration inhibitor factor up-regulates MCP-1 expression in an autocrine manner in hepatocytes during acute mouse liver injury. *Sci. Rep.* **2016**, *6*, 27665. [CrossRef]

30. Peskar, B.M.; Dreyling, K.W.; Peskar, B.A.; May, B.; Goebell, H. Enhanced formation of sulfidopeptide-leukotrienes in ulcerative colitis and Crohn's disease: Inhibition by sulfasalazine and 5-aminosalicylic acid. *Agents Actions Suppl.* **1986**, *18*, 381–383. [CrossRef]

31. Sutherland, L.R.; MacDonald, J.K. Oral 5-aminosalicylic acid for induction of remission in ulcerative colitis. *Cochrane Database Syst. Rev.* **2006**. [CrossRef]

32. Du, B.; Yang, Y.D.; Bian, Z.X.; Xu, B.J. Molecular weight and helix conformation determine intestinal anti-inflammatory effects of exopolysaccharide from *Schizophyllum commune*. *Carbohydr. Polym.* **2017**, *172*, 68–77. [CrossRef] [PubMed]

33. Larrosa, M.; Gascon, M.J.Y.; Selma, M.V.; Sarrias, A.G.; Toti, S.; Ceron, J.J.; Barberan, F.T.; Dolara, P.; Espin, J.C. Effect of a low dose of dietary resveratrol on colon microbiota, inflammation and tissue damage in DSS-induced colitis rat model. *J. Agric. Food Chem.* **2009**, *57*, 2211–2220. [CrossRef] [PubMed]

34. Xiao, H.T.; Lin, C.Y.; Ho, D.H.H.; Peng, J.; Chen, Y.; Tsang, S.W.; Wong, M.; Zhang, X.J.; Zhang, M.; Bian, Z.X. Inhibitory effect of the gallotannin corilagin on dextran sulfate sodium-induced murine ulcerative colitis. *J. Nat. Prod.* **2013**, *76*, 2120–2125. [CrossRef] [PubMed]

35. Markovic, B.S.; Nikolic, A.; Gazdic, M.; Bojic, S.; Vucicevic, L.; Kosic, M.; Mitrovic, S.; Milosavljevic, M.; Besra, G.; Trajkovic, V.; et al. Gal-3 plays an important pro-inflammatory role in the induction phase of acute colitis by promoting activation of NLRP3 inflammasome and production of IL-1β in macrophages. *J. Crohns Colitis* **2016**, *10*, 593–606. [CrossRef] [PubMed]

36. Cani, P.D.; Possemiers, S.; Wiele, T.V.; Guiot, Y.; Everard, A.; Rottier, O.; Geurts, L.; Naslain, D.; Neyrinck, A.; Lambert, D.M.; et al. Changes in gut microbiota control inflammation in obese mice through a mechanism involving GLP-2-driven improvement of gut permeability. *Gut* **2009**, *58*, 1091–1103. [CrossRef]

37. Zhao, H.; Zhang, H.; Wu, H.; Li, H.; Liu, L.; Guo, J.; Li, C.Y.; Shih, D.Q.S.; Zhang, X.L. Protective role of 1, 25 $(OH)_2$ vitamin D_3 in the mucosal injury and epithelial barrier disruption in DSS-induced acute colitis in mice. *BMC Gastroenterol.* **2012**, *12*, 57. [CrossRef] [PubMed]

38. Muthas, D.; Reznichenko, A.; Balendran, C.A.; Böttcher, G.; Clausen, I.G.; Kärrman Mårdh, C.; Ottosson, T.; Uddin, M.; MacDonald, T.T.; Danese, S.; et al. Neutrophils in ulcerative colitis: A review of selected biomarkers and their potential therapeutic implications. *Scand. J. Gastroenterol.* **2017**, *52*, 125–135. [CrossRef]

39. Fukata, M.; Shang, L.; Santaolalla, R.; Sotolongo, J.; Pastorini, C.; España, C.; Ungaro, R.; Harpaz, N.; Cooper, H.S.; Elson, G.; et al. Constitutive activation of epithelial TLR4 augments inflammatory responses to mucosal injury and drives colitis-associated tumorigenesis. *Inflamm. Bowel Dis.* **2010**, *17*, 1464–1473. [CrossRef]

40. Abreu, M.T.; Fukata, M.; Arditi, M. TLR signaling in the gut in health and disease. *J. Immunol.* **2005**, *174*, 4453–4460. [CrossRef]

41. Bantel, H.; Berg, C.; Vieth, M.; Stolte, M.; Kruis, W.; Schulze-Osthoff, K. Mesalazine inhibits activation of transcription factor NF-κB in inflamed mucosa of patients with ulcerative colitis. *Am. J. Gastroenterol.* **2000**, *95*, 3452. [CrossRef]

42. Lu, Y.C.; Yeh, W.C.; Ohashi, P.S. LPS/TLR4 signal transduction pathway. *Cytokine* **2008**, *42*, 145–151. [CrossRef] [PubMed]

43. Endale, M.; Park, S.C.; Kim, S.; Kim, S.H.; Yang, Y.; Cho, J.Y.; Rhee, M.H. Quercetin disrupts tyrosine-phosphorylated phosphatidylinositol 3-kinase and myeloid differentiation factor-88 association, and inhibits MAPK/AP-1 and IKK/NF-κB-induced inflammatory mediators production in RAW 264.7 cells. *Immunobiology* **2013**, *218*, 1452–1467. [CrossRef] [PubMed]

44. Demon, D.; Kuchmiy, A.; Fossoul, A.; Zhu, Q.; Kanneganti, T.D.; Lamkanfi, M. Caspase-11 is expressed in the colonic mucosa and protects against dextran sodium sulfate-induced colitis. *Nature* **2014**, *7*, 1480–1491.

45. Chen, Q.Y.; Duan, X.Y.; Fan, H.; Xu, M.; Tang, Q.; Zhang, L.J.; Shou, Z.X.; Liu, X.X.; Zuo, D.M.; Yang, J.; et al. Oxymatrine protects against DSS-induced colitis via inhibiting the PI3K/AKT signalling pathway. *Int. Immunopharmacol.* **2017**, *53*, 149–157. [CrossRef] [PubMed]

46. Chen, J.L.; Jayachandran, M.; Xu, B.J.; Yu, Z.L. Sea bass (*Lateolabrax maculatus*) accelerates wound healing: A transition from inflammation to proliferation. *J. Ethnopharmacol.* **2019**, *236*, 263–276. [CrossRef] [PubMed]

47. Kirk, P.L. Kjeldahl method for total nitrogen. *Anal. Chem.* **1950**, *22*, 354–358. [CrossRef]

48. Jambrak, A.R.; Mason, T.J.; Lelas, V.; Paniwnyk, L.; Herceg, Z. Effect of ultrasound treatment on particle size and molecular weight of whey proteins. *J. Food Eng.* **2014**, *121*, 15–23. [CrossRef]
49. Winter, S.E.; Winter, M.G.; Xavier, M.N.; Thiennimitr, P.; Poon, V.; Keestra, A.M.; Laughlin, R.C.; Gomez, G.; Wu, J.; Lawhon, S.D.; et al. Host derived nitrate boosts growth of *E. coli* in the inflamed gut. *Science* **2013**, *339*, 708–771. [CrossRef]
50. Xu, X.L.; Yin, P.; Wan, C.R.; Chong, X.L.; Liu, M.J.; Cheng, P.; Chen, J.J.; Liu, F.H.; Xu, J.Q. Punicalagin inhibits inflammation in LPS-induced RAW264.7 macrophages via the suppression of TLR4-mediated MAPKs and NF-κB activation. *Inflammation* **2014**, *37*, 956–965. [CrossRef]
51. Lai, J.L.; Liu, Y.H.; Liu, C.; Qi, M.P.; Liu, R.N.; Zhu, X.F.; Zhou, Q.G.; Chen, Y.Y.; Guo, A.Z. Indirubin inhibits LPS-induced inflammation via TLR4 abrogation mediated by the NF-κB and MAPK signaling pathways. *Inflammation* **2017**, *40*, 1–12. [CrossRef]

International Journal of
Molecular Sciences

Review

Costunolide—A Bioactive Sesquiterpene Lactone with Diverse Therapeutic Potential

Dae Yong Kim [1] and Bu Young Choi [2,*]

[1] Department of Biology Education, Seowon University, Cheongju, Chungbuk 361-742, Korea; kensei@hanmail.net
[2] Department of Pharmaceutical Science & Engineering, Seowon University, Cheongju, Chungbuk 361-742, Korea
* Correspondance: bychoi@seowon.ac.kr; Tel.: +82-043-299-8411

Received: 23 April 2019; Accepted: 12 June 2019; Published: 14 June 2019

Abstract: Sesquiterpene lactones constitute a major class of bioactive natural products. One of the naturally occurring sesquiterpene lactones is costunolide, which has been extensively investigated for a wide range of biological activities. Multiple lines of preclinical studies have reported that the compound possesses antioxidative, anti-inflammatory, antiallergic, bone remodeling, neuroprotective, hair growth promoting, anticancer, and antidiabetic properties. Many of these bioactivities are supported by mechanistic details, such as the modulation of various intracellular signaling pathways involved in precipitating tissue inflammation, tumor growth and progression, bone loss, and neurodegeneration. The key molecular targets of costunolide include, but are not limited to, intracellular kinases, such as mitogen-activated protein kinases, Akt kinase, telomerase, cyclins and cyclin-dependent kinases, and redox-regulated transcription factors, such as nuclear factor-kappaB, signal transducer and activator of transcription, activator protein-1. The compound also diminished the production and/expression of proinflammatory mediators, such as cyclooxygenase-2, inducible nitric oxide synthase, nitric oxide, prostaglandins, and cytokines. This review provides an overview of the therapeutic potential of costunolide in the management of various diseases and their underlying mechanisms.

Keywords: costunolide; antioxidants; anti-inflammatory; anti-allergic; bone regenerating; neuroprotective; antimicrobial; hair growth promoting; anticancer; antidiabetic properties

---

1. Introduction

Drug development from natural sources, particularly from plants, has long been the mainstay in medical management of various human ailments. A wide variety of non-nutritive plant constituents, commonly known as phytochemicals, are being used as therapy for many disease processes, including, but not limited to, infections, diabetes, heart diseases, neurological disorders, and cancer. In fact, it is estimated that about 40% of all medicines are natural compounds or their semisynthetic derivatives [1]. One of the major classes of bioactive phytochemicals is the terpenoids, which are widely present in various plants and marine organisms, and are being examined for developing new antifungal, anticancer, anti-inflammatory, and antiviral agents [2]. For example, artemisinin and paclitaxel are terpenoids used clinically as antimalarial and anticancer agents, respectively. The largest group of sesquiterpene lactones is germacranolides [3], which possesses a 10,5-ring structure and is present in several plant families. Germacranolides are key precursors of other sesquiterpene lactones with various polycyclic skeletons, such as guaianolides, eudesmanolides, etc. [4]. Costunolide, a colorless crystalline powder with a molecular formula of $C_{15}H_{20}O_2$ and a molecular weight of 232.318 g/mol, is a well-known sesquiterpene lactone in the germacranolides series. This compound was first isolated from costus (*Saussurea lappa* Clarke) root and then isolated from various other plant species. [5]. Structurally, costunolide (Figure 1) is a monocarboxylic acid having three double bonds which by catalytic hydrogenation generates hexahydrocostunolide. Partial hydrogenation of costunolide produces dihydrocostunolide [6]. The bioactivity of costunolide is mediated through its functional moiety, α-methylene-γ-lactone, which can react with the cysteine sulfhydryl group of various proteins, thereby altering intracellular redox balance [5]. This review is aimed at summarizing the recent research on costunolide, focusing on its therapeutic potential, underlying mechanisms of action, and the prospect of using costunolide for future drug development.

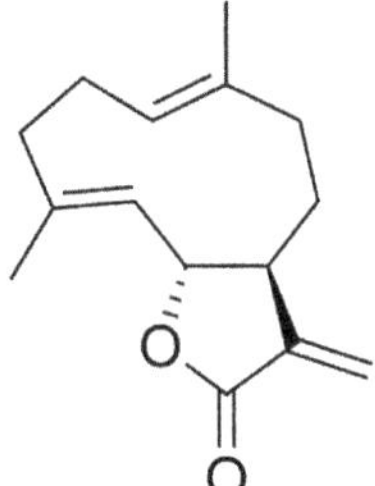

Figure 1. Chemical structure of costunolide.

2. Therapeutic Potential of Costunolide

2.1. Antioxidant and Anti-Inflammatory Effects of Costunolide

Oxidative stress resulting from cellular redox imbalance leads to many diseases, such as diabetes, atherosclerosis, and cardiovascular diseases [7]. The antioxidant activity of costunolide was studied in streptozotocin (STZ)-induced diabetic rat model, which demonstrated marked reduction in the levels of glutathione (GSH) in the brain, heart, liver, pancreas, and kidney. Oral administration of costunolide restored the GSH level in these tissues [8]. Increased levels of GSH may increase the levels of GSH-dependent enzymes, such as glutathione peroxidase (GPx) and glutathione-S-transferase (GST), thereby reducing tissue damage [9]. Oxidative stress oxidizes and damages membrane phospholipid to produce lipid peroxides, such as malondialdehyde (MDA) and hydroxynonenals (HNE), which by forming DNA adducts may cause oxidative tissue damage. Costunolide also decreased lipid peroxidation levels and increased in SOD, catalase, and GPx activity in MCF-7 & MDA-MB-231 cells [10]. In a rat intestinal mucositis (IM) model, administration of costunolide restored 5-floirouracil (5FU)-depleted plasma superoxide dismutase (SOD) levels in rat intestinal mucosa [11]. Costunolide also abrogated hydrogen peroxide (H_2O_2)-induced ROS production in rat pheochromocytoma (PC12) cells [12].

Persistent tissue inflammation plays an important role in the pathogenesis of various infectious and noninfectious diseases, such as rheumatoid arthritis, Alzheimer's disease, and arteriosclerosis [13]. Costunolide exhibited anti-inflammatory properties in a number of preclinical studies. The compound attenuated carrageenan-induced paw edema, myeloperoxidase (MPO) activity and *N*-acetylglucosaminidase (NAG) activity in mice [13]. One of the transcriptional regulators of proinflammatory gene expression is the transcription factor nuclear factor-kappaB (NF-κB). Costunolide negated NF-κB activation via blockade of IκBα phosphorylation in lipopolysaccharide (LPS)-stimulated RAW264.7 cells, thereby reducing the expression of proinflammatory markers, such as inducible nitric oxide synthase (iNOS), and the production of nitric oxide (NO) [14]. Chen et al. also demonstrated that treatment with costunolide inhibited 5-fluorouracil (5-FU)-induced expression of iNOS, cyclooxygenase-2 (COX-2), TNF-α, and the production of nitric oxide (NO) in a mouse model of intestinal mucositis by blocking the activation of NF-κB [11]. Costunolide diminished STAT1 and STAT3 phosphorylation in IL-22 or IFN-γ-induced human keratinocytes [15]. Likewise, treatment of human THP-1 cells with costunolide inhibited interleukin (IL)-6-induced phosphorylation and the

DNA binding activity of signal transducer and activator of transcription (STAT)-3 via downregulation of Janus-activated kinase (JAK)-1 and -2 [16]. Moreover, costunolide showed an anti-inflammatory effect as evidenced by amelioration of ethanol-induced gastric ulcers in mice. This study also reported that the compound suppressed the activation and/or induction of NF-κB, TNF-α, NO, iNOS, and COX-2 [17]. Costunolide inhibited interleukin (IL)-1β protein and mRNA expression in LPS-stimulated RAW264.7 cells by blocking activator protein (AP-1) transcriptional activity via downregulation of mitogen-activated protein kinase (MAPK) phosphorylation [18]. In addition, costunolide alleviated lung inflammation in carrageenan-induced mouse pleurisy model as evidenced by reduced accumulation of polymorphonuclear cells and reduced expression of TNF-α, intracellular adhesion molecule-1 (ICAM-1), P-selectin, and nitrotyrosine [19].

Heme oxygenase-1 (HO-1) has been reported to mediate anti-inflammatory and cytoprotective activity [20]. Pae and colleagues [21] have reported that the production of TNF-α and IL-6 in LPS-stimulated RAW264.7 cells was decreased by treatment with costunolide, which increased the expression and activity of HO-1 via enhanced nuclear accumulation of a redox-regulated transcription factor, nuclear factor erythroidrelated factor-2 (Nrf2). Pretreatment with a HO-1 inhibitor abrogated the inhibitory effect of costunolide on LPS-induced TNF-α and IL-6 production [21]. CD4$^+$ T cell activation and proper differentiation into T helper (Th) cells are important for establishing an adaptive immune response against foreign pathogens. However, an excessive activation of Th cells leads to inflammation and autoimmune diseases [22]. When CD4$^+$ T cells were induced to differentiate, costunolide markedly reduced the differentiation into a population of Th1, Th2, and Th17 subsets. Costunolide also inhibited the expression level of Th subset-polarizing master genes such as T-bet, GATA3, and RORγt. The compound reduced the level of CD4$^+$ T cell activation marker CD69 and attenuated T cell proliferation by blocking phosphorylation of extracellular signal-regulated kinase (ERK) and p38 MAPK [23].

2.2. Anti-Allergic Effects of Costunolide

Chemokines play an important role in inducing various allergic and inflammatory skin diseases, such as atopic dermatitis, psoriasis, and eczema. Keratinocytes are known to respond to various chemokines, such as chemokine (C-C motif) ligand (CCL)-17 (also known as TARC), CCL-22 (alternatively known as MDC), CCL-5 (synonym RANTES), and IL-8, which are involved in precipitating atopic dermatitis [24]. Costunolide significantly reduced mRNA expression of various chemokines including TARC/CCL17, MDC/CCL22, RANTES/CCL5, and IL-8 in HaCaT cells stimulated with TNF-α and IFN-γ [25]. In the OVA-induced asthmatic mouse model, costunolide reduced eosinophil infiltration, inflammation score, and mucin secretion in the lungs. In particular, the increase of eosinophil count in BALF (bronchoalveolar lavage fluid) by OVA was significantly inhibited by costunolide. Moreover, the compound decreased the expression and secretion of Th2 cytokines (IL-4 and IL-13) in BALF and lung tissue [26]. Costunolide reduced the activity of β-hexosaminidase, an enzyme involved in mast cell degranulation, and decreased IL-4 mRNA transcript in IgE-sensitized rat basophilic leukemia (RBL-2H3) cells. In addition, the inhibition of IL-5-dependent growth of Y16 pro-B cells suggests the potential of costunolide or its derivatives to be developed as mast cell stabilizers and pro-B cell proliferation inhibitors in allergic diseases [26,27].

2.3. Costunolide in Bone Remodeling

Osteoporosis, a disease of the bone generally characterized by excessive bone resorption due to poor osteoblastic and enhanced osteoclastic activity, is very common among the elderly population. Although few therapeutic interventions, such as the use of calcium and vitamin D3, parathyroid hormone analogs, bisphosphonates, and monoclonal antibodies, are current clinical recommendations, there is emerging need of developing new drugs [28]. Several studies have demonstrated the potential of costunolide in improving bone health. Costunolide stimulated the growth and differentiation of murine osteoblastic cells (MC3T3-E1) cells, as characterized by increased alkaline phosphatase (ALP) activity, collagen deposition and mineralization. These osteoblastic activity of costunolide was abrogated by cotreatment with pharmacological inhibitors of either estrogen receptor (ER) or phoaphatidylinositol-3-kinase (PI3K), suggesting that the increased mineralization by the compound was associated with increased activation of ER and PI3K [29]. Likewise, costunolide increased the ALP activity and matrix mineralization, and elevated the transcription of a number of differentiation factors, such as distal-less homeobox 5 (Dlx5), runt-related transcription factor 2 (Runx2), and osteocalcin (OC) in mouse mesenchymal stem cell (C3H10T1/2) by stimulating activated transcription factor 4 (ATF4)-dependent increased expression and activity of HO-1. The blockade of HO-1 by treating cells with tin (IV) protoporphyrin IX dichloride (SnPP) blocked costunolide-induced Runx2 expression, suggesting that costunolide-induced osteoblast differentiation is regulated by ATF4-dependent HO-1 expression [30]. The receptor activator of nuclear factor kappa-B ligand (RANKL) induces the differentiation of bone marrow-derived macrophages into osteoclasts, a key mediator of bone resorption. Treatment with costunolide inhibited osteoclast differentiation by blocking the expression of nuclear factor of activated T cells, cystoplasic-1 (NFATc1) through the inhibition of c-Fos transcriptional activity, without affecting c-Fos expression. The compound also attenuated the mRNA expression of tartrate-resistant acid phosphatase (TRAP) and osteoclast-associated receptor (OSCAR) (Figure 2). Thus, costunolide inhibited RANKL-induced osteoclast differentiation by inhibiting c-Fos transcriptional activity [31].

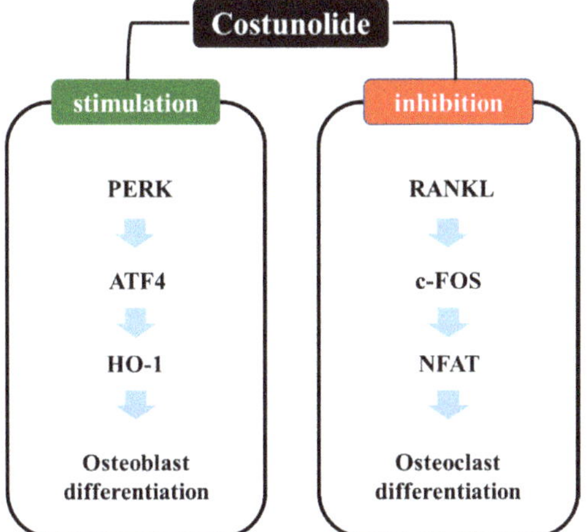

Figure 2. Effect of costunolide on differentiation of osteoblast and osteoclast. Costunolide induces osteoblast differentiation through ATF-4-induced HO-1 expression in mesenchymal stem cells. On the other hand, costunolide suppressed RANKL-mediated osteoclast differentiation via inhibiting RANKL-mediated c-Fos transcriptional activity in bone marrow cells.

2.4. Costunolide as a Neuroprotective Agent

Parkinson disease (PD) is one of the neurodegenerative diseases characterized by reduced dopaminergic (DAergic) neuronal transmission in the substantia nigra (SN). One of the key regulators of DAergic nerve transmission, especially the regeneration of synaptic vesicles, and the storage, metabolism, and release of DA at nerve endings is α-synuclein (ASYN), which is transcriptionally regulated by nuclear receptor related-1 (Nurr1). In PD patients with Nurr1 mutations, Nurr1 expression was decreased and ASYN expression was increased. Whereas the expression of the Nurr1 gene is essential for the development and maintenance of nigral DAergic neurons, overexpression of ASYN causes selective degeneration and toxicity of DAergic neurons. Nurr1 also participates in DA metabolism by regulating vesicular monoamine transporter type 2 (VMAT2) and dopamine transporter (DAT) [32]. Ham et al demonstrated that costunolide inhibited DA-induced apoptosis of human neuroblastoma (SH-SY5Y) cells which was associated with decreased ASYN expression and the restoration of DA-mediated reduced Nurr1, VMAT2, and DAT level [33]. These results suggest the potential of costunolide in the management of PD. Since oxidative stress-mediated neuronal cell death is a well-known cause of many neurodegenerative diseases, the reduction of reactive oxygen species (ROS) can be a pragmatic approach to delay the disease progression. Treatment with costunolide inhibited H_2O_2-induced apoptosis of PC12 cells by reducing intracellular ROS, stabilizing mitochondrial membrane potential (MMP), and decreasing the caspase-3 activity (Figure 3). Moreover, costunolide reduced H_2O_2-induced cell death by blocking the phosphorylation of p38 MAPK and ERK [12]. Besides oxidative stress, persistent inflammation often leads to neurodegeneration. By virtue of its anti-inflammatory properties costunolide inhibited LPS-induced apoptosis of BV2 microglial cells by decreasing the expression of a series of neuroinflammatory mediators, such as TNF-α, IL-1, IL-6, iNOS, macrophage chemoattractant protein-1 (MCP-1), and COX-2 via the inhibition of NF-κB and MAPK activation [34].

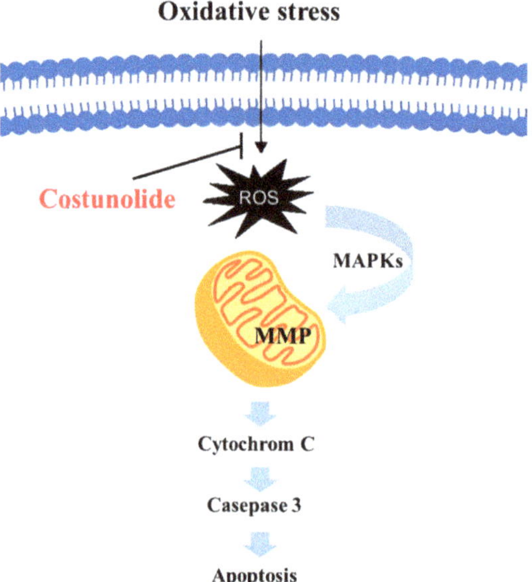

Figure 3. The effect of costunolide on apoptosis of neurons. Costunolide reduced intracellular ROS caused by oxidative stress. As a result, mitochondrial membrane potential (MMP) stabilized and apoptosis-related proteins such as caspase 3 decreased.

2.5. Antimicrobial Properties of Costunolide

Several studies have demonstrated the antimicrobial activity of costunolide (Table 1). The antibacterial activity of costunolide against *Mycobacterium tuberculosis* H37Rv (*M. tuberculosis*) [35] and *Mycobacterium avium* (*M. avium*) [36] in fluorometric Alamar Blue microassay and radiorespirometric bioassay, respectively, suggest that the compound may be considered for developing antitubercular drugs. In addition, in vitro agar diffusion test showed that costunolide exhibited antimicrobial activity against *Staphylococcus aureus* (*S. aureus*), *Escherichia coli* (*E. coli*), and *Pseudomonas aeruginosa* (*P. aeruginosa*) [37]. Costunolide also inhibited the growth of *H. pylori* [38], which is causally linked with gastric and duodenal ulcers. In vitro disc diffusion assay revealed that costunolide inhibited the growth of various pathogenic fungi, such as *Trichophyton mentagrophytes*, *T. simum*, *T. rubrum*, *Epidermophyton floccosum*, *Scopulariopsis* sp., *Aspergillus niger*, *Curvulari lunata*, *Magnaporthe grisea*, and *Candida albicans* [39]. Costunolide also showed antifungal activity against *Botauttis cinereal*, *Colletotrichum acutatum*, *Colletotrichum fragariae* and *Colletotrichum gloeosporioides* [40], and *C. echinulata* [41]. The antiviral property of costunolide was evident from the inhibition of hepatitis B surface antigen (HBsAg) expression in human hepatoma Hep3B cells and that of hepatitis B e antigen (HBeAg), a hepatitis B virus genome replication marker, in human hepatocytes and HepA2 cells [42].

Table 1. Antimicrobial activity of costunolide.

Effect	Tested Organisms		Concentration	Reference
Antibacterial activity	*M. tuberculosis*	MIC (mg/L)	12.5	[35]
	S. aureus *E. coli* *P. aeruginosa*	MIDZ (mm)	18 19 14	[37]
	M. avium *M. tuberculosis*	MIC (μg/mL)	128 32	[36]
	H. pylori	MIC (μg/mL)	100–200	[38]
Antifungal activity	*Trichophyton mentagrophytes* *T. simum* *T. rubrum* *Epidermophyton floccosum* *Scopulariopsis sp.* *Aspergillus niger* *Curvulari lunata* *Magnaporthe grisea*	MIC (μg/mL)	62.5 62 31 or 62 125 250 125 250	[39]
	Colletotrichum acutatum *Colletotrichum fragariae*	MDIZ (mm)	4 6	[40]
	C. echinulata	EC$_{50}$ (μg/mL)	6	[41]
Antiviral activity	*Hepatitis B virus (HBV)*	IC$_{50}$ (μM)	1	[42]

MIC: minimum inhibitory ceoncetration, MDIZ: mean diameter of inhibition zone, EC50: effective concentration, IC50: inhibiton concentration.

Lipoteichoic acid (LTA)-induced acute lung injury (ALI) in mice is a model to represent experimental pneumonia. Treatment with costunolide significantly reduced LTA-induced inflammatory cell infiltration and lung tissue damage by decreasing the production of various cytokines and chemokines. Moreover, the compound inhibited LTA-induced iNOS expression in mouse bone marrow-derived macrophages by blocking phosphorylation of TAK1, p38 MAPK, and ERK, without affecting the activation of NF-κB [43]. Thus, costunolide may be considered as a lead compound for developing novel antimicrobial agents.

2.6. Costunolide in the Treatment of Alopecia

The cosmetic use of herbal products, especially for preventing hair loss or promoting hair growth, have long been practiced throughout the world. The herbal therapies used as hair growth promoters are expected to have low toxicity, be easy to use, low cost, and have high patient compliance. As the physiological and biochemical pathways in hair follicle dermal papillary cells (hHFDPCs) are unfolded, the mechanistic basis of hair growth promotion by many natural products is being explored [44,45]. It has been reported that topical application of costunolide significantly improved hair growth in C57BL/6 mice in vivo and promoted the proliferation of hHFDPCs in vitro [46]. Mechanistically, costunolide inhibited 5α-reductase activity and suppressed transforming growth factor (TGF-β1) induced phosphorylation of Smad-1/5 (mothers against decapentaplegic-1/5) in hHFDPCs, whereas the compound increased the levels of β-catenin and Gli1 mRNA and protein [46]. Thus, the development of costunolide-based formulation for the treatment of alopecia would be an interesting approach pending further studies on the toxicity and pharmacokinetic properties of the compound.

2.7. Costunolide as an Anticancer Agent

The search for anticancer agents from natural sources, especially plants, has led to the discovery of many clinically useful drugs. Extensive investigation of the anticancer effects of costunolide have shown that the compound induces apoptosis and inhibits proliferation of various cancer cells in vitro, and suppresses angiogenesis and metastasis. The following section will shed light on the biochemical processes and molecular targets of costunolide in exerting its anticancer effects.

2.7.1. Inhibition of Cell Proliferation

Costunolide decreased the proliferation of various cancer cells including those of the colon, breast, prostate, liver, gastric, and blood cancer cells [47–51]. Treatment of HCT-116 cells with costunolide decreased cell proliferation by inhibiting phosphorylation of mammalian target of rapamycin (mTOR) and its downstream kinases p70S6K and 4E-BP1, and increasing the phosphorylation and nuclear localization of p53 [52]. The antiproliferative effect of costunolide was mediated, at least in part, through suppression of cellular glutaminolysis via blockade of the promoter activity of glutaminase 1 (GLS1). The compound also decreased GLS1 mRNA and protein expression in p53-dependent manner since pretreatment with a p53 inhibitor reversed costunolide-mediated suppression of GLS1 activity and expression [52]. The antiproliferative effect of costunolide in MCF-7 cells was associated with microtubule polymerization and alteration of spindle morphology. [47].

Several other studies have reported that costunolide inhibits various tumor cell proliferation by blocking G2/M phase of the cell cycle and modulating the effect of cyclins and cyclin-dependent kinases (Cdk) [51,53,54]. The growth inhibition of SW-480 cells upon treatment with costunolide was associated with the downregulation of cyclin D1 and survivin, which was mediated via inhibition of nuclear translocation of β-catenin and its co-activator molecule galectin-3 [49]. Peng et al. [55] reported that costunolide induced cell cycle arrest at G2/M phase in MCF-7 and MDA-MB-231 cells via activation of p53 and p-14-3-3 expression and inhibition of c-Myc, p-AKT, and p-BID expression. Moreover, the ratio of BAX/BCL-2 was significantly increased upon costunolide treatment, which led to the induction of apoptosis in these cells [55]. Another study showed that costunolide-induced G2/M cell cycle arrest in MDA-MB-231 cells, which was mediated through the inhibition of Cdc2 and cyclin B1, and the elevation of p21^{WAF1} expression was independent of p53 activation [56]. Roy and colleagues have demonstrated that the G2/M phase of cell cycle arrest in MCF-7 and MDA-MB-231 cells incubated with costunolide was mediated through the downregulation of cell cycle regulatory proteins, such as cyclin D1, D3, CDK-4, CDK-6, p18^{INK4c}, p21$^{CIP1/Waf-1}$, and p27^{KIP1}. However, the compound did not affect the proliferation of normal mammary epithelial (MCF-10A) cells [57]. Likewise, the increase level of p21^{WAF1} and reduced expression of cyclin B1 and CDK2 by costunolide led to the G2/M phase arrest in K562 cells. According to this study, the compound enhanced imatinib-induced apoptosis in K562 cells via modulation of B cell receptor (Bcr)/Abl and STAT5 signaling pathways. In another study, these authors reported that costunolide sensitized K562 cells to doxorubicin via inhibition of the PI3K/Akt activity [58]. In another study, costunolide arrested the cell cycle at the G2/M phase through the downregulation of Chk2/Cdc25c/Cdk1/cyclin B1 signaling in human hepatoma HA22T and VGH cells [51]. Incubation of human prostate cancer (PC-3, DU-145, and LNCaP) cells with costunolide arrested the cell cycle at the G1 phase, which was associated with the inhibition of the CDK2 activity and Rb phosphorylation [50]. Moreover, costunolide upregulated p53 and p21 expression in human esophageal squamous Eca-109 cells, thereby inducing G1/S phase arrest [59].

2.7.2. Induction of Apoptosis

Mitochondria-Mediated Apoptosis

Costunolide induced mitochondria-mediated apoptosis as evidenced by the inhibition of Bcl-2, induction of Bax, and release of cytochrome c in human prostate (PC3 and DU-145) [60], leukemia (K562) [48], oral cancer (Eca-109) [59], gastric cancer (SGC-7901) [54], lung squamous carcinoma

(SK-MES1) [53], and bladder cancer (T24) [61] cells. Treatment of PC3 and DU-145 cells with costunolide led to the generation of ROS, the phosphorylation of c-Jun-N terminal kinase (JNK) and p38 MAPK, the inhibition of Bcl-2 and Bcl-xl, and the induction of Bax, thereby leading to reduced mitochondrial membrane potential and cytochrome c release and caspase 3 activation. The apoptosis induction by costunolide resulted in the reduced growth of PC3 cells xenograft tumors in nude mice [60]. Likewise, costunolide induced mitochondria-mediated apoptosis by upregulation of Bax, downregulation of Bcl-2, and activation of caspase-3 and poly ADP-ribose polymerase via ROS production and loss of mitochondrial membrane potential in oral cancer Eca-109 cells [59]. The compound also activated JNK in human leukemic U937 cells, thereby leading to mitochondrial cell death via phosphorylation of Bcl-2 and translocation of Bax to mitochondria [62]. Hua et al. demonstrated that the apoptosis of SK-MES-1 cells upon treatment with costunolide was mediated through upregulation of p53 and Bax expression, downregulation of Bcl-2 expression, and caspase-3 activation [53]. In addition to inducing Bax, caspase-3 and PARP cleavage in T24 bladder cancer cells, costunolide also attenuated expression of survivin and Bcl2 as a mechanism of apoptosis induction [61]. Costunolide induced mitochondrial-mediated apoptosis through activation of caspase-3, -8, and -9 in ovarian cancer cell lines (MPSC1PT, A2780PT, and SKOV3PT) and the human endometriotic epithelial cells (11Z, 12Z) [63,64]. Furthermore, incubation of multidrug resistant ovarian cancer cells (OAW42-A) with costunolide attenuated cell growth with an IC_{50} of 25 µM, and induced apoptosis, which was mediated through induction of Bax, decreased the expression of Bcl-2 and cleavage of caspase-3 and 9. Moreover, the compound induced autophagy as evidenced by the elevated expression of LC3 II and Beclin 1 [65].

Endoplasmic Reticulum (ER) Stress-Mediated Apoptosis

Apoptosis may result from continued ER stress that activates unfolded protein response (UPR) signaling pathways. Costunolide activated the ionositol requiring enzyme (IRE)-1α, a resident ER membrane protein, which further activated JNK by recruiting adapter molecules TRAF2 and ASK1 in cultured lung adenocarcinoma cell line A549 cells [66]. Costunolide-activated JNK led to Bcl-2 phosphorylation at serine 70, a mechanism to convert antiapoptotic Bcl-2 to play proapoptotic functions, thereby causing cytochrome c release, caspase-3 activation, and PARP cleavage, leading to induction of apoptosis. Authors have further demonstrated that costunolide-induced ROS generation played a critical role in this process as pretreatment of cells with ROS scavenger *N*-acetyl cysteine abrogated costunolide-induced ER stress and apoptosis [66]. A similar mechanism of ROS-mediated ER stress induction by costunolide led to the expression of Bip and IREα, and the activation of the JNK pathway, leading to apoptosis in human osteosarcoma U2OS cells [67]. Recent studies have shown that the thioredoxin/thioredoxin reductase (TrxR) system causes tumor cell resistance to oxidative stress-induced apoptosis. Surface plasmon resonance analysis and molecular docking study revealed that costunolide directly interacted with TrxR1 via its lactone oxygen atom with Gln-494 of TrxR1 and inhibited the activity of TrxR1, thereby increasing the production of ROS and inducing ROS-dependent ER stress and apoptosis in colon cancer cells (HCT-116, SW-620, and HT-29 cells). This study also demonstrated that the compound arrested G2/M phase of cell cycle and attenuated the expression of cyclin B1, CDC2, MDM2, and Bcl2 and increased the expression of Bax and cleavage of caspase 3, which was reversed by cotreatment with *N*-acetyl cysteine, suggesting the involvement of ROS in costunolide-induced retardation of tumor cell growth. Furthermore, costunolide treatment of mice transplanted with colon cancer cells inhibited tumor growth and decreased TrxR1 activity and ROS levels [68].

Death Receptor-Mediated Apoptosis

Extrinsic mechanisms of apoptosis induction by costunolide have also been reported. The induction of apoptosis in estrogen receptor-negative human breast cancer (MDA-MB-231) cells by costunolide involves the activation of Fas, caspase-8, caspase-3, and the degradation of PARP [56]. Costunolide also increased the phosphorylation of Fas-associated death domain (FADD) at serine 194, leading to apoptotic cell death in human B cell lymphoma. [69].

2.7.3. Telomerase Reverse Transcriptase (TERT) Inhibition

Telomeres, which maintain genomic integrity in normal cells, are shortened upon each cell division, thus leading to chromosomal instability, cellular senescence, and aging. However, the length of the telomeres is maintained by high levels of telomerase enzyme present in tumor cells, thereby allowing cancer cells to be immortal. Therefore, telomerase has been considered as a possible target for cancer treatment [69]. It has been reported that costunolide caused significant inhibition of telomerase activity in human B cell leukemia (NALM-6) cells by decreasing the mRNA and protein expression of human telomerase reverse transcriptase (hTERT), which controls the enzymatic activity of telomerase, and induced apoptosis in these cells [68]. Likewise, costunolide showed strong inhibition of telomerase activity in MCF-7 and MDM-23-231 cells through the downregulation of hTERT mRNA via inactivation of c-Myc and Sp1 transcription factors [69]. The antiproliferative and apoptosis-inducing effects of costunolide was also associated with inhibition of hTERT in human glioma cells [70,71] and human hepatocellular carcinoma (HepG2/C3A, PLC/PRF/5) (A172, U87MG) cells [72].

2.7.4. Inhibition of Angiogenesis

The persistent growth and spread of a tumor require a constant supply of nutrients and oxygen to the cancer cells. The formation of new blood vessels, a process known as angiogenesis, is therefore an essential step in tumor invasion and metastasis. The discovery of angiogenesis inhibitors (e.g., avastin) helps reduce the morbidity and mortality from various cancers. A key angiogenic molecule is vascular endothelial growth factor (VEGF), which by binding with VEGF receptors (VEGFR) on vascular endothelial cells promotes formation of new blood vessels [73].

In a murine cannulated sponge implant angiogenesis model, administration of costunolide to Swiss albino mice implanted with polyester polyurethane sponges used as a framework for fibroblast tissue growth reduced the levels of VEGF and hemoglobin content in fibrovascular tissue, suggesting the antiangiogenic property of the compound [74]. Jeong et al. reported that costunolide attenuated VEGF-induced proliferation and chemotaxis of human umbilical vein endothelial cells (HUVECs), and blocked VEGF-induced phosphorylation of KDR/Flk-1 in NIH 3T3 cells overexpressing KDR/Flk-1. In addition, VEGF-stimulated neovascularization in mouse corneal micropocket analysis was reduced by costunolide treatment [75]. In another study, costunolide significantly reduced VEGF secretion and decreased VEGF mRNA levels in human gastric cancer (AGS), colon cancer (Caco-2), and liver cancer (HepG2/C3A) cells. This study also reported that costunolide significantly reduced VEGFR1 and VEGFR2 expression at both mRNA and protein levels [76].

2.7.5. Inhibition of Tumor Metastasis

Cancer metastasis refers to the spread of tumor cells from their site of origin to other distant parts of the body. Metastasis consists of multistep processes including tumor cell spread, extracellular matrix (ECM) degradation, tumor cell invasion in ECM, angiogenesis, and secondary metastatic tumor growth [77]. Costunolide inhibited TNF-α-induced migration and invasion of MDA-MB-231 breast cancer cells by downregulating the expression of matrix metalloproteinase (MMP)-9 gene via blockade of NF-κB activation. Moreover, the xenograft tumor growth of MBA-MB-231 cells in athymic nude mice was diminished upon treatment with costunolide [77]. The MMP-2 and MMP-9 are key molecules involved in tumor invasion and metastasis. Costunolide significantly inhibited invasion and decreased MMP-2 expression in human neuroblastoma (NB-39) cells [78]. The invasion of soft tissue sarcoma (TE-671, SW-872, and SW982) cells was also inhibited by costunolide via modulation of MMPs expression [79]. Of the various forms of metastasis, lymphatic metastasis is an important determinant in cancer therapy and staging. Costunolide inhibited the proliferation and capillary formation of TR-LE (temperature-sensitive mouse lymphoid endothelial cells) cells, suggesting that the compound can provide clinical benefits as an inhibitor of lymphoproliferative growth during tumor metastasis [80]. Epithelial–mesenchymal transition (EMT) is critical step in tumor invasion and metastasis. One of the mechanisms that the EMT process initiates in tumor invasion is the detyrosination of tubulin via inhibition of tyrosine ligase, and the detyrosinated tubulin forms microtentacles (McTN) which promotes tumor cell reattachment to the endothelial layer during tumor invasion. Costunolide significantly reduced detyrosinated tubulin and the frequency of McTN in multiple invasive breast tumors, thereby preventing tumor cells attachment with endothelial tissue and blocking invasion [81].

2.8. Antidiabetic Effects of Costunolide

An in vitro assay has revealed that the methanol extract of leaves of *Costus speciosus* inhibited α-glucosidase activity with an IC_{50} value of 67.5 µg/ml and attenuated α-amylase activity with an IC_{50} value of 5.88 mg/ml, which is lower than the reference compound acarbose [82]. Since costunolide is abundantly present in leaves of *Costus speciosus*, this study indicates the potential of costunolide in managing glycemic control. A subsequent study demonstrated that costunolide significantly reduced blood glucose level, glycosylated hemoglobin (HbA1c), serum total cholesterol, triglyceride, and LDL cholesterol level in streptozotocin (STZ)-induced diabetic rats [83]. Moreover, the compound remarkably increased plasma insulin, tissue glycogen, HDL cholesterol, and serum protein level [83]. Since oxidative stress affect the pathogenesis and progression of diabetic tissue injury, the induction of antioxidant enzymes, such as glutathione peroxidase, catalase, and superoxide dismutase in STZ-induced diabetic rat's pancreas indicates the role of costunolide in improving glycemic control in diabetes [8]. However, additional studies are warranted to ascertain the antidiabetic property of this compound.

3. Pharmacokinetics and Toxicity Profile

Pharmacokinetic studies are an integral part of the drug discovery process. The understanding of the absorption, distribution, metabolism, and elimination of the drug-to-be is an essential step in new drug development. There have been several studies reporting the pharmacokinetic profile of costunolide. The maximum plasma concentration (C_{max}) and time required to attain highest plasma level of the molecule (T_{max}) after oral administration of costunolide to Wistar rats were found as 0.024 ± 0.004 mg/L and 9.0 ± 1.5 h, respectively. The half-life ($t_{1/2}$) and area under the curve (AUC) were 4.97 h and 0.33 ± 0.03 mg·h/mL, respectively [84]. However, a subsequent study reported that after oral administration of costunolide to Wistar rats, the C_{max} and T_{max} were 19.84 ng/mL and 10.46 h, respectively, and the half-life ($t_{1/2}$) and AUC were 5.54 h and 308.83 ng·h/mL, respectively [85]. According to a recent study, oral administration of costunolide to SD rats showed C_{max}, T_{max}, $t_{1/2}$, and AUC as 0.106 ± 0.045 µg/mL, 8.00 h, 14.62 ± 3.21 h, and 1.23 ± 0.84 µg·h/mL, respectively [86]. The large variation in pharmacokinetic parameters between these studies may be due to the use of different assay techniques and/or animal models. In addition, intravenous administration of costunolide to Sprague–Dawley rats revealed the C_{max} as 12.28 ± 1.47 µg/mL, and the half-life ($t_{1/2}$) and AUC were detected as 1.16 ± 0.06 h and 3.11 ± 0.13 µg·h/mL, respectively [87]. These results would have immense importance in further development of costunolide-based therapy.

Although costunolide has been examined extensively for its therapeutic potential in various animal models as discussed in the previous sections of this review, acute and chronic toxicity studies are scarce. A recent study demonstrated that the compound induced apoptosis in normal Chinese hamster ovarian cells by inducing clastogenic and genotoxic effects as evidenced by micronuclei formation and chromosomal breaks [88]. Thus, more rigorous toxicity studies to determine the lethal dose (LD_{50}) and ensure safety of the compounds is of paramount importance in further progressing the development of costunolide as a drug candidate.

4. Conclusions

Sesquiterpene lactones form a large, structurally diverse group of natural products found almost universally in plants. Extensive investigation of the therapeutic potential of sesquiterpene lactones has yielded important candidates for pharmaceutical development [89]. Costunolide is a well-known sesquiterpene lactone, which has been isolated from various plant species. As has been discussed in earlier sections, costunolide has been reported to possess antioxidant, anti-inflammatory, antiallergic, bone remodeling, neuroprotective, antimicrobial, hair growth promoting, anticancer, and antidiabetic properties (Figure 4 and Table 2). Limited pharmacokinetic studies have also shown that the compound can be bioavailable. However, the majority of these studies have been conducted in cultured cells or using an in vitro system. Considering the therapeutic value of the compound, it would be interesting to further examine the effects of costunolide in various other animal models to reveal the subacute and chronic toxicities, detailed elucidation of molecular mechanisms of action, and structural modifications to develop new therapeutics based on costunolide or its derivatives.

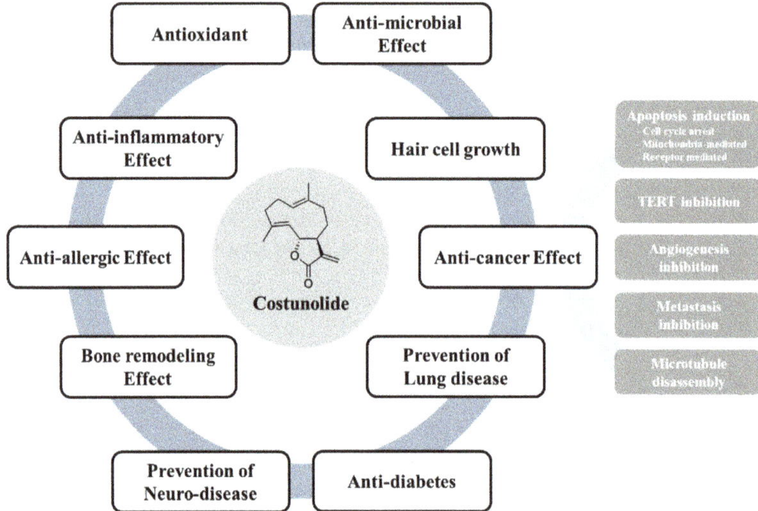

Figure 4. Bioactivities of costunolide. Costunolide could exert its therapeutic potential including antioxidant, anti-inflammatory effect, anti-allergic effect, bone remodeling effect, prevention of neurodegenerative disease, anti-microbial effect, inhibition of alopecia, prevention of lung disease and anti-diabetic effect. In particular, costunolide elicits anti-cancer activities partly through induction of apoptosis, Inhibition of cell proliferation, TERT, angiogenesis, metastasis and microtubule disassembly.

Table 2. Molecular mechanisms underlying bioactivities of costunolide.

Type	Experimental Model	Dose/Concentration	Mechanism of Action	Ref.
Antioxidant effect	STZ-induced diabetic rats	20 mg/kg day	Decreased in TBARS level; increased in GSH content	[8]
	MCF-7, MDA-MB-231	20, 40 μM	Decreased in TBARS level; increased in SOD, catalase, GPx activity	[10]
	5-FU-induced IM	5, 20 mg/kg	Increased in SOD level	[11]
	H_2O_2-stimulated PC12 cells	50, 100 μM	Decreased intracellular ROS	[12]
Anti-inflammatory effect	Cg-induced edema; LPS-induced fever	0.015, 0.15, 0.3 mg/kg	Inhibited edema formation; Reduced the fever index	[13]
	LPS-stimulated RAW264.7 cells	0.5, 1.5, 3 μg/ml	Inhibited NF-κB activity, phosphorylation of IκBα and NO production; suppressed iNOS mRNA expression	[14]
	5-FU-induced IM	5, 20 mg/kg	Decreased the expression of iNOS, COX-2, TNF-α and NO	[11]
	IL-22 or IFN-γ-stimulated keratinocytes	12.5 μM	Inhibited STAT1/3 phosphorylation	[15]
	IL-6-stimulated THP-1 cells	6, 12, 25 ng/ml	Inhibited STAT3 and JAK1/2 phosphorylation	[16]
	Ethanol-induced gastric ulcer	5, 20 mg/kg	Suppressed the activation of NF-κB, TNF-α, COX-2, NO and iNOS	[17]
	LPS-stimulated RAW264.7 cells	0.1, 0.3, 1, 3 μM	Suppressed the protein and mRNA expression of IL-1β; inhibited the activity of AP-1 and the phosphorylation of MAPKs	[18]
	Carrageenan-induced pleurisy	5, 10, 15 mg/kg	Reduced accumulation of PMNs and expression of T TNF-α, ICAM-1, P-selectin and nitrotyrosine	[19]
	LPS-stimulated RAW264.7 cells	0.1, 0.5, 1 μM	Induced HO-1 expression and Nrf2 nuclear accumulation; inhibited production of TNF-α and IL-6	[21]
	CD3/CD28-stimulated CD4$^+$ T cells	0.5, 1, 2 μM	Inhibited the expression of T-bet, GATA3 and RORγt; suppressed the proliferation of CD4$^+$ T cells and expression of CD69; decreased the phosphorylation of ERK and p38	[23]
Antiallergic effect	TNF-α/IFN-γ-stimulated HaCaT cells	2.5, 5, 10 μM	Inhibited the expression of TARC, MDC, RANTES and IL-8	[25]
	IgE-sensitized RBL-2H3	10 μM	Inhibited the expression of β-hexosaminidase	[26]
	OVA-induced mouse asthma model	10 mg/kg	Reduced eosinophil filtration, inflammation score and mucin secretion; decreased the expression of IL-4 and IL-13	
	Ketotifen-stimulated RBL-2H3	0.32, 1.6, 8, 40 μM	Inhibited the release of β-hexosaminidase	[27]
	IL-5-stimulated Y16 cells	0.16, 0.8, 4, 20, 40 μM	Inhibited the proliferation Y16 cells	
Bone remodeling	MC3T3-E1 cells differentiation	10 μM	Increased ALP activity, collagen deposition and mineralization	[29]
	C3H10T1/2 cells differentiation	1, 10, 10^2, 10^3, 10^4 ng/ml	Increased the expression of Dlx5, Runx2, ALP, and OC; reduced the activity of ATF4 and expression of HO-1	[30]
	RANKL-induced osteoclast differentiation	5 μM	Suppressed NFATc1 expression and c-Fos activity	[31]

Table 2. *Cont.*

Type	Experimental Model	Dose/Concentration	Mechanism of Action	Ref.
Neuroprotective agent	DA-stimulated SH-SY5Y	0.8, 4, 2 μM	Decreased the expression of ASYN; increased the expression of Nurr1, VMAT2 and DAT	[33]
	LPS-stimulated BV2 microglial cells	1 μM	Attenuated the expression of TNF-α, IL-1,6, iNOS, MCP-1 and COX-2; inhibited the activation of NF-κB	[34]
Treatment of alopecia	Testosterone-stimulated hHFDPCs	3 μM	Promotes the growth of hHFDPCs; inhibits the 5α-reductase activity	[46]
	Hair growth in mice	3 μM/L	Improved the hair growth	
Inhibition of proliferation	MCF-7 breast cancer cells	10, 100 nM	Inhibited the cell growth; stimulated tubulin assembly	[47]
	K562 leukemia cells	15 μM	Induced cell cycle arrest; induced apoptosis	[48]
	S480 colon cancer cells	5 μM	Suppressed cyclin D1, survivin, β-catenin, and galectin-3; inhibited proliferation and survival of cells	[49]
	LNCaP, PC-3, DU-145 prostate cancer cells	1.3 μM	Inhibited cell proliferation; induced cell cycle arrest at the G1phase	[50]
	HA22T/VGH hepatocellular carcinoma cells	5 μM	Caused G2/M arrest; up-regulated phosphorylation of Chk2, Cdc25c, Cdk1, and cyclin B1	[51]
	HCT-116 colorectal cancer cells	10, 20, 40 μM	Inhibited proliferation; suppressed mTOR phosphorylation and GLS1 activity	[52]
	SK-MES-1 lung squamous carcinoma cells	40, 80 μM	Inhibited growth of cells; induced cell cycle arrest at G1/S phase; upregulated expression of p53 and Bax; downregulated Bcl-2 expression; activated caspase-3	[53]
	SGC-7901 gastric adenocarcinoma cells	20, 40 μM	Arrested cell cycle at G2/M phase; activated caspase-3	[54]
	MCF-7, MDA-MB-231 breast cancer cells	0.9, 1.3, 2.2 μg/mL	Arrested cell cycle at G2/M phase; induced p53 and 14-3-3 expression; inhibited c-Myc, p-Akt and p-BID expression	[55]
	MDA-MB-231 breast cancer cells	15 μM	Induced G2/M cell cycle arrest; upregulated p21WAF1 expression; inhibited cdc2 and cyclin B1 expression	[56]
	MCF-7, MDA-MB-231 breast cancer cells	40 μM	Arrested cell cycle arrest at G2/M phase; inhibited the expression of cyclin D1, D3, CDK-4, CDK-6, p18 INK4c, p21 CIP1/Waf-1 and p27 KIP1	[57]
	K562/ADR chronic myeloid leukemia cells	0.1, 1, 10, 100 μM	Sensitized K562 cells to doxorubicin; inhibited PI3K/Akt activity	[58]
	Eca-109 human esophageal cancer cells	40, 80 μM	Induced cell cycle arrest in G1/S phase; upregulated the expression of p53, p21, Bax and caspase-3; downregulated Bcl-2	[59]

Table 2. *Cont.*

Type	Experimental Model	Dose/Concentration	Mechanism of Action	Ref.
Mitochondria-mediated apoptosis	PC-3, DU-145 prostate cancer cells	20 μM	Enhanced doxorubicin to change of MMP; increased Bax expression and cytochrome c release	[60]
	T24 human bladder cancer cells	25, 50 μM	Increased expression of Bax, downregulated Bcl-2 and surviving; activated caspase-3 and PARP	[48]
	U937 human promonocytic leukemia cells	5, 10	Increased the activation of JNK; inhibited the expression of Bcl-2; induced DNA fragmentation	[61]
	SKOV3, A2780, MPSC1 ovarian cancer cells	10, 20, 30 μM	Triggered the activation of caspase-3, -8, and -9; down-regulated Bcl-2 expression,	[62]
	11Z human epithelial endometriotic cells	IC_{50} 14.21 μM	Induced the activation of caspase-3, -8, and -9; inhibited the activation of Akt and NF-κB	[63]
	ovarian cancer cell line, OAW42-A	12.5, 25, 50 μM	Reduced the mitochondrial membrane potential; increased protein expression of LC3 II and beclin 1	[64]
ER stress-mediated apoptosis	A549 lung adenocarcinoma cells	10, 20, 30 μM	Activated UPR signaling pathways; upregulated GRP78 and IRE1α expression; induced ASK1 and JNK activation	[65]
	U2OS human osteosarcoma cells, A549 human alveolar adenocarcinoma cells, Hela cells	10, 20, 30 μM	Increased expressions of Bip and IREa; increased expressions of p-ASK1, p-JNK and p-ERK; induced generation of Ca^{2+}	[66]
	HCT-116, HT-29, SW620 colon cancer cells	10, 20, 30 μM	Inhibited the activity of TrxR1; induced the expression of p-eIF2a, ATF4 and CHOP	[67]
Death receptor mediated apoptosis	NALM-6 human B cell leukemia cell	10 μM	Increased the phosphorylation of FADD; activated caspase-8	[68]
TERT inhibition	NALM-6 human B cell leukemia cell	10 μM	Suppressed telomerase activity; inhibited the expression of hTERT mRNA and protein	
	MCF-7, MDA-MB-231 breast cancer cells	10, 50, 80, 100 μM	Inhibited the cell growth, telomerase activity and hTERT mRNA expression; inhibited bindings of hTERT promoters; inhibited the expression of c-Myc and Sp1	[69]
	A172, U87MG, T98G glioma cells	10, 20, 30, 40 μM	Decreases Nrf2 levels; Suppressed telomerase activity; decreased expression of G6PD and TKT	[70]
	A172, U87MG glioma cells	30 μM	Inhibited hTERT expression	[71]
	HepG2/C3A, PLC/PRF/5 HCC cells	5, 10, 50 μM	Inhibited AFP secretion and mRNA expression; decreased cell migration	[72]

Table 2. *Cont.*

Type	Experimental Model	Dose/Concentration	Mechanism of Action	Ref.
Inhibition of angiogenesis	subcutaneous murine sponge model	5, 10, 20 mg/kg	Reduced hemoglobin concentration and VEGF levels	[74]
	VEGF-stimulated HUVECs	IC_{50} 5.7 µM	Inhibited VEGF-induced proliferation and migration; inhibited the VEGF-induced autophosphorylation of KDR/Flk-1	[75]
	AGS, Caco-2, HepG2/C3A cancer cells	10 µM	Decreased VEGF secretion and mRNA levels	[76]
Inhibition of tumor metastasis	MDA-MB-231 breast cancer cells	20 µM	Inhibited TNFα-induced cells migration and invasion; reduced phosphorylation of IKK and IκBα; inhibited p65 NF-κB subunit	[77]
	IMR-32, LA-N-1, SK-N-SH neuroblastoma cell	0.1, 1, 10 µM	Inhibited migration and invasion; suppressed MMP2 expression	[78]
	SW-872, SW-982, TE-671 soft tissue sarcomas	3, 10, 20 µg/mL	Inhibited the invasion potential; changed the expression of MMPs	[79]
	TR-LE (temperature-sensitive rat lymphatic endothelial) cells	IC_{50} 1.37 µM	Suppressed cell proliferation; inhibited capillary-like tube formation	[80]
	MDA-MB-157, MDA-MB-436, Bt-549 breast cancer cells	10, 25 µM	Reduced detyrosinated tubulin; decreased microtentacle (McTN) frequency; reduced tumor cell attachment	[81]
Antidiabetic effect	α-Amylase, α-Glucosidase, fructosamine formation, glycation	IC_{50} 5.88 or 67.5 µM	Inhibited the activity of α-Amylase, α-Glucosidase; inhibited fructosamine formation;	[82]
	streptozotocin-induced diabetic rats	5, 10, 20 mg/kg	Reduced glucose levels and HbA_{1c}; increased insulin levels; reduced cholesterol, TG, LDL; increased HDL	[8]

Funding: This research received no external funding.

Conflicts of Interest: The author declares no conflict of interest.

Abbreviations

Akt: protein kinase B; **ALP**: alkaline phosphatase; **AP-1**: activator protein-1; **ASYN**: α-synuclein; **ATF4**: transcription factor 4; **BAX:** BCL2 Associated X; **BCL-2**: B-cell lymphoma 2; **Bcl-xL**: B-cell lymphoma-extra large; **BID BH3**: interacting-domain death agonist; **Bip**: binding immunoglobulin protein; **CCL**: C-C motif ligand; **Cdc2**: cell division cycle protein 2 homolog; **Cdk**: cyclin-dependent kinases; **Chk2**: checkpoint kinase 2; **COX**: cyclooxygenase; **DAT**: dopamine transporter; **Dlx5**: distal-less homeobox 5; **ECM**: extracellular matrix; **EMT**: Epithelial-messenchymal transition; **ER**: estrogen receptor; **ERK**: extracellular signal-regulated kinase; **GATA-3**: GATA-Binding Protein 3; **GLS1**: glutaminase 1; **GPx**: glutathione peroxidase; **GSH**: glutathione; **GST**: glutathione-S-transferase; **HBeAg**: hepatitis B e antigen; **HBsAg**: hepatitis B surface antigen; **HO-1:** heme oxygenase-1; **hTERT**: human telomerase reverse transcriptase; **ICAM-1**: intracellular adhesion molecule-1; **IFN-γ**: Interferon gamma; **IM**: intestinal mucositis; **iNOS**: inducible nitric oxide synthase; **JAK**: Janus-activated kinase; **JNK**: c-Jun N-terminal kinases; **KDR/Flk-1:** kinase insert domain receptor; **LPS**: lipopolysaccharide; **MAPK**: mitogen-activated protein kinase; **MCP-1:** macrophage chemoattractant protein-1; **McTN** : microtentacles; **MDC**: Macrophage-Derived Chemokine; **MDM2**: Mouse double minute 2 homolog; **MMP**: matrix metalloproteinase; **MMP**: mitochondrial membrane potential; **MPO**: myeloperoxidase; **mTOR**: mammalian target of rapamycin; **NAG**: *N*-acetylglucosaminidase; **NF-κB**: nuclear factor-kappaB; **NFATc1**: nuclear factor of activated T cells, cystoplasic-1; **NO**: nitric oxide; **Nrf2**: nuclear factor erythroidrelated factor-2; **Nurr1**: nuclear receptor related-1; **OC**: osteocalcin; **OSCAR**: osteoclast-associated receptor; **OVA**: ovalbumin; **PERK**: pancreatic ER kinase; **PI3K**: phoaphatidylinositol-3-kinase; **RANKL**: receptor activator of nuclear factor kappa-B ligand; **RANTES**: regulated on activation, normal T cell expressed and secreted; **Rb**: retinoblastoma protein; **RORγt**: RAR-related orphan receptor γt; **ROS**: reactive oxygen species; **Runx2**: runt-related transcription factor 2; **SN**: substantia nigra; **SNPP**: Simple Network Paging Protocol; **SOD**: superoxide dismutase; **STAT**: signal transducer and activator of transcription; **T-bet**: T-box transcription factor; **TAK1**: transforming growth factor-β-activated kinase 1; **TARC**: thymus and activation regulated chemokine; **TGF-β1**: transforming growth factor; **TNF-α**: tumor necrosis factor alpha; **TRAP**: tartrate-resistant acid phosphatase; **TrxR**: thioredoxin/thioredoxin reductase; **VEGF**: vascular endothelial growth factor; **VMAT2**: vesicular monoamine transporter type 2.

References

1. Lahlou, M. The success of natural products in drug discovery. *Pharmacol. Pharm.* **2013**, *4*, 17–31. [CrossRef]

2. Wang, G.; Tang, W.; Bidigare, R.R. Terpenoids as therapeutic drugs and pharmaceutical agents. In *Natural Products*; Springer: Berlin, Germany, 2005; pp. 197–227.

3. De Kraker, J.-W.; Franssen, M.C.; Dalm, M.C.; de Groot, A.; Bouwmeester, H.J. Biosynthesis of germacrene A carboxylic acid in chicory roots. Demonstration of a cytochrome P450 (+)-germacrene A hydroxylase and NADP+-dependent sesquiterpenoid dehydrogenase (s) involved in sesquiterpene lactone biosynthesis. *Plant Physiol.* **2001**, *125*, 1930–1940. [CrossRef] [PubMed]

4. Yang, Z.-J.; Ge, W.-Z.; Li, Q.-Y.; Lu, Y.; Gong, J.-M.; Kuang, B.-J.; Xi, X.; Wu, H.; Zhang, Q.; Chen, Y. Syntheses and biological evaluation of costunolide, parthenolide, and their fluorinated analogues. *J. Med. Chem.* **2015**, *58*, 7007–7020. [CrossRef] [PubMed]

5. Rasul, A.; Parveen, S.; Ma, T. Costunolide: A novel anti Costunolide: A novel anti-cancer sesquiterpene lactone cancer sesquiterpene lactone cancer sesquiterpene lactone. *Bangladesh J. Pharmacol.* **2012**, *7*, 6–13. [CrossRef]

6. Rao, A.S.; Kelkar, G.; Bhattacharyya, S. Terpenoids—XXI: The structure of costunolide, a new sesquiterpene lactone from costus root oil. *Tetrahedron* **1960**, *9*, 275–283. [CrossRef]

7. Rice-Evans, C. Flavonoids and isoflavones: Absorption, metabolism, and bioactivity. *Free Radic. Biol. Med.* **2004**, *7*, 827–828. [CrossRef] [PubMed]

8. Eliza, J.; Daisy, P.; Ignacimuthu, S. Antioxidant activity of costunolide and eremanthin isolated from Costus speciosus (Koen ex. Retz) Sm. *Chem. Biol. Interact.* **2010**, *188*, 467–472. [CrossRef]

9. Wu, G.; Fang, Y.-Z.; Yang, S.; Lupton, J.R.; Turner, N.D. Glutathione metabolism and its implications for health. *J. Nutr.* **2004**, *134*, 489–492. [CrossRef]

10. Rajalakshmi, M.; Anita, R. In vitro and in silico evaluation of antioxidant activity of a sesquiterpene lactone, costunolide, isolated from costus specious rhizome on MCF-7 and MDA-MB-231 human breast cancer cell lines. *World J. Pharm. Pharm. Sci.* **2014**, *3*, 1334–1347.

11. Chen, Y.; Zheng, H.; Zhang, J.; Wang, L.; Jin, Z.; Gao, W. Intestinal mucositis repaired activity of costunolide and dehydrocostus in 5-fluorouracil-induced mice model. *RSC. Adv.* **2016**, *6*, 5249–5258. [CrossRef]

12. Cheong, C.-U.; Yeh, C.-S.; Hsieh, Y.-W.; Lee, Y.-R.; Lin, M.-Y.; Chen, C.-Y.; Lee, C.-H. Protective effects of Costunolide against hydrogen peroxide-induced injury in PC12 cells. *Molecules* **2016**, *21*, 898. [CrossRef] [PubMed]

13. Kassuya, C.A.L.; Cremoneze, A.; Barros, L.F.L.; Simas, A.S.; da Rocha Lapa, F.; Mello-Silva, R.; Stefanello, M.É.A.; Zampronio, A.R. Antipyretic and anti-inflammatory properties of the ethanolic extract, dichloromethane fraction and costunolide from Magnolia ovata (Magnoliaceae). *J. Ethnopharmacol.* **2009**, *124*, 369–376. [CrossRef] [PubMed]

14. Koo, T.H.; Lee, J.-H.; Park, Y.J.; Hong, Y.-S.; Kim, H.S.; Kim, K.-W.; Lee, J.J. A sesquiterpene lactone, costunolide, from Magnolia grandiflora inhibits NF-κB by targeting IκB phosphorylation. *Planta Med.* **2001**, *67*, 103–107. [CrossRef] [PubMed]

15. Scarponi, C.; Butturini, E.; Sestito, R.; Madonna, S.; Cavani, A.; Mariotto, S.; Albanesi, C. Inhibition of inflammatory and proliferative responses of human keratinocytes exposed to the sesquiterpene lactones dehydrocostuslactone and costunolide. *PLoS ONE* **2014**, *9*, e107904. [CrossRef] [PubMed]

16. Butturini, E.; Cavalieri, E.; de Prati, A.C.; Darra, E.; Rigo, A.; Shoji, K.; Murayama, N.; Yamazaki, H.; Watanabe, Y.; Suzuki, H. Two naturally occurring terpenes, dehydrocostuslactone and costunolide, decrease intracellular GSH content and inhibit STAT3 activation. *PLoS ONE* **2011**, *6*, e20174. [CrossRef] [PubMed]

17. Zheng, H.; Chen, Y.; Zhang, J.; Wang, L.; Jin, Z.; Huang, H.; Man, S.; Gao, W. Evaluation of protective effects of costunolide and dehydrocostuslactone on ethanol-induced gastric ulcer in mice based on multi-pathway regulation. *Chem. Biol. Interact.* **2016**, *250*, 68–77. [CrossRef] [PubMed]

18. Kang, J.S.; Yoon, Y.D.; Lee, K.H.; Park, S.-K.; Kim, H.M. Costunolide inhibits interleukin-1β expression by down-regulation of AP-1 and MAPK activity in LPS-stimulated RAW 264.7 cells. *Biochem. Biophys. Res. Commun.* **2004**, *313*, 171–177. [CrossRef] [PubMed]

19. Butturini, E.; Di Paola, R.; Suzuki, H.; Paterniti, I.; Ahmad, A.; Mariotto, S.; Cuzzocrea, S. Costunolide and Dehydrocostuslactone, two natural sesquiterpene lactones, ameliorate the inflammatory process associated to experimental pleurisy in mice. *Eur. J. Pharmacol.* **2014**, *730*, 107–115. [CrossRef] [PubMed]

20. Prawan, A.; Kundu, J.K.; Surh, Y.J. Molecular basis of heme oxygenase-1 induction: Implications for chemoprevention and chemoprotection. *Antioxid. Redox Signal.* **2005**, *7*, 1688–1703. [CrossRef]

21. Pae, H.-O.; Jeong, G.-S.; Kim, H.-S.; Woo, W.; Rhew, H.; Kim, H.; Sohn, D.; Kim, Y.-C.; Chung, H.-T. Costunolide inhibits production of tumor necrosis factor-α and interleukin-6 by inducing heme oxygenase-1 in RAW264. 7 macrophages. *Inflamm. Res.* **2007**, *56*, 520–526. [CrossRef]

22. Rahimi, K.; Ahmadi, A.; Hassanzadeh, K.; Soleimani, Z.; Sathyapalan, T.; Mohammadi, A.; Sahebkar, A. Targeting the balance of T helper cell responses by curcumin in inflammatory and autoimmune states. *Autoimmun. Rev.* **2019**. [CrossRef] [PubMed]

23. Park, E.; Song, J.H.; Kim, M.S.; Park, S.-H.; Kim, T.S. Costunolide, a sesquiterpene lactone, inhibits the differentiation of pro-inflammatory CD4+ T cells through the modulation of mitogen-activated protein kinases. *Int. Immunopharmacol.* **2016**, *40*, 508–516. [CrossRef] [PubMed]

24. Kim, J.E.; Kim, J.S.; Cho, D.H.; Park, H.J. Molecular Mechanisms of Cutaneous Inflammatory Disorder: Atopic Dermatitis. *Int. J. Mol. Sci.* **2016**, *17*, 1234. [CrossRef] [PubMed]

25. Seo, C.S.; Lim, H.S.; Jeong, S.J.; Shin, H.K. Anti-allergic effects of sesquiterpene lactones from the root of Aucklandia lappa Decne. *Mol. Med. Rep.* **2015**, *12*, 7789–7795. [CrossRef] [PubMed]

26. Lee, B.-K.; Park, S.-J.; Nam, S.-Y.; Kang, S.; Hwang, J.; Lee, S.-J.; Im, D.-S. Anti-allergic effects of sesquiterpene lactones from Saussurea costus (Falc.) Lipsch. determined using in vivo and in vitro experiments. *J. Ethnopharmacol.* **2018**, *213*, 256–261. [CrossRef] [PubMed]

27. Kim, T.J.; Nam, K.W.; Kim, B.; Lee, S.J.; Oh, K.B.; Kim, K.H.; Mar, W.; Shin, J. Inhibitory Effects of Costunolide Isolated from Laurus nobilis on IgE-induced Degranulation of Mast Cell-like RBL-2H3 Cells and the Growth of Y16 pro-B Cells. *Phytother. Res.* **2011**, *25*, 1392–1397. [CrossRef]

28. Khosla, S. Increasing options for the treatment of osteoporosis. *N. Engl. J. Med.* **2009**, *361*, 818–820. [CrossRef]

29. Lee, Y.S.; Choi, E.M. Costunolide stimulates the function of osteoblastic MC3T3-E1 cells. *Int. Immunopharmacol.* **2011**, *11*, 712–718. [CrossRef]

30. Jeon, W.-J.; Kim, K.-M.; Kim, E.-J.; Jang, W.-G. Costunolide increases osteoblast differentiation via ATF4-dependent HO-1 expression in C3H10T1/2 cells. *Life Sci.* **2017**, *178*, 94–99. [CrossRef]

31. Cheon, Y.H.; Song, M.J.; Kim, J.Y.; Kwak, S.C.; Park, J.H.; Lee, C.H.; Kim, J.J.; Kim, J.Y.; Choi, M.K.; Oh, J. Costunolide Inhibits Osteoclast Differentiation by Suppressing c-Fos Transcriptional Activity. *Phytother. Res.* **2014**, *28*, 586–592. [CrossRef]

32. Hermanson, E.; Joseph, B.; Castro, D.; Lindqvist, E.; Aarnisalo, P.; Wallen, A.; Benoit, G.; Hengerer, B.; Olson, L.; Perlmann, T. Nurr1 regulates dopamine synthesis and storage in MN9D dopamine cells. *Exp. Cell Res.* **2003**, *288*, 324–334. [CrossRef]

33. Ham, A.; Lee, S.-J.; Shin, J.; Kim, K.-H.; Mar, W. Regulatory effects of costunolide on dopamine metabolism-associated genes inhibit dopamine-induced apoptosis in human dopaminergic SH-SY5Y cells. *Neurosci. Lett.* **2012**, *507*, 101–105. [CrossRef] [PubMed]

34. Rayan, N.; Baby, N.; Pitchai, D.; Indraswari, F.; Ling, E.; Lu, J.; Dheen, T. Costunolide inhibits proinflammatory cytokines and iNOS in activated murine BV2 microglia. *Front. Biosci. (Elite Ed.)* **2011**, *3*, 1079–1091. [CrossRef] [PubMed]

35. Luna-Herrera, J.; Costa, M.; Gonzalez, H.; Rodrigues, A.; Castilho, P. Synergistic antimycobacterial activities of sesquiterpene lactones from Laurus spp. *J. Antimicrob. Chemother.* **2007**, *59*, 548–552. [CrossRef] [PubMed]

36. Fischer, N.H.; Lu, T.; Cantrell, C.L.; Castañeda-Acosta, J.; Quijano, L.; Franzblau, S.G. Antimycobacterial evaluation of germacranolides in honour of professor GH Neil Towers 75th birthday. *Phytochemistry* **1998**, *49*, 559–564. [CrossRef]

37. Alaagib, R.M.O.; Ayoub, S.M.H. On the chemical composition and antibacterial activity of Saussurea lappa (Asteraceae). *Pharma Innov.* **2015**, *4 Pt C*, 73–76.

38. Park, J.-B.; Lee, C.-K.; Park, H.J. Anti-Helicobacter pylori effect of costunolide isolated from the stem bark ofMagnolia sieboldii. *Arch. Pharmacal Res.* **1997**, *20*, 275. [CrossRef] [PubMed]

39. Duraipandiyan, V.; Al-Harbi, N.A.; Ignacimuthu, S.; Muthukumar, C. Antimicrobial activity of sesquiterpene lactones isolated from traditional medicinal plant, Costus speciosus (Koen ex. Retz.) Sm. *BMC Complement. Altern. Med.* **2012**, *12*, 13. [CrossRef]

40. Wedge, D.; Galindo, J.; Macıas, F. Fungicidal activity of natural and synthetic sesquiterpene lactone analogs. *Phytochemistry* **2000**, *53*, 747–757. [CrossRef]

41. Barrero, A.F.; Oltra, J.E.; Álvarez, M.R.; Raslan, D.S.; Saúde, D.A.; Akssira, M. New sources and antifungal activity of sesquiterpene lactones. *Fitoterapia* **2000**, *71*, 60–64. [CrossRef]

42. Chen, H.-C.; Chou, C.-K.; Lee, S.-D.; Wang, J.-C.; Yeh, S.-F. Active compounds from Saussurea lappa Clarks that suppress hepatitis B virus surface antigen gene expression in human hepatoma cells. *Antivir. Res.* **1995**, *27*, 99–109. [CrossRef]

43. Chen, Z.; Zhang, D.; Li, M.; Wang, B. Costunolide ameliorates lipoteichoic acid-induced acute lung injury via attenuating MAPK signaling pathway. *Int. Immunopharmacol.* **2018**, *61*, 283–289. [CrossRef] [PubMed]

44. Choi, B.Y. Hair-Growth Potential of Ginseng and Its Major Metabolites: A Review on Its Molecular Mechanisms. *Int. J. Mol. Sci* **2018**, *19*, 2703. [CrossRef] [PubMed]

45. Madaan, A.; Verma, R.; Singh, A.T.; Jaggi, M. Review of Hair Follicle Dermal Papilla cells as in vitro screening model for hair growth. *Int. J. Cosmet. Sci.* **2018**, *40*, 429–450. [CrossRef] [PubMed]

46. Kim, Y.E.; Choi, H.C.; Nam, G.; Choi, B.Y. Costunolide promotes the proliferation of human hair follicle dermal papilla cells and induces hair growth in C57 BL/6 mice. *J. Cosmet. Dermatol.* **2018**, *18*, 414–421. [CrossRef] [PubMed]

47. Bocca, C.; Gabriel, L.; Bozzo, F.; Miglietta, A. A sesquiterpene lactone, costunolide, interacts with microtubule protein and inhibits the growth of MCF-7 cells. *Chem. Biol. Interact.* **2004**, *147*, 79–86. [CrossRef] [PubMed]

48. Cai, H.; He, X.; Yang, C. Costunolide promotes imatinib-induced apoptosis in chronic myeloid leukemia cells via the Bcr/Abl-Stat5 pathway. *Phytother. Res.* **2018**, *32*, 1764–1769. [CrossRef] [PubMed]

49. Dong, G.Z.; Shim, A.R.; Hyeon, J.S.; Lee, H.J.; Ryu, J.H. Inhibition of Wnt/β-Catenin Pathway by Dehydrocostus Lactone and Costunolide in Colon Cancer Cells. *Phytother. Res.* **2015**, *29*, 680–686. [CrossRef] [PubMed]

50. Hsu, J.-L.; Pan, S.-L.; Ho, Y.-F.; Hwang, T.-L.; Kung, F.-L.; Guh, J.-H. Costunolide induces apoptosis through nuclear calcium2+ overload and DNA damage response in human prostate cancer. *J. Urol.* **2011**, *185*, 1967–1974. [CrossRef] [PubMed]

51. Liu, C.-Y.; Chang, H.-S.; Chen, I.-S.; Chen, C.-J.; Hsu, M.-L.; Fu, S.-L.; Chen, Y.-J. Costunolide causes mitotic arrest and enhances radiosensitivity in human hepatocellular carcinoma cells. *Radiat. Oncol.* **2011**, *6*, 56. [CrossRef] [PubMed]

52. Hu, M.; Liu, L.; Yao, W. Activation of p53 by costunolide blocks glutaminolysis and inhibits proliferation in human colorectal cancer cells. *Gene* **2018**, *678*, 261–269. [CrossRef] [PubMed]

53. Hua, P.; Zhang, G.; Zhang, Y.; Sun, M.; Cui, R.; Li, X.; Li, B.; Zhang, X. Costunolide induces G1/S phase arrest and activates mitochondrial-mediated apoptotic pathways in SK-MES 1 human lung squamous carcinoma cells. *Oncol. Lett.* **2016**, *11*, 2780–2786. [CrossRef] [PubMed]

54. Rasul, A.; Yu, B.; Yang, L.-F.; Arshad, M.; Khan, M.; Ma, T.; Yang, H. Costunolide, a sesquiterpene lactone induces G2/M phase arrest and mitochondria-mediated apoptosis in human gastric adenocarcinoma SGC-7901 cells. *J. Med. Plants Res.* **2012**, *6*, 1191–1200.

55. Peng, Z.; Wang, Y.; Fan, J.; Lin, X.; Liu, C.; Xu, Y.; Ji, W.; Yan, C.; Su, C. Costunolide and dehydrocostuslactone combination treatment inhibit breast cancer by inducing cell cycle arrest and apoptosis through c-Myc/p53 and AKT/14-3-3 pathway. *Sci. Rep.* **2017**, *7*, 41254. [CrossRef] [PubMed]

56. Choi, Y.K.; Seo, H.S.; Choi, H.S.; Choi, H.S.; Kim, S.R.; Shin, Y.C.; Ko, S.-G. Induction of Fas-mediated extrinsic apoptosis, p21WAF1-related G2/M cell cycle arrest and ROS generation by costunolide in estrogen receptor-negative breast cancer cells, MDA-MB-231. *Mol. Cell. Biochem.* **2012**, *363*, 119–128. [CrossRef] [PubMed]

57. Roy, A.; Manikkam, R. Cytotoxic Impact of Costunolide Isolated from Costus speciosus on Breast Cancer via Differential Regulation of Cell Cycle—An In-vitro and In-silico Approach. *Phytother. Res.* **2015**, *29*, 1532–1539. [CrossRef] [PubMed]

58. Cai, H.; Li, L.; Jiang, J.; Zhao, C.; Yang, C. Costunolide enhances sensitivity of K562/ADR chronic myeloid leukemia cells to doxorubicin through PI3K/Akt pathway. *Phytother. Res.* **2019**. [CrossRef] [PubMed]

59. Hua, P.; Sun, M.; Zhang, G.; Zhang, Y.; Song, G.; Liu, Z.; Li, X.; Zhang, X.; Li, B. Costunolide Induces Apoptosis through Generation of ROS and Activation of P53 in Human Esophageal Cancer Eca-109 Cells. *J. Biochem. Mol. Toxicol.* **2016**, *30*, 462–469. [CrossRef]

60. Chen, J.; Chen, B.; Zou, Z.; Li, W.; Zhang, Y.; Xie, J.; Liu, C. Costunolide enhances doxorubicin-induced apoptosis in prostate cancer cells via activated mitogen-activated protein kinases and generation of reactive oxygen species. *Oncotarget* **2017**, *8*, 107701–107705. [CrossRef]

61. Rasul, A.; Bao, R.; Malhi, M.; Zhao, B.; Tsuji, I.; Li, J.; Li, X. Induction of apoptosis by costunolide in bladder cancer cells is mediated through ROS generation and mitochondrial dysfunction. *Molecules* **2013**, *18*, 1418–1433. [CrossRef]

62. Choi, J.-H.; Lee, K.-T. Costunolide-induced apoptosis in human leukemia cells: Involvement of c-jun N-terminal kinase activation. *Biol. Pharm. Bull.* **2009**, *32*, 1803–1808. [CrossRef] [PubMed]

63. Yang, Y.-I.; Kim, J.-H.; Lee, K.-T.; Choi, J.-H. Costunolide induces apoptosis in platinum-resistant human ovarian cancer cells by generating reactive oxygen species. *Gynecol. Oncol.* **2011**, *123*, 588–596. [CrossRef] [PubMed]

64. Kim, J.-H.; Yang, Y.-I.; Lee, K.-T.; Park, H.-J.; Choi, J.-H. Costunolide induces apoptosis in human endometriotic cells through inhibition of the prosurvival Akt and nuclear factor kappa B signaling pathway. *Biol. Pharm. Bull.* **2011**, *34*, 580–585. [CrossRef] [PubMed]

65. Fang, Y.; Li, J.; Wu, Y.; Gui, J.; Shen, Y. Costunolide Inhibits the Growth of OAW42-A Multidrug-Resistant Human Ovarian Cancer Cells by Activating Apoptotic and Autophagic Pathways, Production of Reactive Oxygen Species (ROS), Cleaved Caspase-3 and Cleaved Caspase-9. *Med. Sci. Monit.* **2019**, *25*, 3231–3237. [CrossRef] [PubMed]

66. Wang, Z.; Zhao, X.; Gong, X. Costunolide induces lung adenocarcinoma cell line A549 cells apoptosis through ROS (reactive oxygen species)—Mediated endoplasmic reticulum stress. *Cell Biol. Int.* **2016**, *40*, 289–297. [CrossRef] [PubMed]

67. Zhang, C.; Lu, T.; Wang, G.-D.; Ma, C.; Zhou, Y.-F. Costunolide, an active sesquiterpene lactone, induced apoptosis via ROS-mediated ER stress and JNK pathway in human U2OS cells. *Biomed. Pharmacother.* **2016**, *80*, 253–259. [CrossRef] [PubMed]

68. Zhuge, W.; Chen, R.; Vladimir, K.; Dong, X.; Zia, K.; Sun, X.; Dai, X.; Bao, M.; Shen, X.; Liang, G. Costunolide specifically binds and inhibits thioredoxin reductase 1 to induce apoptosis in colon cancer. *Cancer Lett.* **2018**, *412*, 46–58. [CrossRef]

69. Kanno, S.-I.; Kitajima, Y.; Kakuta, M.; Osanai, Y.; Kurauchi, K.; Ujibe, M.; Ishikawa, M. Costunolide-induced apoptosis is caused by receptor-mediated pathway and inhibition of telomerase activity in NALM-6 cells. *Biol. Pharm. Bull.* **2008**, *31*, 1024–1028. [CrossRef]

70. Choi, S.-H.; Im, E.; Kang, H.K.; Lee, J.-H.; Kwak, H.-S.; Bae, Y.-T.; Park, H.-J.; Kim, N.D. Inhibitory effects of costunolide on the telomerase activity in human breast carcinoma cells. *Cancer Lett.* **2005**, *227*, 153–162. [CrossRef]

71. Ahmad, F.; Dixit, D.; Sharma, V.; Kumar, A.; Joshi, S.; Sarkar, C.; Sen, E. Nrf2-driven TERT regulates pentose phosphate pathway in glioblastoma. *Cell Death Dis.* **2017**, *7*, e2213. [CrossRef]

72. Tahtouh, R.; Azzi, A.-S.; Alaaeddine, N.; Chamat, S.; Bouharoun-Tayoun, H.; Wardi, L.; Raad, I.; Sarkis, R.; Antoun, N.A.; Hilal, G. Telomerase inhibition decreases alpha-fetoprotein expression and secretion by hepatocellular carcinoma cell lines: In vitro and in vivo study. *PLoS ONE* **2015**, *10*, e0119512. [CrossRef] [PubMed]

73. Nishida, N.; Yano, H.; Nishida, T.; Kamura, T.; Kojiro, M. Angiogenesis in cancer. *Vasc. Health Risk Manag.* **2006**, *2*, 213–219. [CrossRef]

74. Saraswati, S.; Alhaider, A.A.; Abdelgadir, A.M. Costunolide suppresses an inflammatory angiogenic response in a subcutaneous murine sponge model. *Apmis* **2018**, *126*, 257–266. [CrossRef] [PubMed]

75. Jeong, S.-J.; Itokawa, T.; Shibuya, M.; Kuwano, M.; Ono, M.; Higuchi, R.; Miyamoto, T. Costunolide, a sesquiterpene lactone from Saussurea lappa, inhibits the VEGFR KDR/Flk-1 signaling pathway. *Cancer Lett.* **2002**, *187*, 129–133. [CrossRef]

76. Mahfouz, N.; Tahtouh, R.; Alaaeddine, N.; El Hajj, J.; Sarkis, R.; Hachem, R.; Raad, I.; Hilal, G. Gastrointestinal cancer cells treatment with bevacizumab activates a VEGF autoregulatory mechanism involving telomerase catalytic subunit hTERT via PI3K-AKT, HIF-1α and VEGF receptors. *PLoS ONE* **2017**, *12*, e0179202. [CrossRef] [PubMed]

77. Choi, Y.K.; Cho, S.-G.; Woo, S.-M.; Yun, Y.J.; Jo, J.; Kim, W.; Shin, Y.C.; Ko, S.-G. Saussurea lappa Clarke-derived costunolide prevents TNFα-induced breast cancer cell migration and invasion by inhibiting NF-κB activity. *Evid. Based Complement. Altern. Med.* **2013**, *2013*, 936257. [CrossRef] [PubMed]

78. Tabata, K.; Nishimura, Y.; Takeda, T.; Kurita, M.; Uchiyama, T.; Suzuki, T. Sesquiterpene lactones derived from Saussurea lappa induce apoptosis and inhibit invasion and migration in neuroblastoma cells. *J. Pharmacol. Sci.* **2015**, *127*, 397–403. [CrossRef]

79. Lohberger, B.; Rinner, B.; Stuendl, N.; Kaltenegger, H.; Steinecker-Frohnwieser, B.; Bernhart, E.; Rad, E.B.; Weinberg, A.M.; Leithner, A.; Bauer, R. Sesquiterpene lactones downregulate G2/M cell cycle regulator proteins and affect the invasive potential of human soft tissue sarcoma cells. *PLoS ONE* **2013**, *8*, e66300. [CrossRef]

80. Jeong, D.; Watari, K.; Shirouzu, T.; Ono, M.; Koizumi, K.; Saiki, I.; Kim, Y.-C.; Tanaka, C.; Higuchi, R.; Miyamoto, T. Studies on lymphangiogenesis inhibitors from Korean and Japanese crude drugs. *Biol. Pharm. Bull.* **2013**, *36*, 152–157. [CrossRef]

81. Whipple, R.A.; Vitolo, M.I.; Boggs, A.E.; Charpentier, M.S.; Thompson, K.; Martin, S.S. Parthenolide and costunolide reduce microtentacles and tumor cell attachment by selectively targeting detyrosinated tubulin independent from NF-κB inhibition. *Breast Cancer Res.* **2013**, *15*, R83. [CrossRef]

82. Perera, H.K.; Premadasa, W.K.; Poongunran, J. alpha-glucosidase and glycation inhibitory effects of costus speciosus leaves. *BMC Complement. Altern Med.* **2016**, *16*, 2.

83. Eliza, J.; Daisy, P.; Ignacimuthu, S.; Duraipandiyan, V. Normo-glycemic and hypolipidemic effect of costunolide isolated from Costus speciosus (Koen ex. Retz.) Sm. in streptozotocin-induced diabetic rats. *Chem. Biol. Interact.* **2009**, *179*, 329–334. [CrossRef] [PubMed]

84. Hu, F.; Feng, S.; Wu, Y.; Bi, Y.; Wang, C.; Li, W. Quantitative analysis of costunolide and dehydrocostuslactone in rat plasma by ultraperformance liquid chromatography–electrospray ionization–mass spectrometry. *Biomed. Chromatogr.* **2011**, *25*, 547–554. [CrossRef] [PubMed]

85. Zhang, J.; Hu, X.; Gao, W.; Qu, Z.; Guo, H.; Liu, Z.; Liu, C. Pharmacokinetic study on costunolide and dehydrocostuslactone after oral administration of traditional medicine Aucklandia lappa Decne. by LC/MS/MS. *J. Ethnopharmacol.* **2014**, *151*, 191–197. [CrossRef] [PubMed]

86. Dong, S.; Ma, L.-Y.; Liu, Y.-T.; Yu, M.; Jia, H.-M.; Zhang, H.-W.; Yu, C.-Y.; Zou, Z.-M. Pharmacokinetics of costunolide and dehydrocostuslactone after oral administration of Radix aucklandiae extract in normal and gastric ulcer rats. *J. Asian Nat. Prod. Res.* **2018**, *11*, 1055–1063. [CrossRef] [PubMed]

87. Peng, Z.; Wang, Y.; Gu, X.; Guo, X.; Yan, C. Study on the pharmacokinetics and metabolism of costunolide and dehydrocostus lactone in rats by HPLC-UV and UPLC-Q-TOF/MS. *Biomed. Chromatogr.* **2014**, *28*, 1325–1334. [CrossRef] [PubMed]

88. Singireesu, S.; Misra, S.; Mondal, S.K.; Yerramsetty, S.; Sahu, N.; Katragadda, S.B. Costunolide induces micronuclei formation, chromosomal aberrations, cytostasis, and mitochondrial-mediated apoptosis in Chinese hamster ovary cells. *Cell Biol. Toxicol.* **2018**, *34*, 125–142. [CrossRef]
89. Muschietti, L.V.; Ulloa, J.L. Natural Sesquiterpene Lactones as Potential Trypanocidal Therapeutic Agents: A Review. *Nat. Prod. Commun.* **2016**, *11*, 1569–1578. [CrossRef]

International Journal of
Molecular Sciences

Review

Phytochemicals as Novel Therapeutic Strategies for NLRP3 Inflammasome-Related Neurological, Metabolic, and Inflammatory Diseases

Carolina Pellegrini [1,*], Matteo Fornai [2], Luca Antonioli [2], Corrado Blandizzi [2] and Vincenzo Calderone [1]

[1] Department of Pharmacy, University of Pisa, 56126 Pisa, Italy; vincenzo.calderone@unipi.it
[2] Department of Clinical and Experimental Medicine, University of Pisa, 56126 Pisa, Italy;
    mfornai74@gmail.com (M.F.); lucaant@gmail.com (L.A.); c.blandizzi@gmail.com (C.B.)
* Correspondence: carolina.pellegrini87@gmail.com; Tel.: +39-050-2218757; Fax: +39-050-2218758

Received: 6 May 2019; Accepted: 11 June 2019; Published: 13 June 2019

Abstract: Several lines of evidence point out the relevance of nucleotide-binding oligomerization domain leucine-rich repeat and pyrin domain-containing protein 3 (NLRP3) inflammasome as a pivotal player in the pathophysiology of several neurological and psychiatric diseases (i.e., Parkinson's disease (PD), Alzheimer's disease (AD), multiple sclerosis (MS), amyotrophic lateral sclerosis, and major depressive disorder), metabolic disorders (i.e., obesity and type 2 diabetes) and chronic inflammatory diseases (i.e., intestinal inflammation, arthritis, and gout). Intensive research efforts are being made to achieve an integrated view about the pathophysiological role of NLRP3 inflammasome pathways in such disorders. Evidence is also emerging that the pharmacological modulation of NLRP3 inflammasome by phytochemicals could represent a promising molecular target for the therapeutic management of neurological, psychiatric, metabolic, and inflammatory diseases. The present review article has been intended to provide an integrated and critical overview of the available clinical and experimental evidence about the role of NLRP3 inflammasome in the pathophysiology of neurological, psychiatric, metabolic, and inflammatory diseases, including PD, AD, MS, depression, obesity, type 2 diabetes, arthritis, and intestinal inflammation. Special attention has been paid to highlight and critically discuss current scientific evidence on the effects of phytochemicals on NLRP3 inflammasome pathways and their potential in counteracting central neuroinflammation, metabolic alterations, and immune/inflammatory responses in such diseases.

Keywords: NLRP3 inflammasome; neurological diseases; psychiatric diseases; metabolic diseases; inflammatory diseases; phytochemicals

---

1. Introduction

A growing body of evidence highlights the relevance of nucleotide-binding oligomerization domain leucine-rich repeat and pyrin domain-containing protein 3 (NLRP3) inflammasome in the pathophysiology of several autoinflammatory syndromes (i.e., cryopyrin-associated autoinflammatory syndromes (CAPS)), neurological and psychiatric diseases (i.e., Parkinson's disease (PD), Alzheimer's disease (AD), multiple sclerosis (MS), amyotrophic lateral sclerosis (ALS), and major depressive disorder (MDD)), metabolic disorders (i.e., obesity and type 2 diabetes), and chronic inflammatory diseases (i.e., arthritis and intestinal inflammation) [1–4]. In particular, the NLRP3 inflammasome complex, including NLRP3, adaptor protein apoptosis-associated speck-like protein (ASC), and pro-caspase-1, through the processing and release of interleukin (IL)-1β and IL-18, acts as a key player both in coordinating the host physiology and shaping the central and/or peripheral immune/inflammatory responses in neurological, metabolic, and inflammatory diseases [5]. Indeed, an overactivation of NLRP3 inflammasome signaling has been observed in the brain and blood of patients with neurological disorders, adipose tissue macrophages from obese and diabetic patients, as well as patients with rheumatoid arthritis (RA) and inflammatory bowel diseases (IBDs) [2,6–9]. In keeping with this knowledge, it is becoming increasingly appreciated that drugs targeting the NLRP3 pathway could represent suitable therapeutic options for the management of a large variety of diseases [10]. For instance, the pharmacological blockade of NLRP3 signaling has been found to exert beneficial effects in animal models of PD, MS, obesity, type 2 diabetes, arthritis, and colitis [10]. Of note, given the great interest in the therapeutic potential of phytochemicals, in terms of prevention, cure, and maintenance of remission, intensive efforts are being made to characterize the effects of natural compounds targeting NLRP3 pathways in neurological, metabolic, and inflammatory diseases [11]. In this regard, several studies have shown that various phytochemicals, including polyphenols and glucosinolates, counteract neuroinflammation, metabolic alterations, and immune/inflammatory responses in experimental models of diseases via inhibition of NLRP3 signaling [12–14].

Based on the above background, the present review article intents to provide an integrated and critical overview of the available clinical and experimental evidence about the role of NLRP3 inflammasome in the pathophysiology of neurological, metabolic, and inflammatory diseases, including obesity, type 2 diabetes, PD, AD, MS, depression, arthritis, and intestinal inflammation. Special attention has been paid to highlight and critically discuss current scientific evidence on the effects of phytochemicals on NLRP3 inflammasome pathways and their ability of counteracting central neuroinflammation, metabolic alterations, and immune/inflammatory responses in such diseases.

2. Mechanisms of NLRP3 Inflammasome Activation

NLRP3 inflammasome, the most characterized inflammasome sensor molecule, is a tripartite protein of the nucleotide-binding domain and leucine-rich repeat (NLR) family, containing an amino-terminal pyrin domain (PYRIN) domain, a nucleotide-binding NACHT domain with ATPase activity, and a carboxy-terminal leucine-rich repeat (LRR) domain [15]. NLRP3 is a key sensor of cellular stress, which senses changes in homeostatic cellular state. Currently, two modes of NLRP3 activation have been characterized: Canonical and non-canonical inflammasome activation.

Canonical NLRP3 inflammasome activation requires two parallel and independent steps: Priming (transcription) and activation (oligomerization) (Figure 1) [16]. In the first step, innate immune signaling via toll-like receptor (TLR)-adaptor molecule myeloid differentiation primary response 88 (MyD88) and/or cytokine receptors, such as the tumor necrosis factor (TNF) receptor, promote pro-IL-1β and NLRP3 transcription via nuclear factor-κB (NF-κB) activation. In the second step, the oligomerization and activation of NLRP3 inflammasome lead to caspase-1 activation and, in turn, IL-1β and IL-18 processing and release [16,17]. Different stimuli, including viral RNA, inhibition of glycolytic or mitochondrial metabolism, extracellular osmolarity, α-synuclein (α-syn) and β-amyloid (Aβ) protein accumulation, degradation of extracellular matrix components, and post-translational NLRP3 modification (i.e., phosphorylation and ubiquitination) can initiate NLRP3 inflammasome oligomerization and activation. In addition, the permeabilization of cell membranes to potassium efflux (i.e., mixed lineage kinase domain-like protein (MLKL) activation, exposure to pore-forming gasdermin D, P2X7 purinergic receptor activation by extracellular adenosine triphosphate (ATP), lysosomal damage, and cathepsin release), the consequent release of oxidized mitochondrial DNA, the increase in mitochondrial reactive oxygen species (ROS), and the cardiolipin externalization can activate NLRP3 inflammasome assembly [1,10]. Independently from IL-1β maturation, caspase-1 activation promotes also pyroptosis, a key defense mechanism against microbial infections, which blocks the replication of intracellular pathogens, induces phagocytosis of surviving bacteria, and promotes the release of additional cytosolic proteins, such as high-mobility group box 1 (HMGB1) alarmin, a pro-inflammatory mediator significantly involved in the pathogenesis of several inflammatory chronic diseases [18,19].

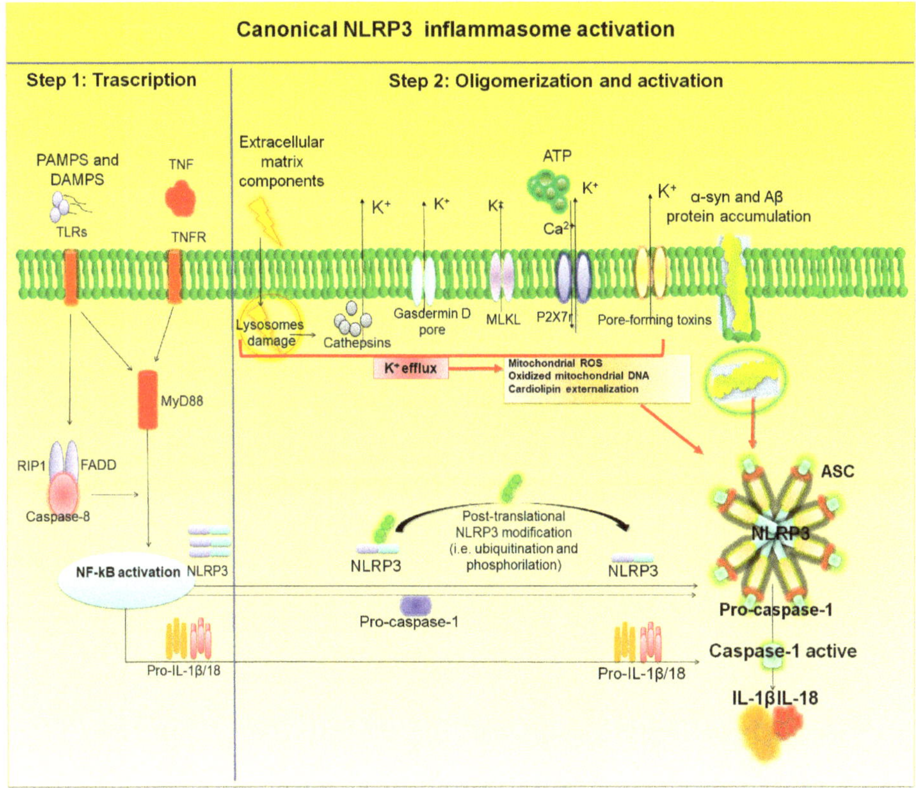

Figure 1. Diagram showing the different molecular mechanisms of canonical nucleotide-binding oligomerization domain leucine rich repeat and pyrin domain-containing protein 3 (NLRP3) inflammasome activation. The first step is regulated by toll-like receptors (TLRs)–adaptor molecule myeloid differentiation primary response 88 (MyD88) pathway and/or tumor necrosis factor receptor (TNFR), which activate pro-IL-1β and NLRP3 transcription via nuclear factor kB (NF-kB) activation. TLR4 stimulation by pathogen-associated molecular pattern molecules (PAMPs) and/or damage-associated molecular pattern molecules (DAMPs) can activate also the receptor-interacting protein 1 (RIP1)–FAS-associated death domain protein (FADD)-caspase-8 protein complex, which, in turn, promotes NF-kB transcription. The second step results in NLRP3 inflammasome oligomerization, leading to caspase-1 activation as well as IL-1β and IL-18 release. Permeabilization of cell membranes to potassium efflux (i.e., mixed lineage kinase domain-like protein (MLKL) activation, exposure to pore-forming Gasdermin D, P2X7 receptor activation by extracellular ATP, lysosomal damage, and cathepsin release) leads to a massive release of oxidized mitochondrial DNA, increase in mitochondrial reactive oxygen species (ROS) and cardiolipin externalization, which, in turn, promote NLRP3 inflammasome oligomerization and activation. α-Syn and Aβ protein accumulation, and post-translational NLRP3 modifications (i.e., phosphorylation and ubiquitination) can also promote the second step of NLRP3 inflammasome activation.

Besides canonical NLRP3 inflammasome activation, a non-canonical activation, which depends on caspase-11 in mice (caspase 4 and caspase 5 in humans), has been characterized (Figure 2) [1]. In this setting, in the first transcription step, gram-negative bacteria (i.e., *Citrobacter rodentium*, *Escherichia coli*, *Legionella pneumophila*, *Salmonella typhimurium*, and *Vibrio cholerae*) activate the TLR4–MyD88 and toll/IL-1 receptor homology-domain-containing adapter-inducing interferon-β (TRIF) pathways which, in turn, promote the transcription of IL-1β, IL-18, and NLRP3, as well as interferon regulatory factor (IRF)-3 and IRF7 genes via NF-κB activation. The IRF3–IRF7 complex promotes the expression of interferon (IFN)-α/β which, in turn, activates the IFN-α/β receptor 1 (IFNAR)/IFNAR2- the janus kinase/signal transducers and activators of transcription (JAK/STAT) pathway leading to transcription of caspase-11 gene. In the second step, unidentified scaffold proteins or receptors, induced by Gram-negative bacteria, cleave and activate caspase-11, which induces pyroptosis, HMGB1, and IL-1α release, and promotes IL-1β processing and release through activation of the canonical NLRP3-ASC-caspase-1 pathway (Figure 2) [1].

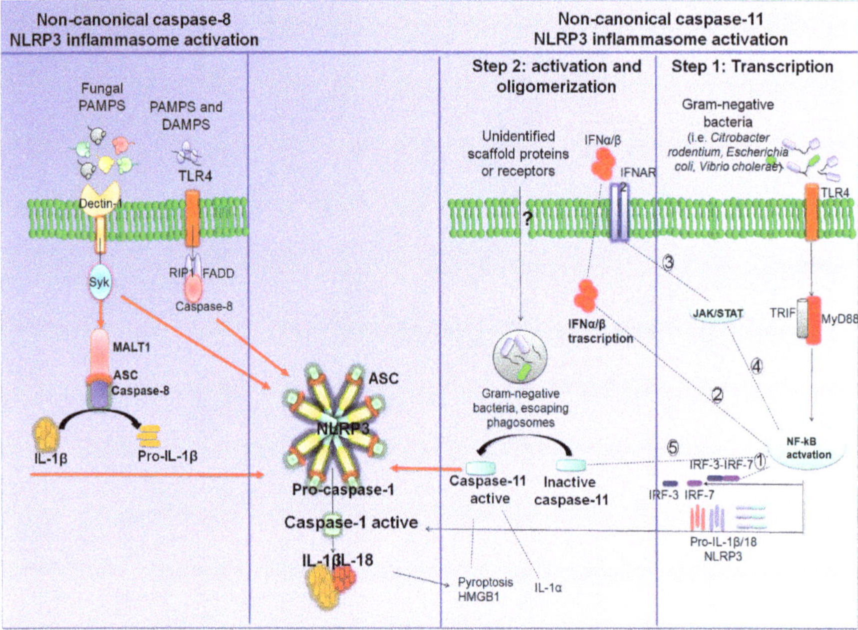

Figure 2. Diagram showing the different molecular mechanisms of non-canonical NLRP3 inflammasome activations. **Left panel**: Non-canonical caspase-8-dependent NLRP3 activation. TLR4 stimulation by PAMPs and/or DAMPs activates RIP1–FADD-caspase-8 intracellular signaling, which, besides promoting the NF-kB transcription step, can activate directly canonical NLRP3 oligomerization and assembly. In addition, fungi PAMPS (i.e., *Candida albicans*, fungal cell wall component β-glucans, and mycobacteria), via dectin-1 stimulation, can promote IL-1β transcription as well as the formation and activation of a mucosa-associated lymphoid tissue lymphoma translocation protein 1 (MALT1)–caspase-8– adaptor protein (ASC) complex, which contributes to processing and release of IL-1β. **Right panel**: Non-canonical caspase-11-dependent NLRP3 activation. In the first step, Gram-negative bacteria activate the TLR4–MyD88 and tumor necrosis factor receptor (TRIF) pathways, with consequent nuclear translocation of NF-κB, which promotes the transcription of IL-1β, IL-18, and NLRP3 as well as interferon regulatory factor (IRF)-3 and IRF7 genes. The IRF3–IRF7 complex (1) elicits the expression of IFN-α/β (2) that binds the interferon (IFN)-α/β receptor 1 IFNAR1/IFNAR2 receptor (3), leading to activation of the janus kinase/signal transducers and activators of transcription (JAK/STAT) pathway (4) and transcription of caspase-11 gene (5). In the second step, unidentified scaffold proteins or receptors induced by Gram-negative bacteria cleave and activate caspase-11, which induces pyroptosis as well as high-mobility group box 1 (HMGB1) and IL-1α release, and promotes the activation of NLRP3–ASC–caspase-1 pathway. Caspase-1 activation promotes also pyroptosis and HMGB1 release.

Of interest, an additional non-canonical caspase-8 dependent NLRP3 activation has been recently characterized (Figure 2) [20–24]. In particular, TLR4 stimulation by pathogen-associated molecular pattern molecules (PAMPs) and/or damage-associated molecular pattern molecules (DAMPs) can activate caspase-8 and its receptor-interacting protein 1 (RIP1)–fatty acid synthase (FAS)-associated death domain protein (FADD) protein, which, in turn, promote both the transcription step and canonical NLRP3 oligomerization and activation [24]. Moreover, fungi (i.e., *Candida albicans*), fungal cell wall component β-glucans, and mycobacteria, via dectin-1 stimulation, have been found to promote IL-1β transcription as well as the formation and activation of a mucosa-associated lymphoid tissue lymphoma translocation protein 1 (MALT1)–caspase-8–ASC complex that contributes to the processing and release of IL-1β (Figure 2) [22]. In particular, caspase-8 can act as a direct IL-1β-converting enzyme, instead of caspase-1. Indeed, IL-1β processing and caspase-8 activation were not evident in NLRP3$^{-/-}$ or ASC$^{-/-}$ bone marrow-derived dendritic cells (BMDCs) [20,23]. Moreover, the release of IL-1β from dendritic cells stimulated with a fungal infection occurred independently of caspase-1, through the association of ASC protein with caspase-8 [22].

Overall, this body of evidence points out that the activation of NLRP3 inflammasome is regulated by several molecular processes, which can be closely interconnected or occur independently. Nevertheless, further in vitro experiments on cultured cells are required to clarify the molecular mechanisms underlying the interplay between caspase-1, -8, and -11 in promoting the canonical and/or non-canonical NLRP3 activations.

3. Effects of Phytochemicals in NLRP3 Inflammasome-Related Diseases

The involvement of inflammasome pathways in the pathophysiology of central nervous system (CNS) diseases, metabolic disorders, and chronic inflammatory diseases is fostering research on the potential therapeutic benefits resulting from the pharmacological targeting of NLRP3 inflammasome. In this field, besides specific programs of novel drug discovery, the scientific community is making intensive efforts to characterize the effects of natural compounds. Current evidence on the effects of different phytochemicals (see chemical structures in Figure 3) on NLRP3 inflammasome pathways and central neuroinflammation, metabolic alterations, and immune/inflammatory responses is addressed in the following sections and summarized in Tables 1–3.

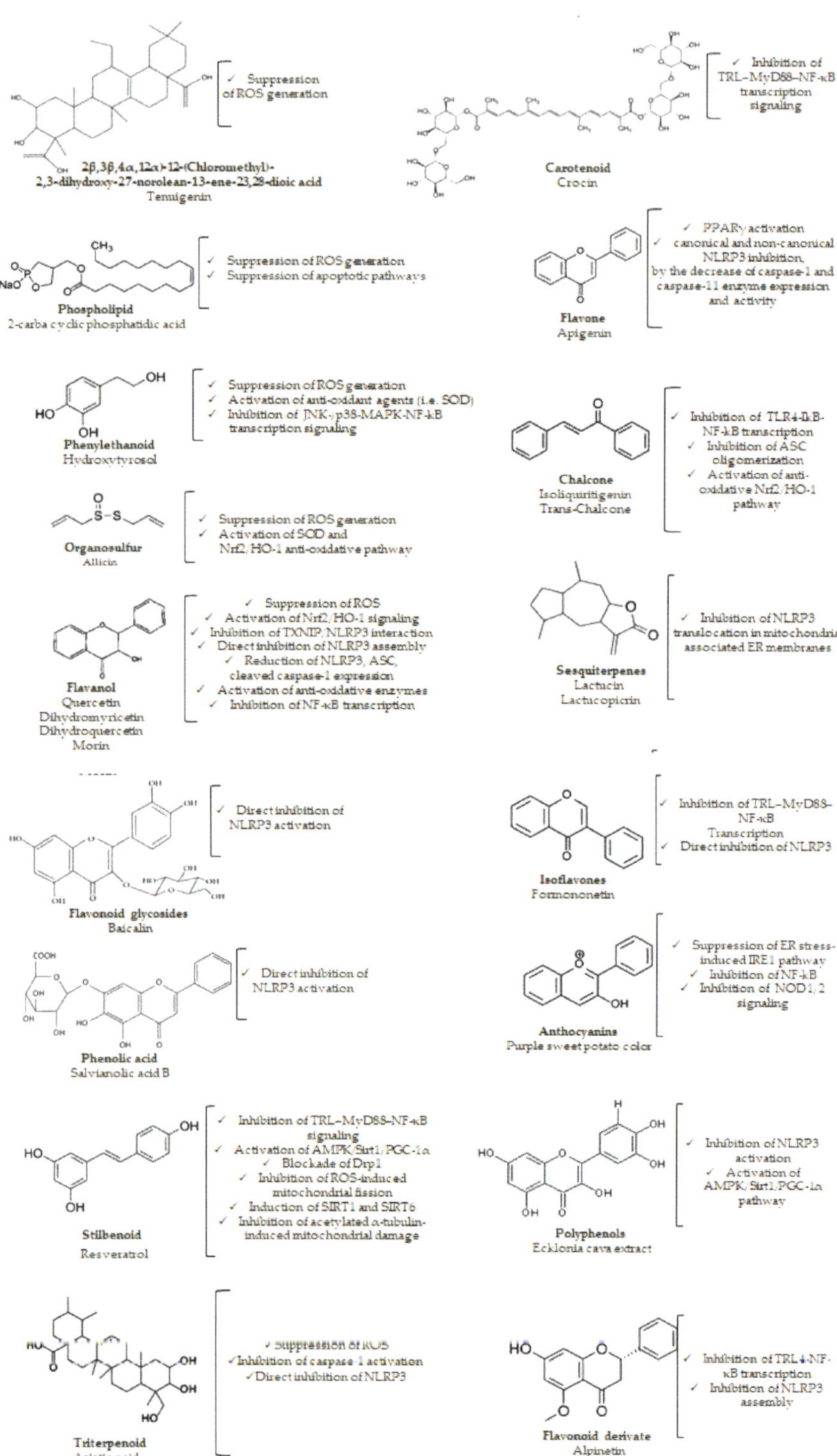

Figure 3. *Cont.*

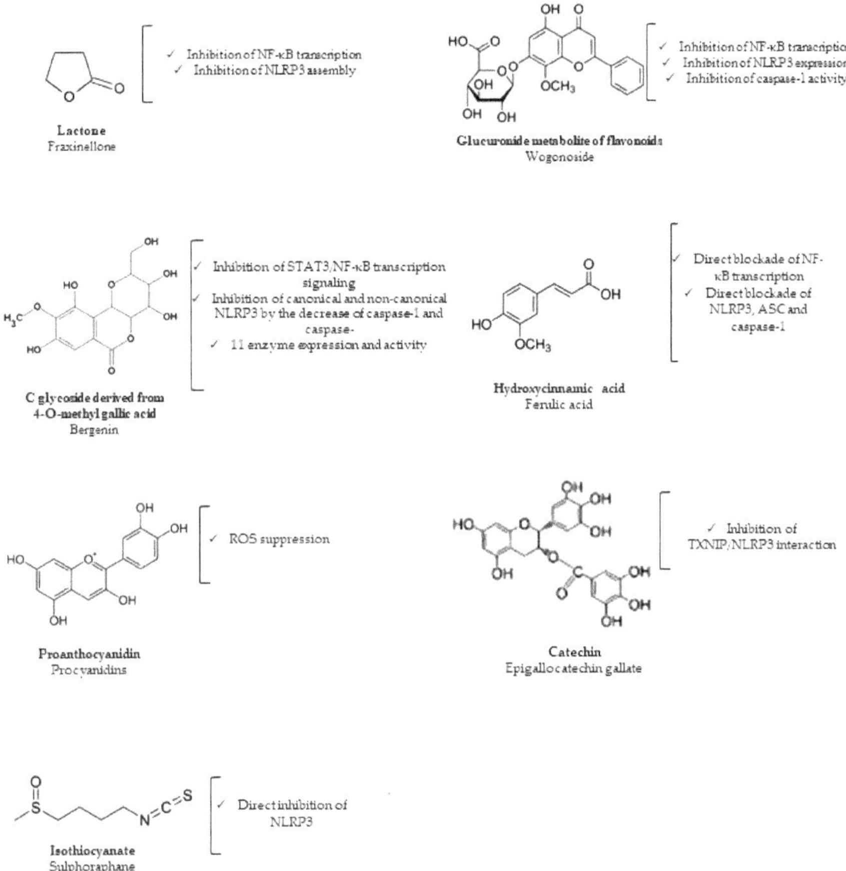

Figure 3. Chemical structures of different phytochemicals acting on NLRP3 inflammasome pathways and related molecular mechanisms through which these compounds modulate NLRP3 activation.

Table 1. Phytochemicals inhibiting nucleotide-binding oligomerization domain leucine rich repeat and pyrin domain-containing protein 3 (NLRP3) inflammasome activation in animal models of central nervous system (CNS) disorders.

Phytochemical	Category	Molecular Mechanisms	Dose and Treatment Time	Experimental Models	Ref.
Tenu genin	*2β,3β,4α,12α)-12-(Chloromethyl)-2,3-dihydroxy-27-norolean-13-ene-23,28-dioic acid*	✓ Suppression of ROS generation	✓ 25 and 50 mg/Kg/day p.o. for 10 days	✓ MPTP-induced central dopaminergic neurodegeneration (toxin-induced model of PD)	[25]
Cyclic phosphatidic acid	*Phospholipid*	✓ Suppression of ROS generation ✓ Suppression apoptotic pathways	✓ 1.6 mg/Kg/day i.p. for 5 weeks	✓ Cuprizone-induced demyelination (animal model of MS) ✓ EAE mice (animal model of MS)	[26]
Hydroxytyrosol	*Phenylethanoid*	✓ Suppression of oxidative stress ✓ Activation of anti-oxidant agents (i.e., SOD) ✓ Inhibition of JNK-/p38-MAPK-NF-kB transcription signaling	✓ 5 mg/Kg/day p.o. for 42 weeks	✓ APP/PS1 mice (genetic model of AD)	[27]
Allicin	*Organosulfur*	✓ Suppression of ROS ✓ Activation of SOD and Nrf2/HO-1 anti-oxidative pathway	✓ 2, 10, 50 mg/Kg/day p.o. for 10 days	✓ CSDS mice (animal model of depression)	[28]
Dihydromyricetin	*Flavonolol*	✓ Reduction of NLRP3, ASC, cleaved caspase-1 expression	✓ 1 mg/Kg/day i.p. for 4 weeks	✓ APP/PS1 mice (genetic model of AD)	[29]
Ginkgo biloba extracts	*n.a*	✓ Direct blockade of caspase-1 activation	✓ 600 mg/day for 20 weeks	✓ TgCRND8 mice (a transgenic model of AD)	[30]
Baicalin	*Flavonoid glycosides*	✓ Direct inhibition of NLRP3 activation	✓ 20 and 40 mg/Kg7day p.o. for 3 weeks	✓ Chronic unpredictable mild stress (CUMS) rats	[31]
Salvianolic acid B	*Phenolic acid*	✓ Direct inhibition of NLRP3 activation	✓ 20 mg/Kg/day i.p. for 7 days	✓ LPS-induced depression rats	[32]
Crocin	*Carotenoid*	✓ Inhibition of TRL–MyD88–NF-κB transcription signaling	✓ 20 mg/Kg7day i.p. for 7 days	✓ LPS-induced depressive mice	[33]
Resveratrol	*Stilbenoid*	✓ Inhibition of TRL–MyD88–NF-κB transcription signaling	✓ 0.02 and 0.2 mg/Kg/day p.o. for 10 days	✓ Intracerebroventricular injection of Aβ1-42 (animal model of AD)	[34]
Resveratrol	*Stilbenoid*	✓ Inhibition of TRL–MyD88–NF-κB transcription signaling ✓ Activation of AMPK/Sirt1/PGC-1α pathway	✓ 20 mg/Kg/day i.p. for 14 days	✓ Ovariectomy-induced anxiety-and depression-like behaviors	[35]
Apigenin	*Flavone*	✓ PPARγ activation	✓ 3 mg/Kg/day p.o. for 3 weeks	✓ CUMS rats (animal model of depression)	[36]

Table 2. Phytochemicals inhibiting NLRP3 inflammasome activation in animal models of metabolic disorders and related comorbidities.

Phytochemical	Category	Molecular Mechanisms	Dose and Treatment Time	Experimental Models	Ref.
Isoliquiritigenin	Chalcone	✓ Inhibition of TLR4-IkB-NF-kB transcription ✓ Inhibition of ASC oligomerization	✓ 0.5% w/w for 20 weeks	✓ HFD-induced obesity	[13]
Lactucin and Lactucopicrin	Sesquiterpene	✓ Inhibition of NLRP3 translocation in mitochondria-associated ER membranes	✓ 50 mg/Kg/twice a week p.o. for 6 weeks	✓ HFD-induced type-2 diabetes	[37]
Resveratrol	Stilbenoid	✓ AMPK stimulation ✓ Blockade of Drp1 ✓ Inhibition of ROS-induced mitochondrial fission	✓ 50 mg/Kg7day p.o. for 7 days	✓ Streptozotocin-induced diabetes	[38]
Quercetin	Flavonol	✓ Inhibition of TXNIP/NLRP3 interaction	✓ 25 mg/Kg/day p.o. for 7 days	✓ Hepatic inflammation by streptozocin-induced type 1 diabetic rats	[39]
Purple sweet potato color	Anthocyanins	✓ Suppression of ER stress-induced IRE1 pathway ✓ Inhibition of NF-kB translocation ✓ Inhibition of NOD1/2 signaling	✓ 700 mg/Kg/day p.o. for 20 weeks	✓ Hepatic inflammation by HFD-induced diabetes	[40]
Resveratrol	Stilbenoid	✓ Induction of SIRT1 and SIRT6	✓ 8 mg/Kg/day p.o. for 4 weeks	✓ Hepatic inflammation associated with HFD-induced diabetes	[41]
Ecklonia cava extract	Polyphenol	✓ Inhibition of NLRP3 activation ✓ Activation of AMPK/Sirt1/PGC-1α pathway	✓ 100 and 500 mg/Kg/day p.o. for 12 weeks	✓ HFD-induced diabetic nephropathy	[42]
Dihydroquercetin	Dihydroflavone	✓ Suppression of ROS ✓ Direct inhibition of NLRP3 assembly	✓ 25, 50, and 100 mg/Kg/day p.o. for 12 weeks	✓ HFD/streptozotocin-induced diabetic nephropathy	[29]
Formononetin	Isoflavone	✓ Inhibition of TRL–MyD88–NF-κB transcription signaling	✓ 25, 50 mg/Kg/day p.o. for 6 weeks	✓ Cognitive dysfunctions associated with streptozocin-induced diabetic	[43]

Table 3. Phytochemicals inhibiting NLRP3 inflammasome activation in animal models of chronic inflammatory diseases (colitis, arthritis, and gout).

Phytochemicals	Category	Molecular Mechanisms	Dose and Treatment Time	Experimental Models	Ref.
Asiatic acid	*Triterpenoid*	✓ Suppression of mitochondrial ROS ✓ Inhibition of caspase-1 activation ✓ Direct inhibition of NLRP3	✓ 3, 10, and 30 mg/Kg/days p.o. for 14 days	✓ DSS-induced colitis	[44]
Fraxinellone	*Lactone*	✓ Inhibition of NF-κB transcription ✓ Inhibition of NLRP3 assembly	✓ 7.5, 15, and 30 mg/Kg/day p.o. for 10 days	✓ DSS-induced colitis	[45]
Wogonoside	*Glucuronide metabolite of the bioactive flavonoid wogonin*	✓ Inhibition of NF-κB transcription ✓ Inhibition of NLRP3 expression ✓ Inhibition of caspase-1 activity	✓ 12.5, 25, and 50 mg/Kg/day p.o. for 10 days	✓ DSS-induced colitis	[46]
Alpinetin	*Flavonid derivate*	✓ Inhibition of TRL4-NF-κB transcription signalling ✓ Inhibition of NLRP3 assembly	✓ 25, 50, and 100 mg/Kg/day i.p. for 3 days before DSS treatment	✓ DSS-induced colitis	[47]
Formononetin	*Isoflavone*	✓ Direct inhibition of NLRP3	✓ 25, 50, and 100 mg/Kg/day i.p. for 7 days	✓ DSS-induced colitis	[48]
Apigenin	*Flavon*	✓ Inhibition of canonical and non-canonical NLRP3 inflammasome pathways, by the decrease of caspase-1 and caspase-11 enzyme expression and activity	✓ 3 g/day of diet resulting approximately to 125 mg/Kg/day for 30 days before DSS treatment	✓ DSS-induced colitis	[14]
Bergenin	*C glycoside derived from 4-O-methyl gallic acid*	✓ Inhibition of STAT3/NF-κB transcription signaling ✓ Inhibition of canonical and non-canonical NLRP3 activation by the decrease of caspase-1 and caspase-11 enzyme expression and activity	✓ 12, 25, and 50 mg/Kg/day p.o. for 3 days before the induction of colitis	✓ TNBS-induced colitis	[49]
Quercetin	*Flavonol*	✓ Activation of Nrf2/HO-1 signaling	✓ 150 mg/Kg/day p.o. for 14 days	✓ collagen-induced arthritis	[50]
Procyanidins	*Proanthocyanidin*	✓ Suppression of ROS	✓ 15, 30, and 60 mg/Kg p.o. 30 minutes before MSU treatment	✓ MSU-induced gouty arthritis	[51]
Epigallocatechin gallate	*Catechin*	✓ Inhibition of TXNIP/NLRP3 interaction	✓ 50 mg/Kg p.o. and s.c. before MSU treatment	✓ MSU-induced gouty arthritis	[52]
Resveratrol	*Stilbenoid*	✓ Inhibition of acetylated α-tubulin-induced mitochondrial damage	✓ 20 µg/Kg 60 min before MSU administration	✓ MSU-induced gouty arthritis	[53]

Table 3. *Cont.*

Phytochemicals	Category	Molecular Mechanisms	Dose and Treatment Time	Experimental Models	Ref.
Trans-Chalcone	*Chalcone*	✓ Activation of anti-oxidative Nrf2/HO-1 pathway ✓ Inhibition of NF-κB transcription signaling	✓ 30 mg/Kg p.o. 30 min before MSU administration	✓ MSU-induced gouty arthritis	[54]
Morin	*Flavonol*	✓ Activation of anti-oxidative enzymes (i.e., CAT and SOD) ✓ Inhibition of NF-κB transcription signaling	✓ 30 mg/Kg/day i.p. for 3 days	✓ MSU-induced acute gout arthritis	[55]
Ferulic acid	*Hydroxycinnamic acid*	✓ Direct blockade of NF-κB transcription signaling ✓ Direct blockade of NLRP3, ASC and caspase-1	✓ 30 mg/Kg i.p. 60 min before MSU administration	✓ MSU-induced acute gout arthritis	[56]
Sulforaphane	*Isothiocyanate*	✓ Direct inhibition of NLRP3	✓ 30 mg/Kg p.o. 60 min before MSU administration	✓ MSU-induced acute gout arthritis	[57]

3.1. Central Nervous System (CNS) Disorders

Increasing evidence supports the contention that central chronic neuroinflammation represents the main process implicated in the pathogenesis of several neurological and psychiatric disorders. Indeed, neuroinflammatory processes in the CNS have been documented in AD, PD, MS, ALS, and MDD patients at different stages of the disease. In this context, NLRP3 inflammasome is emerging as a pivotal driver in the onset of central neuroinflammation and consequent neurodegeneration [58,59], and several studies have shown an overactivation of NLRP3 inflammasome pathways in patients affected by these disorders. [6,60–71].

In support of this view, several studies in animal models of accelerated senescence, AD, PD, MS, ALS, and psychiatric disorders have shown a pivotal role of NLRP3 inflammasome in the onset and progression of central neuroinflammation and neurodegeneration [26]. Gordon et al. [6] observed an increase in IL-1β, caspase-1, NLRP3, and ASC expression in substantia nigra from PD mice induced by intranigral injection of 6-hydroxydopamine or α-syn preformed fibril, and that the pharmacological inhibition of inflammasome with MCC950, a recognized selective NLRP3 inhibitor, counteracted microglia activation, nigrostriatal degeneration, α-syn accumulation, and motor deficits in PD animals. Likewise, Zhang et al. [72] showed that mice subjected to chronic mild stress (a model of depression) displayed an increase in IL-1β, caspase-1, NLRP3, and ASC in hippocampus tissues, and that the pharmacological blockade of NLRP3 attenuated central and peripheral inflammation and ameliorated depressive-like symptoms. These results suggest that central NLRP3 overactivation shapes immune/inflammatory responses that contribute to neuroinflammation, microglial activation, neuronal loss, and cognitive and motor impairments in different CNS disorders, and that the pharmacological modulation of this enzymatic complex could represent a suitable way for treatment of such diseases.

Of interest, in an attempt of understanding the role of NLRP3 inflammasome in the pathophysiology of CNS diseases, several efforts have been made to implement research on the effects of NLRP3 gene deletion and its components in preclinical models of brain disorders. A recent pioneering study by Haneka et al. [73] showed that NLRP3 gene deletion in amyloid precursor protein (APP)/presenilin 1 (PS1) mice (animal model of AD) attenuated learning and memory deficits, neocortex neuronal loss and brain Aβ accumulation, by increasing Aβ phagocytic clearance capacity and insulin-degrading enzyme (IDE) expression (an enzyme able to degrade extracellular Aβ). Bellezza et al. [74] reported that caspase-1 gene deletion in α-synA53T mice (a transgenic model of PD) counteracted central neuroinflammation and microglial activation [75]. These findings further support the view that NLRP3 inflammasome plays a key role in the pathophysiology of CNS disorders. In this regard, a recent and pioneering study by Venegas et al. [76] showed that central NLRP3 activation represent an early event in AD that contributes to Aβ deposition and disease progression. In particular, they observed that NLRP3 activation in the brain from AD mice promoted the release of ASC speck proteins in the extracellular space, which, in turn, rapidly bounded Aβ peptides, leading to an increase in its aggregation and accumulation [76]. These findings represent a point of novelty, since, for the first time, they demonstrate that, in AD, NLRP3 activation contributes to Aβ aggregation and accumulation. However, the molecular mechanisms underlying the interplay between Aβ and NLRP3 as well as its role in the pathophysiology of central neurodegeneration remain poorly understood. In addition, despite these interesting results, no distinction between canonical and/or non-canonical NLRP3 inflammasome activation in CNS diseases has been made. Indeed, at present, only a study by Martin et al. [77] has shown an involvement of caspase-8-dependent non-canonical NLRP3 inflammasome signaling in the pathophysiology of MS. In particular, the authors reported that the activation of non-canonical caspase-8-dependent inflammasome and the consequent massive release of IL-1β promoted Th17 cell differentiation and infiltration in brain tissues of experimental autoimmune encephalomyelitis (EAE) mice (a model of MS), thus contributing to diseases progression [77].

Of interest, various phytochemicals have been found to inhibit NLRP3 activation acting on different steps of the inflammasome cascade and to exert beneficial effects in experimental models of CNS diseases, including PD, AD, MS, and psychiatric disorders (Table 1 and Figure 4). A pioneering study by Fan et al. [25] showed that tenuigenin, a natural extract from *Polygala tenuifolia* root, endowed with antioxidant, anti-aging, and anti-inflammatory properties, and able to cross the blood brain barrier (BBB), exerted beneficial effects in animals with MPTP-induced PD, through inhibition of NLRP3 activation and consequent decrease in IL-1β release [25]. The molecular mechanism underlying this effect was proposed to depend on the ability of tenuigenin to suppress ROS generation, thus suggesting that the blockade of NLRP3 upstream signaling in the CNS could represent a suitable therapeutic target for treatment of PD. Likewise, Yamamoto et al. [78] observed that in vivo treatment with cyclic phosphatidic acid (2ccPA), a natural phospholipid, counteracted demyelination, microglial activation, and motor dysfunctions in mice with cuprizone-induced MS, through NLRP3 inhibition via suppression of mitochondrial oxidative stress and apoptotic pathways. In addition, 2ccPA reduced the density of CD4+ T cells as well as macrophage infiltration in brain tissues from EAE mice. These results support the view that the inhibition of upstream NLRP3 signaling can also counteract the infiltration of CNS T cells and macrophages. Therefore, the blockade of NLRP3 activation in the CNS and the consequent decrease of immune/inflammatory cell infiltration could represent a suitable pharmacological target in the setting of CNS disorders.

Consistent with the above data, Peng et al. [27] observed that treatment with hydroxytyrosol (3,4-dihydroxyphenylethanol), a main polyphenol metabolite of oleuropein, attenuated neuronal impairment, central inflammation, and apoptotic activation in brain tissues from APP/PS1 mice, through the inhibition of NLRP3 inflammasome activation via suppression of ROS, known to activate NLRP3 inflammasome assembly [1]. In particular, hydroxytyrosol has been found to counteract ROS formation through the activation of antioxidant agents, including glutathione and superoxide dismutase. In addition, hydroxytyrosol decreased the expression of C-JunNH$_2$-terminal kinase (JNK)-/p38-mitogen-activated protein kinase (MAPK)-NF-kB pathway, a molecular pathway that regulates the first step of NLRP3 activation. Therefore, the NLRP3 blockade by hydroxytyrosol could be dependent on ROS suppression and inhibition of JNK-/p38-MAPK-NF-kB transcription.

The inhibition of ROS generation as a suitable pharmacological target for inhibiting NLRP3 inflammasome assembly has been confirmed by a subsequent study, showing that in vivo administration of allicin, one of the main active compounds from garlic, attenuated depressive-like behaviors, CNS neuroinflammation, abnormal iron accumulation, and neuronal apoptosis in mice with chronic social defeat stress (CSDS), through NLRP3 inflammasome signaling inhibition by activation of antioxidant pathways and suppression of ROS generation. In particular, allicin increased SOD and nuclear factor (erythroid-derived 2)-like 2 (Nrf2)/heme oxygenase 1 (HO-1) anti-oxidative activities, that, in turn, suppressed ROS levels, thus inhibiting NLRP3 assembly and activation [28].

Besides upstream targeting the inflammasome pathway, a direct blockade of NLRP3 inflammasome assembly has been shown to exert anti-inflammatory effects in CNS disorders. Feng et al. [12] reported that dihydromyricetin, a flavonoid compound derived from the medicinal plant *Ampelopsis grossedentata* and able to cross the BBB, ameliorated memory and cognition deficits, increased neprilysin (NEP) levels (enzyme involved in Aβ clearance), reduced Aβ accumulation, and promoted the shift of migroglial cells towards the anti-inflammatory M$_2$ phenotype in hippocampus and cortex in APP/PS1 mice, through direct inhibition of NLRP3 activation. In particular, dihydromyricetin decreased NLRP3, ASC, and cleaved caspase-1 expression in microglial cells from AD mice. These findings suggest that dihydromyricetin, by counteracting CNS neuroinflammation and Aβ accumulation, could represent a suitable therapeutic approach for the management of AD. Likewise, Liu et al. [30] observed that in vivo treatment of TgCRND8 mice (a transgenic model of AD) with *Ginkgo biloba* extracts improved cognitive functions, attenuated the loss of synaptic structural proteins, counteracted microglial activation, and decreased CNS neuroinflammation, through a direct blockade of caspase-1 activation.

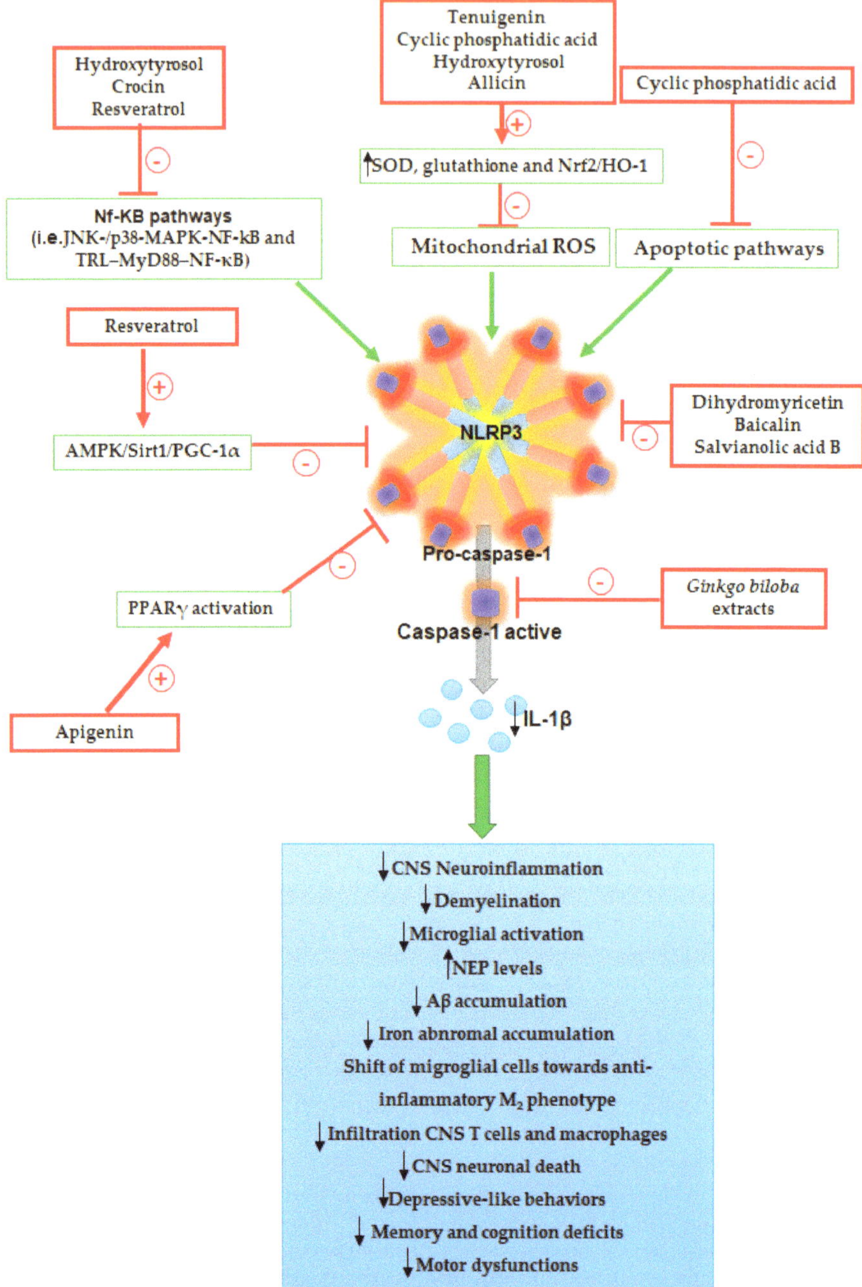

Figure 4. Diagram showing the molecular mechanisms through which phytochemicals can inhibit NLRP3 inflammasome signaling in the setting of CNS disorders.

Consistent with the above data, two recent papers by Liu et al. [31] and Jiang et al. [32] reported that the direct inhibition of NLRP3 activation with two different phytochemicals attenuated

CNS neuroinflammation in animal models of depression. These authors observed that baicalin (5, 6-dihydroxy-7-O-glucuronide flavonoid glycoside), a major polyphenol compound extracted from *Scutellaria radix* roots, and salvianolic acid B, a natural compound extracted from *Salvia miltiorrhiza*, counteracted depressive-like behaviors and central neurogenic/inflammatory responses in rats with chronic unpredictable mild stress (CUMS) and LPS-induced depression, respectively, via direct blockade of NLRP3 inflammasome assembly. Therefore, the direct inhibition of NLRP3 activation with phytochemicals might represent a suitable pharmacological strategy for treatment of CNS disorders.

Recent observations have shown that several polyphenols exerted beneficial effects on CNS disorders via inhibition of the primary TRL–MyD88–NF-κB step of NLRP3 inflammasome activation. Zhang et al. [33] showed that crocin, a carotenoid isolated from *Gardenia jasminoides* and *Crocus sativus*, counteracted CNS neuroinflammation, locomotor disability, and immobility time, in concomitance with a shift of M_1 pro-inflammatory microglia towards M_2 anti-inflammatory phenotype, in mice with LPS-induced depression, through the inhibition of NLRP3 transcription step. Qi et al. [34] and Liu et al. [79] reported that treatment with resveratrol (3, 5, 4′-trihydroxystilbene), a stilbenoid endowed with anti-aging, anti-inflammatory, antioxidant, and anti-apoptotic activities [35,80,81], improved cognitive and behavioral functions in mice with AD induced by intracerebroventricular injection of Aβ1-42 as well as in animals with ovariectomy-induced anxiety- and depression-like behaviors, by inhibiting the NF-κB/NLRP3 signaling. Resveratrol has been found to activate the anti-inflammatory 5′-adenosine monophosphate-activated protein kinase (AMPK)/sirtuin1 (Sirt1)/peroxisome proliferator-activated receptor-gamma coactivator-1alpha (PGC-1α) pathway. Such an effect could contribute further to the inhibition of NLRP3 activation. However, the authors did not demonstrate whether resveratrol influenced NF-κB/NLRP3/IL-1β and AMPK/Sirt1/PGC-1α pathways in independent ways, or whether the inhibition/activation of one can regulate the other. Thus, further investigations are needed to characterize the molecular mechanisms underlying the interplay between AMPK/Sirt1/PGC-1α and NF-κB/NLRP3/IL-1β pathways.

A pioneering study by Li et al. [36] has shown that treatment with apigenin (4′,5,7-trihydroxyflavone), a bioflavonoid with anti-inflammatory and antioxidant activities, exerted beneficial effects in CUMS rats, by counteracting behavioral alterations, CNS inflammation, and oxidative stress, through the inhibition of NLRP3 inflammasome via activation of peroxisome proliferator-activated receptor gamma (PPARγ). Indeed, the concomitant administration of GW9662, a PPARγ antagonist, to CUMS rats counteracted the inhibitory effects of apigenin on NLRP3, thus suggesting that apigenin can exert antidepressant-like effects through the blockade of NLRP3 by activation of PPARγ. These results represent an interesting point of novelty, since they suggest that a dynamic interplay between NLRP3 signaling and PPARγ contributes to neurogenic/inflammatory responses in depression, and that PPARγ blockade could represent a suitable molecular target to modulate NLRP3 activation.

3.2. Metabolic Disorders

Several lines of evidence have shown that obesity is characterized by a chronic low-grade systemic inflammation, that seems to contribute to the development of insulin resistance and the pathogenesis of type 2 diabetes mellitus (T2DM) [82,83]. Indeed, obese patients display an uncontrolled activation of innate immune/inflammatory cells, with consequent massive release of pro-inflammatory cytokines, which, in turn, interfere with several metabolic processes, including insulin synthesis and signaling, and blood glucose levels [83]. In this setting, the NLRP3 inflammasome complex has been found to act as a key immune sensor involved in shaping immune/inflammatory responses [7,84]. Clinical evidence has documented an increase in caspase-1 activity and IL-1β secretion in adipose tissue macrophages from obese and T2D patients, and these patterns were tightly correlated with a condition of insulin resistance [7,85–87]. However, the causal relationship of inflammasome activation with chronic inflammation, obesity, and T2D remains to be clarified. In this context, several research efforts have been made to investigate the effects of NLRP3 gene deletion and its components in pre-clinical models of obesity and diabetes. In particular, several studies have shown that NLRP3$^{-/-}$, caspase-1$^{-/-}$, and ASC$^{-/-}$ mice were less susceptible to the development of obesity induced by high-fat diet (HFD). Others have reported that IL-1β gene depletion in HFD obese mice attenuated adipose tissue inflammation and insulin resistance [88–90].

Of interest, recent studies have shown that NLRP3 inflammasome activation in obese mice, besides sensing innate immune responses, also shapes adaptive immune responses through actions on accumulation and activation of T cells in adipose tissue, that, in turn, contribute to alter insulin sensitivity [7]. In particular, NLRP3 gene depletion in HFD obese mice significantly decreased the number of CD4+ and CD8+ effector memory T cells and increased naive T cells in adipose tissue [7,91,92].

At present, poor evidence is available about the beneficial effects of phytochemicals in animal models of metabolic disorders.

In the setting of obesity, only one study has reported that treatment with isoliquiritigenin, a flavonoid with chalcone structure obtained by *Glycyrrhiza uralensis*, exerted beneficial effects in mice with HFD-induced obesity. Isoliquiritigenin attenuated body weight gain, insulin resistance, hyperglycemia, hypercholesterolemia, and adipose tissues inflammation, via inhibition of both steps of NLRP3 activation. Indeed, isoliquiritigenin inhibited both NF-kB transcription via TLR4/IkB signaling blockade and ASC oligomerization [13].

Two recent studies evaluated the effects of natural compounds acting on NLRP3 intracellular cascade in animal models of T2DM. In the first study, Shim et al. [37] observed that treatment with leaf extracts of *Cichorium intybus*, containing a variety of phytochemicals, such as lactucin and lactucopicrin sesquiterpene lactones, attenuated body weight gain, glucose metabolism, insulin resistance, as well as systemic and adipose tissue inflammation in mice with HFD-induced type-2 diabetes, via inhibition of NLRP3 activation. In addition, the leaf extracts of *Cichorium intybus* promoted the shift of M_1 pro-inflammatory macrophages in adipose tissues towards the M_2 anti-inflammatory phenotype, reducing the inducible nitric oxide synthase (iNOS) and TNF M_1 markers and increasing the expression of Arg-1 and IL-10 M_2 markers. The molecular mechanism was proposed to depend on the ability of lactucin and lactucopicrin of inhibiting the translocation of NLRP3 molecular components in mitochondria-associated endoplasmic reticulum (ER) membranes, a pivotal step in the process of NLRP3 activation [93,94]. In the second study, Li et al. [38] reported that treatment with resveratrol counteracted adipose tissue oxidative stress and inflammation in streptozotocin-induced diabetes through the blockade of NLRP3 activation. In particular, the inhibitory effects of resveratrol were ascribed to its ability to activate the anti-inflammatory AMPK system through the blockade of dynamin related protein-1(Drp1)- and ROS-induced mitochondrial fission and the consequent inhibition of thioredoxin-interacting protein (TXNIP)/NLRP3 interaction. These results, although generated in different animal models of metabolic disorders, show that natural compounds targeting both priming and activation steps of NLRP3 activation could represent suitable therapeutic options for the management of obesity and diabetes.

Of interest, in the setting of metabolic syndromes, the majority of available studies have investigated the effects of NLRP3-targeting phytochemicals in complications associated with obesity and diabetes, including hepatic inflammation, non-alcoholic fatty liver disease (NAFLD), diabetic nephropathy, and cognitive impairment [29,39–43]. Wang et al. [39] observed that the dietary favonol quercetin alleviated hepatic oxidative stress, inflammation, and steatosis in rats with streptozocin-induced diabetes, through the blockade of TXNIP/NLRP3 interaction [39]. Likewise, Wang et al. [40] reported that purple sweet potato color (PSPC), containing anthocyanins, ameliorated the hepatic histopathologic damage associated with HFD-induced diabetes, through NLRP3 inflammasome inhibition. The molecular mechanisms were proposed to depend on the ability of PSPC anthocyanins of inhibiting ER stress-induced inositol-requiring enzyme 1 (IRE1) signaling as well as nucleotide-binding oligomerization domain-containing protein 1/2 (NOD1/2)-NF-kB transcription, thus suggesting that this class of flavonoids can act on both steps required for NLRP3 activation. In a subsequent paper,

Yang et al. [41] showed that resveratrol counteracted hepatic inflammation in HFD mice, through the inhibition of NLRP3 assembly, by inducing the expression of Sirt1 and Sirt6 anti-inflammatory proteins. In particular, treatment with resveratrol decreased the levels of hepatic triglyceride along with TNF, IL-1β, and IL-6 pro-inflammatory cytokines. However, the authors did not investigate the molecular mechanisms underlying the interplay among SIRT1, SIRT6, and NLRP3 inflammasome.

Two recent papers reported that phytochemicals acting on NLRP3 activation alleviated obesity-induced renal damage and diabetic nephropathy. In the first paper, Eo et al. [42] observed that *Ecklonia cava* polyphenol extract (ECPE) ameliorated the renal histopathological damage and inflammation in HFD-induced obese mice. In particular, treatment with ECPE decreased systemic and tissue parameters related to kidney inflammation, including body and kidney weight, sterol regulatory element-binding protein (SREBP-1), acetyl-coA carboxylase (ACC), as well as fatty acid synthase (FAS) and renal NFκB, MCP-1, TNF, and C-reactive protein expression, through the activation of the anti-inflammatory AMPK/SIRT1/PGC-1α pathway and the inhibition of NLRP3 activation. However, the authors did not clarify whether ECPE blocked directly the inflammasome assembly, or indirectly through the activation of anti-inflammatory pathways related to NLRP3 activation. In the second paper, Tao et al. [29] observed that the dihydroflavone dihydroquercetin exerted protective effects in rats with HFD/streptozotocin-induced diabetic nephropathy, by attenuating urine microalbumin excretion and renal histopathological lesions, through NLRP3 blockade via both inhibition of upstream NLRP3 signaling and acting directly on NLRP3 assembly.

Besides the beneficial effects of phytochemicals targeting NLRP3 signaling on liver and kidney inflammation associated with diabetes, a recent study has shown that the isoflavone formononetin alleviated cognitive dysfunctions associated with diabetes via NLRP3 inhibition. In this setting, treatment with formononetin attenuated learning and memory deficiencies and decreased circulating and hippocampus levels of malondialdehyde (MDA), TNF, IL-1β, and IL-6 in mice with streptozocin-induced diabetes. The molecular mechanism underlying NLRP3 inflammasome blockade was proposed to depend on the ability of formononetin to block the priming TLR4/MyD88/NF-κB step involved in inflammasome activation through the inhibition of extracellular HMGB1. These findings represent a point of novelty, since they highlight a novel mechanism of inhibition of the first step of NLRP3 activation. [43].

3.3. Chronic Inflammatory Diseases

An increasing number of phytochemicals, including phenols, polyphenols, triterpenoids, and isothiocyanates, have been found to exert beneficial effects in several animal models of chronic inflammatory diseases through the inhibition of NLRP3 inflammasome pathways.

3.3.1. Inflammatory Bowel Diseases

IBDs, including Crohn's disease (CD) and ulcerative colitis (UC), comprise chronic and relapsing inflammatory disorders that affect the gastrointestinal tract [95]. Recent studies have shown that the NLRP3 inflammasome complex, besides acting as a key player in the maintenance of intestinal homeostasis, shapes innate immune responses during bowel inflammation, thus contributing to sustain the ongoing inflammatory processes, the disruption of enteric epithelial barrier through a deregulation of tight junction proteins (i.e., claudin-1, claudin-2, and junctional adhesion molecule-A), as well as epithelial cell apoptosis [1,96]. Recent clinical evidence has documented an increased IL-1β secretion from colonic tissues and macrophages of IBD patients, these patterns being correlated with the severity of the disease [8]. In addition, Liu et al. [97] showed an increased expression of NLRP3, ASC and IL-1β in the colonic mucosa from IBD patients, as compared with healthy controls.

In an attempt to better understand the role of NLRP3 inflammasome in the pathophysiology of bowel inflammation, several studies have investigated the effects of gene deletion and in vivo pharmacological modulation of NLRP3 inflammasome signaling in preclinical models of colitis [1]. In particular, it has been shown that the gene deletion of molecular components involved in both canonical and non-canonical NLRP3 inflammasome activation had both protective and detrimental roles in bowel inflammation, depending of the choice of experimental model. This led to postulate that, in the first phase of enteric inflammation, NLRP3 inflammasome contributes to tissue repair and maintenance of epithelial barrier integrity, while, in the chronic phase of inflammation, an overactivation of NLRP3 pathways leads to a massive release of IL-1β and IL-18, that contribute to immune/inflammatory responses and impairment of the intestinal epithelial barrier [1]. In support of this view, recent studies have shown that the in vivo pharmacological modulation of NLRP3 with drugs targeting different steps of NLRP3 activation, including inhibition of NF-kB transcription, protection against mitochondrial damage, activation of the Keap-1/NFE-related factor 2 (Nrf2) antioxidant pathway, inhibition of pro-caspase-1 cleavage, direct blockade of canonical and non-canonical NLRP3 activations, or IL-1β receptor exerted beneficial effects on bowel inflammation [98–103].

There is a large body of evidence that natural compounds targeting NLRP3 are able to exert anti-inflammatory effects on dextran sulphate sodium (DSS)-induced colitis in mice and colitis induced by 2,4,6-trinitrobenzenesulfonic acid (TNBS) in rats [1,11]. Guo et al. [44] observed that oral administration of asiatic acid, a natural triterpenoid compound, dose-dependently attenuated body weight loss, histological damage, myeloperoxidase activity, as well as colonic TNF, IL-1β, IL-6, and IFN-γ levels in mice with DSS-induced colitis through the inhibition of NLRP3 inflammasome activation. In particular, asiatic acid inhibited the upstream signaling of inflammasome activation by suppressing mitochondrial ROS generation, caspase-1 activation, and the inflammasome assembly [44].

The inhibition of NF-κB signaling and NLRP3 activation has been shown to also exert anti-inflammatory effects in colitis. Wu et al. [45] observed that treatment with fraxinellone, a natural lactone, reduced the weight loss, diarrhea, colonic macroscopic damage, enteric TNF, IL-1β, IL-6, and IL-18 levels, CD11b+ macrophage infiltration, as well as the mRNA levels of intercellular adhesion molecule 1 (ICAM1), vascular cell adhesion molecule 1 (VCAM1), iNOS, and cyclooxygenase-2 (COX-2) in mice with DSS-induced colitis, through NF-κB signaling and NLRP3 blockade. Likewise, treatment with wogonoside, a glucuronide metabolite of the bioactive flavonoid wogonin, exerted beneficial effects on bowel inflammation via direct inhibition of NF-κB and NLRP3 expression, as well as caspase-1 expression and activity [46]. These results indicate that phytochemicals targeting both steps of NLRP3 activation could represent a suitable and promising pharmacological target for treatment of bowel inflammation.

Consistent with the above data, He et al. [47] showed that the alpinetin, a flavonoid isolated from *Alpinia katsumadai Hayata*, attenuated diarrhea, colonic shortening, histological damage, and myeloperoxidase activity as well as colonic TNF and IL-1β expression in mice with DSS-induced colitis, likely by suppressing TRL4-NF-κB and NLRP3-ASC-caspase-1 signaling. However, the authors documented the ability of alpinetin of inhibiting NLRP3 activation in in vitro THP-1 cells, omitting the evaluation of alpinetin effects on NLRP3 activation in DSS mice. A recent study by Wu et al. [48] reported that formononetin alleviated the colonic shortening, histological damage, macrophage infiltration, colonic epithelial cell injury, and restored colonic tight junction (zonulin-1, claudin-1, and occludin) protein expression, by counteracting the increased expression of NLRP3 components (NLRP3, ASC, IL-1β) in DSS mice. These findings show, for the first time, that the isoflavone formononetin, via NLRP3 blockade, exerts beneficial effects on colitis both by counteracting colonic inflammation and restoring the integrity of epithelial barrier.

Of interest, two pioneering papers by Marquez-Flores et al. [14] and Oficjalska et al. [49] have shown that the inhibition of canonical and non-canonical NLRP3 activation with polyphenols counteracted bowel inflammation in animals with DSS- and TNBS-induced colitis. In the first paper, Marquez-Flores et al. [14] observed that a dietary apigenin enrichment decreased the macroscopic and microscopic signs of colitis, and reduced colonic PGE, COX-2 and iNOS expression as well as serum matrix metalloproteinase (MMP-3) levels in DSS mice, through the inhibition of both canonical and non-canonical NLRP3 inflammasome pathways, through a decrease in caspase-1 and caspase-11 expression and activity. In the second study, treatment with bergenin, a polyphenolic compound, classified as a C glycoside derived from 4-O-methyl gallic acid, in rats with TNBS-induced colitis alleviated colonic macroscopic and microscopic damage, decreased neutrophilic infiltration, as well as colonic COX- 2 and iNOS expression and IL-1β, IFN-γ, and IL-10 levels, by modulation of STAT3 and NF-κB signaling and blockade of canonical and non-canonical NLRP3 inflammasome pathways.

3.3.2. Rheumatoid Arthritis

RA is a chronic autoinflammatory disease characterized by synovial inflammation and irreversible joint destruction, leading to joint functional impairment and often premature mortality [104]. Several lines of evidence support the view that NLRP3 inflammasome plays a pivotal role in the pathophysiology of RA [104]. Indeed, polymorphisms in different regions of NLRP3 gene have been associated with increased RA susceptibility and disease severity [105–107]. In addition, recent papers showed that RA patients are characterized by an increased mRNA and protein expression of inflammasome components, including NLRP3, ASC, caspase-1, and IL-1β levels in the synovia as well as circulating monocytes/macrophages, dendritic cells, and neutrophils, thus suggesting that the activation of NLRP3 inflammasome contributes both to tissues and systemic inflammation in RA [108–112]. However, current clinical evidence does not clarify whether NLRP3 inflammasome contributes to the pathophysiology of RA, or whether its activation occurs rather as a consequence of the initiation of synovial inflammatory processes.

To better understand the pathophysiological role of NLRP3 in RA, several investigations have been performed in animal models of arthritis. Wei et al. [113] showed that IL-18$^{-/-}$ mice were less susceptible to develop collagen-induced arthritis (CIA), as compared with wild-type animals, thus indicating a role of the pro-inflammatory cytokine IL-18 in the pathophysiology of RA. Guo et al. [9] observed an increased protein expression of NLRP3 and active caspase-1, along with an increment of IL-1β levels in the knee joint synovia and sera from CIA animals. In addition, they reported that the pharmacological blockade of NLRP3 with the selective inhibitor MCC950, through a decrease in the production of IL-1β, attenuated disease severity counteracting joint inflammation and bone destruction, thus suggesting that NLRP3-induced IL-1β release contributes to shaping and sustaining the immune/inflammatory processes in RA, and, most importantly, that inflammasome could represent a suitable pharmacological target for the management of RA [9].

However, there is very limited evidence that a natural compound targeting NLRP3 inflammasome can exert anti-inflammatory effects on experimental arthritis. Yang et al. [50] observed that oral administration of quercetin attenuated arthritic scores and paw edema decreased the joint levels of TNF, IL-6, PGE2, COX-2, iNOS, and Th17 cells and increased the number of Treg cells in mice with collagen-induced arthritis, through the inhibition of NLRP3 inflammasome activation. In particular, the authors found that quercetin inhibited the upstream signaling of NLRP3 activation via activation of anti-oxidant Nrf2/HO-1 signaling. As a confirmation, the application of HO-1 siRNA to fibroblast-like synoviocytes abolished the inhibitory effects of quercetin on NLRP3 activation [50].

3.3.3. Gout

Gout is a chronic inflammatory arthritis characterized by an increase in urate concentrations and deposition of monosodium urate (MSU) crystals in joints [114,115]. In this context, increasing evidence supports the contention that NLRP3 inflammasome activation and the consequent massive release of IL-1β and IL-18 following MSU deposition promote mast cell, monocyte, and neutrophil influx into the synovium and joint fluids, thus contributing to the pathophysiology of gout [116–118]. In support of this view, several studies have shown that MSU crystals activated NLRP3 assembly in PBMCs from gout patients [119]. In addition, clinical trials have observed that the blockade of NLRP3 downstream signaling with IL-1 inhibitors, including rilonacept, canakinumab, and anakinra counteracted joint inflammation and attenuated the disease severity in patients with acute and chronic gout [120–123]. However, no clinical studies are available about the effects of direct blockade of IL-18 or its receptor in gout.

Of interest, the implementation of gouty arthritis animal models has allowed to better clarify the pathophysiological role of NLRP3 in gout. In particular, Nomura et al. [124] observed that NLRP3-/- mice were less susceptible to the development of MSU-induced inflammation [105]. Likewise, Martinon et al. [125] showed that MSU-induced gouty arthritis mice with gene deletion of inflammasome components, including caspase-1, ASC, and NLRP3, displayed a decrease in joint inflammation and neutrophil influx.

Several lines of evidence have shown that various NLRP3-targeting natural compounds can exert anti-inflammatory effects on MSU-induced gouty arthritis. Liu et al. [51] reported that administration of procyanidins, grape seed-derived natural flavonoids, attenuated the ankle circumference increase and joint inflammation in mice with acute MSU-induced gouty arthritis, through the inhibition of oxidative stress-induced upstream signaling of NLRP3 activation. Likewise, Jhang et al. [52] observed that epigallocatechin gallate, a bioactive polyphenol from green tea endowed with antioxidant activities, attenuated peritoneal inflammation in MSU animals via a blockade of TXNIP/NLRP3 interaction. In this study, epigallocatechin gallate decreased peritoneal neutrophil infiltration as well as the levels of neutrophil cytosolic factor 1, IL-6, IL-1β, monocyte chemoattractant protein-1 (MCP-1) in peritoneal lavage fluid, and amyloid A levels in serum, thus suggesting that NLRP3 blockade could represent a viable pharmacological strategy for the management of peritoneal inflammation associated with gout.

A pioneering study by Misawa et al. [53] showed that resveratrol administration to MSU mice exerted anti-inflammatory effects through the blockade of NLRP3 inflammasome assembly by reducing acetylated α-tubulin-induced mitochondrial damage. In particular, resveratrol, through the decrease in α-tubulin acetylation, prevented the optimal spatial conformation of NLRP3 and ASC, thus inhibiting inflammasome oligomerization and activation. These results show, for the first time, a novel mechanism of inhibition of upstream NLRP3 signaling by natural compounds.

Several studies have reported that different flavonoids exerted anti-inflammatory effects on experimental gout arthritis through the inhibition of both steps of NLRP3 activation. *Trans*-chalcone (1,3-diphenyl-2-propen-1-one), an open-chain flavonoid, attenuated knee joint inflammation and pain in MSU mice, via activation of antioxidative Nrf2/HO-1 pathway and inhibition of NF-κB/NLRP3 signaling [54]. Likewise, the administration of morin, a dietary bioflavonol, ameliorated ankle swelling, synovial hyperplasia, inflammatory cell infiltration, and cartilage degeneration as well as ankle levels of monocyte chemoattractant protein (MCP-1), iNOS, COX-2, and TNF, and IL-6 and IL-1β in rats with MSU-induced acute gouty arthritis, through the inhibition of NLRP3 activation by both increasing the activity of anti-oxidative enzymes, such as catalase (CAT) and SOD, and suppressing the NF-κB transcription step [55].

In further support of the above observations, a pioneering study by Doss et al. [56] showed that the intraperitoneal administration of ferulic acid, a phenolic phytochemical endowed with antioxidant, antihyperlipidemic, hypotensive, antimicrobial, anticarcinogenic, anti-inflammatory, and hepatoprotective activities, attenuated paw edema, elastase levels, lysosomal enzymes, inflammatory cell infiltration, and TNF and IL-1β levels in ankle joints in MSU rats. Such anti-inflammatory effects were ascribed to the ability of ferulic acid to inhibit directly both steps of NLRP3 activation. Indeed, the molecular docking analysis showed that ferulic acid displayed significant ligand efficiency towards pro-caspase-1, NF-κB, ASC, and NLRP3. These findings provide the first demonstration of direct molecular interactions between a phenolic phytochemical and NLRP3 inflammasome signaling, thus paving the way to the identification of novel phenolic compounds acting directly on NF-κB and NLRP3 components for the treatment of gouty arthritis.

One study has shown that sulforaphane (1-isothiocyanato-4-methylsulfinylbutane), a natural dietary isothiocyanate derivative, abundant in cruciferous vegetables, exerted anti inflammatory effects on gouty arthritis via direct inhibition of NLRP3 inflammasome signaling [57]. Indeed, the oral administration of sulforaphane dose-dependently attenuated foot swelling and neutrophil recruitment while decreasing foot Il-1β levels and caspase-1 activity in animals with acute gout induced by MSU and air pouch. In vitro experiments indicated that sulforaphane suppressed MSU,

ATP, and nigericin-induced NLRP3 inflammasome activation in bone-marrow-derived macrophages (BMDMs) [57], thus suggesting that sulforaphane could represent a promising phytochemical entity for prevention or treatment of gouty inflammation.

4. Discussion

Current data from human studies suggest that NLRP3 inflammasome activation represents a common path to a variety of diseases, including neurological, psychiatric, metabolic, and chronic inflammatory disorders. Indeed, even though each disease displays distinct clinical, pathological, and genetic features, patients with PD, AD, MS, ALS, depression, obesity, diabetes, IBD, RA, and gout are characterized by an overactivation of the inflammasome signaling that contributes to central neuroinflammation, metabolic alterations, and immune/inflammatory responses in such disorders. However, human studies do not provide a clear causal relationship of NLRP3 activation with CNS, metabolic, or inflammatory diseases.

The development of experimental models of neurological and psychiatric diseases, metabolic disorders, and chronic inflammatory diseases has allowed us to better understand the pathophysiological role of NLRP3 in these pathological conditions. Indeed, even though each experimental model displays distinct pathophysiological features, NLRP3 activation has been shown to contribute to central neuroinflammation, metabolic dysfunctions, and immune/inflammatory responses. In support of this view, gene depletion of inflammasome components or in vivo NLRP3 modulation with drugs acting at different levels of the inflammasome cascade in animal models have been found to counteract the progression of central neuroinflammation, metabolic alterations, and immune/inflammatory responses.

Based on the above considerations, and pooling together the available human and pre-clinical evidence, it is conceivable that the NLRP3 inflammasome complex represents a pivotal node for immune sensing in the innate immune system and that its activation in CNS, metabolic, and chronic inflammatory disorders contributes to shape the immune/inflammatory responses as well as to sustain the pathophysiological events underlying these diseases. In addition, NLRP3 activation has been shown to promote the adaptive immune responses, influencing the differentiation of T cells to pro-inflammatory Th17 phenotypes, thus suggesting that NLRP3 is not only one of the early immune sensors for innate immune response, but also for shaping of adaptive immune signals. Nevertheless, the exact role of NLRP3 in the pathophysiology of CNS, metabolic, and inflammatory disorders remains to be elucidated, with particular regard for the molecular mechanisms through which NLRP3 inflammasome can influence the adaptive immune system. Moreover, the majority of current human and pre-clinical studies have focused their attention on the role of canonical NLRP3 inflammasome activation, thus disregarding the evaluation of possible contributions by non-canonical caspase-8 and caspase-11-dependent NLRP3 activation pathways. This is a point of high interest, since an involvement of non-canonical caspase-8- and caspase-11-dependent inflammasome activation in the onset and progression of demyielinization and bowel inflammation, respectively, has been documented only in the setting of experimental MS and colitis. In addition, the role of other inflammasomes, including NLRP6, NLRP1, AIM2, and NLRC4, in the pathophysiology of CNS, metabolic, and inflammatory disorders remains unclear and poorly investigated [126–128]. Accordingly, there is a strong need for further studies aimed at characterizing the role of non-canonical NLRP3 inflammasome pathways and other inflammasomes in the pathophysiology of CNS, metabolic, and inflammatory diseases.

Another crucial issue concerns the rick of adverse reactions following full inhibition of NLRP3, such as cancer observed in NLRP3-/- mice [129]. Several lines of evidence have shown that NLRP3-/- mice are more prone to develop colorectal cancer induced by azoxymethane/DSS, suggesting a protective role of NLRP3 inflammasome in tumorigenesis [130]. Conversely, others reported that NLRP3 inflammasome activation and the consequent release of IL-1β and IL-18 promoted tumor growth, proliferation, invasion, and metastasis in lung cancer, melanoma, breast cancer, and head and neck squamous cell carcinoma [129]. These findings suggest that NLRP3 inflammasome can play both protective and detrimental roles in tumors, likely depending on the tissue context. Therefore, further studies are needed to clarify the role of NLRP3 inflammasome in tumorigenesis, and to characterize the putative detrimental/protective effects associated with pharmacological agents acting as NLRP3 inflammasome inhibitors.

Of note, despite several issues about the role of NLRP3 in the pathophysiology of CNS, metabolic, and inflammatory diseases that still need to be addressed, the beneficial effects resulting from NLRP3 modulation by phytochemicals in various experimental models allow us to postulate that the inhibition of NLRP3 with natural compounds could represent a viable pharmacological approach for the management of a variety of diseases. In this respect, current evidence shows that a number of polyphenols (i.e., flavonoids, stilbenoids, and phenols), triterpenoids, isothiocyanates, and carotenoids, acting on different steps of canonical or non-canonical NLRP3 inflammasome activation signaling, can counteract central neuroinflammation and neurodegeneration, metabolic alterations and related comorbidities, and immune/inflammatory responses in animal models of CNS, metabolic, and inflammatory diseases. However, in this area, a number of issues remain to be clarified: (1) What are the exact mechanisms through which 2ccPA and hydroxytyrosol inhibit NLRP3 activation in MS and AD, respectively? (2) What are the molecular mechanisms underlying NLRP3 inhibition by sulforaphane in gout? (3) How do some polyphenols block both canonical and non-canonical NLRP3 assembly? (4) Can phytochemicals targeting both canonical and non-canonical inflammasome activation be regarded as the most reliable pharmacological approaches for the management of CNS, metabolic, and inflammatory diseases? (5) Can a dietary supplementation with phytochemicals prevent or counteract neuroinflammatory and neurodegenerative processes, metabolic alterations, and immune/inflammatory responses? (6) Can phytochemicals exert beneficial effects in CNS, metabolic, and inflammatory diseases through the inhibition of other inflammasomes, including NRLP1, NLRP6, AIM2, and NLRC4? To clarify these points, intensive research efforts should be addressed to investigate, by means of molecular, biochemical, and pharmacological approaches, the effects of phytochemicals targeting NLRP3 inflammasome pathways and other inflammasomes in in vitro experiments on cultured cells and different experimental models.

5. Conclusions and Future Perspectives

Current pre-clinical studies show that several phytochemical compounds can exert beneficial effects in CNS, metabolic, and inflammatory disorders through different modes of inhibition of NLRP3 inflammasome activation. In particular, polyphenols (i.e., flavonoids, stilbenoids, and phenols), triterpenoids, isothiocyanates, and carotenoids, acting on different steps of inflammasome activation, have been found to counteract central neuroinflammation and neurodegeneration, metabolic alterations, and immune/inflammatory responses in various experimental models. The main molecular mechanisms underlying NLRP3 blockade by phytochemicals have been ascribed to their ability of: (1) Inhibiting upstream NLRP3 activation by suppression of ROS generation; (2) directly blocking NF-kB-mediated transcription steps and/or NLRP3 oligomerization; (3) activating anti-inflammatory pathways, including AMPK/SIRT1/PGC-1α, that, in turn, could increase the expression of several ROS-detoxifying enzymes and inhibit directly NLRP3 assembly.

Based on this body of evidence, it can be proposed that phytochemicals acting at different steps of NLRP3 signaling could represent suitable pharmacological approaches for the management of a variety of diseases sharing the presence of chronic and persistent inflammatory conditions.

One considerable gap in our knowledge concerns whether these natural compounds can also interfere with non-canonical caspase-8 and/or caspase-11-dependent NLRP3 inflammasome activations. Another relevant issue stems from the lack of translational evidence about the effects of phytochemicals targeting NLRP3 in patients with CNS, metabolic, or inflammatory diseases. Despite some clinical trials that have shown the beneficial effects of phytochemicals in patients with CNS, metabolic, and inflammatory disorders, no evidence is currently available about the effects of natural compounds targeting NLRP3 in humans. In this respect, a translation of preclinical evidence into clinical practice could allow a better understanding of the protective effects of phytochemicals acting on NLRP3 in patients with CNS, metabolic, and inflammatory disorders. Unraveling these aspects could pave the way to novel therapeutic options for both the prevention and clinical management of neurological, psychiatric, metabolic, and inflammatory diseases.

Author Contributions: C.P., M.F., and L.A. conceived and wrote the review manuscript; C.B. and V.C. revised the reviewed manuscript.

Funding: The present work has been supported by the PRA_2018_31 granted by the University of Pisa.

Conflicts of Interest: The authors declare no conflict of interest.

Abbreviations

α-syn	α-synuclein
Aβ	β amyloid
AD	Alzheimer's disease
ALS	amyotrophic lateral sclerosis
AMPK	5' adenosine monophosphate-activated protein kinase
AOM	azoxymethane
ASC	adaptor protein
ATP	adenosine triphosphate
CAPS	cryopyrin-associated autoinflammatory syndromes
CAT	catalase
DAMPs	damage-associated molecular pattern molecules
Drp-1	dynamin related protein-1
DSS	dextran sulphate sodium
ER	endoplasmic reticulum
FADD	FAS-associated death domain protein
HFD	high fat diet
IKb	IkappaB kinase
IRE1	Inositol-requiring enzyme 1
HMGB1	high mobility group box 1
HO-1	heme oxygenase 1
ip	intraperitoneal
IRF	interferon regulatory factor
IFN	interferon
IFNAR	interferon-α/β receptor
IL	interleukin
JAK/ STAT	janus kinase/signal transducers and activators of transcription
JNK	JunNH$_2$-terminal kinase
MLKL	mixed lineage kinase domain-like protein
MALT1	mucosa-associated lymphoid tissue lymphoma translocation protein 1
MAPK	mitogen-activated protein kinase
MDD	major depressive disorder
MPTP	l-methyl-4-phenyl-l,2,3,6-tetrahydropyridine
MS	multiple sclerosis
MSU	monosodium urate
MyD88	myeloid differentiation primary response 88
NF-kB	nuclear factor kB
NLR	nucleotide-binding domain and leucine-rich repeat
NLRP3	nucleotide-binding oligomerization domain leucine rich repeat and pyrin domain-containing protein 3
NOD1/2	nucleotide-binding oligomerization domain-containing protein 1/2
NRF2/ARE	nuclear factor (erythroid-derived 2)-like 2/antioxidant response element

P2X	purinergic receptor 7
PAMPs	pathogen-associated molecular pattern molecules
PD	Parkinson's disease
PGC-1α	peroxisome proliferator-activated receptor-gamma coactivator-1alpha
p.o.	oral
PPARγ	peroxisome proliferator-activated receptor gamma
PYRIN	amino-terminal pyrin domain domain
RIP1	receptor-interacting protein 1
ROS	reactive oxygen species
s.c.	subcutaneous
Sirt1	sirtuin1
SOD	superoxide dismutase
STAT3	Signal transducer and activator of transcription 3
TLR	toll like receptor
TNBS	2,4,6-trinitrobenzenesulfonic acid
TNFR	tumor necrosis factor receptor
TRIF	toll/IL-1 receptor homology (TIR)-domain-containing adapter-inducing interferon-β
TXNIP	thioredoxin-interacting protein

References

1. Pellegrini, C.; Antonioli, L.; Lopez-Castejon, G.; Blandizzi, C.; Fornai, M. Canonical and Non-Canonical Activation of NLRP3 Inflammasome at the Crossroad between Immune Tolerance and Intestinal Inflammation. *Front. Immunol.* **2017**, *8*, 36. [CrossRef] [PubMed]
2. Jiang, D.; Chen, S.; Sun, R.; Zhang, X.; Wang, D. The NLRP3 inflammasome: Role in metabolic disorders and regulation by metabolic pathways. *Cancer Lett.* **2018**, *419*, 8–19. [CrossRef] [PubMed]
3. Broderick, L.; De Nardo, D.; Franklin, B.S.; Hoffman, H.M.; Latz, E. The inflammasomes and autoinflammatory syndromes. *Annu. Rev. Pathol.* **2015**, *10*, 395–424. [CrossRef] [PubMed]
4. Song, L.; Pei, L.; Yao, S.; Wu, Y.; Shang, Y. NLRP3 Inflammasome in Neurological Diseases, from Functions to Therapies. *Front. Cell. Neurosci.* **2017**, *11*, 63. [CrossRef] [PubMed]
5. Strowig, T.; Henao-Mejia, J.; Elinav, E.; Flavell, R. Inflammasomes in health and disease. *Nature* **2012**, *481*, 278–286. [CrossRef] [PubMed]
6. Gordon, R.; Albornoz, E.A.; Christie, D.C.; Langley, M.R.; Kumar, V.; Mantovani, S.; Robertson, A.A.B.; Butler, M.S.; Rowe, D.B.; O'Neill, L.A.; et al. Inflammasome inhibition prevents alpha-synuclein pathology and dopaminergic neurodegeneration in mice. *Sci. Transl. Med.* **2018**, *10*, eaah4066. [CrossRef] [PubMed]
7. Vandanmagsar, B.; Youm, Y.H.; Ravussin, A.; Galgani, J.E.; Stadler, K.; Mynatt, R.L.; Ravussin, E.; Stephens, J.M.; Dixit, V.D. The NLRP3 inflammasome instigates obesity-induced inflammation and insulin resistance. *Nat. Med.* **2011**, *17*, 179–188. [CrossRef] [PubMed]
8. Coccia, M.; Harrison, O.J.; Schiering, C.; Asquith, M.J.; Becher, B.; Powrie, F.; Maloy, K.J. IL-1beta mediates chronic intestinal inflammation by promoting the accumulation of IL-17A secreting innate lymphoid cells and CD4$^+$ Th17 cells. *J. Exp. Med.* **2012**, *209*, 1595–1609. [CrossRef] [PubMed]
9. Guo, C.; Fu, R.; Wang, S.; Huang, Y.; Li, X.; Zhou, M.; Zhao, J.; Yang, N. NLRP3 inflammasome activation contributes to the pathogenesis of rheumatoid arthritis. *Clin. Exp. Immunol.* **2018**, *194*, 231–243. [CrossRef] [PubMed]
10. Mangan, M.S.J.; Olhava, E.J.; Roush, W.R.; Seidel, H.M.; Glick, G.D.; Latz, E. Targeting the NLRP3 inflammasome in inflammatory diseases. *Nat. Rev. Drug Discov.* **2018**, *17*, 588–606. [CrossRef] [PubMed]
11. Jahan, S.; Kumar, D.; Chaturvedi, S.; Rashid, M.; Wahajuddin, M.; Khan, Y.A.; Goyal, S.N.; Patil, C.R.; Mohanraj, R.; Subramanya, S.; et al. Therapeutic Targeting of NLRP3 Inflammasomes by Natural Products and Pharmaceuticals: A Novel Mechanistic Approach for Inflammatory Diseases. *Curr. Med. Chem.* **2017**, *24*, 1645–1670. [CrossRef] [PubMed]
12. Feng, J.; Wang, J.X.; Du, Y.H.; Liu, Y.; Zhang, W.; Chen, J.F.; Liu, Y.J.; Zheng, M.; Wang, K.J.; He, G.Q. Dihydromyricetin inhibits microglial activation and neuroinflammation by suppressing NLRP3 inflammasome activation in APP/PS1 transgenic mice. *CNS Neurosci. Ther.* **2018**, *24*, 1207–1218. [CrossRef] [PubMed]

13. Honda, H.; Nagai, Y.; Matsunaga, T.; Okamoto, N.; Watanabe, Y.; Tsuneyama, K.; Hayashi, H.; Fujii, I.; Ikutani, M.; Hirai, Y.; et al. Isoliquiritigenin is a potent inhibitor of NLRP3 inflammasome activation and diet-induced adipose tissue inflammation. *J. Leukoc. Biol.* **2014**, *96*, 1087–1100. [CrossRef] [PubMed]

14. Marquez-Flores, Y.K.; Villegas, I.; Cardeno, A.; Rosillo, M.A.; Alarcon-de-la-Lastra, C. Apigenin supplementation protects the development of dextran sulfate sodium-induced murine experimental colitis by inhibiting canonical and non-canonical inflammasome signaling pathways. *J. Nutr. Biochem.* **2016**, *30*, 143–152. [CrossRef] [PubMed]

15. Duncan, J.A.; Bergstralh, D.T.; Wang, Y.; Willingham, S.B.; Ye, Z.; Zimmermann, A.G.; Ting, J.P. Cryopyrin/NALP3 binds ATP/dATP, is an ATPase, and requires ATP binding to mediate inflammatory signaling. *Proc. Natl. Acad. Sci. USA* **2007**, *104*, 8041–8046. [CrossRef] [PubMed]

16. Latz, E.; Xiao, T.S.; Stutz, A. Activation and regulation of the inflammasomes. *Nat. Rev. Immunol.* **2013**, *13*, 397–411. [CrossRef] [PubMed]

17. Bauernfeind, F.G.; Horvath, G.; Stutz, A.; Alnemri, E.S.; MacDonald, K.; Speert, D.; Fernandes-Alnemri, T.; Wu, J.; Monks, B.G.; Fitzgerald, K.A.; et al. Cutting edge: NF-kappaB activating pattern recognition and cytokine receptors license NLRP3 inflammasome activation by regulating NLRP3 expression. *J. Immunol.* **2009**, *183*, 787–791. [CrossRef]

18. Aachoui, Y.; Sagulenko, V.; Miao, E.A.; Stacey, K.J. Inflammasome-mediated pyroptotic and apoptotic cell death, and defense against infection. *Curr. Opin. Microbiol.* **2013**, *16*, 319–326. [CrossRef]

19. Lu, B.; Wang, H.; Andersson, U.; Tracey, K.J. Regulation of HMGB1 release by inflammasomes. *Protein Cell* **2013**, *4*, 163–167. [CrossRef]

20. Antonopoulos, C.; Russo, H.M.; El Sanadi, C.; Martin, B.N.; Li, X.; Kaiser, W.J.; Mocarski, E.S.; Dubyak, G.R: Caspase-8 as an Effector and Regulator of NLRP3 Inflammasome Signaling. *J. Biol. Chem.* **2015**, *290*, 20167–20184. [CrossRef]

21. Chen, M.; Xing, Y.; Lu, A.; Fang, W.; Sun, B.; Chen, C.; Liao, W.; Meng, G. Internalized Cryptococcus neoformans Activates the Canonical Caspase-1 and the Noncanonical Caspase-8 Inflammasomes. *J. Immunol.* **2015**, *195*, 4962–4972. [CrossRef] [PubMed]

22. Gringhuis, S.I.; Kaptein, T.M.; Wevers, B.A.; Theelen, B.; van der Vlist, M.; Boekhout, T.; Geijtenbeek, T.B. Dectin-1 is an extracellular pathogen sensor for the induction and processing of IL-1beta via a noncanonical caspase-8 inflammasome. *Nat. Immunol.* **2012**, *13*, 246–254. [CrossRef] [PubMed]

23. Chung, H.; Vilaysane, A.; Lau, A.; Stahl, M.; Morampudi, V.; Bondzi-Simpson, A.; Platnich, J.M.; Bracey, N.A.; French, M.C.; Beck, P.L.; et al. NLRP3 regulates a non-canonical platform for caspase-8 activation during epithelial cell apoptosis. *Cell Death Differ.* **2016**, *23*, 1331–1346. [CrossRef] [PubMed]

24. Gurung, P.; Anand, P.K.; Malireddi, R.K.; Vande Walle, L.; Van Opdenbosch, N.; Dillon, C.P.; Weinlich, R.; Green, D.R.; Lamkanfi, M.; Kanneganti, T.D. FADD and caspase-8 mediate priming and activation of the canonical and noncanonical Nlrp3 inflammasomes. *J. Immunol.* **2014**, *192*, 1835–1846. [CrossRef] [PubMed]

25. Fan, Z.; Liang, Z.; Yang, H.; Pan, Y.; Zheng, Y.; Wang, X. Tenuigenin protects dopaminergic neurons from inflammation via suppressing NLRP3 inflammasome activation in microglia. *J. Neuroinflamm.* **2017**, *14*, 256. [CrossRef] [PubMed]

26. Tha, K.K.; Okuma, Y.; Miyazaki, H.; Murayama, T.; Uehara, T.; Hatakeyama, R.; Hayashi, Y.; Nomura, Y. Changes in expressions of proinflammatory cytokines IL-1beta, TNF-alpha and IL-6 in the brain of senescence accelerated mouse (SAM) P8. *Brain Res.* **2000**, *885*, 25–31. [CrossRef]

27. Peng, Y.; Hou, C.; Yang, Z.; Li, C.; Jia, L.; Liu, J.; Tang, Y.; Shi, L.; Li, Y.; Long, J.; et al. Hydroxytyrosol mildly improve cognitive function independent of APP processing in APP/PS1 mice. *Mol. Nutr. Food Res.* **2016**, *60*, 2331–2342. [CrossRef] [PubMed]

28. Gao, W.; Wang, W.; Liu, G.; Zhang, J.; Yang, J.; Deng, Z. Allicin attenuated chronic social defeat stress induced depressive-like behaviors through suppression of NLRP3 inflammasome. *Metab. Brain Dis.* **2019**, *34*, 319–329. [CrossRef] [PubMed]

29. Ding, T.; Wang, S.; Zhang, X.; Zai, W.; Fan, J.; Chen, W.; Bian, Q.; Luan, J.; Shen, Y.; Zhang, Y.; et al. Kidney protection effects of dihydroquercetin on diabetic nephropathy through suppressing ROS and NLRP3 inflammasome. *PhytoMed. Int. J. Phytother. Phytopharm.* **2018**, *41*, 45–53. [CrossRef]

30. Liu, X.; Hao, W.; Qin, Y.; Decker, Y.; Wang, X.; Burkart, M.; Schotz, K.; Menger, M.D.; Fassbender, K.; Liu, Y. Long-term treatment with Ginkgo biloba extract EGb 761 improves symptoms and pathology in a transgenic mouse model of Alzheimer's disease. *Brain Behav. Immun.* **2015**, *46*, 121–131. [CrossRef]

31. Liu, X.; Liu, C. Baicalin ameliorates chronic unpredictable mild stress-induced depressive behavior: Involving the inhibition of NLRP3 inflammasome activation in rat prefrontal cortex. *Int. Immunopharmacol.* **2017**, *48*, 30–34. [CrossRef] [PubMed]

32. Jiang, P.; Guo, Y.; Dang, R.; Yang, M.; Liao, D.; Li, H.; Sun, Z.; Feng, Q.; Xu, P. Salvianolic acid B protects against lipopolysaccharide-induced behavioral deficits and neuroinflammatory response: Involvement of autophagy and NLRP3 inflammasome. *J. Neuroinflamm.* **2017**, *14*, 239. [CrossRef] [PubMed]

33. Zhang, L.; Previn, R.; Lu, L.; Liao, R.F.; Jin, Y.; Wang, R.K. Crocin, a natural product attenuates lipopolysaccharide-induced anxiety and depressive-like behaviors through suppressing NF-kB and NLRP3 signaling pathway. *Brain Res. Bull.* **2018**, *142*, 352–359. [CrossRef] [PubMed]

34. Qi, Y.; Shang, L.; Liao, Z.; Su, H.; Jing, H.; Wu, B.; Bi, K.; Jia, Y. Intracerebroventricular injection of resveratrol ameliorated Abeta-induced learning and cognitive decline in mice. *Metab. Brain Dis.* **2019**, *34*, 257–266. [CrossRef] [PubMed]

35. Csiszar, A. Anti-inflammatory effects of resveratrol: Possible role in prevention of age-related cardiovascular disease. *Ann. N. Y. Acad. Sci.* **2011**, *1215*, 117–122. [CrossRef]

36. Li, R.; Wang, X.; Qin, T.; Qu, R.; Ma, S. Apigenin ameliorates chronic mild stress-induced depressive behavior by inhibiting interleukin-1beta production and NLRP3 inflammasome activation in the rat brain. *Behav. Brain Res.* **2016**, *296*, 318–325. [CrossRef] [PubMed]

37. Shim, D.W.; Han, J.W.; Ji, Y.E.; Shin, W.Y.; Koppula, S.; Kim, M.K.; Kim, T.K.; Park, P.J.; Kang, T.B.; Lee, K.H. Cichorium intybus Linn. Extract Prevents Type 2 Diabetes Through Inhibition of NLRP3 Inflammasome Activation. *J. Med. Food* **2016**, *19*, 310–317. [CrossRef]

38. Li, A.; Zhang, S.; Li, J.; Liu, K.; Huang, F.; Liu, B. Metformin and resveratrol inhibit Drp1-mediated mitochondrial fission and prevent ER stress-associated NLRP3 inflammasome activation in the adipose tissue of diabetic mice. *Mol. Cell. Endocrinol.* **2016**, *434*, 36–47. [CrossRef]

39. Wang, W.; Wang, C.; Ding, X.Q.; Pan, Y.; Gu, T.T.; Wang, M.X.; Liu, Y.L.; Wang, F.M.; Wang, S.J.; Kong, L.D. Quercetin and allopurinol reduce liver thioredoxin-interacting protein to alleviate inflammation and lipid accumulation in diabetic rats. *Br. J. Pharmacol.* **2013**, *169*, 1352–1371. [CrossRef]

40. Wang, X.; Zhang, Z.F.; Zheng, G.H.; Wang, A.M.; Sun, C.H.; Qin, S.P.; Zhuang, J.; Lu, J.; Ma, D.F.; Zheng, Y.L. The Inhibitory Effects of Purple Sweet Potato Color on Hepatic Inflammation Is Associated with Restoration of NAD$^+$ Levels and Attenuation of NLRP3 Inflammasome Activation in High-Fat-Diet-Treated Mice. *Molecules* **2017**, *22*, 1315. [CrossRef]

41. Yang, S.J.; Lim, Y. Resveratrol ameliorates hepatic metaflammation and inhibits NLRP3 inflammasome activation. *Metab. Clin. Exp.* **2014**, *63*, 693–701. [CrossRef] [PubMed]

42. Eo, H.; Park, J.E.; Jeon, Y.J.; Lim, Y. Ameliorative Effect of Ecklonia cava Polyphenol Extract on Renal Inflammation Associated with Aberrant Energy Metabolism and Oxidative Stress in High Fat Diet-Induced Obese Mice. *J. Agric. Food Chem.* **2017**, *65*, 3811–3818. [CrossRef] [PubMed]

43. Wang, J.; Wang, L.; Zhou, J.; Qin, A.; Chen, Z. The protective effect of formononetin on cognitive impairment in streptozotocin (STZ)-induced diabetic mice. *Biomed. Pharmacother. Biomed. Pharmacother.* **2018**, *106*, 1250–1257. [CrossRef] [PubMed]

44. Guo, W.; Liu, W.; Jin, B.; Geng, J.; Li, J.; Ding, H.; Wu, X.; Xu, Q.; Sun, Y.; Gao, J. Asiatic acid ameliorates dextran sulfate sodium-induced murine experimental colitis via suppressing mitochondria-mediated NLRP3 inflammasome activation. *Int. Immunopharmacol.* **2015**, *24*, 232–238. [CrossRef] [PubMed]

45. Wu, X.F.; Ouyang, Z.J.; Feng, L.L.; Chen, G.; Guo, W.J.; Shen, Y.; Wu, X.D.; Sun, Y.; Xu, Q. Suppression of NF-kappaB signaling and NLRP3 inflammasome activation in macrophages is responsible for the amelioration of experimental murine colitis by the natural compound fraxinellone. *Toxicol. Appl. Pharmacol.* **2014**, *281*, 146–156. [CrossRef]

46. Sun, Y.; Zhao, Y.; Yao, J.; Zhao, L.; Wu, Z.; Wang, Y.; Pan, D.; Miao, H.; Guo, Q.; Lu, N. Wogonoside protects against dextran sulfate sodium-induced experimental colitis in mice by inhibiting NF-kappaB and NLRP3 inflammasome activation. *Biochem. Pharmacol.* **2015**, *94*, 142–154. [CrossRef]

47. He, X.; Wei, Z.; Wang, J.; Kou, J.; Liu, W.; Fu, Y.; Yang, Z. Alpinetin attenuates inflammatory responses by suppressing TLR4 and NLRP3 signaling pathways in DSS-induced acute colitis. *Sci. Rep.* **2016**, *6*, 28370. [CrossRef]

48. Wu, D.; Wu, K.; Zhu, Q.; Xiao, W.; Shan, Q.; Yan, Z.; Wu, J.; Deng, B.; Xue, Y.; Gong, W.; et al. Formononetin Administration Ameliorates Dextran Sulfate Sodium-Induced Acute Colitis by Inhibiting NLRP3 Inflammasome Signaling Pathway. *Mediat. Inflamm.* **2018**, *2018*, 3048532.

49. Lopes de Oliveira, G.A.; Alarcon de la Lastra, C.; Rosillo, M.A.; Castejon Martinez, M.L.; Sanchez-Hidalgo, M.; Rolim Medeiros, J.V.; Villegas, I. Preventive effect of bergenin against the development of TNBS-induced acute colitis in rats is associated with inflammatory mediators inhibition and NLRP3/ASC inflammasome signaling pathways. *Chem.-Biol. Interact.* **2019**, *297*, 25–33. [CrossRef]

50. Yang, Y.; Zhang, X.; Xu, M.; Wu, X.; Zhao, F.; Zhao, C. Quercetin attenuates collagen-induced arthritis by restoration of Th17/Treg balance and activation of Heme Oxygenase 1-mediated anti-inflammatory effect. *Int. Immunopharmacol.* **2018**, *54*, 153–162. [CrossRef]

51. Liu, H.J.; Pan, X.X.; Liu, B.Q.; Gui, X.; Hu, L.; Jiang, C.Y.; Han, Y.; Fan, Y.X.; Tang, Y.L.; Liu, W.T. Grape seed-derived procyanidins alleviate gout pain via NLRP3 inflammasome suppression. *J. Neuroinflamm.* **2017**, *14*, 74. [CrossRef] [PubMed]

52. Jhang, J.J.; Lu, C.C.; Yen, G.C. Epigallocatechin gallate inhibits urate crystals-induced peritoneal inflammation in C57BL/6 mice. *Mol. Nutr. Food Res.* **2016**, *60*, 2297–2303. [CrossRef] [PubMed]

53. Misawa, T.; Saitoh, T.; Kozaki, T.; Park, S.; Takahama, M.; Akira, S. Resveratrol inhibits the acetylated alpha-tubulin-mediated assembly of the NLRP3-inflammasome. *Int. Immunol.* **2015**, *27*, 425–434. [CrossRef] [PubMed]

54. Staurengo-Ferrari, L.; Ruiz-Miyazawa, K.W.; Pinho-Ribeiro, F.A.; Fattori, V.; Zaninelli, T.H.; Badaro-Garcia, S.; Borghi, S.M.; Carvalho, T.T.; Alves-Filho, J.C.; Cunha, T.M.; et al. Trans-Chalcone Attenuates Pain and Inflammation in Experimental Acute Gout Arthritis in Mice. *Front. Pharmacol.* **2018**, *9*, 1123. [CrossRef] [PubMed]

55. Dhanasekar, C.; Rasool, M. Morin, a dietary bioflavonol suppresses monosodium urate crystal-induced inflammation in an animal model of acute gouty arthritis with reference to NLRP3 inflammasome, hypo-xanthine phospho-ribosyl transferase, and inflammatory mediators. *Eur. J. Pharmacol.* **2016**, *786*, 116–127. [CrossRef] [PubMed]

56. Doss, H.M.; Dey, C.; Sudandiradoss, C.; Rasool, M.K. Targeting inflammatory mediators with ferulic acid, a dietary polyphenol, for the suppression of monosodium urate crystal-induced inflammation in rats. *Life Sci.* **2016**, *148*, 201–210. [CrossRef] [PubMed]

57. Yang, G.; Yeon, S.H.; Lee, H.E.; Kang, H.C.; Cho, Y.Y.; Lee, H.S.; Lee, J.Y. Suppression of NLRP3 inflammasome by oral treatment with sulforaphane alleviates acute gouty inflammation. *Rheumatology* **2018**, *57*, 727–736. [CrossRef] [PubMed]

58. Heneka, M.T.; McManus, R.M.; Latz, E. Inflammasome signalling in brain function and neurodegenerative disease. *Nat. Rev. Neurosci.* **2018**, *19*, 610–621. [CrossRef]

59. Kaufmann, F.N.; Costa, A.P.; Ghisleni, G.; Diaz, A.P.; Rodrigues, A.L.S.; Peluffo, H.; Kaster, M.P. NLRP3 inflammasome-driven pathways in depression: Clinical and preclinical findings. *Brain Behav. Immun.* **2017**, *64*, 367–383. [CrossRef]

60. Wang, W.; Nguyen, L.T.; Burlak, C.; Chegini, F.; Guo, F.; Chataway, T.; Ju, S.; Fisher, O.S.; Miller, D.W.; Datta, D.; et al. Caspase-1 causes truncation and aggregation of the Parkinson's disease-associated protein alpha-synuclein. *Proc. Natl. Acad. Sci. USA* **2016**, *113*, 9587–9592. [CrossRef]

61. Zendedel, A.; Johann, S.; Mehrabi, S.; Joghataei, M.T.; Hassanzadeh, G.; Kipp, M.; Beyer, C. Activation and Regulation of NLRP3 Inflammasome by Intrathecal Application of SDF-1a in a Spinal Cord Injury Model. *Mol. Neurobiol.* **2016**, *53*, 3063–3075. [CrossRef] [PubMed]

62. Keane, R.W.; Dietrich, W.D.; de Rivero Vaccari, J.P. Inflammasome Proteins As Biomarkers of Multiple Sclerosis. *Front. Neurol.* **2018**, *9*, 135. [CrossRef]

63. Saresella, M.; La Rosa, F.; Piancone, F.; Zoppis, M.; Marventano, I.; Calabrese, E.; Rainone, V.; Nemni, R.; Mancuso, R.; Clerici, M. The NLRP3 and NLRP1 inflammasomes are activated in Alzheimer's disease. *Mol. Neurodegener.* **2016**, *11*, 23. [CrossRef] [PubMed]

64. Zhou, Y.; Lu, M.; Du, R.H.; Qiao, C.; Jiang, C.Y.; Zhang, K.Z.; Ding, J.H.; Hu, G. MicroRNA-7 targets Nod-like receptor protein 3 inflammasome to modulate neuroinflammation in the pathogenesis of Parkinson's disease. *Mol. Neurodegener.* **2016**, *11*, 28. [CrossRef] [PubMed]

65. Tan, F.C.; Hutchison, E.R.; Eitan, E.; Mattson, M.P. Are there roles for brain cell senescence in aging and neurodegenerative disorders? *Biogerontology* **2014**, *15*, 643–660. [CrossRef]

66. Qin, X.Y.; Zhang, S.P.; Cao, C.; Loh, Y.P.; Cheng, Y. Aberrations in Peripheral Inflammatory Cytokine Levels in Parkinson Disease: A Systematic Review and Meta-analysis. *JAMA Neurol.* **2016**, *73*, 1316–1324. [CrossRef] [PubMed]

67. Ming, X.; Li, W.; Maeda, Y.; Blumberg, B.; Raval, S.; Cook, S.D.; Dowling, P.C. Caspase-1 expression in multiple sclerosis plaques and cultured glial cells. *J. Neurol. Sci.* **2002**, *197*, 9–18. [CrossRef]

68. Cao, Y.; Goods, B.A.; Raddassi, K.; Nepom, G.T.; Kwok, W.W.; Love, J.C.; Hafler, D.A. Functional inflammatory profiles distinguish myelin-reactive T cells from patients with multiple sclerosis. *Sci. Transl. Med.* **2015**, *7*, 287ra74. [CrossRef]

69. Italiani, P.; Carlesi, C.; Giungato, P.; Puxeddu, I.; Borroni, B.; Bossu, P.; Migliorini, P.; Siciliano, G.; Boraschi, D. Evaluating the levels of interleukin-1 family cytokines in sporadic amyotrophic lateral sclerosis. *J. Neuroinflamm.* **2014**, *11*, 94. [CrossRef]

70. Alcocer-Gomez, E.; de Miguel, M.; Casas-Barquero, N.; Nunez-Vasco, J.; Sanchez-Alcazar, J.A.; Fernandez-Rodriguez, A.; Cordero, M.D. NLRP3 inflammasome is activated in mononuclear blood cells from patients with major depressive disorder. *Brain Behav. Immun.* **2014**, *36*, 111–117. [CrossRef]

71. Kim, H.K.; Andreazza, A.C.; Elmi, N.; Chen, W.; Young, L.T. Nod-like receptor pyrin containing 3 (NLRP3) in the post-mortem frontal cortex from patients with bipolar disorder: A potential mediator between mitochondria and immune-activation. *J. Psychiatr. Res.* **2016**, *72*, 43–50. [CrossRef] [PubMed]

72. Zhang, Y.; Liu, L.; Liu, Y.Z.; Shen, X.L.; Wu, T.Y.; Zhang, T.; Wang, W.; Wang, Y.X.; Jiang, C.L. NLRP3 Inflammasome Mediates Chronic Mild Stress-Induced Depression in Mice via Neuroinflammation. *Int. J. Neuropsychopharmacol.* **2015**, *18*(8), pyv006. [CrossRef]

73. Heneka, M.T.; Kummer, M.P.; Stutz, A.; Delekate, A.; Schwartz, S.; Vieira-Saecker, A.; Griep, A.; Axt, D.; Remus, A.; Tzeng, T.C.; et al. NLRP3 is activated in Alzheimer's disease and contributes to pathology in APP/PS1 mice. *Nature* **2013**, *493*, 674–678. [CrossRef]

74. Bellezza, I.; Grottelli, S.; Costanzi, E.; Scarpelli, P.; Pigna, E.; Morozzi, G.; Mezzasoma, L.; Peirce, M.J.; Moresi, V.; Adamo, S.; et al. Peroxynitrite Activates the NLRP3 Inflammasome Cascade in SOD1(G93A) Mouse Model of Amyotrophic Lateral Sclerosis. *Mol. Neurobiol.* **2018**, *55*, 2350–2361. [CrossRef] [PubMed]

75. Codolo, G.; Plotegher, N.; Pozzobon, T.; Brucale, M.; Tessari, I.; Bubacco, L.; de Bernard, M. Triggering of inflammasome by aggregated alpha-synuclein, an inflammatory response in synucleinopathies. *PLoS ONE* **2013**, *8*, e55375. [CrossRef]

76. Venegas, C.; Kumar, S.; Franklin, B.S.; Dierkes, T.; Brinkschulte, R.; Tejera, D.; Vieira-Saecker, A.; Schwartz, S.; Santarelli, F.; Kummer, M.P.; et al. Microglia-derived ASC specks cross-seed amyloid-beta in Alzheimer's disease. *Nature* **2017**, *552*, 355–361. [CrossRef] [PubMed]

77. Martin, B.N.; Wang, C.; Zhang, C.J.; Kang, Z.; Gulen, M.F.; Zepp, J.A.; Zhao, J.; Bian, G.; Do, J.S.; Min, B.; et al. T cell-intrinsic ASC critically promotes T(H)17-mediated experimental autoimmune encephalomyelitis. *Nat. Immunol.* **2016**, *17*, 583–592. [CrossRef] [PubMed]

78. Yamamoto, S.; Yamashina, K.; Ishikawa, M.; Gotoh, M.; Yagishita, S.; Iwasa, K.; Maruyama, K.; Murakami-Murofushi, K.; Yoshikawa, K. Protective and therapeutic role of 2-carba-cyclic phosphatidic acid in demyelinating disease. *J. Neuroinflamm.* **2017**, *14*, 142. [CrossRef]

79. Liu, T.; Ma, Y.; Zhang, R.; Zhong, H.; Wang, L.; Zhao, J.; Yang, L.; Fan, X. Resveratrol ameliorates estrogen deficiency-induced depression- and anxiety-like behaviors and hippocampal inflammation in mice. *Psychopharmacology* **2019**, 1–15. [CrossRef] [PubMed]

80. Hsu, S.C.; Huang, S.M.; Chen, A.; Sun, C.Y.; Lin, S.H.; Chen, J.S.; Liu, S.T.; Hsu, Y.J. Resveratrol increases anti-aging Klotho gene expression via the activating transcription factor 3/c-Jun complex-mediated signaling pathway. *Int. J. Biochem. Cell Biol.* **2014**, *53*, 361–371. [CrossRef]

81. Yousuf, S.; Atif, F.; Ahmad, M.; Hoda, N.; Ishrat, T.; Khan, B.; Islam, F. Resveratrol exerts its neuroprotective effect by modulating mitochondrial dysfunctions and associated cell death during cerebral ischemia. *Brain Res.* **2009**, *1250*, 242–253. [CrossRef] [PubMed]

82. Ota, T. Obesity-induced inflammation and insulin resistance. *Front. Endocrinol.* **2014**, *5*, 204. [CrossRef] [PubMed]

83. Esser, N.; Legrand-Poels, S.; Piette, J.; Scheen, A.J.; Paquot, N. Inflammation as a link between obesity, metabolic syndrome and type 2 diabetes. *Diabetes Res. Clin. Pract.* **2014**, *105*, 141–150. [CrossRef]

84. Rheinheimer, J.; de Souza, B.M.; Cardoso, N.S.; Bauer, A.C.; Crispim, D. Current role of the NLRP3 inflammasome on obesity and insulin resistance: A systematic review. *Metab. Clin. Exp.* **2017**, *74*, 1–9. [CrossRef] [PubMed]

85. Stienstra, R.; van Diepen, J.A.; Tack, C.J.; Zaki, M.H.; van de Veerdonk, F.L.; Perera, D.; Neale, G.A.; Hooiveld, G.J.; Hijmans, A.; Vroegrijk, I.; et al. Inflammasome is a central player in the induction of obesity and insulin resistance. *Proc. Natl. Acad. Sci. USA* **2011**, *108*, 15324–15329. [CrossRef] [PubMed]

86. Goossens, G.H.; Blaak, E.E.; Theunissen, R.; Duijvestijn, A.M.; Clement, K.; Tervaert, J.W.; Thewissen, M.M. Expression of NLRP3 inflammasome and T cell population markers in adipose tissue are associated with insulin resistance and impaired glucose metabolism in humans. *Mol. Immunol.* **2012**, *50*, 142–149. [CrossRef]

87. Lee, H.M.; Kim, J.J.; Kim, H.J.; Shong, M.; Ku, B.J.; Jo, E.K. Upregulated NLRP3 inflammasome activation in patients with type 2 diabetes. *Diabetes* **2013**, *62*, 194–204. [CrossRef]

88. Donath, M.Y.; Boni-Schnetzler, M.; Ellingsgaard, H.; Halban, P.A.; Ehses, J.A. Cytokine production by islets in health and diabetes: Cellular origin, regulation and function. *Trends Endocrinol. Metab. TEM* **2010**, *21*, 261–267. [CrossRef]

89. Wen, H.; Gris, D.; Lei, Y.; Jha, S.; Zhang, L.; Huang, M.T.; Brickey, W.J.; Ting, J.P. Fatty acid-induced NLRP3-ASC inflammasome activation interferes with insulin signaling. *Nat. Immunol.* **2011**, *12*, 408–415. [CrossRef]

90. McGillicuddy, F.C.; Harford, K.A.; Reynolds, C.M.; Oliver, E.; Claessens, M.; Mills, K.H.; Roche, H.M. Lack of interleukin-1 receptor I (IL-1RI) protects mice from high-fat diet-induced adipose tissue inflammation coincident with improved glucose homeostasis. *Diabetes* **2011**, *60*, 1688–1698. [CrossRef]

91. Zhou, R.; Tardivel, A.; Thorens, B.; Choi, I.; Tschopp, J. Thioredoxin-interacting protein links oxidative stress to inflammasome activation. *Nat. Immunol.* **2010**, *11*, 136–140. [CrossRef] [PubMed]

92. Stienstra, R.; Joosten, L.A.; Koenen, T.; van Tits, B.; van Diepen, J.A.; van den Berg, S.A.; Rensen, P.C.; Voshol, P.J.; Fantuzzi, G.; Hijmans, A.; et al. The inflammasome-mediated caspase-1 activation controls adipocyte differentiation and insulin sensitivity. *Cell Metab.* **2010**, *12*, 593–605. [CrossRef] [PubMed]

93. Zhou, R.; Yazdi, A.S.; Menu, P.; Tschopp, J. A role for mitochondria in NLRP3 inflammasome activation. *Nature* **2011**, *469*, 221–225. [CrossRef] [PubMed]

94. Hara, H.; Tsuchiya, K.; Kawamura, I.; Fang, R.; Hernandez-Cuellar, E.; Shen, Y.; Mizuguchi, J.; Schweighoffer, E.; Tybulewicz, V.; Mitsuyama, M. Phosphorylation of the adaptor ASC acts as a molecular switch that controls the formation of speck-like aggregates and inflammasome activity. *Nat. Immunol.* **2013**, *14*, 1247–1255. [CrossRef] [PubMed]

95. Neurath, M.F. Cytokines in inflammatory bowel disease. *Nat. Rev. Immunol.* **2014**, *14*, 329–342. [CrossRef] [PubMed]

96. Lissner, D.; Schumann, M.; Batra, A.; Kredel, L.I.; Kuhl, A.A.; Erben, U.; May, C.; Schulzke, J.D.; Siegmund, B. Monocyte and M1 Macrophage-induced Barrier Defect Contributes to Chronic Intestinal Inflammation in IBD. *Inflamm. Bowel Dis.* **2015**, *21*, 1297–1305. [CrossRef] [PubMed]

97. Liu, L.; Dong, Y.; Ye, M.; Jin, S.; Yang, J.; Joosse, M.E.; Sun, Y.; Zhang, J.; Lazarev, M.; Brant, S.R.; et al. The Pathogenic Role of NLRP3 Inflammasome Activation in Inflammatory Bowel Diseases of Both Mice and Humans. *J. Crohn's Colitis* **2017**, *11*, 737–750. [CrossRef]

98. Lazaridis, L.D.; Pistiki, A.; Giamarellos-Bourboulis, E.J.; Georgitsi, M.; Damoraki, G.; Polymeros, D.; Dimitriadis, G.D.; Triantafyllou, K. Activation of NLRP3 Inflammasome in Inflammatory Bowel Disease: Differences Between Crohn's Disease and Ulcerative Colitis. *Dig. Dis. Sci.* **2017**, *62*, 2348–2356. [CrossRef]

99. Sutterwala, F.S.; Haasken, S.; Cassel, S.L. Mechanism of NLRP3 inflammasome activation. *Ann. N. Y. Acad. Sci.* **2014**, *1319*, 82–95. [CrossRef]

100. Cocco, M.; Pellegrini, C.; Martinez-Banaclocha, H.; Giorgis, M.; Marini, E.; Costale, A.; Miglio, G.; Fornai, M.; Antonioli, L.; Lopez-Castejon, G.; et al. Development of an Acrylate Derivative Targeting the NLRP3 Inflammasome for the Treatment of Inflammatory Bowel Disease. *J. Med. Chem.* **2017**, *60*, 3656–3671. [CrossRef]

101. Perera, A.P.; Kunde, D.; Eri, R. NLRP3 Inhibitors as Potential Therapeutic Agents for Treatment of Inflammatory Bowel Disease. *Curr. Pharm. Des.* **2017**, *23*, 2321–2327. [CrossRef]

102. Perera, A.P.; Fernando, R.; Shinde, T.; Gundamaraju, R.; Southam, B.; Sohal, S.S.; Robertson, A.A.B.; Schroder, K.; Kunde, D.; Eri, R. MCC950, a specific small molecule inhibitor of NLRP3 inflammasome attenuates colonic inflammation in spontaneous colitis mice. *Sci. Rep.* **2018**, *8*, 8618. [CrossRef] [PubMed]

103. Pellegrini, C.; Fornai, M.; Colucci, R.; Benvenuti, L.; D'Antongiovanni, V.; Natale, G.; Fulceri, F.; Giorgis, M.; Marini, E.; Gastaldi, S.; et al. A Comparative Study on the Efficacy of NLRP3 Inflammasome Signaling Inhibitors in a Pre-clinical Model of Bowel Inflammation. *Front. Pharmacol.* **2018**, *9*, 1405. [CrossRef] [PubMed]

104. Shen, H.H.; Yang, Y.X.; Meng, X.; Luo, X.Y.; Li, X.M.; Shuai, Z.W.; Ye, D.Q.; Pan, H.F. NLRP3: A promising therapeutic target for autoimmune diseases. *Autoimmun. Rev.* **2018**, *17*, 694–702. [CrossRef] [PubMed]

105. Jenko, B.; Praprotnik, S.; Tomsic, M.; Dolzan, V. NLRP3 and CARD8 Polymorphisms Influence Higher Disease Activity in Rheumatoid Arthritis. *J. Med. Biochem.* **2016**, *35*, 319–323. [CrossRef] [PubMed]

106. Mathews, R.J.; Robinson, J.I.; Battellino, M.; Wong, C.; Taylor, J.C.; Biologics in Rheumatoid Arthritis Genetics and Genomics Study Syndicate (BRAGGSS); Eyre, S.; Churchman, S.M.; Wilson, A.G.; Isaacs, J.D.; et al. Evidence of NLRP3-inflammasome activation in rheumatoid arthritis (RA); genetic variants within the NLRP3-inflammasome complex in relation to susceptibility to RA and response to anti-TNF treatment. *Ann. Rheum. Dis.* **2014**, *73*, 1202–1210. [PubMed]

107. Paramel, G.V.; Sirsjo, A.; Fransen, K. Role of genetic alterations in the NLRP3 and CARD8 genes in health and disease. *Mediat. Inflamm.* **2015**, *2015*, 846782. [CrossRef]

108. Rosengren, S.; Hoffman, H.M.; Bugbee, W.; Boyle, D.L. Expression and regulation of cryopyrin and related proteins in rheumatoid arthritis synovium. *Ann. Rheum. Dis.* **2005**, *64*, 708–714. [CrossRef]

109. Choulaki, C.; Papadaki, G.; Repa, A.; Kampouraki, E.; Kambas, K.; Ritis, K.; Bertsias, G.; Boumpas, D.T.; Sidiropoulos, P. Enhanced activity of NLRP3 inflammasome in peripheral blood cells of patients with active rheumatoid arthritis. *Arthritis Res. Ther.* **2015**, *17*, 257. [CrossRef]

110. Ruscitti, P.; Cipriani, P.; Di Benedetto, P.; Liakouli, V.; Berardicurti, O.; Carubbi, F.; Ciccia, F.; Alvaro, S.; Triolo, G.; Giacomelli, R. Monocytes from patients with rheumatoid arthritis and type 2 diabetes mellitus display an increased production of interleukin (IL)-1beta via the nucleotide-binding domain and leucine-rich repeat containing family pyrin 3(NLRP3)-inflammasome activation: A possible implication for therapeutic decision in these patients. *Clin. Exp. Immunol.* **2015**, *182*, 35–44.

111. Dayer, J.M. The pivotal role of interleukin-1 in the clinical manifestations of rheumatoid arthritis. *Rheumatology* **2003**, *42* (Suppl. 2), ii3–ii10. [CrossRef]

112. Zhang, H.; Wang, S.; Huang, Y.; Wang, H.; Zhao, J.; Gaskin, F.; Yang, N.; Fu, S.M. Myeloid-derived suppressor cells are proinflammatory and regulate collagen-induced arthritis through manipulating Th17 cell differentiation. *Clin. Immunol.* **2015**, *157*, 175–186. [CrossRef] [PubMed]

113. Wei, X.Q.; Leung, B.P.; Arthur, H.M.; McInnes, I.B.; Liew, F.Y. Reduced incidence and severity of collagen-induced arthritis in mice lacking IL-18. *J. Immunol.* **2001**, *166*, 517–521. [CrossRef] [PubMed]

114. Choi, H.K.; Mount, D.B.; Reginato, A.M.; American College of, P.; American Physiological, S. Pathogenesis of gout. *Ann. Intern. Med.* **2005**, *143*, 499–516. [CrossRef] [PubMed]

115. Martinon, F. Mechanisms of uric acid crystal-mediated autoinflammation. *Immunol. Rev.* **2010**, *233*, 218–232. [CrossRef] [PubMed]

116. Mitroulis, I.; Kambas, K.; Ritis, K. Neutrophils, IL-1beta, and gout: Is there a link? *Semin. Immunopathol.* **2013**, *35*, 501–512. [CrossRef] [PubMed]

117. Kingsbury, S.R.; Conaghan, P.G.; McDermott, M.F. The role of the NLRP3 inflammasome in gout. *J. Inflamm. Res.* **2011**, *4*, 39–49.

118. Cavalcanti, N.G.; Marques, C.D.; Lins, E.L.T.U.; Pereira, M.C.; Rego, M.J.; Duarte, A.L.; Pitta Ida, R.; Pitta, M.G. Cytokine Profile in Gout: Inflammation Driven by IL-6 and IL-18? *Immunol. Investig.* **2016**, *45*, 383–395. [CrossRef]

119. Mylona, E.E.; Mouktaroudi, M.; Crisan, T.O.; Makri, S.; Pistiki, A.; Georgitsi, M.; Savva, A.; Netea, M.G.; van der Meer, J.W.; Giamarellos-Bourboulis, E.J.; et al. Enhanced interleukin-1beta production of PBMCs from patients with gout after stimulation with Toll-like receptor-2 ligands and urate crystals. *Arthritis Res. Ther.* **2012**, *14*, R158. [CrossRef]

120. Terkeltaub, R.; Sundy, J.S.; Schumacher, H.R.; Murphy, F.; Bookbinder, S.; Biedermann, S.; Wu, R.; Mellis, S.; Radin, A. The interleukin 1 inhibitor rilonacept in treatment of chronic gouty arthritis: Results of a placebo-controlled, monosequence crossover, non-randomised, single-blind pilot study. *Ann. Rheum. Dis.* **2009**, *68*, 1613–1617. [CrossRef]

121. McGonagle, D.; Tan, A.L.; Shankaranarayana, S.; Madden, J.; Emery, P.; McDermott, M.F. Management of treatment resistant inflammation of acute on chronic tophaceous gout with anakinra. *Ann. Rheum. Dis.* **2007**, *66*, 1683–1684. [CrossRef]

122. So, A.; De Smedt, T.; Revaz, S.; Tschopp, J. A pilot study of IL-1 inhibition by anakinra in acute gout. *Arthritis Res. Ther.* **2007**, *9*, R28. [CrossRef] [PubMed]

123. So, A.; De Meulemeester, M.; Pikhlak, A.; Yucel, A.E.; Richard, D.; Murphy, V.; Arulmani, U.; Sallstig, P.; Schlesinger, N. Canakinumab for the treatment of acute flares in difficult-to-treat gouty arthritis: Results of a multicenter, phase II, dose-ranging study. *Arthritis Rheum.* **2010**, *62*, 3064–3076. [CrossRef] [PubMed]

124. Nomura, J.; So, A.; Tamura, M.; Busso, N. Intracellular ATP Decrease Mediates NLRP3 Inflammasome Activation upon Nigericin and Crystal Stimulation. *J. Immunol.* **2015**, *195*, 5718–5724. [CrossRef] [PubMed]

125. Martinon, F.; Petrilli, V.; Mayor, A.; Tardivel, A.; Tschopp, J. Gout-associated uric acid crystals activate the NALP3 inflammasome. *Nature* **2006**, *440*, 237–241. [CrossRef] [PubMed]

126. Yi, Y.S. Role of inflammasomes in inflammatory autoimmune rheumatic diseases. *Korean J. Physiol. Pharmacol.* **2018**, *22*, 1–15. [CrossRef]

127. Normand, S.; Delanoye-Crespin, A.; Bressenot, A.; Huot, L.; Grandjean, T.; Peyrin-Biroulet, L.; Lemoine, Y.; Hot, D.; Chamaillard, M. Nod-like receptor pyrin domain-containing protein 6 (NLRP6) controls epithelial self-renewal and colorectal carcinogenesis upon injury. *Proc. Natl. Acad. Sci. USA* **2011**, *108*, 9601–9606. [CrossRef]

128. Henao-Mejia, J.; Elinav, E.; Thaiss, C.A.; Flavell, R.A. Inflammasomes and metabolic disease. *Annu. Rev. Physiol.* **2014**, *76*, 57–78. [CrossRef] [PubMed]

129. Moossavi, M.; Parsamanesh, N.; Bahrami, A.; Atkin, S.L.; Sahebkar, A. Role of the NLRP3 inflammasome in cancer. *Mol. Cancer* **2018**, *17*, 158. [CrossRef]

130. Zaki, M.H.; Boyd, K.L.; Vogel, P.; Kastan, M.B.; Lamkanfi, M.; Kanneganti, T.D. The NLRP3 inflammasome protects against loss of epithelial integrity and mortality during experimental colitis. *Immunity* **2010**, *32*, 379–391. [CrossRef] [PubMed]

International Journal of
Molecular Sciences

Article

Skullcapflavone II Inhibits Degradation of Type I Collagen by Suppressing MMP-1 Transcription in Human Skin Fibroblasts

Young Hun Lee [1], Eun Kyoung Seo [2] and Seung-Taek Lee [1],*

[1] Department of Biochemistry, College of Life Science and Biotechnology, Yonsei University, Seoul 03722, Korea; isop_tera@naver.com

[2] Graduate school of Pharmaceutical Sciences, College of Pharmacy, Ewha Womans University, Seoul 03790, Korea; yuny@ewha.ac.kr

*   Correspondence: stlee@yonsei.ac.kr; Tel.: +82-2-2123-2703

Received: 27 April 2019; Accepted: 31 May 2019; Published: 4 June 2019

Abstract: Skullcapflavone II is a flavonoid derived from the root of *Scutellaria baicalensis*, a herbal medicine used for anti-inflammatory and anti-cancer therapies. We analyzed the effect of skullcapflavone II on the expression of matrix metalloproteinase-1 (MMP-1) and integrity of type I collagen in foreskin fibroblasts. Skullcapflavone II did not affect the secretion of type I collagen but reduced the secretion of MMP-1 in a dose- and time-dependent manner. Real-time reverse transcription-PCR and reporter gene assays showed that skullcapflavone II reduced MMP-1 expression at the transcriptional level. Skullcapflavone II inhibited the serum-induced activation of the extracellular signal-regulated kinase (ERK) and c-Jun N-terminal kinase (JNK) signaling pathways required for MMP-1 transactivation. Skullcapflavone II also reduced tumor necrosis factor (TNF)-α-induced nuclear factor kappa light chain enhancer of activated B cells (NF-κB) activation and subsequent MMP-1 expression. In three-dimensional culture of fibroblasts, skullcapflavone II down-regulated TNF-α-induced MMP-1 secretion and reduced breakdown of type I collagen. These results indicate that skullcapflavone II is a novel biomolecule that down-regulates MMP-1 expression in foreskin fibroblasts and therefore could be useful in therapies for maintaining the integrity of extracellular matrix.

Keywords: extracellular matrix; fibroblast; inflammation; MMP-1; skullcapflavone II; type I collagen

---

1. Introduction

Skullcapflavone II, a naturally occurring flavonoid compound with a polyphenolic structure also known as neobaicalein, is derived from the roots of *Scutellaria baicalensis*, *S. litwinowii*, and *S. pinnatifida* [1–3]. Biological functions attributed to skullcapflavone II include reducing inflammation, inhibiting osteoclastogenesis, decreasing cell growth, inducing apoptosis, and down-regulating cholesterol [2,4–7]. Skullcapflavone II inhibits ovalbumin-induced airway inflammation via a decrease in transforming growth factor-β1 expression and subsequent decrease in SMAD2/3 activation [4]. Skullcapflavone II inhibits osteoclastogenesis with reduced activation of mitogen-activated protein kinases (MAPKs), Src, and cyclic adenosine monophosphate (cAMP) response element binding protein, and attenuates the survival and bone resorption function of osteoclasts via down-regulation of integrin signaling [5]. Skullcapflavone II inhibits cell proliferation in a variety of cancer cell lines, including LNCaP, PC-3, and HeLa [2,6]. In addition, skullcapflavone II reportedly inhibits the mRNA expression of proprotein convertase subtilisin/kexin type 9, which prevents the recycling of endocytosed low-density-lipoprotein receptors (LDLRs) to the cell surface, thereby increasing cell-surface LDLR levels and lowering plasma cholesterol levels [7].

Collagen is the primary structural protein in extracellular spaces in mammals and functions to strengthen and support various connective tissues, such as tendons, ligaments, bones, and skin [8,9]. The collagen family contain at least one triple-helical domain consisting of three α-chains with a repeating amino acid sequence $(Gly-X-Y)_n$ [10]. A total of 28 types of collagen identified in humans can be divided into subfamilies based on their supramolecular assemblies; fibrils, beaded filaments, anchoring fibrils, and networks [11,12].

Collagen is degraded during various normal physiological processes involving tissue remodeling, such as organ morphogenesis, wound healing, and skin aging. In addition, collagen is degraded during numerous pathological conditions, such as inflammation, arthritis, atherosclerotic cardiovascular disease, and tumorigenesis [10,11]. Matrix metalloproteinases (MMPs) are zinc-dependent endopeptidases that play a major role in tissue remodeling processes by cleaving extracellular matrix components, including collagen [13,14]. MMP-1 cleaves the triple helix of fibril-forming collagens, including types I, II, and III; in type I collagen, it cleaves at $Gly^{775}{\downarrow}Ile^{776}$ of the α1 chain and $Gly^{775}{\downarrow}Leu^{776}$ of the α2 chain to generate 3/4- and 1/4-length fragments [15].

Collagen is also frequently degraded in inflammation lesions. In inflamed acne lesions, for example, collagen degradation is increased as a result of up-regulation of inflammatory cytokine and MMP expression [16]. Among the various collagen types, type I accounts for over 90% of all collagens in the human body [9] and is highly expressed in fibroblasts [17]. In addition, MMP-1 is expressed in unstimulated fibroblasts and is upregulated by inflammation [18,19]. Because skullcapflavone II exhibits anti-inflammatory activity, we investigated the effect of skullcapflavone II on the expression of MMP-1 and the integrity of type I collagen in fibroblasts. Specifically, we examined whether skullcapflavone II affects the production of type I collagen or MMP-1-mediated degradation of type I collagen. In addition, we analyzed the signaling pathways involved in skullcapflavone II-mediated suppression of MMP-1 expression. Finally, using three-dimensional (3D) culture of fibroblasts, we examined the effect of skullcapflavone II on breakdown of type I collagen.

2. Results

2.1. Skullcapflavone II Decreased MMP-1 Expression in Foreskin Fibroblasts

We investigated the effect of skullcapflavone II (Figure S1) on the secretion of MMP-1 and type I collagen in primary human foreskin fibroblasts and primary human buttock dermal fibroblasts. It was reported that human fibroblasts secrete MMP-1 as a pro-form; mostly unglycosylated (52 kDa) and partly glycosylated (57 kDa) [20]. We detected a major band of MMP-1 at 52 kDa in foreskin fibroblasts and buttock dermal fibroblasts (Figure 1A), suggesting that it should be proMMP-1. Skullcapflavone II significantly decreased the secretion of MMP-1 in a dose-dependent manner in both cell types but did not significantly affect the secretion of type I collagen (Figure 1A). Compared with untreated cells, foreskin fibroblasts secreted significantly lower amounts of MMP-1 in the presence of 3 µM skullcapflavone II, with the effect decreasing by 24 h in serum-free Dulbecco's Modified Eagle's Medium (DMEM) and by 48 h in DMEM supplemented with 3% fetal bovine serum (FBS) (Figure 1B).

To examine whether the reduction in MMP-1 secretion by cells treated with skullcapflavone II was associated with a decrease in cell proliferation, we monitored the growth of skullcapflavone II-treated foreskin fibroblasts using the 3-(4,5-dimethyl thiazol-2-yl)-2,5-diphenyltetrazolium bromide (MTT) assay. Cell growth was unaffected by skullcapflavone II at concentrations up to 3 µM but slightly decreased at a concentration of 10 µM (up to 15% decrease in growth at 2 days) (Figure 2A). Flow cytometry analyses indicated that the decrease in the growth of foreskin fibroblasts in the presence of 10 µM skullcapflavone II was not due to cytotoxic effects (Figure 2B). For subsequent experiments, therefore, we used 3 µM skullcapflavone II, a concentration that had no effect on cell growth.

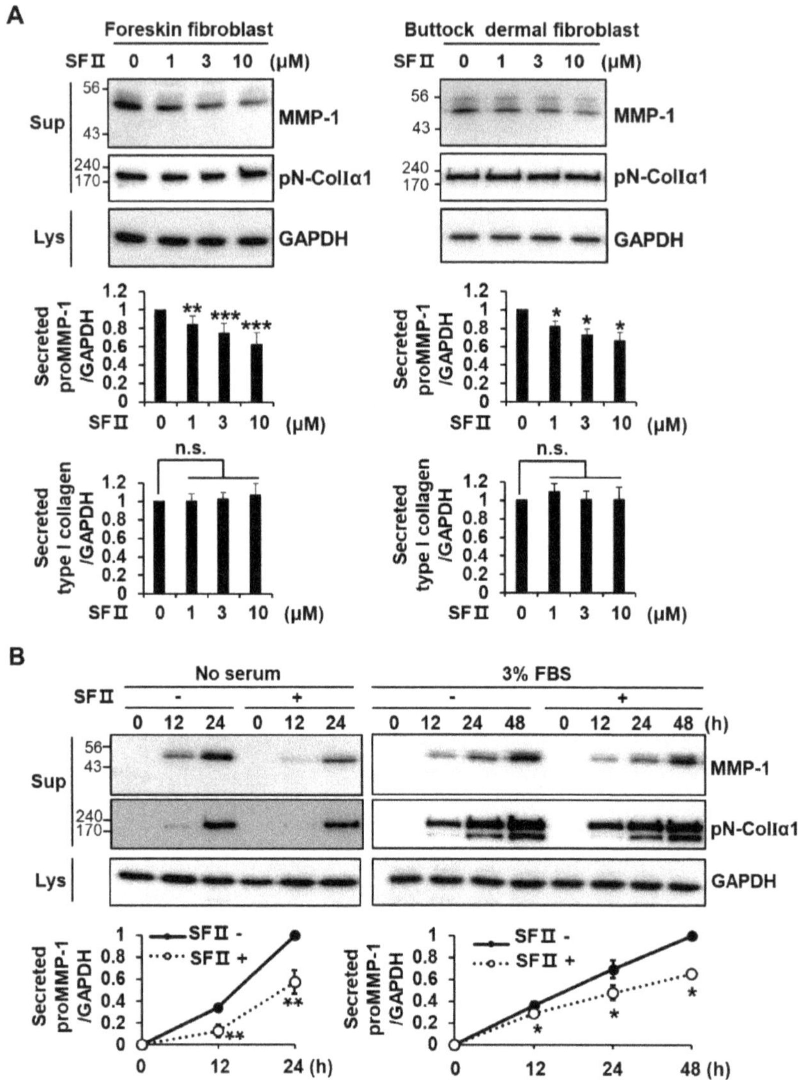

Figure 1. Effect of skullcapflavone II on expression of matrix metalloproteinase-1 (MMP-1) and type I collagen in fibroblasts. (**A**) Primary human foreskin fibroblasts and primary human buttock skin fibroblasts were incubated for 24 h in serum-free Dulbecco's Modified Eagle's Medium (DMEM) with the indicated concentrations of skullcapflavone II. * $p < 0.05$, ** $p < 0.01$, and *** $p < 0.001$ vs. the sample incubated with 0 μM skullcapflavone II; n.s., not significant. (**B**) Foreskin fibroblasts were incubated for the indicated times in serum-free DMEM or DMEM containing 3% fetal bovine serum (FBS) with (+) or without (−) 3 μM skullcapflavone II. Conditioned medium and cell lysates were analyzed by 9% SDS-PAGE and Western blot with anti-MMP-1, anti-pN-Colα1, and anti-glyceraldehyde 3-phosphate dehydrogenase (GAPDH) antibodies. * $p < 0.05$ and ** $p < 0.01$ vs. the sample incubated without skullcapflavone II.

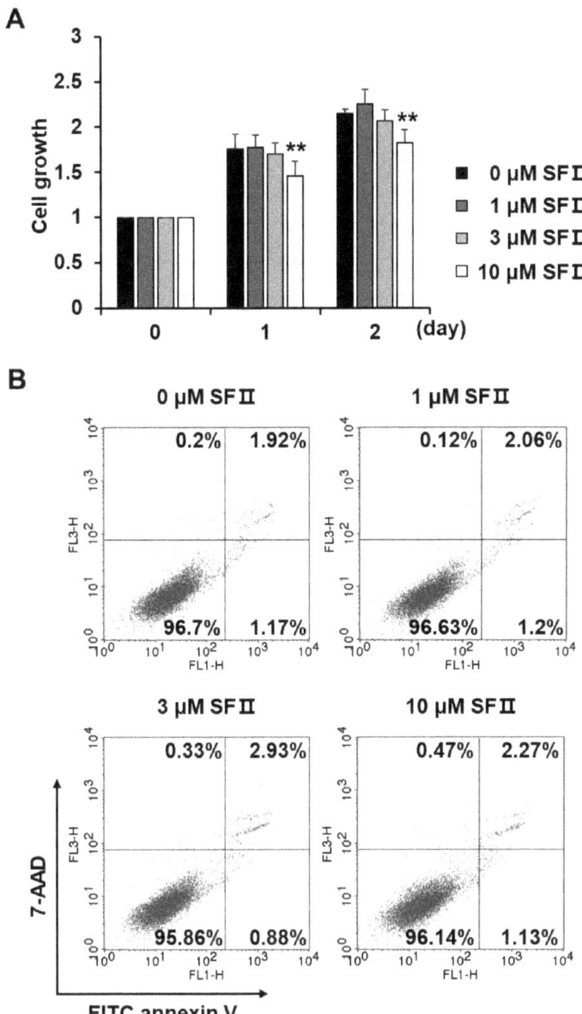

Figure 2. Effect of skullcapflavone II on proliferation and cytotoxicity of foreskin fibroblasts. Sub-confluent foreskin fibroblasts were plated in DMEM containing 10% FBS and incubated with indicated concentrations of skullcapflavone II. (**A**) The number of viable cells was measured based on the absorbance at a wavelength of 565 nm using 3-(4,5-dimethyl thiazol-2-yl)-2,5-diphenyltetrazolium bromide (MTT) reagent. The number of viable cells in the presence of skullcapflavone II is shown as fold change relative to that in the absence of skullcapflavone II. ** $p < 0.01$ vs. the sample incubated with 0 µM skullcapflavone II. (**B**) After 24 h of incubation with skullcapflavone II, apoptotic cells were stained with fluorescein isothiocyanate (FITC) annexin V and 7-aminoactinomycin (7-AAD) and then analyzed by flow cytometry.

2.2. Skullcapflavone II Decreased Transcription of the MMP-1 Gene in Foreskin Fibroblasts

To determine whether the observed skullcapflavone II-mediated inhibition of MMP-1 secretion was due to suppression of *MMP-1* transcription, we performed real-time RT-PCR and reporter gene assays using foreskin fibroblasts. Consistent with MMP-1 protein levels, *MMP-1* mRNA levels declined in a dose-dependent manner in cells treated with skullcapflavone II (Figure 3A). From the reference curve generated by serial dilution of *MMP-1* cDNA, the *MMP-1* mRNA molecules per cell were estimated to be 40.0 at 0 uM skullcapflavone II and 27.2 at 10 uM skullcapflavone II. Treatment with skullcapflavone II also reduced luciferase activity driven by the *MMP-1* promoter to 70% of the control (Figure 3B). These data demonstrate that skullcapflavone II inhibits expression of the *MMP-1* gene at the transcription level.

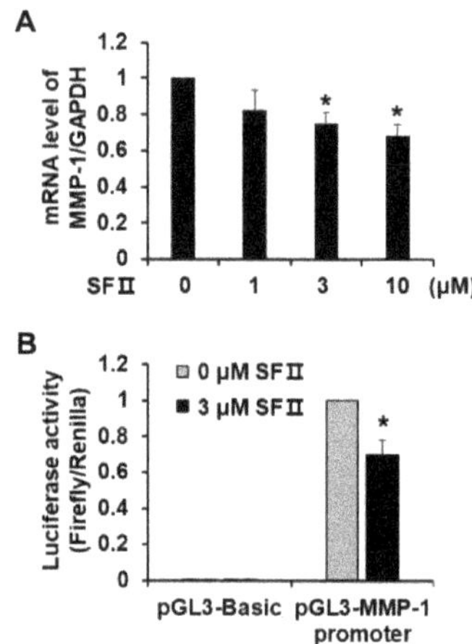

Figure 3. Effect of skullcapflavone II on MMP-1 transcription in foreskin fibroblasts. (**A**) *MMP-1* mRNA levels in DMSO- or skullcapflavone II-treated foreskin fibroblasts were determined using real-time RT-PCR. * $p < 0.05$ vs. the sample incubated with 0 µM skullcapflavone II. (**B**) Foreskin fibroblasts were transiently co-transfected with pGL3-*MMP-1* promoter and pRL-TK as a transfection control. Cells were treated overnight with 3 µM skullcapflavone II. Luciferase activity was determined as the ratio of firefly/*Renilla* luciferase activity. The activity of the *MMP-1* promoter in the presence of skullcapflavone II relative to that in the absence of skullcapflavone II is shown. *$p < 0.05$ vs. the sample incubated with 0 µM skullcapflavone II.

2.3. *Skullcapflavone II Inhibited MMP-1 Expression by Blocking the Activation of Activator Protein-1 (AP-1)*

MMP-1 gene expression is positively regulated by the activation of extracellular signal-regulated kinase (ERK), c-Jun N-terminal kinase (JNK), p38 MAPK, and nuclear factor kappa light chain enhancer of activated B cells (NF-κB) signaling pathways [21,22]. To investigate how skullcapflavone II suppresses MMP-1 transactivation, phosphorylation of various signaling proteins was analyzed in foreskin fibroblasts incubated with and without skullcapflavone II. Serum-induced tyrosine phosphorylation of cellular proteins was significantly decreased in a dose-dependent manner in cells treated with skullcapflavone II (Figure 4A). The presence of FBS strongly enhanced the phosphorylation of ERK1/2, moderately enhanced the phosphorylation of JNK, and weakly enhanced the phosphorylation of p38 MAPK (Figure 4B). Treatment with skullcapflavone II reduced the serum-induced phosphorylation of ERK1/2 and JNK in a dose-dependent manner but had no effect on the phosphorylation of p38 MAPK (Figure 4B). Phosphorylation of NF-κB p65 was not affected by the presence of FBS and did not decrease in the presence of skullcapflavone II. In addition, skullcapflavone II significantly decreased the serum-enhanced phosphorylation of c-Jun (Figure 4C). These findings demonstrate that skullcapflavone II down-regulates *MMP-1* expression by reducing activation of the ERK and JNK pathways and thus activation of the transcription factor AP-1, which plays an important role in MMP-1 transactivation.

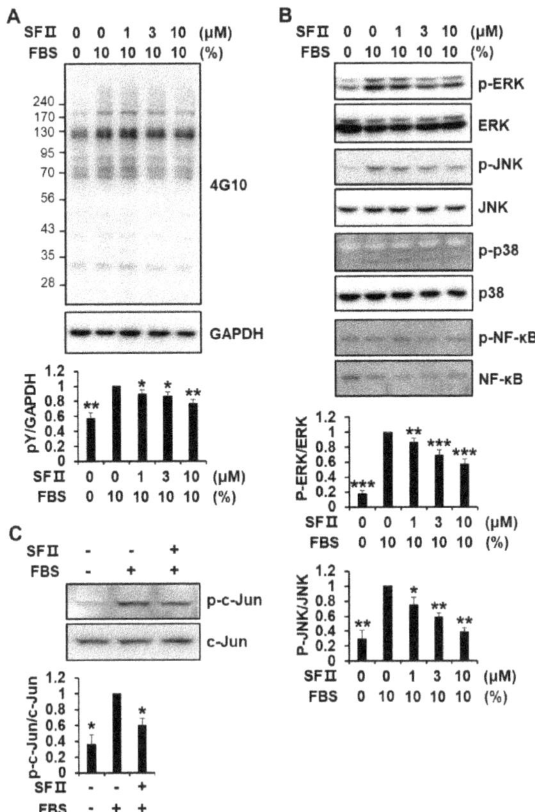

Figure 4. Effect of skullcapflavone II on phosphorylation of signaling molecules in foreskin fibroblasts. Sub-confluent foreskin fibroblasts were starved for 24 h. (**A,B**) Cells were pre-incubated for 30 min with indicated concentrations of skullcapflavone II and then stimulated for 10 min with 10% FBS. Cell lysates were subjected to 9% SDS-PAGE and Western blot analysis with anti-phosphotyrosine (4G10) (A), anti-phospho-extracellular signal-regulated kinase (ERK), anti-phospho-c-Jun N-terminal kinase (JNK), anti-phospho-p38 mitogen-activated protein kinases (MAPK), and anti-phospho-nuclear factor kappa light chain enhancer of activated B cells (NF-κB) p65 antibodies (**B**). (**C**) Cells were pre-incubated for 30 min with (+) or without (−) 3 μM skullcapflavone II, and then FBS was added to a final concentration of 10% and incubated for 30 min. Cell lysates were analyzed by Western blotting with anti-phospho-c-Jun and anti-c-Jun antibodies. * $p < 0.05$, ** $p < 0.01$, and *** $p < 0.001$ vs. the sample incubated with 10% FBS alone.

2.4. Skullcapflavone II Inhibited Tumor Necrosis Factor (TNF)-α-Induced MMP-1 Expression by Blocking Activation of NF-κB

The pro-inflammatory cytokine TNF-α up-regulates MMP-1 expression in diverse cell types, including dermal fibroblasts [23–25]. As expected, stimulation of foreskin fibroblasts with TNF-α significantly up-regulated MMP-1 secretion (Figure 5A). Treatment with skullcapflavone II decreased MMP-1 expression induced by TNF-α. However, consistent with the effect of serum stimulation, skullcapflavone II treatment did not affect the secretion of type I collagen in TNF-α-treated cells (Figure 5A).

Since TNF-α is known to activate NF-κB signaling [26], we investigated whether skullcapflavone II suppresses NF-κB activation for the down-regulation of MMP-1. Treatment with skullcapflavone II also reduced the TNF-α-induced phosphorylation of NF-κB p65 (Figure. 5B). Although stimulation of cells with TNF-α in serum-free medium for 24 h did not lead to a significant increase in phosphorylation of ERK1/2 and JNK, skullcapflavone II reduced the phosphorylation of ERK1/2 with or without TNF-α stimulation (Figure 5B). Moreover, phosphorylation of ERK1/2 and JNK was increased by stimulation with TNF-α for 10 min in serum-free medium but significantly decreased by treatment with skullcapflavone II (Figure S2). These findings suggest that skullcapflavone II down-regulates TNF-α-induced MMP-1 expression by reducing the activation of NF-κB and AP-1.

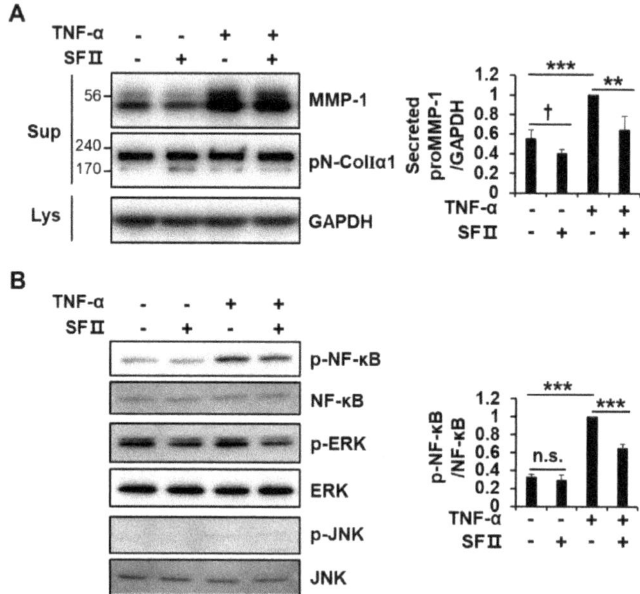

Figure 5. Effect of skullcapflavone II on tumor necrosis factor (TNF)-α-induced MMP-1 expression in foreskin fibroblasts. Foreskin fibroblasts were incubated for 24 h in a serum-free medium with (+) or without (-) 3 μM skullcapflavone II and with (+) or without (−) 1 ng/mL TNF-α. (**A**) The serum-free conditioned medium and cell lysates were analyzed by 9% SDS-PAGE and Western blot with anti-MMP-1, anti-pN-Collα1, and anti-GAPDH antibodies. (**B**) Cell lysates were analyzed by Western blotting using the indicated antibodies. ** $p < 0.01$ and *** $p < 0.001$ vs. the sample incubated with TNF-α alone; † $p < 0.05$ vs. the sample incubated without TNF-α and without skullcapflavone II; n.s., not significant.

2.5. Skullcapflavone II Decreased the Breakdown of Type I Collagen in 3D Culture of Foreskin Fibroblasts

MMP-1 is a central enzyme involved in the degradation of type I collagen. Since proMMP-1 secreted from fibroblasts in 2D culture is not activated into mature active form (Figure 1A), it is difficult to detect cleavage of type I collagen by MMP-1 in the medium. We therefore examined whether skullcapflavone II affects the degradation of type I collagen by down-regulating MMP-1 expression using 3D culture of foreskin fibroblasts. Consistent with two-dimensional culture conditions, skullcapflavone II down-regulated the TNF-α-induced secretion of MMP-1 in 3D culture (Figure 6A). Experiments using collagen type I cleavage-site antibody revealed that TNF-α stimulation increased the generation of cleaved 3/4 fragments of type I collagen. Interestingly, treatment with skullcapflavone II significantly decreased the amount of cleaved 3/4 fragments of type I collagen in 3D culture of foreskin fibroblasts (Figure 6B). These data demonstrate that skullcapflavone II inhibits collagenolysis via down-regulation of MMP-1 expression.

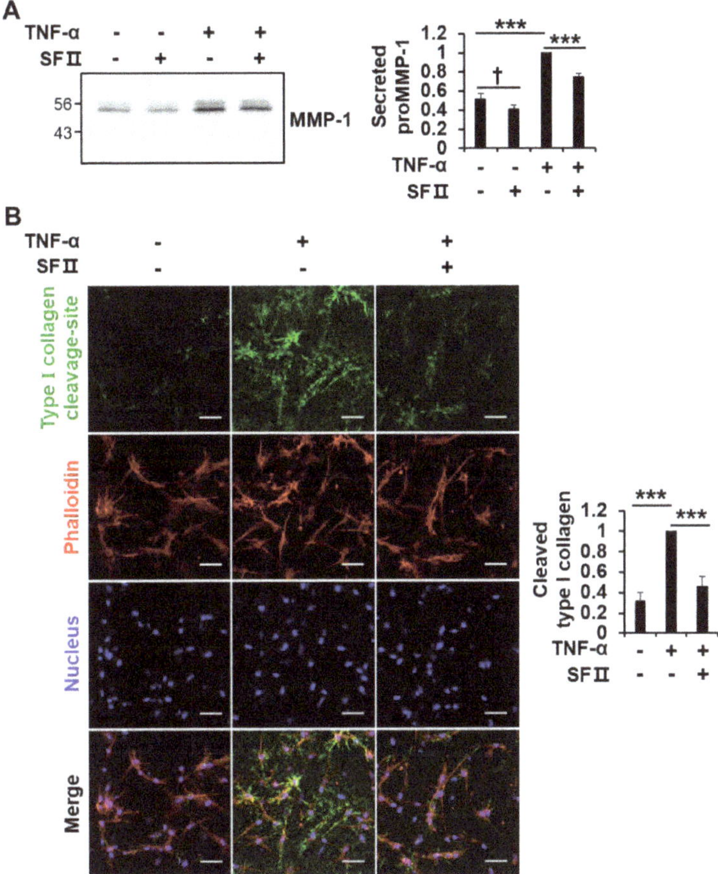

Figure 6. Effect of skullcapflavone II on TNF-α-induced type I collagen degradation in 3D culture of foreskin fibroblasts. Foreskin fibroblasts were embedded within a 3D type I collagen matrix. After polymerization for 1 h at 37 °C, cells embedded in the collagen matrix were incubated for 24 h in serum-free DMEM with (+) or without (-) 3 μM skullcapflavone II and with (+) or without (−) 1 ng/mL TNF-α. (**A**) The conditioned medium in the 3D culture was analyzed by 9% SDS-PAGE and Western blot with anti-MMP-1 antibody. *** $p < 0.001$ vs. the sample incubated with TNF-α alone; † $p < 0.05$ vs. the sample incubated without TNF-α and without skullcapflavone II (**B**) The 3D matrix containing foreskin fibroblasts was stained with anti-type I collagen cleavage-site antibody and Alexa Fluor® 488 goat anti-rabbit IgG (H+L), phalloidin–rhodamine and Hoechst 33258, and cells were analyzed by confocal fluorescence microscopy (200×). Scale bar, 50 μm. The quantification of cleaved type I collagen (Alexa Fluor®488) relative to nuclear staining (Hoechst 33258) obtained from six randomly chosen fields is shown as the mean ± S.D. *** $p < 0.001$ vs. the sample incubated with TNF-α alone.

3. Discussion

Roots of *Scutellaria baicalensis*, *S. litwinowii*, and *S. pinnatifida* are used as herbal medicines due to their anti-tumorigenic, anti-fibrotic, anti-inflammatory, and antioxidant effects [2–6,27,28]. Skullcapflavone II is one of the common components of *Scutellaria* roots, and it is considered a potential therapeutic compound for use in treating cancer, allergic asthma, and bone diseases. In this study, we examined whether skullcapflavone II regulates the expression of type I collagen and the enzyme MMP-1, which degrades type I collagen in fibroblasts.

We found that skullcapflavone II decreased the secretion of MMP-1 but had no effect on the secretion of type I collagen by foreskin fibroblasts and buttock dermal fibroblasts. Skullcapflavone II did not affect the proliferation of foreskin fibroblasts at concentrations up to 3 μM but inhibited proliferation at a concentration of 10 μM. However, at a concentration of 10 μM, skullcapflavone II did not induce apoptosis of foreskin fibroblasts. Our findings thus suggest that, at higher concentrations, skullcapflavone II reduces not only the secretion of MMP-1 by fibroblasts but also their proliferation.

We found that skullcapflavone II inhibits the expression of MMP-1 at the transcriptional level. Induction of MMP-1 transcription depends on activation of the transcription factors AP-1 and NF-κB. The *MMP-1* promoter has three AP-1 binding sites, at −70, −186, and −1602 bp [22,29], and an NF-κB binding site at −2886 bp [30,31], upstream of the transcriptional start site. AP-1 is composed of the polypeptides c-Jun and c-Fos. Activation of JNK subsequently leads to c-Jun phosphorylation, whereas activation of ERK induces c-Fos transactivation [32,33]. In our study, serum (i.e., FBS) stimulation induced tyrosine phosphorylation of cellular proteins and phosphorylation of ERK, JNK, p38, and c-Jun. Skullcapflavone II treatment decreased the serum-induced phosphorylation of cellular proteins, ERK, JNK, and c-Jun. These results demonstrate that skullcapflavone II reduces MMP-1 expression by suppressing AP-1 activation under serum-induced conditions.

In addition to inducing MMP-1 expression, AP-1 also up-regulates cell proliferation [32,33]. It is interesting to note that, in our experiments, skullcapflavone II reduced cell proliferation at higher concentrations but did not induce apoptosis. As we demonstrated here that skullcapflavone II suppresses the activation of AP-1, higher concentrations of skullcapflavone II could slow progression of the cell cycle without causing cell death.

MMP-1 expression is known to be up-regulated by a variety of inflammatory cytokines and growth factors, such as TNF-α, interleukins-1, -4, -5, -6, -8, and -10, fibroblast growth factors-1, -2, -7, and -9, epidermal growth factor (EGF), and platelet-derived growth factor [34,35]. Dermal fibroblasts are often subjected to inflammatory conditions associated with infections involving bacteria, viruses, or fungi or as a result of ultraviolet or ionized radiation exposure. As expected, treatment of fibroblasts with TNF-α to mimic an inflammatory state resulted in up-regulated secretion of MMP-1 and increased phosphorylation of NF-κB p65 as well as ERK1/2 and JNK. ERK and JNK were activated following short-term stimulation with TNF-α, and activation of NF-κB p65 was sustained with long-term TNF-α stimulation. Skullcapflavone II decreased TNF-α-induced MMP-1 secretion and also decreased phosphorylation of NF-κB p65, ERK1/2, and JNK. These results suggest that skullcapflavone II inhibits MMP-1 transactivation by suppressing both the AP-1 and NF-κB signaling pathways under pro-inflammatory conditions.

Skullcapflavone II also reportedly has antioxidant activity [5]. Reactive oxygen species (ROS) such as H_2O_2 and singlet oxygen ($^1O_2$) can activate NF-κB under a variety of circumstances [36–38]. For example, ROS activate IκB kinase by phosphorylation of its subunits on serine and tyrosine residues of the activation loops [39]. ROS also activate NF-κB through tyrosine phosphorylation of IκBα, without its degradation [40]. In addition, ROS can activate AP-1 under a variety of physiological conditions such as inflammation and tumorigenesis [41,42]. ROS activate protein tyrosine kinases by specific oxidation of cysteine SH groups, inducing an activating conformational change in the enzymes [43,44]. Thus, ROS induce autophosphorylation of receptor protein tyrosine kinases such as the EGF receptor [43] and tyrosine phosphorylation of downstream signaling proteins such as Src [45], thus enhancing activation of AP-1. We demonstrated that skullcapflavone II decreases cellular ROS generation (data not shown) and suppresses the TNF-α-induced activation of NF-κB and serum-induced tyrosine phosphorylation of cellular proteins and activation of ERK1/2 and JNK. Therefore, we hypothesized that the decreased activation of NF-κB and AP-1 by skullcapflavone II is mediated, at least in part, by its anti-oxidant activity.

Breakdown of collagen by MMP-1 is a critical process in the regulation of tissue remodeling, development, and morphogenesis [12,46]. Fibroblasts grown in 2D culture secrete the pro-form of MMP-1 which cannot cleave type I collagen. To characterize the effect of skullcapflavone II on the breakdown of type I collagen, we used an in vivo-mimicking 3D culture system of foreskin fibroblasts with an anti-type I collagen cleavage site antibody that enables detection of cleaved 3/4 fragments of type I collagen [47,48]. We found that skullcapflavone II decreased TNF-α-induced MMP-1 expression and degradation of type I collagen in 3D culture of foreskin fibroblasts. Therefore, we assume that proMMP-1 is at least partially processed into mature MMP-1 in a 3D culture system. Based on these results, we believe that skullcapflavone II could be useful in therapies aimed at maintaining the integrity of the extracellular matrix or for treating aging-induced and inflammation-related deterioration of the extracellular matrix.

Some natural flavonoids, such as quercetin, kaempferol, wogonin, apigenin, and luteolin, have also demonstrated efficacy at suppressing MMP-1 expression by reducing AP-1 activation [49,50]. However, relatively high concentrations (>10 μM) of these flavonoids are required to suppress MMP-1 expression in human dermal fibroblasts [49]. In our study, much lower concentrations (≤3 μM) of skullcapflavone II were sufficient to significantly down-regulate the expression of MMP-1. In addition, skullcapflavone II has methoxy (O–CH$_3$) groups at positions 6, 7, and 8 of the A ring and the 6′-position of the B ring, which is functionally important because it has been reported that polymethoxyflavones pass through the cell membrane and are readily transported via the blood circulation [51,52]. Therefore, we hypothesize that these O–CH$_3$ groups play an important role in the greater bioavailability of skullcapflavone II compared to other flavonoids.

In summary, this is the first study demonstrating the inhibitory effect of skullcapflavone II on the expression of MMP-1 and degradation of type I collagen in foreskin fibroblasts (Figure 7). We propose that skullcapflavone II would be a useful chemopreventive compound for treating physiological conditions associated with up-regulation of MMP-1 and the loss of extracellular matrix integrity, such as skin aging.

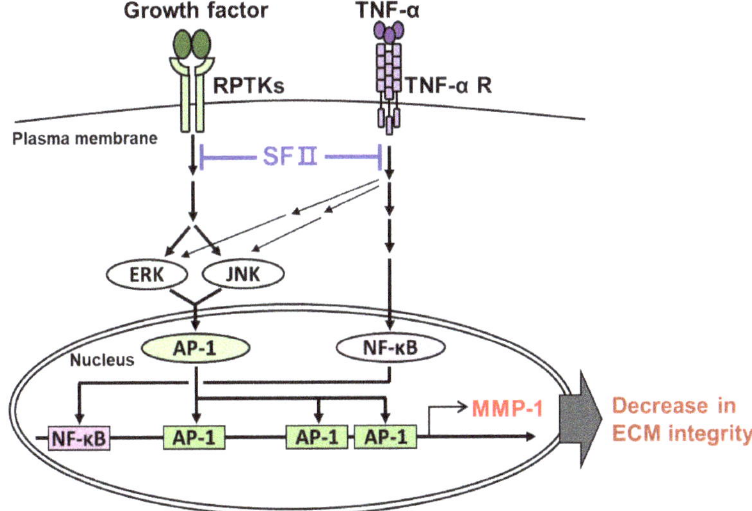

Figure 7. A proposed model describing the role of skullcapflavone II in collagenolysis inhibition. Growth factors in serum activate the transcription factor activator protein-1 (AP-1) through activation of the ERK and JNK pathways. TNF-α activates the transcription factor NF-κB as well as AP-1. Skullcapflavone II inhibits serum- and TNF-α-induced activation of AP-1 and NF-κB, which are required for MMP-1 expression. Skullcapflavone II thus maintains the integrity of extracellular matrix by suppressing MMP-1 expression.

4. Materials and Methods

4.1. Reagents and Antibodies

Skullcapflavone II (5-hydroxy-2-[2-hydroxy-6-methoxyphenyl]-6,7,8-trimethoxychromen-4-one) was purchased from ChemFaces (Wuhan, Hubei, China). Antibodies against phospho-ERK, ERK2, p38, and NF-κB p65 were purchased from Santa Cruz Biotechnology (Santa Cruz, CA, USA). Antibodies against phospho-NF-κB p65 (Ser536), phospho-JNK, JNK, phospho-p38, phospho-c-Jun, and c-Jun were purchased from Cell Signaling Technology (Danvers, MA, USA). Anti-phospho-tyrosine antibody (clone 4G10) was purchased from Millipore (Billerica, MA, USA). Anti-collagen type I cleavage-site antibody was purchased from ImmunoGlobe (Himmelstadt, Germany). Anti-GAPDH antibody was purchased from AbClone (Seoul, Korea). Horseradish peroxidase-conjugated goat anti-mouse IgG and rabbit IgG were obtained from KOMA Biotech (Seoul, Korea). Anti-MMP-1 and pro-collagen α1(I) N-propeptide (pN-Collα1) antibodies were a gift from Dr. Chung, J. H. (Seoul National University College of Medicine, Republic of Korea) [53]. Alexa Fluor® 488 goat anti-rabbit IgG (H+L) and rhodamine-conjugated phalloidin were purchased from Thermo Fisher Scientific (Waltham, MA, USA).

4.2. Cloning of the Human MMP-1 Promoter in a Reporter Plasmid

To generate a reporter construct of the human *MMP-1* promoter (GenBank Accession No. NM_000011.10), a 1938-bp DNA fragment including the promoter of the human *MMP-1* gene (−1880 to +40) was PCR-amplified using genomic DNA from human dermal fibroblasts as a template, PrimeSTAR® GXL DNA polymerase (TaKaRa, Shiga-ken, Japan), and the primer pair 5′-GAA *GCTAGC*TCCCTCACAGTCGAGTATATCTGCCAC-3′, which includes a *Nhe*I site (italicized) and 5′-GAA*AAGCTT*GCAAGGTAAGTGATGGCTTCCCAG-3′, which includes a *Hind*III site (italicized). The PCR product cleaved with *Nhe*I and *Hind*III was cloned into the pGL3-Basic luciferase reporter (Promega, Madison, WI, USA) that was digested with *Nhe*I and *Hind*III to generate the pGL3-*MMP-1* promoter.

4.3. Cell Culture

A primary culture of human foreskin fibroblasts was obtained from Welskin (Seoul, Korea). A primary culture of human dermal fibroblasts derived from a buttock skin of a young individual was provided by Chung, J. H. (Seoul National University College of Medicine, Seoul, Korea). Cells were maintained sub-confluently in DMEM (Gibco/Thermo Fisher Scientific, Waltham, MA, USA) supplemented with 10% FBS (Gibco/Thermo Fisher Scientific), 100 U/mL penicillin, and 100 µg/mL streptomycin at 37 °C in an atmosphere of 5% CO_2 and 95% air. Cells were plated on culture dishes and incubated overnight for attachments. The passage numbers for foreskin fibroblasts or buttock dermal fibroblasts were 13–18 or 8–13, respectively.

4.4. RNA Isolation and Reverse Transcription (RT)-PCR Analysis

Total RNA was isolated from foreskin fibroblasts using TRIzol reagent (Invitrogen, Carlsbad, CA, USA) as described previously [54]. cDNA was synthesized from total RNA (2 µg) using AMV RT system (Promega, Madison, WI, USA) and oligo $(dT)_{15}$ primer. Real-time PCR was conducted using a QuantiTect SYBR Green PCR kit (Qiagen, Hilden, Germany) and the QuantStudio 3 Real-Time PCR system (Applied Biosystems, Foster City, CA, USA). Primer sequences and annealing temperatures are described in Table S1.

4.5. Preparation of Conditioned Media and Cell Lysates, and Western Blot Analysis

Sub-confluent cells were incubated with serum-free medium for 24 h, and the resulting conditioned medium was collected by centrifugation at 2000× *g* for 3 min. Cell pellets were lysed with 1× SDS sample buffer for analysis of GAPDH or radio-immunoprecipitation assay (RIPA) lysis buffer (50 mM

Tris-HCl, pH7.4, 150 mM NaCl, 1% NP-40, 0.5% sodium deoxycholate, and 0.1% SDS) containing 1 mM NaF, 1 mM NA$_3$VO$_4$, and a SIGMAFAST^{TM} protease inhibitor tablet (Sigma-Aldrich, St. Louis, MO, USA) for analysis of signaling proteins. Western blot analysis was performed as described previously [55,56]. The MultiGauge software (Fujifilm, Tokyo, Japan) was used to quantify the band intensities.

4.6. Cell Growth Assay

Cell growth was analyzed as previously described [57]. Foreskin fibroblasts (0.5×10^4 cells/well) were plated in 96-well plates and incubated in medium supplemented with 10% FBS and various concentrations of skullcapflavone II for up to 2 days. Viable cells were stained with MTT, solubilized with dimethyl sulfoxide (DMSO), and the absorbance was measured at 565 nm using a microplate reader (Molecular Devices, San Jose, CA, USA).

4.7. Flow Cytometry

Foreskin fibroblasts (0.5×10^4 cells/well) were seeded in 6-well plates and incubated in DMEM supplemented with 10% FBS and various concentrations of skullcapflavone II for 24 h. Briefly, cells were washed twice with cold PBS and then resuspended in binding buffer (10 mM HEPES, pH 7.4, 140 mM NaCl, and 2.5 mM CaCl$_2$) to a final density of 1×10^6 cells/mL. A total volume of 100 µL of detached cells, including 5 µL of FITC annexin V (BD Biosciences, Bedford, MA, USA) and 5 µL of 50 µg/mL 7-aminoactinomycin D (7-AAD, Invitrogen), was incubated for 15 min at room temperature. Apoptotic cells were then analyzed by flow cytometry (BD FACSCalibur, BD Biosciences, Bedford, MA, USA).

4.8. Dual-Luciferase Reporter Assay

Transfection of reporter genes into foreskin fibroblasts was conducted using Lipofectamine LTX (Thermo Fisher Scientific). Foreskin fibroblasts (5×10^4 cells/well) were seeded in 24-well plates, and the medium was replaced with fresh DMEM supplemented with 10% FBS. Promoterless pGL3-Basic (0.36 µg) or pGL3-*MMP1* promoter (0.5 µg) encoding firefly luciferase driven by *MMP-1* promoter and pRL-TK (0.05 µg, Promega, Madison, WI, USA) encoding *Renilla* luciferase driven by herpes simplex virus thymidine kinase promoter in 25 µL Opti-MEM were incubated with Lipofectamine LTX (0.75 µL, Invitrogen, Carlsbad, CA, USA) and PLUS^{TM} reagent (0.25 µL, Invitrogen, Carlsbad, CA, USA) in 25 µL Opti-MEM (Gibco/Thermo Fisher Scientific, Waltham, MA, USA) for 20 min at room temperature. Cells were treated with this mixture for 5 h and then incubated with DMEM supplemented with 10% FBS for 24 h. The cells were then treated with skullcapflavone II or DMSO in serum-free medium for 12 h. Luciferase activity was measured using the dual-luciferase reporter assay system (Promega), and the firefly luciferase activity in transfected cells was normalized to the *Renilla* luciferase activity.

4.9. Collagenolysis in 3D Culture, Confocal Fluorescence Microscopy, and Image Acquisition

Foreskin fibroblasts (5×10^5 cells/mL) were trypsinized and resuspended in 2.8 mg/mL of rat tail collagen I solution (Corning Inc., Corning, NY, USA):5× DMEM:10× reconstitution buffer (260 mM NaHCO$_3$, 200 mM HEPES, and 50 mM NaOH) = 7:2:1, and then 0.15 mL of the cell mixture was placed in a glass-bottom (35 mm × 10 mm, hole 13 φ) dish (SPL Life Sciences, Gyeonggi-do, Korea). After solidifying for 1 h at 37 °C, 2 mL of phenol red-free DMEM (Hyclone, South Logan, UT, USA) with or without TNF-α and/or skullcapflavone II was added, and collagen-embedded cells were incubated for 24 h at 37 °C in an atmosphere of 5% CO$_2$ and 95% air. For nucleus staining, cells were incubated with Hoechst 33258 (2 µg/mL) for 30 min and then fixed in 3.7% paraformaldehyde for 30 min, permeabilized in 0.2% Triton-X 100 for 10 min, blocked in 3% bovine serum albumin for 30 min, and immunostained overnight at 4 °C with rabbit anti-collagen type I cleavage-site antibody (2.5 µg/mL). The cells were then washed with PBS and incubated with Alexa Fluor$^{®}$ 488 goat anti-rabbit IgG (H+L) (Invitrogen, Carlsbad, CA, USA) and phalloidin-rhodamine (1 U/mL). Images were obtained on a confocal microscope (LSM700; Carl Zeiss, Feldbach, Switzerland) with 20× Plan-Apochromat objective lens and Zen software (Carl Zeiss, Oberkochen, Germany). The excitation wavelengths were 405 nm for Hoechst 33258, 488 nm for Alexa Fluor® 488, and 555 nm for rhodamine. To avoid bias during image acquisition, all images were obtained from randomly selected fields and acquired using the same parameters including exposure time, laser power, and offset settings. The intensity of fluorescence was determined using Image J software (National Institutes of Health, Bethesda, MD, USA).

4.10. Statistical Analyses

All data are expressed as the mean ± S.D. of at least three independent experiments. Statistical significance was analyzed using the paired two-tailed Student's *t*-test, except for the fluorescence image analysis using the unpaired two-tailed Student's *t*-test. A *p*-value <0.05 was considered indicative of statistical significance.

5. Conclusions

We were able to show that skullcapflavone II inhibits the expression of MMP-1 and the degradation of type I collagen in foreskin fibroblasts. Skullcapflavone II suppresses transcription of MMP-1 mRNA through reduced activation of AP-1 and NF-κB. Skullcapflavone II also reduces type I collagen degradation in the 3D fibroblast culture. We propose that skullcapflavone II would be a useful chemopreventive compound for the treatment of physiological conditions associated with the up-regulation of MMP-1 and the loss of extracellular matrix integrity, such as skin aging.

Supplementary Materials: Supplementary materials can be found at http://www.mdpi.com/1422-0067/20/11/2734/s1.

Author Contributions: Y.H.L. and S.-T.L. designed the study, analyzed data, and wrote the manuscript. Y.H.L. performed the experiments. E.K.S. provided natural compounds from medicinal herbs. All authors reviewed and approved the manuscript.

Funding: This work was supported by grants from the National Research Foundation of Korea (no. 2014M3C9A2064597, 2016R1A2B4007904, and 2019R111A2A01061685 to Seung-Taek Lee). Young Hun Lee is a recipient of the 2017 Research Scholarship from the Graduate School of Yonsei University.

Conflicts of Interest: The authors have no conflicts of interest to declare.

Abbreviations

3D	three-dimensional
7-AAD	7-aminoactinomycin D
AP-1	activator protein-1
cAMP	cyclic cyclic adenosine monophosphate
DMEM	Dulbecco's Modified Eagle's medium
DMSO	dimethyl sulfoxide
EGF	epidermal growth factor
ERK	extracellular signal-regulated kinase
FBS	fetal bovine serum
FITC	fluorescein isothiocyanate
GAPDH	glyceraldehyde 3-phosphate dehydrogenase
JNK	c-Jun N-terminal kinase
LDLR	low-density-lipoprotein receptor
MAPK	mitogen-activated protein kinase
MMP	matrix metalloproteinase
MTT	3-(4,5-dimethyl thiazol-2-yl)-2,5-diphenyltetrazolium bromide
NF-κB	nuclear factor kappa light chain enhancer of activated B cells
pN-Col1α1	pro-collagen α1(I) N-propeptide
RIPA	radio-immunoprecipitation assay
ROS	reactive oxygen species
TNF	tumor necrosis factor

References

1. Kimura, Y.; Okuda, H.; Ogita, Z. Effects of flavonoids isolated from *Scutellariae* radix on fibrinolytic system induced by trypsin in human umbilical vein endothelial cells. *J. Nat. Prod.* **1997**, *60*, 598–601. [CrossRef]
2. Tayarani-Najarani, Z.; Asili, J.; Parsaee, H.; Mousavi, S.H.; Mashhadian, N.V.; Mirzaee, A.; Emami, S.A. Wogonin and neobaicalein from *Scutellaria litwinowii* roots are apoptotic for HeLa cells. *Rev. Bras. Farmacogn.* **2012**, *22*, 268–276. [CrossRef]
3. Boozari, M.; Mohammadi, A.; Asili, J.; Emami, S.A.; Tayarani-Najaran, Z. Growth inhibition and apoptosis induction by *Scutellaria pinnatifida* A. Ham. on HL-60 and K562 leukemic cell lines. *Environ. Toxicol. Pharmacol.* **2015**, *39*, 307–312. [CrossRef] [PubMed]
4. Jang, H.Y.; Ahn, K.S.; Park, M.J.; Kwon, O.K.; Lee, H.K.; Oh, S.R. Skullcapflavone II inhibits ovalbumin-induced airway inflammation in a mouse model of asthma. *Int. Immunopharmacol.* **2012**, *12*, 666–674. [CrossRef] [PubMed]
5. Lee, J.; Son, H.S.; Lee, H.I.; Lee, G.R.; Jo, Y.J.; Hong, S.E.; Kim, N.; Kwon, M.; Kim, N.Y.; Kim, H.J.; et al. Skullcapflavone II inhibits osteoclastogenesis by regulating reactive oxygen species and attenuates the survival and resorption function of osteoclasts by modulating integrin signaling. *FASEB J.* **2019**, *33*, 2026–2036. [CrossRef] [PubMed]
6. Bonham, M.; Posakony, J.; Coleman, I.; Montgomery, B.; Simon, J.; Nelson, P.S. Characterization of chemical constituents in *Scutellaria baicalensis* with antiandrogenic and growth-inhibitory activities toward prostate carcinoma. *Clin. Cancer Res.* **2005**, *11*, 3905–3914. [CrossRef] [PubMed]
7. Nhoek, P.; Chae, H.S.; Masagalli, J.N.; Mailar, K.; Pel, P.; Kim, Y.M.; Choi, W.J.; Chin, Y.W. Discovery of flavonoids from *Scutellaria baicalensis* with inhibitory activity against PCSK 9 expression: Isolation, synthesis and their biological evaluation. *Molecules* **2018**, *23*, 504. [CrossRef]
8. Kafienah, W.; Buttle, D.J.; Burnett, D.; Hollander, A.P. Cleavage of native type I collagen by human neutrophil elastase. *Biochem. J.* **1998**, *330*, 897–902. [CrossRef] [PubMed]
9. Gelse, K.; Poschl, E.; Aigner, T. Collagens-structure, function, and biosynthesis. *Adv. Drug. Deliv. Rev.* **2003**, *55*, 1531–1546. [CrossRef] [PubMed]
10. Chang, S.W.; Buehler, M.J. Molecular biomechanics of collagen molecules. *Materials Today* **2014**, *17*, 70–76. [CrossRef]
11. Ricard-Blum, S. The collagen family. *Cold Spring Harb. Perspect. Biol.* **2011**, *3*, a004978. [CrossRef] [PubMed]
12. Cole, M.A.; Quan, T.; Voorhees, J.J.; Fisher, G.J. Extracellular matrix regulation of fibroblast function: redefining our perspective on skin aging. *J. Cell Commun. Signal* **2018**, *12*, 35–43. [CrossRef] [PubMed]
13. Nagase, H.; Visse, R.; Murphy, G. Structure and function of matrix metalloproteinases and TIMPs. *Cardiovasc. Res.* **2006**, *69*, 562–573. [CrossRef] [PubMed]
14. Fanjul-Fernandez, M.; Folgueras, A.R.; Cabrera, S.; Lopez-Otin, C. Matrix metalloproteinases: Evolution, gene regulation and functional analysis in mouse models. *Biochim. Biophys. Acta* **2010**, *1803*, 3–19. [CrossRef] [PubMed]
15. Song, F.; Wisithphrom, K.; Zhou, J.; Windsor, L.J. Matrix metalloproteinase dependent and independent collagen degradation. *Front. Biosci.* **2006**, *11*, 3100–3120. [CrossRef] [PubMed]
16. Kang, S.; Cho, S.; Chung, J.H.; Hammerberg, C.; Fisher, G.J.; Voorhees, J.J. Inflammation and extracellular matrix degradation mediated by activated transcription factors nuclear factor-kappaB and activator protein-1 in inflammatory acne lesions in vivo. *Am. J. Pathol.* **2005**, *166*, 1691–1699. [CrossRef]
17. Ivarsson, M.; McWhirter, A.; Borg, T.K.; Rubin, K. Type I collagen synthesis in cultured human fibroblasts: Regulation by cell spreading, platelet-derived growth factor and interactions with collagen fibers. *Matrix Biol.* **1998**, *16*, 409–425. [CrossRef]
18. Lindner, D.; Zietsch, C.; Becher, P.M.; Schulze, K.; Schultheiss, H.P.; Tschope, C.; Westermann, D. Differential expression of matrix metalloproteases in human fibroblasts with different origins. *Biochem. Res. Int.* **2012**, *2012*, 875742. [CrossRef] [PubMed]
19. Du, G.; Liu, C.; Li, X.; Chen, W.; He, R.; Wang, X.; Feng, P.; Lan, W. Induction of matrix metalloproteinase-1 by tumor necrosis factor-alpha is mediated by interleukin-6 in cultured fibroblasts of keratoconus. *Exp. Biol. Med. (Maywood)* **2016**, *241*, 2033–2041. [CrossRef]

20. Wilhelm, S.M.; Eisen, A.Z.; Teter, M.; Clark, S.D.; Kronberger, A.; Goldberg, G. Human fibroblast collagenase: Glycosylation and tissue-specific levels of enzyme synthesis. *Proc. Natl. Acad. Sci. USA* **1986**, *83*, 3756–3760. [CrossRef]

21. Cortez, D.M.; Feldman, M.D.; Mummidi, S.; Valente, A.J.; Steffensen, B.; Vincenti, M.; Barnes, J.L.; Chandrasekar, B. IL-17 stimulates MMP-1 expression in primary human cardiac fibroblasts via p38 MAPK- and ERK1/2-dependent C/EBP-beta, NF-kappaB, and AP-1 activation. *Am. J. Physiol. Heart Circ. Physiol.* **2007**, *293*, H3356–H3365. [CrossRef] [PubMed]

22. Vincenti, M.P.; Brinckerhoff, C.E. Transcriptional regulation of collagenase (MMP-1, MMP-13) genes in arthritis: Integration of complex signaling pathways for the recruitment of gene-specific transcription factors. *Arthritis Res.* **2002**, *4*, 157–164. [CrossRef] [PubMed]

23. Hozawa, S.; Nakamura, T.; Nakano, M.; Adachi, M.; Tanaka, H.; Takahashi, Y.; Tetsuya, M.; Miyata, N.; Soma, H.; Hibi, T. Induction of matrix metalloproteinase-1 gene transcription by tumour necrosis factor alpha via the p50/p50 homodimer of nuclear factor-kappa B in activated human hepatic stellate cells. *Liver Int.* **2008**, *28*, 1418–1425. [CrossRef] [PubMed]

24. Lin, S.K.; Wang, C.C.; Huang, S.; Lee, J.J.; Chiang, C.P.; Lan, W.H.; Hong, C.Y. Induction of dental pulp fibroblast matrix metalloproteinase-1 and tissue inhibitor of metalloproteinase-1 gene expression by interleukin-1alpha and tumor necrosis factor-alpha through a prostaglandin-dependent pathway. *J. Endod.* **2001**, *27*, 185–189. [CrossRef] [PubMed]

25. Bae, S.; Jung, Y.; Choi, Y.M.; Li, S. Effects of Er-Miao-San extracts on TNF-alpha-induced MMP-1 expression in human dermal fibroblasts. *Biol. Res.* **2015**, *48*, 8. [CrossRef] [PubMed]

26. Ghosh, S.; May, M.J.; Kopp, E.B. NF-kappa B and Rel proteins: Evolutionarily conserved mediators of immune responses. *Annu. Rev. Immunol.* **1998**, *16*, 225–260. [CrossRef] [PubMed]

27. Delazar, A.; Nazemiyeh, H.; Afshar, F.H.; Barghi, N.; Esnaashari, S.; Asgharian, P. Chemical compositions and biological activities of *Scutellaria pinnatifida* A. Hamilt aerial parts. *Res. Pharm. Sci.* **2017**, *12*, 187–195. [PubMed]

28. Shang, X.; He, X.; He, X.; Li, M.; Zhang, R.; Fan, P.; Zhang, Q.; Jia, Z. The genus *Scutellaria* an ethnopharmacological and phytochemical review. *J. Ethnopharmacol.* **2010**, *128*, 279–313. [CrossRef]

29. Benbow, U.; Brinckerhoff, C.E. The AP-1 site and MMP gene regulation: What is all the fuss about? *Matrix Biol.* **1997**, *15*, 519–526. [CrossRef]

30. Wang, Y.P.; Liu, I.J.; Chiang, C.P.; Wu, H.C. Astrocyte elevated gene-1 is associated with metastasis in head and neck squamous cell carcinoma through p65 phosphorylation and upregulation of MMP1. *Mol. Cancer* **2013**, *12*, 109. [CrossRef]

31. Rowan, A.D.; Young, D.A. Collagenase gene regulation by pro-inflammatory cytokines in cartilage. *Front. Biosci.* **2007**, *12*, 536–550. [CrossRef] [PubMed]

32. Shaulian, E.; Karin, M. AP-1 in cell proliferation and survival. *Oncogene* **2001**, *20*, 2390–2400. [CrossRef]

33. Trop-Steinberg, S.; Azar, Y. AP-1 expression and its clinical relevance in immune eisorders and cancer. *Am. J. Med. Sci.* **2017**, *353*, 474–483. [CrossRef]

34. Pardo, A.; Selman, M. MMP-1: The elder of the family. *Int. J. Biochem. Cell Biol.* **2005**, *37*, 283–288. [CrossRef] [PubMed]

35. Barchowsky, A.; Frleta, D.; Vincenti, M.P. Integration of the NF-kappaB and mitogen-activated protein kinase/AP-1 pathways at the collagenase-1 promoter: Divergence of IL-1 and TNF-dependent signal transduction in rabbit primary synovial fibroblasts. *Cytokine* **2000**, *12*, 1469–1479. [CrossRef] [PubMed]

36. Gloire, G.; Legrand-Poels, S.; Piette, J. NF-kappaB activation by reactive oxygen species: Fifteen years later. *Biochem. Pharmacol.* **2006**, *72*, 1493–1505. [CrossRef] [PubMed]

37. Volanti, C.; Matroule, J.Y.; Piette, J. Involvement of oxidative stress in NF-kappaB activation in endothelial cells treated by photodynamic therapy. *Photochem. Photobiol.* **2002**, *75*, 36–45. [CrossRef]

38. Schreck, R.; Rieber, P.; Baeuerle, P.A. Reactive oxygen intermediates as apparently widely used messengers in the activation of the NF-kappa B transcription factor and HIV-1. *EMBO J.* **1991**, *10*, 2247–2258. [CrossRef]

39. Kamata, H.; Manabe, T.; Oka, S.; Kamata, K.; Hirata, H. Hydrogen peroxide activates IkappaB kinases through phosphorylation of serine residues in the activation loops. *FEBS Lett.* **2002**, *519*, 231–237. [CrossRef]

40. Takada, Y.; Mukhopadhyay, A.; Kundu, G.C.; Mahabeleshwar, G.H.; Singh, S.; Aggarwal, B.B. Hydrogen peroxide activates NF-kappa B through tyrosine phosphorylation of I kappa B alpha and serine phosphorylation of p65: Evidence for the involvement of I kappa B alpha kinase and Syk protein-tyrosine kinase. *J. Biol. Chem.* **2003**, *278*, 24233–24241. [CrossRef]

41. Morgan, M.J.; Liu, Z.G. Crosstalk of reactive oxygen species and NF-kappaB signaling. *Cell Res.* **2011**, *21*, 103–115. [CrossRef]

42. Qin, Z.; Robichaud, P.; He, T.; Fisher, G.J.; Voorhees, J.J.; Quan, T. Oxidant exposure induces cysteine-rich protein 61 (CCN1) via c-Jun/AP-1 to reduce collagen expression in human dermal fibroblasts. *PLoS ONE* **2014**, *9*, e115402. [CrossRef]

43. Knebel, A.; Rahmsdorf, H.J.; Ullrich, A.; Herrlich, P. Dephosphorylation of receptor tyrosine kinases as target of regulation by radiation, oxidants or alkylating agents. *EMBO J.* **1996**, *15*, 5314–5325. [CrossRef]

44. Giannoni, E.; Buricchi, F.; Raugei, G.; Ramponi, G.; Chiarugi, P. Intracellular reactive oxygen species activate Src tyrosine kinase during cell adhesion and anchorage-dependent cell growth. *Mol. Cell. Biol.* **2005**, *25*, 6391–6403. [CrossRef]

45. Natarajan, V.; Scribner, W.M.; al-Hassani, M.; Vepa, S. Reactive oxygen species signaling through regulation of protein tyrosine phosphorylation in endothelial cells. *Environ. Health Perspect.* **1998**, *106*, 1205–1212.

46. Amar, S.; Smith, L.; Fields, G.B. Matrix metalloproteinase collagenolysis in health and disease. *Biochim. Biophys. Acta. Mol. Cell Res.* **2017**, *1864*, 1940–1951. [CrossRef]

47. Orgaz, J.L.; Pandya, P.; Dalmeida, R.; Karagiannis, P.; Sanchez-Laorden, B.; Viros, A.; Albrengues, J.; Nestle, F.O.; Ridley, A.J.; Gaggioli, C.; et al. Diverse matrix metalloproteinase functions regulate cancer amoeboid migration. *Nat. Commun.* **2014**, *5*, 4255. [CrossRef]

48. Lagoutte, E.; Villeneuve, C.; Lafanechere, L.; Wells, C.M.; Jones, G.E.; Chavrier, P.; Rosse, C. LIMK regulates tumor-cell invasion and matrix degradation through tyrosine phosphorylation of MT1-MMP. *Sci. Rep.* **2016**, *6*, 24925. [CrossRef]

49. Lim, H.; Kim, H.P. Inhibition of mammalian collagenase, matrix metalloproteinase-1, by naturally-occurring flavonoids. *Planta Med.* **2007**, *73*, 1267–1274. [CrossRef]

50. Hwang, Y.P.; Oh, K.N.; Yun, H.J.; Jeong, H.G. The flavonoids apigenin and luteolin suppress ultraviolet A-induced matrix metalloproteinase-1 expression via MAPKs and AP-1-dependent signaling in HaCaT cells. *J. Dermatol. Sci.* **2011**, *61*, 23–31. [CrossRef]

51. Gan, L.S.; Hsyu, P.H.; Pritchard, J.F.; Thakker, D. Mechanism of intestinal absorption of ranitidine and ondansetron: Transport across Caco-2 cell monolayers. *Pharm. Res.* **1993**, *10*, 1722–1725. [CrossRef]

52. Li, S.M.; Pan, M.H.; Lo, C.Y.; Tan, D.; Wang, Y.; Shahidi, F.; Ho, C.T. Chemistry and health effects of polymethoxyflavones and hydroxylated polymethoxyflavones. *J. Funct. Foods* **2009**, *1*, 2–12. [CrossRef]

53. Kim, E.J.; Kim, Y.K.; Kim, M.K.; Kim, S.; Kim, J.Y.; Lee, D.H.; Chung, J.H. UV-induced inhibition of adipokine production in subcutaneous fat aggravates dermal matrix degradation in human skin. *Sci. Rep.* **2016**, *6*, 25616. [CrossRef]

54. Shin, W.S.; Hong, Y.; Lee, H.W.; Lee, S.T. Catalytically defective receptor protein tyrosine kinase PTK7 enhances invasive phenotype by inducing MMP-9 through activation of AP-1 and NF-kappaB in esophageal squamous cell carcinoma cells. *Oncotarget* **2016**, *7*, 73242–73256. [CrossRef]

55. Shin, W.S.; Na, H.W.; Lee, S.T. Biphasic effect of PTK7 on KDR activity in endothelial cells and angiogenesis. *Biochim. Biophys. Acta* **2015**, *1853*, 2251–2260. [CrossRef]

56. Shin, W.S.; Gim, J.; Won, S.; Lee, S.T. Biphasic regulation of tumorigenesis by PTK7 expression level in esophageal squamous cell carcinoma. *Sci. Rep.* **2018**, *8*, 8519. [CrossRef]

57. Lee, Y.H.; Park, J.H.; Cheon, D.H.; Kim, T.; Park, Y.E.; Oh, E.S.; Lee, J.E.; Lee, S.T. Processing of syndecan-2 by matrix metalloproteinase-14 and effect of its cleavage on VEGF-induced tube formation of HUVECs. *Biochem. J.* **2017**, *474*, 3719–3732. [CrossRef]

International Journal of
Molecular Sciences

Review

The Anti-inflammatory Effects of Dietary Anthocyanins against Ulcerative Colitis

Shiyu Li [1], Binning Wu [1,2], Wenyi Fu [1] and Lavanya Reddivari [1,*]

[1]   Department of Food Science, Purdue University, 745 Agriculture Mall Drive, West Lafayette, IN 47907, USA; li3291@purdue.edu (S.L.); wu1515@purdue.edu (B.W.); fu205@purdue.edu (W.F.)
[2]   Department of Plant Science, Penn State University, University Park, PA 16802, USA
*   Correspondence: lreddiva@purdue.edu; Tel.: +1-(765)496-6102; Fax: +1-(765)94-7953

Received: 3 May 2019; Accepted: 23 May 2019; Published: 27 May 2019

Abstract: Ulcerative colitis (UC), which is a major form of inflammatory bowel disease (IBD), is a chronic relapsing disorder of the gastrointestinal tract affecting millions of people worldwide. Alternative natural therapies, including dietary changes, are being investigated to manage or treat UC since current treatment options have serious negative side effects. There is growing evidence from animal studies and human clinical trials that diets rich in anthocyanins, which are pigments in fruits and vegetables, protect against inflammation and increased gut permeability as well as improve colon health through their ability to alter bacterial metabolism and the microbial milieu within the intestines. In this review, the structure and bioactivity of anthocyanins, the role of inflammation and gut bacterial dysbiosis in UC pathogenesis, and their regulation by the dietary anthocyanins are discussed, which suggests the feasibility of dietary strategies for UC mitigation.

Keywords: anthocyanins; anti-inflammatory; colitis; colonic inflammation

1. Anthocyanins

Anthocyanins, which is a clan of flavonoids, are water-soluble polyphenolic pigments that are responsible for the pigmentation of anthocyanin-rich foods including fruits (black plums, blackberries, blueberries, and grapes), vegetables (black plums, blackberries, blueberries, and grapes), and grains (black rice, red rice, and black soybeans) [1–5]. Different crops vary in the composition and the content of anthocyanins ranging from 0.1% to 1.0% [6,7]. Additionally, oxidation, enzymolysis, and environmental factors such as temperature, light, and pH can alter anthocyanin levels [8]. Previous studies showed that malonylation enhanced the stability of anthocyanins in water [9]. Most of the anthocyanins exert better stability under acidic conditions while high pH leads to anthocyanin degradation [10,11]. pH-dependent reversible structure transformation occurs between the following forms: flavylium cation (red), quinonoidal base (blue), carbinol pseudobase (colorless), and chalcone (colorless) [12] in aqueous solution [13]. In plants, anthocyanins aid in pollination and anthocyanin pigments can serve as natural food colorants [11,14].

Anthocyanins are naturally present in plants as glycosides carrying glucose, galactose, arabinose, rhamnose, and xylose [15]. Deglycosylated anthocyanins known as anthocyanidins are unstable and rarely found in nature [16]. The instability of anthocyanidins is due to the presence of flavylium ion and its peculiar electron distribution [17]. To date, a total of 27 aglycones and over 700 anthocyanins have been identified based on their chemical structures [1,18]. Anthocyanins share a basic C-6 (A ring)-C-3 (C ring)-C-6 (B ring) carbon skeleton (Figure 1) with a varying number of hydroxyl groups and sugars with different degrees of methylation [19]. Approximately 665 natural anthocyanins are derived from six commonly found anthocyanidins (Figure 2): cyanidin (Cy), peonidin (Pn), pelargonidin (Pg), malvidin (Mv), delphinidin (Dp), and petunidin (Pt) [13,20].

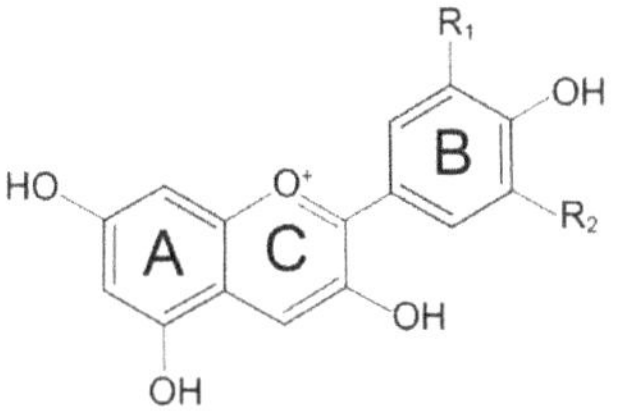

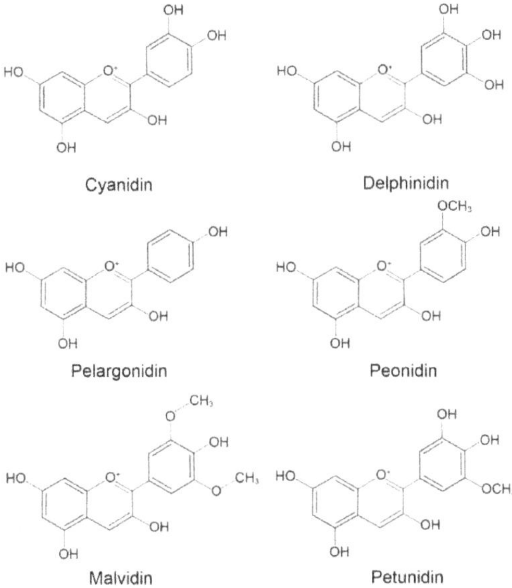

Figure 1. The basic structure of anthocyanin.

Figure 2. Structures of six major anthocyanidins.

Red-colored or blue-colored fruits, vegetables, and grains serve as sources of various anthocyanins. For example, 100 g kokum can provide 1000 to 2400 mg anthocyanins [21], 100 g strawberry contains 13-315 mg anthocyanins [22], and 100 g red wine grapes supply 30-750 mg anthocyanins [23]. As reported by Raul Zamora-Ros et al., daily consumption of anthocyanins varies depending on the region, weather condition, gender, and lifestyle [24]. Among all European regions that are investigated, Italy had the highest daily anthocyanin intake (~43.74 mg/day), with men consuming 49% more anthocyanins daily than women. The opposite pattern was observed in the UK, where daily anthocyanin intake of women is 21% higher than men [24]. The estimated anthocyanin daily intake in the US is about 11.6 mg/day [25].

1.1. Anthocyanin Bioavailability

The structure of anthocyanins is a key factor that determines their bioavailability and bioactivity. Bioavailability is defined as the rate and extent to which a compound is absorbed and utilized by the organism to perform multiple physiological effects [26]. Thus, the bioavailability has been considered as an essential index in evaluating the efficacy of bioactive compounds. Absorption is the main factor that influences the bioavailability of anthocyanins. The absorption rate varies depending on the molecular size, sugar moiety, and acylated groups. Moreover, the interference by other materials within the food matrix is also a considerable factor that affects the absorption. An *in vitro* study conducted by Yi et al. showed that anthocyanins with more free hydroxyl groups and fewer OCH_3 groups had lower bioavailability [27]. Anthocyanidin-glucosides exhibited higher bioavailability than anthocyanidin-galactosides, while non-acylated anthocyanins have better absorption than the acylated ones [28,29]. Studies also found that anthocyanins can be absorbed mainly in their intact glycosidic forms through the stomach and small intestine [19]. Anthocyanins were detected in the plasma within a few minutes after intake, which indicates the rapid absorption in the stomach [30]. Talavera et al. indicated that 19% to 37% of bilberry anthocyanins were absorbed by gastric fluid within 30 min [31]. An *in vivo* study showed that the highest absorption of anthocyanins occurred in the jejunum ($55.3 \pm 7.6\%$) whereas minor absorption occurred in the duodenum ($10.4 \pm 7.6\%$), which supports the role of the small intestine as a major site for anthocyanin absorption [32]. Unabsorbed anthocyanins travel down to the colon. However, both humans [33] and mice studies [34] demonstrated that most of the cyanidin-3-glucosides (C3G) that enter the large intestine was excreted in feces. Although anthocyanins display high absorption in the gastrointestinal tract, the bioavailability of anthocyanins is less than 1% [35–37]. Recent studies suggest that anthocyanins similar to other flavonoids are metabolized by colonic microbiota (Table 1) [38,39] and the metabolic function might be a direct result of metabolomic indicators rather than the bioavailability [40].

Table 1. Bacterial metabolites of major anthocyanidins.

Chemical Class	Bacteria	Major Metabolites	Reference
Cyanidin		Vanillic acid and protocatechuic acid	[41–43]
Peonidin		Vanillic acid and protocatechuic acid	[41,42]
Pelargonidin	*Lachnospiraceae, Bifidobacteria,* and *Lactobacillus.*	4-hydroxybenzoic acid, hydroxycinnamic acid, p-coumaric acid, ferulic acid, and caffeic acid	[41,42]
Malvidin		Syringic acid, gallic acid, and pyrogallol	[44]
Delphinidin		Gallic acid and syringic acid	[41,42,45]
Petunidin		Gallic acid	[42]

1.2. Anthocyanin and Human Health

Anthocyanins have been indicated to be a group of bioactive compounds with numerous health benefits because of their anti-inflammatory, anti-oxidant, anti-obesity, anti-angiogenesis, anti-cancer, anti-diabetes, anti-microbial, neuroprotection, and immunomodulation properties (Table 2) [9]. Studies demonstrated that anthocyanins exhibited a strong attenuating effect against colitis [46] and colon cancer [47]. The anti-angiogenic effect of anthocyanins has been proven on human esophageal and intestinal microvascular endothelial cells [48]. Significant evidence supports the preventive efficacy of anthocyanins against many neurodegenerative diseases such as Parkinson's disease and Alzheimer's disease [49]. Previous studies indicated that middle-aged and older-aged women with a high consumption of anthocyanin-rich foods exhibited 32% and 18% reduction in risk of myocardial infarction, respectively [50,51]. Additionally, human obesity prevention and blood glucose tolerance effects of anthocyanin have also been reported [52,53]. Anthocyanins have been shown to reduce oxidative stress either by scavenging reactive oxygen species or by inducing anti-oxidant enzymes. Anthocyanins in black currant skin induced the anti-oxidant enzymes and eased the oxidative stress through activation of the Nrf2 signaling pathway [54]. Moreover, oxidative stress can increase inflammation by enhanced pro-inflammatory gene expression and inflammation, which, in turn, can lead to oxidative stress (ref-curcumin review). Antioxidative effects of anthocyanins can contribute to the anti-inflammatory properties, but we will not be covering the anti-oxidative effects of anthocyanins. In this review, we will focus on the anti-inflammatory effects of anthocyanins against ulcerative colitis (UC).

Table 2. Sources of anthocyanins and their health benefits.

Chemical Class	Plant Source	Health Benefit	Reference
Cyanidin	Blueberries, bilberries, cranberries, elderberries, raspberry seeds, strawberries, purple corn, tea, purple carrot, purple rice	Anti-inflammatory and anti-cancer activity, prevention of cardiac disease, amelioration of perturbations in mitochondrial energy metabolism, and scavenging of reactive oxygen species as well as the promotion of neuronal plasticity.	[55–59]
Peonidin	Cranberry, blackcurrant, blueberry, huckleberry, bilberry, myrtles, roselle plants, purple-fleshed sweet potatoes, raw black rice, and centella asiatica	Antioxidative, anti-inflammatory, antimicrobial, antidiabetic, and cardioprotective effect.	[55,56,59,60]
Pelargonidin	Cranberry, verbena, strawberry, red corn, red potato	Cardiovascular disease prevention, obesity control, alleviation of diabetes, improvement of vision and memory, and increased immune defenses.	[61–65]
Malvidin	Red grape, blue pimpernel, cranberry, blueberries, saskatoon berries	Antioxidative, anti-inflammatory, and anti-cancer activity.	[66]
Delphinidin	Cranberry, Bilberry, Pomegranate, red potato, purple potato	Anti-inflammatory, prevention of bone loss, and anti-cancer activity.	[61,64,67–70]
Petunidin	Cranberry, grapes, black goji, color-fleshed potato, mango, bluberry, red banana, black bean	Antioxidative, anti-inflammatory, anti-diabetic, and neuroprotective effect.	[55,56,71–77]

2. Ulcerative Colitis Pathogenesis

Ulcerative colitis (UC), which is a chronic and idiopathic inflammatory disease of the colon, is one of the major forms of inflammatory bowel disease (IBD). UC occurs with several clinical symptoms, such as abdominal and/or rectal pain, diarrhea, bloody stool, weight loss, fever, and even rectal prolapse under the severe scenario. UC is also associated with an increased risk of colon cancer [78]. Recent studies have identified various genetic and environmental factors involved in UC pathogenesis. Studies showed that UC is more common in western and northern countries when compared with eastern countries [79]. The peak age for UC occurrence is 30 to 40 years [80] and people with infection history of nontyphoid *Salmonella* or *Campylobacter* exhibit eight to 10 times more risk to develop UC in later years [81]. Moreover, former smoking [82], high fat, and/or sugar diets [83], hormone replacement, and anti-inflammatory therapy have been shown to be closely related to increased risk of UC [83–86]. Collectively, UC is a wide-spread inflammatory disease all over the world and can worsen the quality of a patient's life due to the continuous, serious clinical symptoms, possible complications, and sustained medical intervention [46].

2.1. Impaired Barrier Function and Inflammatory Signaling Pathways

Pathologically, UC is characterized by epithelial ulceration, immune cell infiltration in the lamina propria, crypt abscess, enlarged spleen and liver, and impaired intestinal epithelial barrier function [87,88]. The integrity of the mucus layer, the production, and assembly of tight junction (TJ) proteins are two main factors to evaluate intestinal barrier function. Decreased thickness of the mucus layer and expression of TJ proteins (claudins, occludin, and zonula occluden-1 (ZO-1)) and increased gut permeability against bacterial product have been found in chemical-induced colitis models [89–91]. Weakened epithelium barrier function with increased permeability allows for the translocation of commensal bacteria and microbial products into the bowel wall and, ultimately, activates the innate and adaptive immune response.

Several components involved in the gut immunity have been highly implicated in UC pathogenesis including dendritic cells (DCs), macrophages, eosinophils, neutrophils, T-cells, B-cells, and their secreted cytokines and chemokines. Disturbed responses of effector T-cells, T-helper 2 (Th2), and Th17 were observed in the context of UC. Th2 produces cytokines such as tumor necrosis factor alpha (TNF-α), IL-5, IL-6, and IL-13 while Th17 produces IL-17A, IL-21, and IL-22 to activate multiple target cells and downstream signaling pathways to exert their pro-inflammatory functions by binding to corresponding receptors [92–94]. TNF, IL-6, IL-17A, and IL-22 levels are significantly elevated in experimental colitis and UC patients [95–97]. TNF binds to TNFR1 and TNFR2, followed by the recruitment of TNF receptor-associated factor 2 (TRAF2) and activation of JNK-dependent kinase cascade, MEKK kinase cascade, and the nuclear factor-κB (NF-κB) signaling pathway to induce apoptosis, necroptosis, and production of other pro-inflammatory cytokines [93,98]. IL-6, which is another key cytokine in UC, functions in governing the proliferation and survival of Th1 and Th2 cells by pairing with IL-1β to serve as a signaling molecule for the generation of regulatory B cells and mediate STAT3-dependent T cell production of anti-inflammatory cytokine IL-10 [99,100]. IL-13 is identified to be an important effector cytokine in UC to induce epithelial cell apoptosis and compromise epithelial restitution velocity [101]. Similar to IL-10, IL-22 is an anti-inflammatory cytokine involved in wound healing and production of defensins and mucins against bacterial invasion [102]. Up-regulation of antigen-presenting cells (APCs) expressing Toll-like receptors 4 (TLR4) is another scenario in human UC. Binding of TLR4 to ligand lipopolysaccharide (LPS) triggers activation of NF-κB via protein adaptor MyD88 and allows for transcription of numerous inflammatory genes such as TNF-α, IL-6, IL-1β, and cyclooxygenase-2 (COX-2) [103,104].

2.2. Gut Microbiota Dysbiosis

Gut-commensal bacteria have a profound impact on host health and the pathogenesis of UC. Gut microbiota play an important role in nutrition, immunomodulation, and various metabolic processes to exhibit their beneficial function in maintaining gut homeostasis [105]. Intestinal symbiotic bacteria help in maintaining intestinal stability and prevent the colonization of pathogens. For example, capsular polysaccharide A (PSA) of *Bacteroides fragilis* can be delivered to regulatory T cells (Tregs) to induce interleukin-10 (IL-10) production against experimental colitis [106]. Gut microbial metabolites such as short-chain fatty acids (SCFAs) produced via dietary fiber fermentation also play a key role in maintaining colon health [107,108]. Moreover, utilization of non-pathogenic commensal bacteria *Lactobacillus* and *Bifidobacterium* as probiotics have shown promising results in UC remission [109–111]. Dysbiosis of gut bacteria with respect to diversity and bacterial load might be one of the contributing factors to the pathogenesis of UC because of the overstimulation of mucosal immune response [112]. 16S rRNA sequencing performed on fecal and biopsy samples from UC patients revealed a reduction in bacterial alpha diversity and an increase in total bacterial load compared to healthy subjects [113]. Evident reductions of bacterial phyla in UC patients include *Bacteroidetes* and *Firmicutes*, among which two SCFA producing bacteria from the genus, *Phascolarctobacterium*, and *Roseburia*, were significantly reduced in abundance [114]. Conversely, concentrations of adhesive invasive *E.coli* have increased under the UC condition [115]. The impaired intestinal mucosal barrier in predisposed subjects is marked as one of the early events of UC as the consequence of gut microbial dysbiosis. Gut bacterial dysbiosis-induced release of enterotoxins lead to increased intestinal permeability and immune dysfunction [116,117].

3. Anthocyanin and Ulcerative Colitis

The rapidly rising incidence of UC makes the prevention, therapy, and control of this disease important. Current standard UC therapies utilize aminosalicylates, immunosuppressants, and biologicals to interfere with the inflammatory cascade. However, the long-term use of these therapeutic agents may result in undesirable side effects such as vomiting, nausea, headache, and fatigue [91]. Hence, there is an urgent demand for developing effective and evidence-based therapeutic strategies with minimal side effects. Bioactive compounds such as anthocyanins might be potential candidates against UC [92]. There is extensive evidence from laboratory animal studies and human clinical trials that dietary anthocyanins derived from fruits and vegetables protect against intestinal inflammation and provide health benefits to the colon [48,118–120]. Anthocyanins exert its anti-inflammatory effects against UC through effective protection of intestinal mucosal integrity, restoration of epithelial barrier function, immunomodulation, and regulation of gut microbiota [90,121].

3.1. Anthocyanins: Mucosal Integrity and Intestinal Epithelial Barrier Function

The integrity of the mucus layer and tight junction proteins are two key factors to maintain regular intestinal epithelial barrier function. The mucus layer provides a physiochemical barrier to protect the epithelial cell surface. Previous studies indicated that anthocyanins-rich food consumption significantly increased the secretion of membrane-associated mucins and wound-enclosure proteins including MUC1, MUC2, MUC3, Cdc42, Rac1, GAL2, GAL3, GAL4, and RELMβ, which play a vital role in the mucus injury repair process [121,122]. Tight junctions establish the paracellular barrier that controls the flow of molecules in the intercellular space between epithelial cells. As the building blocks of epithelial tight junction, different TJ proteins play different roles. Claudin 1 and Claudin 4 contribute to the tightening of the epithelium, whereas Claudin 2 may be partially responsible for the luminal uptake of antigenic macromolecules because of induction of TJ strand discontinuities [123–125]. Occludin involved in cellular adhesion regulates paracellular permeability [126]. ZO-1, which is a classic TJ marker, functions as an "anchor" and is responsible for linking occludin, claudin, and actin cytoskeleton to enhance the epithelial barrier [127,128]. Anthocyanins from a purple-fleshed potato reduced the cell permeability *in vitro* using a Caco-2 cells [129]. In another study, mice were supplemented with 100 mg/kg black rice extract via oral gavage, and then provided with 2% DSS in their drinking water for five days to induce colitis. Mice on black rice supplementation showed a reduced histological score, which suggests alleviated mucosal injury and edema compared to DSS treatment [90]. In a DSS-induced murine colitis model, the cooked black bean diet (20%) consumption for two weeks significantly inhibited the colon shortening and spleen enlargement in mice [130]. Shima Bibi et al. evaluated the intestinal barrier protective activity of anthocyanins from red raspberries and reported that the red raspberries supplementation observably suppressed the elevation of claudin-2 protein and enhanced the expression of claudin-3 and ZO-1 under DSS treatment [122]. These above results indicate that anthocyanins can protect the tight junctions by modulating the ratio of TJ-positive and negative proteins and confirm the protective effect of anthocyanins from different fruits and vegetables against colonic inflammation [131].

3.2. Anthocyanins and Immunomodulation

Anthocyanin-rich bilberry extract (ARBE) and single anthocyanin cyanidin-3-O-glycoside (C3G) application significantly inhibited the expression and secretion of TNF-α in stimulated human colon epithelial T84 cells [132]. Blueberry supplementation in an obesity-associated chronic inflammation rat model showed elevated production of acetate and reduced expression levels of TNF-α and IL-1β compared to control rats [133]. The protective effect of blueberry anthocyanin extract has also been confirmed in trinitrobenzene sulfonic acid (TNBS)-induced colitis mice model, where researchers found that anthocyanin treatment restored not only IL-10 secretion but also reduced serum levels of IL-12, TNF-α, and IFN-γ. In the same study, anthocyanin supplementation showed amelioration of morphological and histological symptoms of colitis in a dose-dependent manner [134]. In a recent study by Lei Zhao et al., mice supplemented with 100 mg/kg black rice extract via oral gavage showed a reduction in DSS-induced colonic IL-6, IL-1β, and TNF-α expression levels and MPO levels that are linearly related to the neutrophil infiltration [90]. Anthocyanin fraction from the tubers of purple yam down-regulated TNF-α, IFN-γ, and inflammation-associated ROS-producing enzyme myeloperoxidase (MPO) in mice treated with TNBS to induce colitis [135]. Similar observations are reported in a study using grapes, where anthocyanin-rich grape pomace extracts were found to prevent a DSS-induced increase of IL-6, MPO, and nitric oxide synthase (iNOS), whose production is triggered by bacterial products and pro-inflammatory cytokines [136]. Administration of purple-fleshed potatoes rich in malvidin and petunidin have shown to reduce the secretion of pro-inflammatory cytokines and, thereby, attenuate dextran sodium sulfate (DSS)-induced colitis in mice [88]. Anthocyanins also play a role in inhibiting chemokine release and the subsequent NF-κB signaling pathway (Figure 3). Cyanidin and C3G displayed a clear inhibitory effect on macrophage migration and pro-inflammatory chemokines monocyte chemoattractant protein-1 (MCP-1) and macrophage inflammatory protein-related protein-2 (MRP-2) *in vitro* [137]. The p-Coumaroyl anthocyanin mixture (contains petanin, peonanin, malvanin, and pelanin) extracted from a dark purple-fleshed potato cultivar Jayoung displayed an inhibitory effect on the transcriptional activity and translocation of NF-κB in RAW264.7 macrophages [138]. Another *in vitro* study reported that a pure sour cherry anthocyanin extract addition to human Caco-2 cells receded the translocation of a p65 subunit from the cytosol to nuclei [139]. Studies also linked the anti-inflammatory activity of anthocyanins to the inhibition of the COX-2 cascade. Both *in vivo* and *in vitro* evidence show that anthocyanins can suppress the expression level of COX-2 as well as the transactivation of AP-1, which is a transcription factor that regulates COX-2 gene expression [140,141]. Moreover, C3G can reduce COX-2 producing prostaglandin E2 (PGE2) production in human intestine HT-29 cells [142]. Additionally, a six-week ARBE treatment on UC patients revealed decreased serum levels of TNF-α, IFN-γ, and activated NF-κB subunit p65 and increased serum levels of IL-10 and IL-22 [143]. These results suggest that anthocyanins act as anti-inflammatory agents by their transcriptional and translational regulation of cytokines to inhibit/suppress pro-inflammatory cytokines and elevate the anti-inflammatory cytokines.

3.3. Anthocyanins and Gut Microbiota

The health-promoting effects of individual anthocyanins and their mixtures have been attributed not only to their direct effects in the colon but also to their metabolism by intestinal microbiota and their alteration of intestinal microbial populations. Anthocyanins and gut microbiota exhibit a two-way interaction to impact host physiology. Intestinal microbiota as a "metabolizing organ" plays a critical role in maintaining gastrointestinal health [144] and host metabolism [145,146]. Gut microbiota is a crucial determinant of anthocyanin bioavailability.

In the lumen of the large intestine, unabsorbed anthocyanins are exposed to microbiota-mediated biotransformation, which includes three significant conditions: hydrolysis (breaking glycosidic linkages), fission (cleaving heterocycle), and demethylation. Bacterial species that carry corresponding β-glucosidase, β-glucuronidase, α-rhamnosidase, or demethylase such as *Clostridium* spp., *Butyrivibrio* spp., *Lactobacillus* spp., *B. fragilis*, and *B. ovatus*, etc., are actively involved in this process [147,148]. Anthocyanin biotransformation also produces glucose, which is an essential energy source required for bacterial growth [144]. Primary anthocyanin-derived metabolites are phenolic acids, whose anti-inflammatory effects have been verified by substantial studies. For example, the predominant metabolite of cyanidin and protocatechuic acid (PCA) has been shown to suppress COX-2 and iNOS protein expression and attenuate DSS-induced UC in mice [149]. Gallic acid as another anthocyanin-derived metabolite was shown to reduce the growth of potentially harmful bacteria such as *Clostridium histolyticum* and *Bacteroides* spp. without any negative effect on measured beneficial bacteria [150].

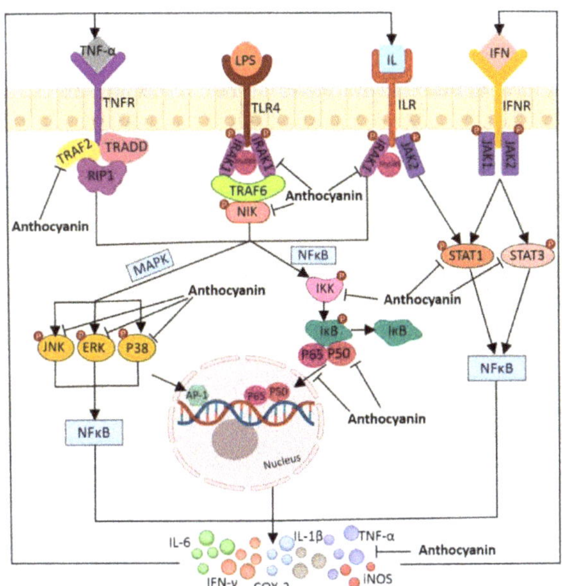

Figure 3. The mechanisms through which anthocyanins act as anti-inflammatory agents. Inflammatory signaling pathways including NF-kB, MAPKs (P38, ERK, JNK), and STATs were activated by ligand binding of the pro-inflammatory cytokines TNF-α, LPS, IL, and IFN, which eventually leads to the translocation of transcription factors to the nucleus, transcriptional activation, and cytokine production. Anthocyanins attenuated the cascade of inflammatory responses by inhibiting the translocation of transcription factors (P50 and P65), the phosphorylation of IRAK1, NIK, IKK, STAT1, STAT3, P38, ERK, and JNK, the secretion of inflammatory cytokines (IL-6, IL-1β, TNF-α, iNOS, COX-2, and IFN-γ), and activation of NF-kB, MAPK, and STAT inflammatory signaling pathways.

There is broad agreement that dietary anthocyanins and their metabolites have potential health benefits via modulation of the gut microbiota [44,150]. Increasing evidence supports the idea that anthocyanins can function as prebiotics, which contributes to the growth of certain commensal bacteria [44,151,152]. Both *in vitro* and *in vivo* studies have shown an elevated growth of potentially beneficial bacteria such as *Lactobacillus* spp. and *Bifidobacterium* spp. after administration of anthocyanin-rich products [44,151,152]. Anthocyanins can also interact with starch, SCFAs, and ferric iron to indirectly modulate gut microbiota. Anthocyanins exert the beneficial effect by increasing the levels of SCFAs, which has the antimicrobial impact on pathogens [153]. Moreover, it was found that anthocyanins were able to affect the digestion of starch by inhibiting digestive enzymes, such as α-amylase [154,155]. The indigestible starch goes down to the large intestine, where it can act as an energy source for several probiotic bacteria such as *lactobacilli*, *bifidobacteria*, and *streptococci*, which are beneficial to human health [155,156]. Another impressive result showed that indigestible dietary fiber components, such as β-glucans and resistant starch, can significantly increase the production of SCFAs [157,158]. Evidence indicated that the dysbiosis of the gut microbiota and impaired intestinal barrier function could be induced by Fe deficiency [159]. However, this situation can be alleviated with anthocyanin supplementation. Studies reported that C3G, cyanidin-3-5-diglucoside, petunidin-3-glucoside, and delphinidin-3-glucoside exerted substantial ferric ion chelating activities. Ferric ion chelation increases its solubility and bioavailability and may contribute to the intestinal homeostasis [160–162].

The above evidence demonstrated the anti-inflammatory properties of anthocyanins and the potential of anthocyanin to be used as novel therapeutic agents in UC treatment. Even though the mechanism behind anthocyanin-induced UC mitigation is not entirely known, it is highly likely that anthocyanin and bacteria interplay while anthocyanin-derived metabolites play a crucial role. There is no proven consensus regarding the bioavailability of anthocyanins, and minimal research has been done to elucidate the bioactivity of anthocyanins *in vivo*. Majority of studies focusing on the anti-colitis effect of anthocyanins utilize fruit or grain extract containing other bioactive compounds that are known to have an anti-oxidant effect. Thus, it is challenging to ascribe the observed UC relief to anthocyanins solely. Moreover, the possible synergistic effect of anthocyanins with other phytochemicals and fiber is a topic that requires more attention and effort to address the need for searching for a natural and safe anti-colitis strategy.

Author Contributions: Conceptualization, L.R.; Writing, S.L. and B.W.; Formatting, S.L. and W.F.; Review & Editing, L.R.

Funding: An Agriculture and Food Research Initiative competitive grant 2016-67017-24512 from the USDA National Institute of Food and Agriculture supported the research.

Acknowledgments: We acknowledge Jairam K. P. Vanamala, Ph.D., for reviewing the article.

Conflicts of Interest: The authors declare no conflict of interest.

Abbreviations

APCs	Antigen-presenting cells
ARBE	Anthocyanin-rich bilberry extract
C3G	Cyanidin-3-glucoside
Cdc	Cell division control protein
COX-2	Cyclooxygenase-2
DCs	Dendritic cells
DSS	Dextran sodium sulfate
ERK	Extracellular signal-regulated kinase
GAL	Galectin
IBD	Inflammatory bowel disease
IFN-γ	Interferon gamma
IL	Interleukin
iNOS	Nitric oxide synthase
JNK	c-Jun N-terminal kinase
LPS	Lipopolysaccharide
MAPK	Mitogen-activated protein kinase
MCP-1	Chemoattractant protein-1
MPO	Myeloperoxidase
MRP-2	Macrophage inflammatory protein-related protein-2
MUC	Mucin
NF-κB	Nuclear factor-κB
PCA	Protocatechuic acid
PGE2	Prostaglandin E2
PSA	Polysaccharide A
RELMβ	Resistin-Like Molecule-beta
ROS	Reactive oxygen species
SCFA	Short chain fatty acid
STAT	Signal transducer and activator of transcription
Th	T-helper
TJ	Tight junction
TLR4	Toll-like receptors 4

TNBS	Trinitrobenzene sulfonic acid
TNFR	Tumor necrosis factor receptor
TNF-α	Tumor necrosis factor alpha
TRAF	TNF receptor-associated factor
Tregs	Regulatory T cells
UC	Ulcerative colitis
ZO-1	Zonula occludens-1

References

1. Wallace, T.C.; Giusti, M.M. Anthocyanins. *Adv. Nutr.* **2015**, *6*, 620–622. [CrossRef]
2. Andersen, O.M.; Markham, K.R. *Flavonoids: Chemistry, biochemistry and applications*; CRC Press: Boca Raton, FL, USA, 2005.
3. McGhie, T.K.; Walton, M.C. The bioavailability and absorption of anthocyanins: Towards a better understanding. *Mol. Nutr. Food Res.* **2007**, *51*, 702–713. [CrossRef]
4. Borges, G.D.S.C.; Vieira, F.G.K.; Copetti, C.; Gonzaga, L.V.; Zambiazi, R.C.; Mancini Filho, J.; Fett, R. Chemical characterization, bioactive compounds, and antioxidant capacity of jussara (euterpe edulis) fruit from the atlantic forest in southern brazil. *Food Res. Int.* **2011**, *44*, 2128–2133. [CrossRef]
5. Sui, X.; Zhang, Y.; Zhou, W. Bread fortified with anthocyanin-rich extract from black rice as nutraceutical sources: Its quality attributes and in vitro digestibility. *Food Chem.* **2016**, *196*, 910–916. [CrossRef] [PubMed]
6. Morais, C.A.; de Rosso, V.V.; Estadella, D.; Pisani, L.P. Anthocyanins as inflammatory modulators and the role of the gut microbiota. *J. Nutr. Biochem.* **2016**, *33*, 1–7. [CrossRef]
7. Pojer, E.; Mattivi, F.; Johnson, D.; Stockley, C.S. The case for anthocyanin consumption to promote human health: A review. *Compr. Rev. Food Sci. Food Saf.* **2013**, *12*, 483–508. [CrossRef]
8. Welch, C.R.; Wu, Q.; Simon, J.E. Recent advances in anthocyanin analysis and characterization. *Curr. Anal. Chem.* **2008**, *4*, 75–101. [CrossRef] [PubMed]
9. Pérez-Gregorio, R.M.; García-Falcón, M.S.; Simal-Gándara, J.; Rodrigues, A.S.; Almeida, D.P. Identification and quantification of flavonoids in traditional cultivars of red and white onions at harvest. *J. Food Compos. Anal.* **2010**, *23*, 592–598. [CrossRef]
10. Woodward, G.; Kroon, P.; Cassidy, A.; Kay, C. Anthocyanin stability and recovery: Implications for the analysis of clinical and experimental samples. *J. Agric. Food Chem.* **2009**, *57*, 5271–5278. [CrossRef] [PubMed]
11. Khoo, H.E.; Azlan, A.; Tang, S.T.; Lim, S.M. Anthocyanidins and anthocyanins: Colored pigments as food, pharmaceutical ingredients, and the potential health benefits. *Food Nutr. Res.* **2017**, *61*, 1361779. [CrossRef]
12. Brouillard, R. *Chemical structure of anthocyanins*; Academic Press: New York, NY, USA, 1982; Volume 1.
13. He, J.; Giusti, M.M. Anthocyanins: Natural colorants with health-promoting properties. *Annu. Rev. Food Sci. Technol.* **2010**, *1*, 163–187. [CrossRef] [PubMed]
14. Wrolstad, R.E.; Durst, R.W.; Lee, J. Tracking color and pigment changes in anthocyanin products. *Trends Food Sci. Technol.* **2005**, *16*, 423–428. [CrossRef]
15. Samadi, A.K.; Bilsland, A.; Georgakilas, A.G.; Amedei, A.; Amin, A.; Bishayee, A.; Azmi, A.S.; Lokeshwar, B.L.; Grue, B.; Panis, C. Seminars in cancer biology. In *A Multi-Targeted Approach to Suppress Tumor-Promoting Inflammation*; Elsevier: Amsterdam, The Netherlands, 2015; pp. S151–S184.
16. Andersen, Ø.M.; Jordheim, M. Basic anthocyanin chemistry and dietary sources. *Anthocyanins Health Dis.* **2013**, *1*, 13–89.
17. Smeriglio, A.; Barreca, D.; Bellocco, E.; Trombetta, D. Chemistry, pharmacology and health benefits of anthocyanins. *Phytother. Res.* **2016**, *30*, 1265–1286. [CrossRef]
18. de Pascual-Teresa, S.; Sanchez-Ballesta, M.T. Anthocyanins: From plant to health. *Phytochem. Rev.* **2008**, *7*, 281–299. [CrossRef]
19. Fang, J. Bioavailability of anthocyanins. *Drug Metab. Rev.* **2014**, *46*, 508–520. [CrossRef] [PubMed]
20. Kong, J.-M.; Chia, L.-S.; Goh, N.-K.; Chia, T.-F.; Brouillard, R. Analysis and biological activities of anthocyanins. *Phytochemistry* **2003**, *64*, 923–933. [CrossRef]
21. Nayak, C.A.; Srinivas, P.; Rastogi, N.K. Characterisation of anthocyanins from garcinia indica choisy. *Food Chem.* **2010**, *118*, 719–724. [CrossRef]

22. da Silva, F.L.; Escribano-Bailón, M.T.; Alonso, J.J.P.; Rivas-Gonzalo, J.C.; Santos-Buelga, C. Anthocyanin pigments in strawberry. *Lwt-Food Sci. Technol.* **2007**, *40*, 374–382. [CrossRef]

23. Böhm, H.G. Mazza und E. Miniati: Anthocyanins in Fruits, Vegetables and Grains. 362 Seiten, zahlr. Abb. und Tab. CRC Press: Boca Raton, Ann Arbor, London, Tokyo 1993. Preis: 144—£. *Food Nahrung* **1994**, *38*, 343. [CrossRef]

24. Zamora-Ros, R.; Knaze, V.; Luján-Barroso, L.; Slimani, N.; Romieu, I.; Touillaud, M.; Kaaks, R.; Teucher, B.; Mattiello, A.; Grioni, S. Estimation of the intake of anthocyanidins and their food sources in the european prospective investigation into cancer and nutrition (epic) study. *Br. J. Nutr.* **2011**, *106*, 1090–1099. [CrossRef] [PubMed]

25. Sebastian, R.S.; Wilkinson Enns, C.; Goldman, J.D.; Martin, C.L.; Steinfeldt, L.C.; Murayi, T.; Moshfegh, A.J. A new database facilitates characterization of flavonoid intake, sources, and positive associations with diet quality among us adults. *J. Nutr.* **2015**, *145*, 1239–1248. [CrossRef] [PubMed]

26. Yousuf, B.; Gul, K.; Wani, A.A.; Singh, P. Health benefits of anthocyanins and their encapsulation for potential use in food systems: A review. *Crit. Rev. Food Sci. Nutr.* **2016**, *56*, 2223–2230. [CrossRef]

27. Yi, W.; Akoh, C.C.; Fischer, J.; Krewer, G. Absorption of anthocyanins from blueberry extracts by caco-2 human intestinal cell monolayers. *J. Agric. Food Chem.* **2006**, *54*, 5651–5658. [CrossRef] [PubMed]

28. Tsuda, T.; Shiga, K.; Ohshima, K.; Kawakishi, S.; Osawa, T. Inhibition of lipid peroxidation and the active oxygen radical scavenging effect of anthocyanin pigments isolated from phaseolus vulgaris l. *Biochem. Pharmacol.* **1996**, *52*, 1033–1039. [CrossRef]

29. Zhang, Y.; Vareed, S.K.; Nair, M.G. Human tumor cell growth inhibition by nontoxic anthocyanidins, the pigments in fruits and vegetables. *Life Sci.* **2005**, *76*, 1465–1472. [CrossRef] [PubMed]

30. Milbury, P.E.; Cao, G.; Prior, R.L.; Blumberg, J. Bioavailablility of elderberry anthocyanins. *Mech. Ageing Dev.* **2002**, *123*, 997–1006. [CrossRef]

31. Talavera, S.; Felgines, C.; Texier, O.; Besson, C.; Lamaison, J.-L.; Rémésy, C. Anthocyanins are efficiently absorbed from the stomach in anesthetized rats. *J. Nutr.* **2003**, *133*, 4178–4182. [CrossRef]

32. Matuschek, M.C.; Hendriks, W.H.; McGhie, T.K.; Reynolds, G.W. The jejunum is the main site of absorption for anthocyanins in mice. *J. Nutr. Biochem.* **2006**, *17*, 31–36. [CrossRef]

33. Czank, C.; Cassidy, A.; Zhang, Q.; Morrison, D.J.; Preston, T.; Kroon, P.A.; Botting, N.P.; Kay, C.D. Human metabolism and elimination of the anthocyanin, cyanidin-3-glucoside: A 13c-tracer study. *Am. Clin. Nutr.* **2013**, *97*, 995–1003. [CrossRef]

34. Felgines, C.; Krisa, S.; Mauray, A.; Besson, C.; Lamaison, J.-L.; Scalbert, A.; Mérillon, J.-M.; Texier, O. Radiolabelled cyanidin 3-o-glucoside is poorly absorbed in the mouse. *Br. J. Nutr.* **2010**, *103*, 1738–1745. [CrossRef]

35. Bub, A.; Watzl, B.; Heeb, D.; Rechkemmer, G.; Briviba, K. Malvidin-3-glucoside bioavailability in humans after ingestion of red wine, dealcoholized red wine and red grape juice. *Eur. J. Nutr.* **2001**, *40*, 113–120. [CrossRef] [PubMed]

36. Matsumoto, H.; Inaba, H.; Kishi, M.; Tominaga, S.; Hirayama, M.; Tsuda, T. Orally administered delphinidin 3-rutinoside and cyanidin 3-rutinoside are directly absorbed in rats and humans and appear in the blood as the intact forms. *J. Agric. Food Chem.* **2001**, *49*, 1546–1551. [CrossRef] [PubMed]

37. Manach, C.; Williamson, G.; Morand, C.; Scalbert, A.; Rémésy, C. Bioavailability and bioefficacy of polyphenols in humans. I. Review of 97 bioavailability studies. *Am. J. Clin. Nutr.* **2005**, *81*, 230S–242S. [CrossRef] [PubMed]

38. Aura, A.-M.; Martin-Lopez, P.; O'Leary, K.A.; Williamson, G.; Oksman-Caldentey, K.-M.; Poutanen, K.; Santos-Buelga, C. In vitro metabolism of anthocyanins by human gut microflora. *Eur. J. Nutr.* **2005**, *44*, 133–142. [CrossRef]

39. Keppler, K.; Humpf, H.-U. Metabolism of anthocyanins and their phenolic degradation products by the intestinal microflora. *Bioorganic Med. Chem.* **2005**, *13*, 5195–5205. [CrossRef]

40. Vamanu, E.; Gatea, F.; Sârbu, I.; Pelinescu, D. An in vitro study of the influence of curcuma longa extracts on the microbiota modulation process, in patients with hypertension. *Pharmaceutics* **2019**, *11*, 191. [CrossRef] [PubMed]

41. Fleschhut, J.; Kratzer, F.; Rechkemmer, G.; Kulling, S.E. Stability and biotransformation of various dietary anthocyanins in vitro. *Eur. J. Nutr.* **2006**, *45*, 7–18. [CrossRef]

42. Forester, S.C.; Waterhouse, A.L. Identification of cabernet sauvignon anthocyanin gut microflora metabolites. *J. Agric. Food Chem.* **2008**, *56*, 9299–9304. [CrossRef]

43. Salyer, J.; Park, S.; Ricke, S.; Lee, S. Analysis of microbial populations and metabolism of anthocyanins by mice gut microflora fed with blackberry powder. *J. Nutr. Food Sci.* **2013**, *3*, 1–5. [CrossRef]

44. Hidalgo, M.; Oruna-Concha, M.J.; Kolida, S.; Walton, G.E.; Kallithraka, S.; Spencer, J.P.; de Pascual-Teresa, S. Metabolism of anthocyanins by human gut microflora and their influence on gut bacterial growth. *J. Agric. Food Chem.* **2012**, *60*, 3882–3890. [CrossRef]

45. Chen, Y.; Li, Q.; Zhao, T.; Zhang, Z.; Mao, G.; Feng, W.; Wu, X.; Yang, L. Biotransformation and metabolism of three mulberry anthocyanin monomers by rat gut microflora. *Food Chem.* **2017**, *237*, 887–894. [CrossRef] [PubMed]

46. Zielińska, M.; Lewandowska, U.; Podsędek, A.; Cygankiewicz, A.I.; Jacenik, D.; Sałaga, M.; Kordek, R.; Krajewska, W.M.; Fichna, J. Orally available extract from brassica oleracea var. Capitata rubra attenuates experimental colitis in mouse models of inflammatory bowel diseases. *J. Funct. Foods* **2015**, *17*, 587–599. [CrossRef]

47. Sugata, M.; Lin, C.-Y.; Shih, Y.-C. Anti-inflammatory and anticancer activities of taiwanese purple-fleshed sweet potatoes (ipomoea batatas l. Lam) extracts. *Biomed. Res. Int.* **2015**, *2015*, 768093. [CrossRef]

48. Medda, R.; Lyros, O.; Schmidt, J.L.; Jovanovic, N.; Nie, L.; Link, B.J.; Otterson, M.F.; Stoner, G.D.; Shaker, R.; Rafiee, P. Anti inflammatory and anti angiogenic effect of black raspberry extract on human esophageal and intestinal microvascular endothelial cells. *Microvasc. Res.* **2015**, *97*, 167–180. [CrossRef] [PubMed]

49. Youdim, K.A.; Shukitt-Hale, B.; Joseph, J.A. Flavonoids and the brain: Interactions at the blood–brain barrier and their physiological effects on the central nervous system. *Free Radic. Biol. Med.* **2004**, *37*, 1683–1693. [CrossRef] [PubMed]

50. Cassidy, A.; Mukamal, K.J.; Liu, L.; Franz, M.; Eliassen, A.H.; Rimm, E.B. High anthocyanin intake is associated with a reduced risk of myocardial infarction in young and middle-aged women. *Circulation* **2013**, *127*, 188–196. [CrossRef]

51. Mink, P.J.; Scrafford, C.G.; Barraj, L.M.; Harnack, L.; Hong, C.-P.; Nettleton, J.A.; Jacobs, D.R., Jr. Flavonoid intake and cardiovascular disease mortality: A prospective study in postmenopausal women. *Am. J. Clin. Nutr.* **2007**, *85*, 895–909. [CrossRef]

52. Vendrame, S.; Del Bo, C.; Ciappellano, S.; Riso, P.; Klimis-Zacas, D. Berry fruit consumption and metabolic syndrome. *Antioxidants* **2016**, *5*, 34. [CrossRef] [PubMed]

53. Overall, J.; Bonney, S.; Wilson, M.; Beermann, A.; Grace, M.; Esposito, D.; Lila, M.; Komarnytsky, S. Metabolic effects of berries with structurally diverse anthocyanins. *Int. J. Mol. Sci.* **2017**, *18*, 422. [CrossRef] [PubMed]

54. J Thoppil, R.; Bhatia, D.; F Barnes, K.; Haznagy-Radnai, E.; Hohmann, J.; S Darvesh, A.; Bishayee, A. Black currant anthocyanins abrogate oxidative stress through nrf2-mediated antioxidant mechanisms in a rat model of hepatocellular carcinoma. *Curr. Cancer Drug Targets* **2012**, *12*, 1244–1257.

55. Wu, X.; Prior, R.L. Systematic identification and characterization of anthocyanins by hplc-esi-ms/ms in common foods in the united states: Fruits and berries. *J. Agric. Food Chem.* **2005**, *53*, 2589–2599. [CrossRef]

56. Khoo, C.; Falk, M. Cranberry polyphenols: Effects on cardiovascular risk factors. In *Polyphenols in human health and disease*; Elsevier: Amsterdam, The Netherlands, 2014; pp. 1049–1065.

57. Rothenberg, D.O.; Yang, H.; Chen, M.; Zhang, W.; Zhang, L. Metabolome and transcriptome sequencing analysis reveals anthocyanin metabolism in pink flowers of anthocyanin-rich tea (camellia sinensis). *Molecules* **2019**, *24*, 1064. [CrossRef]

58. Tsutsumi, A.; Horikoshi, Y.; Fushimi, T.; Saito, A.; Koizumi, R.; Fujii, Y.; Hu, Q.Q.; Hirota, Y.; Aizawa, K.; Osakabe, N. Acylated anthocyanins derived from purple carrot (daucus carota l.) induce elevation of blood flow in rat cremaster arteriole. *Food Funct.* **2019**, *10*, 1726–1735. [CrossRef] [PubMed]

59. Wongwichai, T.; Teeyakasem, P.; Pruksakorn, D.; Kongtawelert, P.; Pothacharoen, P. Anthocyanins and metabolites from purple rice inhibit il-1beta-induced matrix metalloproteinases expression in human articular chondrocytes through the nf-kappab and erk/mapk pathway. *Biomed. Pharmacother.* **2019**, *112*, 108610. [CrossRef] [PubMed]

60. Jayaprakasam, B.; Vareed, S.K.; Olson, L.K.; Nair, M.G. Insulin secretion by bioactive anthocyanins and anthocyanidins present in fruits. *J. Agric. Food Chem.* **2005**, *53*, 28–31. [CrossRef] [PubMed]

61. Amini, A.M.; Muzs, K.; Spencer, J.P.; Yaqoob, P. Pelargonidin-3-o-glucoside and its metabolites have modest anti-inflammatory effects in human whole blood cultures. *Nutr. Res.* **2017**, *46*, 88–95. [CrossRef]

62. Tsuda, T. Dietary anthocyanin-rich plants: Biochemical basis and recent progress in health benefits studies. *Mol. Nutr. Food Res.* **2012**, *56*, 159–170. [CrossRef] [PubMed]

63. Rodriguez-Mateos, A.; Heiss, C.; Borges, G.; Crozier, A. Berry (poly) phenols and cardiovascular health. *J. Agric. Food Chem.* **2013**, *62*, 3842–3851. [CrossRef]

64. Stushnoff, C.; Holm, D.; Thompson, M.D.; Jiang, W.; Thompson, H.J.; Joyce, N.I.; Wilson, P. Antioxidant properties of cultivars and selections from the colorado potato breeding program. *Am. J. Potato Res.* **2008**, *85*, 267. [CrossRef]

65. Nankar, A.N.; Dungan, B.; Paz, N.; Sudasinghe, N.; Schaub, T.; Holguin, F.O.; Pratt, R.C. Quantitative and qualitative evaluation of kernel anthocyanins from southwestern united states blue corn. *J. Sci Food Agric.* **2016**, *96*, 4542–4552. [CrossRef]

66. Bognar, E.; Sarszegi, Z.; Szabo, A.; Debreceni, B.; Kalman, N.; Tucsek, Z.; Sumegi, B.; Gallyas, F., Jr. Antioxidant and anti-inflammatory effects in raw264. 7 macrophages of malvidin, a major red wine polyphenol. *PLoS ONE* **2013**, *8*, e65355. [CrossRef] [PubMed]

67. Moriwaki, S.; Suzuki, K.; Muramatsu, M.; Nomura, A.; Inoue, F.; Into, T.; Yoshiko, Y.; Niida, S. Delphinidin, one of the major anthocyanidins, prevents bone loss through the inhibition of excessive osteoclastogenesis in osteoporosis model mice. *PLoS ONE* **2014**, *9*, e97177. [CrossRef] [PubMed]

68. Hafeez, B.B.; Siddiqui, I.A.; Asim, M.; Malik, A.; Afaq, F.; Adhami, V.M.; Saleem, M.; Din, M.; Mukhtar, H. A dietary anthocyanidin delphinidin induces apoptosis of human prostate cancer pc3 cells in vitro and in vivo: Involvement of nuclear factor-κb signaling. *Cancer Res.* **2008**, *68*, 8564–8572. [CrossRef]

69. Spilmont, M.; Léotoing, L.; Davicco, M.J.; Lebecque, P.; Miot-Noirault, E.; Pilet, P.; Rios, L.; Wittrant, Y.; Coxam, V. Pomegranate peel extract prevents bone loss in a preclinical model of osteoporosis and stimulates osteoblastic differentiation in vitro. *Nutrients* **2015**, *7*, 9265–9284. [CrossRef] [PubMed]

70. Lao, F.; Sigurdson, G.T.; Giusti, M.M. Health benefits of purple corn (zea mays l.) phenolic compounds. *Compr. Rev. Food Sci. Food Saf.* **2017**, *16*, 234–246. [CrossRef]

71. Muche, B.M.; Speers, R.A.; Rupasinghe, H.P.V. Storage temperature impacts on anthocyanins degradation, color changes and haze development in juice of "merlot" and "ruby" grapes (vitis vinifera). *Front. Nutr.* **2018**, *5*, 100. [CrossRef]

72. Tang, P.; Giusti, M.M. Black goji as a potential source of natural color in a wide ph range. *Food Chem.* **2018**, *269*, 419–426. [CrossRef] [PubMed]

73. Rocha-Parra, D.; Chirife, J.; Zamora, C.; de Pascual-Teresa, S. Chemical characterization of an encapsulated red wine powder and its effects on neuronal cells. *Molecules* **2018**, *23*, 842. [CrossRef]

74. Kalita, D.; Holm, D.G.; LaBarbera, D.V.; Petrash, J.M.; Jayanty, S.S. Inhibition of alpha-glucosidase, alpha-amylase, and aldose reductase by potato polyphenolic compounds. *PLoS ONE* **2018**, *13*, e0191025. [CrossRef] [PubMed]

75. Lopez-Cobo, A.; Verardo, V.; Diaz-de-Cerio, E.; Segura-Carretero, A.; Fernandez-Gutierrez, A.; Gomez-Caravaca, A.M. Use of hplc- and gc-qtof to determine hydrophilic and lipophilic phenols in mango fruit (mangifera indica l.) and its by-products. *Food Res. Int.* **2017**, *100*, 423–434. [CrossRef]

76. Fu, X.; Cheng, S.; Liao, Y.; Huang, B.; Du, B.; Zeng, W.; Jiang, Y.; Duan, X.; Yang, Z. Comparative analysis of pigments in red and yellow banana fruit. *Food Chem.* **2018**, *239*, 1009–1018. [CrossRef] [PubMed]

77. Aguilera, Y.; Mojica, L.; Rebollo-Hernanz, M.; Berhow, M.; de Mejia, E.G.; Martin-Cabrejas, M.A. Black bean coats: New source of anthocyanins stabilized by beta-cyclodextrin copigmentation in a sport beverage. *Food Chem.* **2016**, *212*, 561–570. [CrossRef] [PubMed]

78. Adams, S.M.; Bornemann, P.H. Ulcerative colitis. *Am. Fam. Physician* **2013**, *87*, 699–705.

79. Burisch, J.; Pedersen, N.; Čuković-Čavka, S.; Brinar, M.; Kaimakliotis, I.; Duricova, D.; Shonová, O.; Vind, I.; Avnstrøm, S.; Thorsgaard, N. East–west gradient in the incidence of inflammatory bowel disease in europe: The ecco-epicom inception cohort. *Gut* **2014**, *63*, 588–597. [CrossRef] [PubMed]

80. Cosnes, J.; Gower–Rousseau, C.; Seksik, P.; Cortot, A. Epidemiology and natural history of inflammatory bowel diseases. *Gastroenterology* **2011**, *140*, 1785–1794. e4. [CrossRef] [PubMed]

81. Jess, T.; Simonsen, J.; Nielsen, N.M.; Jørgensen, K.T.; Bager, P.; Ethelberg, S.; Frisch, M. Enteric salmonella or campylobacter infections and the risk of inflammatory bowel disease. *Gut* **2011**, *60*, 318–324. [CrossRef]

82. Sahami, S.; Kooij, I.; Meijer, S.; Van den Brink, G.; Buskens, C.; Te Velde, A. The link between the appendix and ulcerative colitis: Clinical relevance and potential immunological mechanisms. *Am. J. Gastroenterol.* **2016**, *111*, 163. [CrossRef] [PubMed]

83. Hou, J.K.; Abraham, B.; El-Serag, H. Dietary intake and risk of developing inflammatory bowel disease: A systematic review of the literature. *Am. J. Gastroenterol.* **2011**, *106*, 563. [CrossRef]

84. Ananthakrishnan, A.N.; Higuchi, L.M.; Huang, E.S.; Khalili, H.; Richter, J.M.; Fuchs, C.S.; Chan, A.T. Aspirin, nonsteroidal anti-inflammatory drug use, and risk for crohn disease and ulcerative colitis: A cohort study. *Ann. Intern. Med.* **2012**, *156*, 350–359. [CrossRef]

85. Khalili, H.; Higuchi, L.M.; Ananthakrishnan, A.N.; Manson, J.E.; Feskanich, D.; Richter, J.M.; Fuchs, C.S.; Chan, A.T. Hormone therapy increases risk of ulcerative colitis but not crohn's disease. *Gastroenterology* **2012**, *143*, 1199–1206. [CrossRef]

86. Ungaro, R.; Bernstein, C.N.; Gearry, R.; Hviid, A.; Kolho, K.-L.; Kronman, M.P.; Shaw, S.; Van Kruiningen, H.; Colombel, J.-F.; Atreja, A. Antibiotics associated with increased risk of new-onset crohn's disease but not ulcerative colitis: A meta-analysis. *Am. J. Gastroenterol.* **2014**, *109*, 1728. [CrossRef] [PubMed]

87. Chen, L.; Zhou, Z.; Yang, Y.; Chen, N.; Xiang, H. Therapeutic effect of imiquimod on dextran sulfate sodium-induced ulcerative colitis in mice. *PLoS ONE* **2017**, *12*, e0186138. [CrossRef] [PubMed]

88. Reddivari, L.; Wang, T.; Wu, B.; Li, S. Potato: An Anti-Inflammatory Food. *Am. J. Potato Res.* **2019**, *96*, 164–169. [CrossRef]

89. Johansson, M.E.; Gustafsson, J.K.; Sjöberg, K.E.; Petersson, J.; Holm, L.; Sjövall, H.; Hansson, G.C. Bacteria penetrate the inner mucus layer before inflammation in the dextran sulfate colitis model. *PLoS ONE* **2010**, *5*, e12238. [CrossRef] [PubMed]

90. Zhao, L.; Zhang, Y.; Liu, G.; Hao, S.; Wang, C.; Wang, Y. Black rice anthocyanin-rich extract and rosmarinic acid, alone and in combination, protect against dss-induced colitis in mice. *Food Funct.* **2018**, *9*, 2796–2808. [CrossRef]

91. Minaiyan, M.; Ghannadi, A.; Mahzouni, P.; Jaffari-Shirazi, E. Comparative study of berberis vulgaris fruit extract and berberine chloride effects on acetic acid-induced colitis in rats. *Iran. J. Pharm. Res. Ijpr* **2011**, *10*, 97. [PubMed]

92. Atreya, R.; Mudter, J.; Finotto, S.; Müllberg, J.; Jostock, T.; Wirtz, S.; Schütz, M.; Bartsch, B.; Holtmann, M.; Becker, C. Blockade of interleukin 6 trans signaling suppresses t-cell resistance against apoptosis in chronic intestinal inflammation: Evidence in crohn disease and experimental colitis in vivo. *Nat. Med.* **2000**, *6*, 583. [CrossRef]

93. Neurath, M.F. Cytokines in inflammatory bowel disease. *Nat. Rev. Immunol.* **2014**, *14*, 329. [CrossRef]

94. Su, L.; Nalle, S.C.; Shen, L.; Turner, E.S.; Singh, G.; Breskin, L.A.; Khramtsova, E.A.; Khramtsova, G.; Tsai, P.Y.; Fu, Y.X. Tnfr2 activates mlck-dependent tight junction dysregulation to cause apoptosis-mediated barrier loss and experimental colitis. *Gastroenterology* **2013**, *145*, 407–415. [CrossRef]

95. Hernández-Chirlaque, C.; Aranda, C.J.; Ocón, B.; Capitán-Cañadas, F.; Ortega-González, M.; Carrero, J.J.; Suárez, M.D.; Zarzuelo, A.; Sánchez de Medina, F.; Martínez-Augustin, O. Germ-free and antibiotic-treated mice are highly susceptible to epithelial injury in dss colitis. *J. Crohn's Colitis* **2016**, *10*, 1324–1335. [CrossRef]

96. Bernardo, D.; Vallejo-Díez, S.; Mann, E.R.; Al-Hassi, H.O.; Martínez-Abad, B.; Montalvillo, E.; Tee, C.T.; Murananthan, A.U.; Núñez, H.; Peake, S.T. Il-6 promotes immune responses in human ulcerative colitis and induces a skin-homing phenotype in the dendritic cells and t cells they stimulate. *Eur. J. Immunol.* **2012**, *42*, 1337–1353. [CrossRef]

97. Ono, Y.; Kanai, T.; Sujino, T.; Nemoto, Y.; Kanai, Y.; Mikami, Y.; Hayashi, A.; Matsumoto, A.; Takaishi, H.; Ogata, H. T-helper 17 and interleukin-17–producing lymphoid tissue inducer-like cells make different contributions to colitis in mice. *Gastroenterology* **2012**, *143*, 1288–1297. [CrossRef]

98. Wu, Y.-D.; Zhou, B. Tnf-α/nf-κb/snail pathway in cancer cell migration and invasion. *Br. J. Cancer* **2010**, *102*, 639. [CrossRef]

99. Hunter, C.A.; Jones, S.A. Il-6 as a keystone cytokine in health and disease. *Nat. Immunol.* **2015**, *16*, 448. [CrossRef]

100. Stumhofer, J.S.; Silver, J.S.; Laurence, A.; Porrett, P.M.; Harris, T.H.; Turka, L.A.; Ernst, M.; Saris, C.J.; O'Shea, J.J.; Hunter, C.A. Interleukins 27 and 6 induce stat3-mediated t cell production of interleukin 10. *Nat. Immunol.* **2007**, *8*, 1363. [CrossRef]

101. Heller, F.; Florian, P.; Bojarski, C.; Richter, J.; Christ, M.; Hillenbrand, B.; Mankertz, J.; Gitter, A.H.; Bürgel, N.; Fromm, M. Interleukin-13 is the key effector th2 cytokine in ulcerative colitis that affects epithelial tight junctions, apoptosis, and cell restitution. *Gastroenterology* **2005**, *129*, 550–564. [CrossRef]

102. Pickert, G.; Neufert, C.; Leppkes, M.; Zheng, Y.; Wittkopf, N.; Warntjen, M.; Lehr, H.-A.; Hirth, S.; Weigmann, B.; Wirtz, S. Stat3 links il-22 signaling in intestinal epithelial cells to mucosal wound healing. *J. Exp. Med.* **2009**, *206*, 1465–1472. [CrossRef]

103. Ordas, I.; Eckmann, L.; Talamini, M.; Baumgart, D.C.; Sandborn, W.J. Ulcerative colitis. *Lancet* **2012**, *380*, 1606–1619. [CrossRef]

104. Liu, T.; Zhang, L.; Joo, D.; Sun, S.C. Nf-kappab signaling in inflammation. *Signal. Transduct Target.* **2017**, 2.

105. Quigley, E.M. Gut bacteria in health and disease. *Gastroenterol. Hepatol.* **2013**, *9*, 560.

106. Mazmanian, S.K.; Round, J.L.; Kasper, D.L. A microbial symbiosis factor prevents intestinal inflammatory disease. *Nature* **2008**, *453*, 620. [CrossRef] [PubMed]

107. Lindemann, R.K.; Gabrielli, B.; Johnstone, R.W. Histone-deacetylase inhibitors for the treatment of cancer. *Cell Cycle* **2004**, *3*, 777–786. [CrossRef]

108. Smith, P.M.; Howitt, M.R.; Panikov, N.; Michaud, M.; Gallini, C.A.; Bohlooly-y, M.; Glickman, J.N.; Garrett, W.S. The microbial metabolites, short-chain fatty acids, regulate colonic treg cell homeostasis. *Science* **2013**, *341*, 569–573. [CrossRef]

109. Matsuoka, K.; Uemura, Y.; Kanai, T.; Kunisaki, R.; Suzuki, Y.; Yokoyama, K.; Yoshimura, N.; Hibi, T. Efficacy of bifidobacterium breve fermented milk in maintaining remission of ulcerative colitis. *Dig. Dis. Sci.* **2018**, *63*, 1910–1919. [CrossRef]

110. Tamaki, H.; Nakase, H.; Inoue, S.; Kawanami, C.; Itani, T.; Ohana, M.; Kusaka, T.; Uose, S.; Hisatsune, H.; Tojo, M. Efficacy of probiotic treatment with bifidobacterium longum 536 for induction of remission in active ulcerative colitis: A randomized, double-blinded, placebo-controlled multicenter trial. *Dig. Endosc.* **2016**, *28*, 67–74. [CrossRef]

111. Zocco, M.; Dal Verme, L.Z.; Cremonini, F.; Piscaglia, A.; Nista, E.; Candelli, M.; Novi, M.; Rigante, D.; Cazzato, I.; Ojetti, V. Efficacy of lactobacillus gg in maintaining remission of ulcerative colitis. *Aliment. Pharmacol. Ther.* **2006**, *23*, 1567–1574. [CrossRef]

112. Shen, Z.-H.; Zhu, C.-X.; Quan, Y.-S.; Yang, Z.-Y.; Wu, S.; Luo, W.-W.; Tan, B.; Wang, X.-Y. Relationship between intestinal microbiota and ulcerative colitis: Mechanisms and clinical application of probiotics and fecal microbiota transplantation. *World J. Gastroenterol.* **2018**, *24*, 5. [CrossRef]

113. Ott, S.; Musfeldt, M.; Wenderoth, D.; Hampe, J.; Brant, O.; Fölsch, U.; Timmis, K.; Schreiber, S. Reduction in diversity of the colonic mucosa associated bacterial microflora in patients with active inflammatory bowel disease. *Gut* **2004**, *53*, 685–693. [CrossRef]

114. Morgan, X.C.; Tickle, T.L.; Sokol, H.; Gevers, D.; Devaney, K.L.; Ward, D.V.; Reyes, J.A.; Shah, S.A.; LeLeiko, N.; Snapper, S.B. Dysfunction of the intestinal microbiome in inflammatory bowel disease and treatment. *Genome Biol.* **2012**, *13*, R79. [CrossRef]

115. Sokol, H.; Lepage, P.; Seksik, P.; Dore, J.; Marteau, P. Temperature gradient gel electrophoresis of fecal 16s rrna reveals active escherichia coli in the microbiota of patients with ulcerative colitis. *J. Clin. Microbiol.* **2006**, *44*, 3172–3177. [CrossRef]

116. Obiso, R.; Azghani, A.O.; Wilkins, T.D. The bacteroides fragilis toxin fragilysin disrupts the paracellular barrier of epithelial cells. *Infect. Immun.* **1997**, *65*, 1431–1439.

117. Wells, C.L.; Van de Westerlo, E.; Jechorek, R.P.; Feltis, B.; Wilkins, T.; Erlandsen, S. Bacteroides fragilis enterotoxin modulates epithelial permeability and bacterial internalization by ht-29 enterocytes. *Gastroenterology* **1996**, *110*, 1429–1437. [CrossRef]

118. Akiyama, S.; Nesumi, A.; Maeda-Yamamoto, M.; Uehara, M.; Murakami, A. Effects of anthocyanin-rich tea "sunrouge" on dextran sodium sulfate-induced colitis in mice. *BioFactors* **2012**, *38*, 226–233. [CrossRef]

119. Biedermann, L.; Mwinyi, J.; Scharl, M.; Frei, P.; Zeitz, J.; Kullak-Ublick, G.A.; Vavricka, S.R.; Fried, M.; Weber, A.; Humpf, H.-U. Bilberry ingestion improves disease activity in mild to moderate ulcerative colitis—an open pilot study. *J. Crohn's Colitis* **2013**, *7*, 271–279. [CrossRef]

120. Kim, J.-M.; Kim, J.-S.; Yoo, H.; Choung, M.-G.; Sung, M.-K. Effects of black soybean [glycine max (l.) merr.] seed coats and its anthocyanidins on colonic inflammation and cell proliferation in vitro and in vivo. *J. Agric. Food Chem.* **2008**, *56*, 8427–8433. [CrossRef]

121. Monk, J.M.; Wu, W.; Hutchinson, A.L.; Pauls, P.; Robinson, L.E.; Power, K.A. Navy and black bean supplementation attenuates colitis-associated inflammation and colonic epithelial damage. *J. Nutr. Biochem.* **2018**, *56*, 215–223. [CrossRef]

122. Bibi, S.; Kang, Y.; Du, M.; Zhu, M.-J. Dietary red raspberries attenuate dextran sulfate sodium-induced acute colitis. *J. Nutr. Biochem.* **2018**, *51*, 40–46. [CrossRef]

123. Turksen, K.; Troy, T.-C. Barriers built on claudins. *J. Cell Sci.* **2004**, *117*, 2435–2447. [CrossRef]

124. Morita, K.; Furuse, M.; Fujimoto, K.; Tsukita, S. Claudin multigene family encoding four-transmembrane domain protein components of tight junction strands. *Proc. Natl. Acad. Sci. USA* **1999**, *96*, 511–516. [CrossRef]

125. Al-Asmakh, M.; Hedin, L. Microbiota and the control of blood-tissue barriers. *Tissue Barriers* **2015**, *3*, e1039691. [CrossRef]

126. Feldman, G.J.; Mullin, J.M.; Ryan, M.P. Occludin: Structure, function and regulation. *Adv. Drug Deliv. Rev.* **2005**, *57*, 883–917. [CrossRef]

127. Umeda, K.; Matsui, T.; Nakayama, M.; Furuse, K.; Sasaki, H.; Furuse, M.; Tsukita, S. Establishment and characterization of cultured epithelial cells lacking expression of zo-1. *J. Biol. Chem.* **2004**, *279*, 44785–44794. [CrossRef]

128. Groschwitz, K.R.; Hogan, S.P. Intestinal barrier function: Molecular regulation and disease pathogenesis. *J. Allergy Clin. Immunol.* **2009**, *124*, 3–20. [CrossRef]

129. Sun, X.; Du, M.; Navarre, D.A.; Zhu, M.J. Purple potato extract promotes intestinal epithelial differentiation and barrier function by activating amp-activated protein kinase. *Mol. Nutr. Food Res.* **2018**, *62*, 1700536. [CrossRef]

130. Zhang, C.; Monk, J.M.; Lu, J.T.; Zarepoor, L.; Wu, W.; Liu, R.; Pauls, K.P.; Wood, G.A.; Robinson, L.; Tsao, R. Cooked navy and black bean diets improve biomarkers of colon health and reduce inflammation during colitis. *Br. J. Nutr.* **2014**, *111*, 1549–1563. [CrossRef]

131. Shan, Q.; Zheng, Y.; Lu, J.; Zhang, Z.; Wu, D.; Fan, S.; Hu, B.; Cai, X.; Cai, H.; Liu, P. Purple sweet potato color ameliorates kidney damage via inhibiting oxidative stress mediated nlrp3 inflammasome activation in high fat diet mice. *Food Chem. Toxicol.* **2014**, *69*, 339–346. [CrossRef]

132. Triebel, S.; Trieu, H.-L.; Richling, E. Modulation of inflammatory gene expression by a bilberry (vaccinium myrtillus l.) extract and single anthocyanins considering their limited stability under cell culture conditions. *J. Agric. Food Chem.* **2012**, *60*, 8902–8910. [CrossRef]

133. Fischer, J.G.; Keirsey, K.I.; Kirkland, R.; Lee, S.; Grunewald, Z.I.; de La Serre, C.B. Blueberry supplementation influences the gut microbiota, inflammation, and insulin resistance in high-fat-diet–fed rats. *J. Nutr.* **2018**, *148*, 209–219.

134. Wu, L.H.; Xu, Z.L.; Dong, D.; He, S.A.; Yu, H. Protective effect of anthocyanins extract from blueberry on tnbs-induced ibd model of mice. *Evid.-Based Complementary Altern. Med. Ecam* **2011**, *2011*, 525462. [CrossRef]

135. Chen, T.; Hu, S.; Zhang, H.; Guan, Q.; Yang, Y.; Wang, X. Anti-inflammatory effects of dioscorea alata l. Anthocyanins in a tnbs-induced colitis model. *Food Funct.* **2017**, *8*, 659–669. [CrossRef]

136. Boussenna, A.; Cholet, J.; Goncalves-Mendes, N.; Joubert-Zakeyh, J.; Fraisse, D.; Vasson, M.P.; Texier, O.; Felgines, C. Polyphenol-rich grape pomace extracts protect against dextran sulfate sodium-induced colitis in rats. *J. Sci. Food Agric.* **2016**, *96*, 1260–1268. [CrossRef]

137. Choe, M.-R.; Ji Hye, K.; Yoo, H.; Yang, C.-H.; Kim, M.-O.; Yu, R.-N.; Choe, S.-Y. Cyanidin and cyanidin-3-o-β-d-glucoside suppress the inflammatory responses of obese adipose tissue by inhibiting the release of chemokines mcp-1 and mrp-2. *J. Food Sci. Nutr.* **2007**, *12*, 148–153.

138. Lee, H.H.; Lee, S.G.; Shin, J.S.; Lee, H.Y.; Yoon, K.; Ji, Y.W.; Jang, D.S.; Lee, K.T. P-coumaroyl anthocyanin mixture isolated from tuber epidermis of solanum tuberosum attenuates reactive oxygen species and pro-inflammatory mediators by suppressing nf-kappab and stat1/3 signaling in lps-induced raw264.7 macrophages. *Biol. Pharm. Bull.* **2017**, *40*, 1894–1902. [CrossRef]

139. Le Phuong Nguyen, T.; Fenyvesi, F.; Remenyik, J.; Homoki, J.R.; Gogolak, P.; Bacskay, I.; Feher, P.; Ujhelyi, Z.; Vasvari, G.; Vecsernyes, M.; et al. Protective effect of pure sour cherry anthocyanin extract on cytokine-induced inflammatory caco-2 monolayers. *Nutrients* **2018**, *10*, 861. [CrossRef]

140. Jung, S.K.; Lim, T.-G.; Seo, S.G.; Lee, H.J.; Hwang, Y.-S.; Choung, M.-G.; Lee, K.W. Cyanidin-3-o-(2″-xylosyl)-glucoside, an anthocyanin from siberian ginseng (acanthopanax senticosus) fruits, inhibits uvb-induced cox-2 expression and ap-1 transactivation. *Food Sci. Biotechnol.* **2013**, *22*, 507–513. [CrossRef]

141. Li, L.; Wang, L.; Wu, Z.; Yao, L.; Wu, Y.; Huang, L.; Liu, K.; Zhou, X.; Gou, D. Anthocyanin-rich fractions from red raspberries attenuate inflammation in both raw264.7 macrophages and a mouse model of colitis. *Sci. Rep.* **2014**, *4*, 6234. [CrossRef]

142. Pereira, S.R.; Pereira, R.; Figueiredo, I.; Freitas, V.; Dinis, T.C.; Almeida, L.M. Comparison of anti-inflammatory activities of an anthocyanin-rich fraction from portuguese blueberries (vaccinium corymbosum l.) and 5-aminosalicylic acid in a tnbs-induced colitis rat model. *PLoS ONE* **2017**, *12*, e0174116. [CrossRef] [PubMed]

143. Roth, S.; Spalinger, M.R.; Gottier, C.; Biedermann, L.; Zeitz, J.; Lang, S.; Weber, A.; Rogler, G.; Scharl, M. Bilberry-derived anthocyanins modulate cytokine expression in the intestine of patients with ulcerative colitis. *PLoS ONE* **2016**, *11*, e0154817. [CrossRef]

144. Faria, A.; Fernandes, I.; Norberto, S.; Mateus, N.; Calhau, C.A.O. Interplay between anthocyanins and gut microbiota. *J. Agric. Food Chem.* **2014**, *62*, 6898–6902. [CrossRef] [PubMed]

145. Ley, R.E.; Peterson, D.A.; Gordon, J.I. Ecological and evolutionary forces shaping microbial diversity in the human intestine. *Cell* **2006**, *124*, 837–848. [CrossRef] [PubMed]

146. Pan, P.; Lam, V.; Salzman, N.; Huang, Y.-W.; Yu, J.; Zhang, J.; Wang, L.-S. Black raspberries and their anthocyanin and fiber fractions alter the composition and diversity of gut microbiota in f-344 rats. *Nutr. Cancer* **2017**, *69*, 943–951. [CrossRef]

147. Cassidy, A.; Minihane, A.M. The role of metabolism (and the microbiome) in defining the clinical efficacy of dietary flavonoids. *Am. J. Clin. Nutr* **2017**, *105*, 10–22. [CrossRef]

148. Selma, M.V.; Espin, J.C.; Tomas-Barberan, F.A. Interaction between phenolics and gut microbiota: Role in human health. *J. Agric. Food Chem.* **2009**, *57*, 6485–6501. [CrossRef]

149. Farombi, E.O.; Adedara, I.A.; Awoyemi, O.V.; Njoku, C.R.; Micah, G.O.; Esogwa, C.U.; Owumi, S.E.; Olopade, J.O. Dietary protocatechuic acid ameliorates dextran sulphate sodium-induced ulcerative colitis and hepatotoxicity in rats. *Food Funct.* **2016**, *7*, 913–921. [CrossRef] [PubMed]

150. Parkar, S.G.; Trower, T.M.; Stevenson, D.E. Fecal microbial metabolism of polyphenols and its effects on human gut microbiota. *Anaerobe* **2013**, *23*, 12–19. [CrossRef] [PubMed]

151. Molan, A.-L.; Liu, Z.; Kruger, M. The ability of blackcurrant extracts to positively modulate key markers of gastrointestinal function in rats. *World J. Microbiol. Biotechnol.* **2010**, *26*, 1735–1743. [CrossRef]

152. Bialonska, D.; Ramnani, P.; Kasimsetty, S.G.; Muntha, K.R.; Gibson, G.R.; Ferreira, D. The influence of pomegranate by-product and punicalagins on selected groups of human intestinal microbiota. *Int. J. Food Microbiol.* **2010**, *140*, 175–182. [CrossRef]

153. Gibson, G.; Wang, X. Regulatory effects of bifidobacteria on the growth of other colonic bacteria. *J. Appl. Bacteriol.* **1994**, *77*, 412–420. [CrossRef] [PubMed]

154. Miao, M.; Jiang, H.; Jiang, B.; Zhang, T.; Cui, S.W.; Jin, Z. Phytonutrients for controlling starch digestion: Evaluation of grape skin extract. *Food Chem.* **2014**, *145*, 205–211. [CrossRef]

155. Camelo-Méndez, G.A.; Agama-Acevedo, E.; Sanchez-Rivera, M.M.; Bello-Pérez, L.A. Effect on in vitro starch digestibility of mexican blue maize anthocyanins. *Food Chem.* **2016**, *211*, 281–284. [CrossRef]

156. Magallanes-Cruz, P.A.; Flores-Silva, P.C.; Bello-Perez, L.A. Starch structure influences its digestibility: A review. *J. Food Sci.* **2017**, *82*, 2016–2023. [CrossRef]

157. Immerstrand, T.; Andersson, K.E.; Wange, C.; Rascon, A.; Hellstrand, P.; Nyman, M.; Cui, S.W.; Bergenståhl, B.; Trägårdh, C.; Öste, R. Effects of oat bran, processed to different molecular weights of β-glucan, on plasma lipids and caecal formation of scfa in mice. *Br. J. Nutr.* **2010**, *104*, 364–373. [CrossRef]

158. Immerstrand, T. *Cholesterol-lowering properties of oats: Effects of processing and the role of oat components*; Division of Applied Nutrition and Food Chemistry, Lund University: Lund, Sweden, 2010.

159. Dostal, A.; Fehlbaum, S.; Chassard, C.; Zimmermann, M.B.; Lacroix, C. Low iron availability in continuous in vitro colonic fermentations induces strong dysbiosis of the child gut microbial consortium and a decrease in main metabolites. *Fems Microbiol. Ecol.* **2013**, *83*, 161–175. [CrossRef]

160. Xie, Y.; Zhu, X.; Li, Y.; Wang, C. Analysis of the ph-dependent fe (iii) ion chelating activity of anthocyanin extracted from black soybean [glycine max (l.) merr.] coats. *J. Agric. Food Chem.* **2018**, *66*, 1131–1139. [CrossRef]

Int. J. Mol. Sci. **2019**, *20*, 2588

161. Buchweitz, M.; Brauch, J.; Carle, R.; Kammerer, D. Application of ferric anthocyanin chelates as natural blue food colorants in polysaccharide and gelatin based gels. *Food Res. Int.* **2013**, *51*, 274–282. [CrossRef]
162. Serobatse, K.R.; Kabanda, M.M. Antioxidant and antimalarial properties of butein and homobutein based on their ability to chelate iron (ii and iii) cations: A dft study in vacuo and in solution. *Eur. Food Res. Technol.* **2016**, *242*, 71–90. [CrossRef]

 International Journal of
Molecular Sciences

Review

MicroRNA-Mediated Health-Promoting Effects of Phytochemicals

Hara Kang

Division of Life Sciences, College of Life Sciences and Bioengineering, Incheon National University, Incheon 22012, Korea; harakang@inu.ac.kr; Tel.: +82-32-835-8238; Fax: +82-32-835-0763

Received: 29 April 2019; Accepted: 21 May 2019; Published: 23 May 2019

Abstract: Phytochemicals are known to benefit human health by modulating various cellular processes, including cell proliferation, apoptosis, and inflammation. Due to the potential use of phytochemicals as therapeutic agents against human diseases such as cancer, studies are ongoing to elucidate the molecular mechanisms by which phytochemicals affect cellular functions. It has recently been shown that phytochemicals may regulate the expression of microRNAs (miRNAs). MiRNAs are responsible for the fine-tuning of gene expression by controlling the expression of their target mRNAs in both normal and pathological cells. This review summarizes the recent findings regarding phytochemicals that modulate miRNA expression and promote human health by exerting anticancer, photoprotective, and anti-hepatosteatosis effects. Identifying miRNAs modulated by phytochemicals and understanding the regulatory mechanisms mediated by their target mRNAs will facilitate the efforts to maximize the therapeutic benefits of phytochemicals.

Keywords: phytochemicals; microRNA; health-promoting effects

1. Introduction

Phytochemicals are plant-derived compounds abundant in a variety of fruits, vegetables, herbs, and many other plants [1]. More than 10,000 phytochemicals have been identified to date, and their medicinal properties including anti-inflammatory and anticancer effects, are being investigated [2]. Polyphenols, alkaloids, terpenoids, and organosulfur compounds are well-known phytochemical groups with anticancer properties [3]. For example, polyphenols such as resveratrol and curcumin modulate oxidative stress and inflammatory signaling, and perform antioxidant and anti-inflammatory activities [4,5]. In addition, they inhibit angiogenesis by regulating the gene expression of key regulators, such as vascular endothelial growth factor and hypoxia inducible factor 1 subunit alpha, and promote apoptosis by modulating levels of Bcl2, BCL2-associated X (Bax), and p53 [6,7]. These activities cooperatively lead to the prevention of carcinogenesis. Due to the medicinal properties of phytochemicals and their availability as therapeutic agents, their underlying molecular mechanisms are being investigated.

Emerging data suggest that some phytochemicals regulate the expression of various microRNAs (miRNAs). MiRNAs are small, noncoding RNAs involved in a wide variety of cellular processes, such as development, proliferation, differentiation, and apoptosis [8]. MiRNAs regulate gene expression by binding to target mRNAs via base pairing with the 3′-untranslated regions (3′UTRs) of the target mRNAs [9]. The miRNA-target mRNA interaction usually results in the degradation or translational repression of the target mRNA. As a single miRNA can target multiple mRNAs implicated in various cellular phenomena, miRNAs are crucial post-transcriptional regulators for the fine-tuning of normal cellular physiology [8]. Expression of miRNAs is tissue- or developmental stage-specific, and their aberrant expression is associated with the development of pathogenic conditions [10]. For example, abnormal expression of several miRNAs is associated with the initiation and progression of carcinogenesis and metastasis [11]. Therefore, miRNA profiles in a wide range of diseases have been analyzed for their potential use as diagnostic or prognostic indicators.

Extensive emerging evidence suggests that phytochemicals affect the expression profiles of miRNAs in pathological conditions, especially cancer. Moreover, several studies have identified novel target mRNAs of phytochemical-modulated miRNAs and investigated the underlying mechanism for the miRNA-mediated therapeutic activities of a few phytochemicals. Herein, we review the current information on phytochemicals that benefit human health by modulating miRNA expression. Specifically, we discuss phytochemicals that exhibit anticancer, photoprotective, and anti-hepatosteatosis effects. This not only helps to understand health-promoting effects of phytochemicals at the molecular level, but also allows us to think about further research and therapeutic development.

2. Phytochemicals and miRNAs

Although emerging data clearly suggest that miRNA expression is specifically regulated by phytochemicals, relatively little is known about their molecular relationships to date. Baselga-Escudero et al. suggested that polyphenols directly bind to mature miRNAs and that the chemical structure of polyphenols influences the expression of miRNAs [12]. In 1 H NMR spectroscopy studies, direct binding of resveratrol and (-)-epigallocatechin-3-gallate (EGCG) to miR-33a and miR-122 was observed [12]. While resveratrol binds miR-33a and miR-122 through an A ring interaction and increases the expression levels of these miRNAs, EGCG decreases miR-33a and miR-122 expression by direct binding through an interaction with all rings in the molecules. Additionally, phytochemicals have been shown to affect miRNA levels by regulating molecules associated with controlling miRNA expression [13,14]. For example, EGCG binds the HIF-1α protein, which is a transcriptional activator of miR-210, and interferes with the hydroxylation of Pro residues in the oxygen-dependent degradation domain [14]. As the hydroxylation of Pro residues is essential for proteasome-mediated degradation of HIF-1α [15], EGCG binding increases HIF-1α expression and consequently enhances miR-210 expression levels. Pan el al. demonstrated that resveratrol regulates the expression of c-Myc, which is a transcriptional activator of miR-17, resulting in the suppression of oncogenic miR-17 levels [13].

3. Phytochemicals with Anticancer Effects

The anticancer effect is a well-known phytochemical function that is important for promoting human health [16]. We summarize the recent reports on phytochemicals that exert their intracellular anticancer effects by altering miRNA expression (Figure 1).

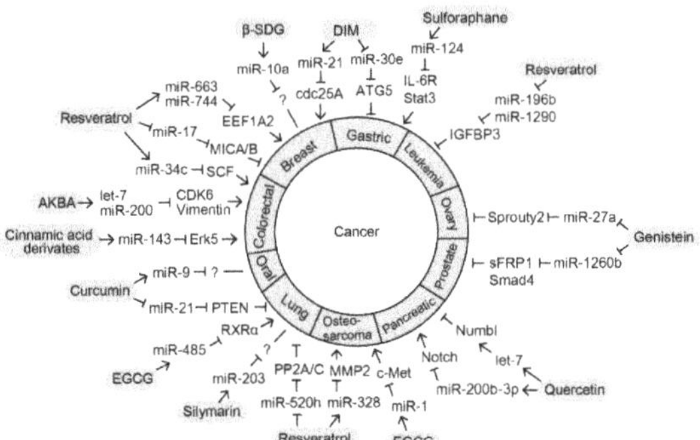

Figure 1. MicroRNA (miRNA)-mediated anticancer functions of phytochemicals. Phytochemicals regulate the expression levels of miRNAs implicated in the development of various cancers. Arrow indicates activation at the levels of expression; lines with a perpendicular line at the end indicates inhibition of expression; question mark indicates an unknown target molecule.

3.1. Resveratrol

Resveratrol (*trans*-3, 5, 4′-trihydroxystilbene) is a polyphenol found in various plants, including berries and grapes [17]. Resveratrol is known for its beneficial properties in human health and has been intensively studied for its anticancer function in multiple cancers such as breast, prostate, stomach, pancreas, and thyroid cancers since 1997 [18,19]. Recently, resveratrol was tested in clinical trials for colon cancer [20]. Several studies provide evidence that the anticancer function of resveratrol is mediated by modulating miRNA expression [13,21–26].

Yang et al. analyzed changes in the profiles of 754 human miRNAs following resveratrol treatment in osteosarcoma cells [24]. In response to resveratrol, miR-328 is the most highly upregulated miRNA. MiR-328 targets the 3′UTR of matrix metalloproteinase 2 (MMP2), an essential regulator of metastasis. Therefore, resveratrol exerts its anticancer effects, such as inhibition of migration, invasion, and adhesion of osteosarcoma cells, by regulating the expression of miR-328 and MMP2.

In lung cancer cells, miR-520h is downregulated following resveratrol treatment, and consequently its target genes, PP2A/C, are derepressed [25]. PP2A/C inhibits AKT/NF-κB-mediated FOXC2 expression. Since the FOXC2 is a key regulator that activates epithelial-mesenchymal transition (EMT) and metastasis, suppression of FOXC2 by resveratrol-mediated regulation of the miR-520h-PP2A/C axis induces mesenchymal-epithelial transition (MET) and exerts anticancer effects in lung cancer cells.

In the breast cancer cell line MCF7, resveratrol treatment increases the expression levels of miR-663 and miR-744, and inhibits cell proliferation [21]. Vislovukh et al. demonstrated that the putative oncogene, isoform A2 of eukaryotic translation elongation factor 1A (EEF1A2), is a direct target of both miR-663 and miR-744, and downregulation of EEF1A2 negatively affects cell proliferation [21]. Therefore, the anticancer effect of resveratrol on breast cancer progression could be mediated by regulating the expression of miR-663 and miR-744 that suppress EEF1A2. Resveratrol also promotes natural killer (NK) cell-mediated anticancer immune responses in breast cancer [13]. Oncogenic miR-17, which belongs to the miR-17-92 cluster, is transcriptionally upregulated by the binding of c-Myc and promotes proliferation, migration, and angiogenesis of breast cancer cells [13]. Resveratrol downregulates c-Myc expression and consequently miR-17 expression is suppressed. Suppression of miR-17 upregulates its target genes, such as major histocompatibility complex class I chain-related proteins A and B (MICA and MICB). Ultimately, MICA and MICB increase the susceptibility of breast cancer cells to lysis by NK cells.

Resveratrol also affects the proliferation, migration, invasion, and apoptosis of colorectal cancer cells by regulating miR-34c-mediated stem cell factor (SCF) expression [23]. In HT-29 cells in vitro and in mouse xenografts in vivo, exposure to resveratrol increases the level of miR-34c while decreasing the level of SCF, a known target of miR-34c.

It has been reported that the anticancer effect of resveratrol in acute lymphoblastic leukemia (ALL) is mediated by miRNAs [26]. Zhou et al. observed that miR-196b and miR-1290 are highly expressed in T-cell ALL and B-cell ALL, respectively, and their expression is downregulated by resveratrol [26]. Both miR-196b and miR-1290 directly bind to the 3′UTR of insulin-like growth factor binding protein 3 (IGFBP3) and the downregulation of IGFBP3 is associated with the development of leukemia. Therefore, the resveratrol-induced decrease in miR-196b and miR-1290 levels and consequent recovery of IGFBP3 expression might ultimately inhibit the proliferation and migration of ALL cells.

Mitochondrial dynamics are involved in the development of human diseases including cancer; increased fission or decreased fusion leads to the formation of fragmented mitochondria that triggers apoptosis [27]. Resveratrol may exert an anticancer effect by regulating mitochondrial dynamics via modulating miRNA levels [22]. In several cancer cell lines, such as DLD1, HeLa, and MCF-7, resveratrol increases the expression of miR-326, which in turn directly targets a regulator of mitochondrial fusion, pyruvate kinase M2 (PKM2) [28]. In this way, resveratrol promotes cancer cell apoptosis by decreasing mitochondrial fusion induced by the miR-326-PKM2 axis.

3.2. (-)-Epigallocatechin-3-Gallate

EGCG is a major polyphenol in green tea and has been shown to have anticancer effects [29]. In a variety of cancer cells, miRNA profiling studies have been performed to determine the role of miRNAs in regulating EGCG-mediated anticancer functions [14,30,31]. In human hepatocellular carcinoma HepG2 cells, 13 miRNAs were upregulated, while 48 miRNAs were downregulated by EGCG [30]. Tsang et al. demonstrated that miR-16 potently promoted apoptosis [30]. EGCG-induced miR-16 expression suppresses the anti-apoptotic protein Bcl-2 level, leading to the apoptosis of HepG2 cells.

Wang et al. found that the expression of miR-210 was prominently increased by EGCG in both human and mouse lung cancer cells in vitro [14]. Upregulation of miR-210 inhibits the proliferation and anchorage-independent growth of lung cancer cells. Zhou et al. further supported the significant role of miRNAs in the anticancer function of EGCG in vivo [31]. EGCG inhibits tobacco carcinogen-induced lung tumors in A/J mice by modulating the expression of 21 miRNAs, including miR-210. Additionally, Jiang et al. showed that EGCG enhances the expression levels of miR-485 which directly targets the retinoid X receptor (RXRα) in non-small cell lung cancer (NSCLC) cells [32]. EGCG-induced miR-485-RXRα axis represses cancer stem cells (CSC)-like characteristics of NSCLC cells, resulting in the inhibition of cell growth and promotion of apoptosis.

The anticancer function of EGCG in suppressing cell proliferation has been shown to be mediated by miRNAs in osteosarcoma and oral squamous cell carcinomas [33,34]. In an miRNA profiling study, miR-1 was identified as a mediator of EGCG anticancer activities [34]. miR-1 levels are downregulated in osteosarcoma tumor tissues and miR-1 is increased by EGCG to target a proto-oncogene c-MET. In oral squamous cell carcinomas, miR-204 is upregulated by EGCG and targets Slug and Sox4, which are associated with cancer stemness and EMT [33].

Moreover, it has been reported that EGCG affects the expression of several miRNAs known to act as oncogenes or tumor suppressors [35,36]. In prostate cancer cells of mice treated with EGCG, oncogenic miR-21 is downregulated and tumor suppressive miR-330 is upregulated [36]. In human malignant neuroblastoma SH-SY5Y and SK-N-DZ cells, oncogenic miR-92, miR-93, and miR-105b are downregulated and tumor suppressive miR-7-1, miR-34a, and miR-99a are upregulated by EGCG [35]. Therefore, EGCG may exert its anticancer effects by regulating miRNAs involved in tumorigenesis.

3.3. Curcumin

Curcumin is a natural phytochemical derived from the root and rhizome of *Curcuma longa* and has been known to have antioxidant, anti-inflammatory, and anticancer properties [37,38]. Recent studies suggest that the anticancer effect of curcumin is mediated by modulation of miRNAs. Zhang et al. demonstrated the anti-proliferative and pro-apoptotic activities of curcumin in NSCLC cells via the miR-21-phosphatase and tensin homolog (PTEN) axis [39]. In NSCLC cells, curcumin downregulates the expression of miR-21, which is a typical oncogenic miRNA implicated in cancer development and progression. Downregulation of miR-21 by curcumin results in the derepression of its target gene, PTEN. Since PTEN is a tumor suppressor, the elevation of PTEN expression might mediate the anticancer effects of curcumin [40].

Aberrant activation of the Wnt signaling pathway affects cell growth, the cell cycle, and invasion in carcinogenesis [41]. Curcumin was shown to suppress cancer cell growth by regulating the Wnt signaling pathway [42]. In oral squamous cell carcinoma, miR-9 is downregulated, and curcumin increases miR-9 expression [43]. The upregulation of miR-9 by curcumin elevates the levels of GSK-3β and phosphorylated GSK-3β, resulting in inhibition of the Wnt signaling pathway. Xiao et al. observed that the expression level of cyclin D1, a putative target of miR-9, was reduced when miR-9 was overexpressed in SCC-9 cells, although whether cyclin D1 is targeted by miR-9 via direct binding was not verified [43]. The direct target gene of miR-9 under curcumin-treated conditions remains to be identified.

3.4. Quercetin

Quercetin is a flavonoid derived from fruits and vegetables, such as berries, apples, onions, and broccoli [44]. Recently, anticancer functions of quercetin through miRNA modulation were reported in pancreatic cancer cells [45]. Nwaeburu et al. performed a miRNA profiling analysis of pancreatic ductal adenocarcinoma following treatment with quercetin [45]. They found 105 differentially expressed miRNAs in quercetin-treated cells; 80 miRNAs including let-7c, miR-200a-3p, and miR-200b-3p were upregulated, while 25 miRNAs including miR-103a-3p, miR-125b, and miR-1202 were downregulated. Nwaeburu et al. then investigated how let-7c, one of the most highly upregulated miRNAs, inhibits pancreatic cancer progression [45]. While most miRNAs inhibit the expression of their target genes, let-7c directly targets the 3′UTR of Numbl, an inhibitor of Notch expression, and increases the expression of Numbl. Induction of Numbl by let-7c subsequently antagonizes Notch signaling, known to be involved in cell proliferation, angiogenesis, and development [46], thus leading to tumor growth inhibition and increased apoptosis of pancreatic ductal adenocarcinoma cells.

In addition, an anti-proliferative function of miR-200b-3p induced by quercetin in cancer stem cells (CSCs) has been reported [47]. Notch and Numb are crucial for the regulation of CSC self-renewal; Notch is required for symmetric division and Numb is a marker for asymmetric division [48]. Nwaeburu et al. demonstrated that miR-200b-3p targets Notch by directly binding its 3′UTR, which in turn inhibits proliferation and self-renewal of CSCs [47]. Together, these results indicate that quercetin exerts its anticancer effects by regulating Notch signaling via miRNAs.

3.5. 3,3′-Diindolylmethane

One of the natural derivatives of indoles from cruciferous vegetables, 3,3′-diindolylmethane (DIM), has been reported to have an anticancer function in various cancer cells [49]. Ye et al. demonstrated that DIM inhibited the proliferation of gastric cancer cells in vitro and tumor growth in vivo in a xenograft mouse model [50]. DIM downregulates miR-30e, which is highly expressed in multiple types of tumors. Autophagy-related gene 5 (ATG5), an essential component for the generation of autophagosomes [51], was validated as a direct target of miR-30e. These observations suggest that the suppression of miR-30e by DIM rescues ATG5 expression and induces autophagy, which in turn inhibits the proliferation of gastric cancer cells. Therefore, autophagy regulation mediated by the miR-30e-ATG5 axis is crucial for the anticancer function of DIM in gastric cancer cells.

In addition, DIM inhibits breast cancer cell growth in vitro and in vivo [52]. DIM treatment inhibits the proliferation of MCF-7 and MDA-MB-468 breast cancer cells, and the growth of transplanted human breast carcinoma cells in a mouse model. The inhibitory effect of DIM on cell proliferation is due to cell cycle arrest. Jin et al. observed that in response to DIM, miR-21 is upregulated and cdc25A, a putative target of miR-21, is downregulated [52]. As cdc25A is a crucial regulator of cell cycle progression, the anticancer function of DIM via cell cycle arrest might be mediated in part by the modulation of miR-21 and cdc25A expression.

3.6. Sulforaphane

Sulforaphane (SFN), a dietary phytochemical converted from cruciferous plants, such as broccoli, carrots, and kale, is known to have anticancer functions [53]. Wang et al. reported that SFN enhances the anticancer function of cisplatin, a DNA-targeting cytotoxic platinum compound, by modulating miRNA expression in gastric cancer cells [54]. Treatment with cisplatin alone causes side effects that increase the CSC-like properties of gastric cancer cells by activating the interleukin-6 (IL-6)-mediated signal transducer and activator of transcription 3 (Stat3) signaling; however, cotreatment with SFN improves the chemotherapy efficacy. SFN increases the expression of miR-124, which targets the IL-6 receptor (IL-6R) and Stat3 by directly binding their 3′UTRs. IL-6R is known to mediate activation of the downstream molecules of IL-6 signaling, such as mitogen-activated protein kinase, phosphatidylinositol

3-kinase (PI3K), and Stat3 [55]. Therefore, SFN inhibits the CSC-like viability of gastric cancer cells by regulating the expression levels of miR-124 and its target genes, IL-6R and Stat3.

In breast cancer cells, SFN affects cell cycle arrest, senescence, apoptosis, and autophagy [56]. Lewinska et al. investigated the changes in the miRNA profiles of three breast cancer cell lines, MCF-7, MDA-MB-231, and SK-BR-3, upon treatment with SFN [56]. Sixty miRNAs were upregulated and 32 were downregulated. Specifically, the levels of miR-23b-3p, miR-92b-3p, miR-381-3p, and miR-382 were significantly reduced in the three cell lines. The target genes recognized by these miRNAs that link them to their anticancer functions remain to be identified.

3.7. Genistein

Genistein is a major isoflavonoid isolated from soybeans [57]. Genistein has been shown to prevent multiple cancers by regulating the expression of specific oncogenic miRNAs, such as miR-27a and miR-1260b [58–63]. Genistein downregulates miR-27a expression in uveal melanoma, ovarian cancer, pancreatic cancer, and lung cancer cells to inhibit cell proliferation, migration, or invasion [60–63]. Zinc finger and BTB domain (Broad-Complex, Tramtrack and Bric a brac) containing 10 and Sprouty2 were validated as targets of miR-27a in uveal melanoma cells and ovarian cancer cells, respectively [60,62]. Oncogenic miR-1260b is downregulated by genistein in renal and prostate cancer cells [58,59]. The Wnt signaling pathway is typically activated in cancer cells to promote tumorigenesis and miR-1260b targets tumor suppressor genes associated with Wnt signaling, such as *sFRP1*, *Dkk2*, and *Smad4* [58,59]. Therefore, genistein-induced downregulation of miR-1260b rescues the expression levels of target genes that antagonize Wnt signaling and ultimately inhibits cancer cell proliferation and invasion.

3.8. Acetyl-11-Keto-β-Boswellic Acid

Boswellic acids, the major components of a gum resin derived from *Boswellia serrata*, have been known to perform anti-inflammatory functions [64]. Takahashi et al. reported that acetyl-11-keto-β-boswellic acid (AKBA), an active component in boswellic acids, has an anticancer function in colorectal cancer cells [64]. AKBA inhibits cell viability, colony formation, proliferation, and migration, and enhances apoptosis in colorectal cancer cells. This anticancer effect is partly mediated by the regulation of well-known tumor suppressive miRNAs, such as the let-7 and miR-200 families [64]. AKBA increases the expression levels of let-7b, let-7i, miR-200b, and miR-200c, and decreases expression of their known target genes implicated in EMT, such as CDK6 and vimentin. Therefore, AKBA-mediated regulation of the let-7 and miR-200 families may play an important role in inhibiting metastasis in colorectal cancer.

3.9. Silymarin

Silymarin is a flavonoid isolated from the milk thistle, *Silybum marianum* L. Gaertn, and has been shown to have anticancer activities [65]. Singh et al. reported that silymarin inhibits the migratory activities of NSCLC cells, including A549, H1299, and H460, by modulating the expression of miR-203, a tumor suppressor [65]. Silymarin enhances miR-203 expression and decreases histone deacetylases and zinc finger E-box binding homeobox 1 (ZEB1) expression. As ZEB1 is implicated in the activation of EMT, downregulation of ZEB1 by silymarin might inhibit the metastasis of cancer cells. Although a novel target gene of miR-203 was not identified in the study, it was shown that miR-203 plays its tumor suppressor role in multiple cancers by suppressing its known target genes, such as annexin A4 and apoptosis inhibitor 4 [66,67].

3.10. β-Sitosterol-D-glucoside

β-Sitosterol-D-glucoside (β-SDG) is a phytosterol found in *Salvia sahendica* and *Arctotis arctotoides* and reported to possess anti-inflammatory, antimicrobial, immunomodulatory, and anti-proliferative functions in multiple cancer cells [68]. Xu et al. demonstrated that β-SDG isolated from sweet potato has an anticancer function in breast cancer cells [69]. β-SDG inhibits the proliferation of breast cancer cells by elevating the levels of pro-apoptotic Bax and BCL2 associated agonist of cell death (Bad), and reducing the levels of anti-apoptotic Bcl-2 and Bcl-xl. β-SDG also suppresses tumor growth in MCF7 cell-injected xenograft models. Moreover, Xu et al. showed that the expression level of miR-10a is increased by β-SDG, and this induction of miR-10a promotes apoptosis and suppresses PI3K and p-Akt levels [69]. Therefore, the anticancer function of β-SDG is probably mediated in part by miR-10a. A direct target gene of miR-10a involved in its anticancer activities needs to be identified to shed light on the specific molecular mechanism involved.

3.11. Arctigenin

Arctigenin is a lignan from the seeds of *Arctium lappa*. Several studies demonstrated the anticancer functions of arctigenin in pancreatic, breast, and lung cancers through the regulation of apoptosis and proliferation [70]. Wang et al. studied the effects of arctigenin on the miRNA profile of mouse prostate tumor tissues; the levels of miR-126 and miR-21 decrease whereas those of miR-135a, miR-205, miR-22-3p, miR-455, and miR-96 increase in response to arctigenin [71]. The molecular mechanism that connects these regulated miRNAs and the anticancer function of arctigenin remains to be elucidated.

3.12. Cinnamic Acid Derivatives

Propolis is a plant mastic that contains many chemical components including flavonoids, benzoic acid, and cinnamic acid derivatives such as artepilin C, baccharin, and drupanin [72]. Chemical components isolated from propolis have been shown to possess anti-inflammatory and anticancer properties [72,73]. Kumazaki et al. demonstrated that artepilin C, baccharin, and drupanin inhibit the proliferation of colon cancer cells, such as DLD-1 and SW480 [74]. Specifically, the combined treatment with baccharin and drupanin has a synergistic effect on apoptosis activation in DLD-1 cells. Moreover, the expression level of miR-143 is increased by cotreatment with baccharin and drupanin, and miR-143 subsequently represses the expression of a target gene, Erk5, resulting in cell cycle arrest [74]. These observations support the idea that the pro-apoptotic anticancer function of propolis cinnamic acid derivatives is due to modulation of miRNAs and their target genes.

4. Phytochemicals with Photoprotective Effects

In addition to their anticancer effect, phytochemicals have been recently reported to inhibit or protect against UV-induced cellular damage [75,76]. Analysis of the altered miRNA profiles in arctiin- or troxerutin-pretreated keratinocytes upon UVB stimulation suggests that the phytochemical-induced regulation of miRNAs is functionally associated with the photoprotective effect.

4.1. Arctiin

Arctiin is a lignan found in plants, such as *Arctium lappa* and *Forsythiae fructus*, and is known to inhibit cell proliferation and inflammation [77,78]. Recently, Cha et al. demonstrated the photoprotective activities of arctiin in keratinocytes, the predominant cell type in the epidermis [75]. Arctiin inhibits cell death and cytotoxicity and enhances DNA repair and wound healing in UVB-exposed keratinocytes. UV radiation induces DNA damage in keratinocytes, which results in cellular senescence, apoptosis, or cancer. MiRNA profile analyses of arctiin-pretreated HaCaT cells under UVB radiation revealed that the expression levels of four miRNAs were increased and those of 62 were decreased (>2-fold) [75]. These results raise the possibility that the protective function of arctiin against UVB-induced cellular damage is mediated by the modulation of miRNA expression.

4.2. Troxerutin

Troxerutin {vitamin P4; 3′, 4′, 7′-Tris[O-(2-hydroxyethyl)]}, is a natural flavonoid rutin found in extracts of *Sophora japonica*, which exhibits antioxidant and anti-inflammatory activities [79,80].

Lee et al. demonstrated that troxerutin promotes cell migration and inhibits cell death in UVB-exposed HaCaT cells [76]. Since the expression of miRNAs involved in apoptosis and cell cycle arrest is aberrantly regulated under UVB radiation, Lee et al. hypothesized that the protective function of troxerutin against UVB-induced DNA damage and apoptosis is mediated by modulating miRNA expression [76]. Indeed, miRNA microarray analysis indicated that five miRNAs were upregulated and 63 were downregulated in troxerutin-pretreated HaCaT cells under UVB exposure, suggesting that the photoprotective effect of troxerutin in UVB-induced cellular damage is mediated by miRNA regulation.

5. A Phytochemical with Anti-Hepatosteatosis Effects

MiRNAs are known to regulate lipid homeostasis, and the aberrant expression of miRNAs has been examined in multiple metabolic diseases, including obesity and diabetes [81]. Jeon et al. demonstrated that fisetin (3, 3′, 4′, 7-tetrahydroxyflavone), a natural flavonol found in various fruits and vegetables, regulates lipid metabolism by modulating miRNA expression [82]. In high-fat mouse liver, the levels of five miRNAs: miR-22*, miR-146a, miR-146b, miR-802, and miR-378, are increased, and fisetin suppresses the expression of miR-378 [82]. In addition, Jeon et al. identified nuclear respiratory factor-1 (NRF-1), a transcription factor implicated in mitochondrial function, as a target of miR-378 [82]. These results suggest that a high-fat diet increases miR-378 levels and the consequent repression of NRF-1 expression leads to mitochondrial dysfunction and development of hepatosteatosis (hepatic fat accumulation). Furthermore, Jeon et al. provide evidence that this metabolic alteration is rescued by fisetin via modulation of miRNA expression [82].

6. Discussion and Conclusions

We provided an overview of the recent research on miRNA-mediated health-promoting effects of phytochemicals (Figure 2). There have been efforts to elucidate the molecular mechanisms underlying the regulation of miRNA expression by phytochemicals. To date, the majority of studies have been focused on the phytochemical-induced changes in miRNA profiles in pathological conditions such as cancer, and more profiling data continue to emerge. Since miRNAs play an important role in maintaining normal cell physiology as well as inducing pathological conditions [83,84], it is perhaps not surprising that phytochemicals exert their medicinal effects through miRNA regulation. As a single miRNA can affect numerous target mRNAs, and a phytochemical can play a wide range of regulatory roles via miRNAs in cells. Elucidating the molecular mechanisms underlying the health-promoting effect of phytochemicals by identifying the crucial regulatory miRNAs and their targets will facilitate the efforts to maximize the therapeutic benefits of phytochemicals.

Understanding the anticancer function of phytochemicals mediated by miRNA regulation is currently an active area of research. As several miRNAs function as either oncogenic miRNAs or tumor suppressors [11], phytochemicals could exert their anticancer effects by directly controlling miRNA expression. In general, miR-21, miR-17, miR-30e, and miR-520h function as oncogenic miRNAs, whereas miR-200, miR-34c, miR-143, and let-7 function as tumor suppressors [11]. One phytochemical could regulate several oncogenic or tumor suppressive miRNAs in various cancers. For example, resveratrol controls miR-520h, miR-17, miR-328, miR-196b, miR-1290, miR-34c, miR-663, and miR-744 in leukemia, osteosarcoma, lung, prostate, breast, and colorectal cancers [13,21,23–26]. On the other hand, a typical oncogenic or tumor suppressive miRNA may be modulated by multiple phytochemicals, as in the case of miR-200, whose expression is regulated by at least two phytochemicals, AKBA and quercetin [64].

The phytochemical SFN enhances the anticancer function of the conventional therapeutic drug cisplatin by modulating miRNAs [54]. Moreover, combined treatment with two phytochemicals, baccharin and drupanin, exerts a synergistic anticancer effect in colorectal cancer cells [74]. These results suggest that phytochemicals possess great potential for development as effective health-promoting and anticancer therapeutic agents. However, since the majority of studies have focused on identifying miRNAs modulated by phytochemicals, future studies are required to explore the direct targets of the miRNAs and verify their associated functions. For example, although SFN, silymarin, β-SDG, and arctigenin exert anticancer effects, and arctiin and troxerutin exert a photoprotective effect through controlling miRNA expression, the direct targets of the miRNAs have yet to be thoroughly investigated [54,65,69,71,75,76].

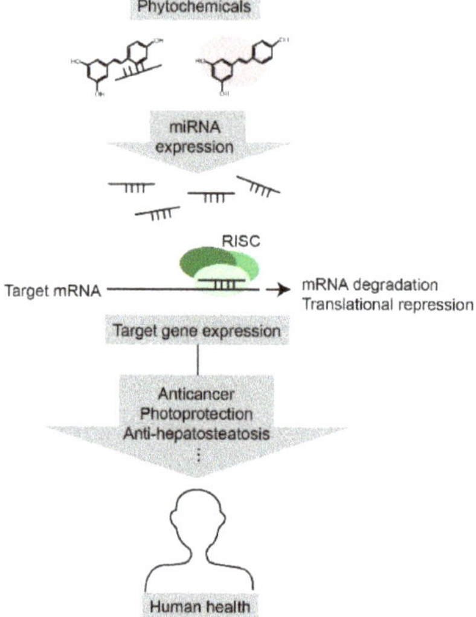

Figure 2. Phytochemicals have a positive impact on human health by regulating miRNA expression. Phytochemicals alter the gene expression profile by modulating miRNA expression, which cause anticancer, photoprotection, and anti-hepatosteatosis effects.

Finally, little is known about the mechanism by which the expression levels of miRNAs are modulated by phytochemicals. A few studies suggested that phytochemicals can bind to miRNAs directly or affect the transcription or processing of miRNAs by regulating a third molecule [12–14]. Interestingly, genistein regulates specific microRNAs, such as miR-27a and miR-1260b, in various cancer cells to elicit anticancer functions [58–63]. A distinct mechanism likely exists to regulate specific microRNA expression. If so, understanding the molecular mechanism by which genistein regulates the expression of specific miRNAs will be useful for controlling multiple types of cancers. Therefore, more extensive studies of the molecular mechanisms responsible for regulating miRNA expression levels are warranted.

Funding: This work was supported by the Incheon National University Research Grant in 2016 and the Basic Science Research Program through the National Research Foundation of Korea (NRF) funded by the Ministry of Education, Science, and Technology (2019R1A2C1004053).

Conflicts of Interest: The author declare no conflict of interest.

Abbreviations

miRNAs	microRNAs
Bax	BCL2 associated X
UTR	untranslated region
MMP2	matrix metalloproteinase 2
EMT	epithelial-mesenchymal transition
MET	mesenchymal-epithelial transition
EEF1A2	isoform A2 of eukaryotic translation elongation factor 1A
NK	natural killer

MICA	major histocompatibility complex class I chain-related protein A
MICB	major histocompatibility complex class I chain-related protein B
ALL	acute lymphoblastic leukemia
IGFBP3	insulin-like growth factor binding protein 3
PKM2	pyruvate kinase M2
NSCLC	non-small cell lung cancer
PTEN	phosphatase and tensin homolog
DIM	3,3′-diindolylmethane
CSC	cancer stem cell
ATG5	autophagy-related gene 5
SFN	sulforaphane
Stat3	signal transducer and activator of transcription 3
IL-6R	interleukin-6 receptor
PI3K	phosphatidylinositol 3-kinase
AKBA	acetyl-11-keto-β-boswellic acid
β-SDG	β-Sitosterol-D-glucoside
NRF-1	nuclear respiratory factor-1

References

1. Manach, C.; Hubert, J.; Llorach, R.; Scalbert, A. The complex links between dietary phytochemicals and human health deciphered by metabolomics. *Mol. Nutr. Food Res.* **2009**, *53*, 1303–1315. [CrossRef]
2. Reddy, L.; Odhav, B.; Bhoola, K.D. Natural products for cancer prevention: A global perspective. *Pharmacol. Ther.* **2003**, *99*, 1–13. [CrossRef]
3. Kaur, V.; Kumar, M.; Kumar, A.; Kaur, K.; Dhillon, V.S.; Kaur, S. Pharmacotherapeutic potential of phytochemicals: Implications in cancer chemoprevention and future perspectives. *Biomed. Pharmacother.* **2018**, *97*, 564–586. [CrossRef]
4. Bishayee, A. Cancer prevention and treatment with resveratrol: From rodent studies to clinical trials. *Cancer Prev. Res.* **2009**, *2*, 409–418. [CrossRef]
5. Kasi, P.D.; Tamilselvam, R.; Skalicka-Wozniak, K.; Nabavi, S.F.; Daglia, M.; Bishayee, A.; Pazoki-Toroudi, H.; Nabavi, S.M. Molecular targets of curcumin for cancer therapy: An updated review. *Tumour Biol.* **2016**, *37*, 13017–13028. [CrossRef]
6. Reuben, S.C.; Gopalan, A.; Petit, D.M.; Bishayee, A. Modulation of angiogenesis by dietary phytoconstituents in the prevention and intervention of breast cancer. *Mol. Nutr. Food Res.* **2012**, *56*, 14–29. [CrossRef] [PubMed]
7. Siddiqui, I.A.; Bharali, D.J.; Nihal, M.; Adhami, V.M.; Khan, N.; Chamcheu, J.C.; Khan, M.I.; Shabana, S.; Mousa, S.A.; Mukhtar, H. Excellent anti-proliferative and pro-apoptotic effects of (-)-epigallocatechin-3-gallate encapsulated in chitosan nanoparticles on human melanoma cell growth both in vitro and in vivo. *Nanomedicine* **2014**, *10*, 1619–1626. [CrossRef]
8. Cai, Y.; Yu, X.; Hu, S.; Yu, J. A brief review on the mechanisms of mirna regulation. *Genom. Proteom. Bioinform.* **2009**, *7*, 147–154. [CrossRef]
9. Gregory, R.I.; Chendrimada, T.P.; Cooch, N.; Shiekhattar, R. Human risc couples microrna biogenesis and posttranscriptional gene silencing. *Cell* **2005**, *123*, 631–640. [CrossRef] [PubMed]
10. Gregory, R.I.; Shiekhattar, R. Microrna biogenesis and cancer. *Cancer Res.* **2005**, *65*, 3509–3512. [CrossRef]
11. Zhang, B.; Pan, X.; Cobb, G.P.; Anderson, T.A. Micrornas as oncogenes and tumor suppressors. *Dev. Biol.* **2007**, *302*, 1–12. [CrossRef]
12. Baselga-Escudero, L.; Blade, C.; Ribas-Latre, A.; Casanova, E.; Suarez, M.; Torres, J.L.; Salvado, M.J.; Arola, L.; Arola-Arnal, A. Resveratrol and egcg bind directly and distinctively to mir-33a and mir-122 and modulate divergently their levels in hepatic cells. *Nucleic. Acids Res.* **2014**, *42*, 882–892. [CrossRef]
13. Pan, J.; Shen, J.; Si, W.; Du, C.; Chen, D.; Xu, L.; Yao, M.; Fu, P.; Fan, W. Resveratrol promotes mica/b expression and natural killer cell lysis of breast cancer cells by suppressing c-myc/mir-17 pathway. *Oncotarget* **2017**, *8*, 65743–65758. [CrossRef]

14. Wang, H.; Bian, S.; Yang, C.S. Green tea polyphenol egcg suppresses lung cancer cell growth through upregulating mir-210 expression caused by stabilizing hif-1alpha. *Carcinogenesis* **2011**, *32*, 1881–1889. [CrossRef]
15. Kaelin, W.G., Jr.; Ratcliffe, P.J. Oxygen sensing by metazoans: The central role of the hif hydroxylase pathway. *Mol. Cell* **2008**, *30*, 393–402. [CrossRef] [PubMed]
16. Amin, A.R.; Kucuk, O.; Khuri, F.R.; Shin, D.M. Perspectives for cancer prevention with natural compounds. *J. Clin. Oncol.* **2009**, *27*, 2712–2725. [CrossRef] [PubMed]
17. Pirola, L.; Frojdo, S. Resveratrol: One molecule, many targets. *IUBMB Life* **2008**, *60*, 323–332. [CrossRef] [PubMed]
18. Jang, M.; Cai, L.; Udeani, G.O.; Slowing, K.V.; Thomas, C.F.; Beecher, C.W.; Fong, H.H.; Farnsworth, N.R.; Kinghorn, A.D.; Mehta, R.G.; et al. Cancer chemopreventive activity of resveratrol, a natural product derived from grapes. *Science* **1997**, *275*, 218–220. [CrossRef]
19. Siddiqui, I.A.; Sanna, V.; Ahmad, N.; Sechi, M.; Mukhtar, H. Resveratrol nanoformulation for cancer prevention and therapy. *Ann. N. Y. Acad. Sci.* **2015**, *1348*, 20–31. [CrossRef] [PubMed]
20. Nguyen, A.V.; Martinez, M.; Stamos, M.J.; Moyer, M.P.; Planutis, K.; Hope, C.; Holcombe, R.F. Results of a phase i pilot clinical trial examining the effect of plant-derived resveratrol and grape powder on wnt pathway target gene expression in colonic mucosa and colon cancer. *Cancer Manag. Res.* **2009**, *1*, 25–37.
21. Vislovukh, A.; Kratassiouk, G.; Porto, E.; Gralievska, N.; Beldiman, C.; Pinna, G.; El'skaya, A.; Harel-Bellan, A.; Negrutskii, B.; Groisman, I. Proto-oncogenic isoform a2 of eukaryotic translation elongation factor eef1 is a target of mir-663 and mir-744. *Br. J. Cancer* **2013**, *108*, 2304–2311. [CrossRef]
22. Wu, H.; Wang, Y.; Wu, C.; Yang, P.; Li, H.; Li, Z. Resveratrol induces cancer cell apoptosis through mir-326/pkm2-mediated er stress and mitochondrial fission. *J. Agric. Food Chem.* **2016**, *64*, 9356–9367. [CrossRef]
23. Yang, S.; Li, W.; Sun, H.; Wu, B.; Ji, F.; Sun, T.; Chang, H.; Shen, P.; Wang, Y.; Zhou, D. Resveratrol elicits anti-colorectal cancer effect by activating mir-34c-kitlg in vitro and in vivo. *BMC Cancer* **2015**, *15*, 969. [CrossRef]
24. Yang, S.F.; Lee, W.J.; Tan, P.; Tang, C.H.; Hsiao, M.; Hsieh, F.K.; Chien, M.H. Upregulation of mir-328 and inhibition of creb-DNA-binding activity are critical for resveratrol-mediated suppression of matrix metalloproteinase-2 and subsequent metastatic ability in human osteosarcomas. *Oncotarget* **2015**, *6*, 2736–2753. [CrossRef]
25. Yu, Y.H.; Chen, H.A.; Chen, P.S.; Cheng, Y.J.; Hsu, W.H.; Chang, Y.W.; Chen, Y.H.; Jan, Y.; Hsiao, M.; Chang, T.Y.; et al. Mir-520h-mediated foxc2 regulation is critical for inhibition of lung cancer progression by resveratrol. *Oncogene* **2013**, *32*, 431–443. [CrossRef]
26. Zhou, W.; Wang, S.; Ying, Y.; Zhou, R.; Mao, P. Mir-196b/mir-1290 participate in the antitumor effect of resveratrol via regulation of igfbp3 expression in acute lymphoblastic leukemia. *Oncol. Rep.* **2017**, *37*, 1075–1083. [CrossRef]
27. Grandemange, S.; Herzig, S.; Martinou, J.C. Mitochondrial dynamics and cancer. *Semin. Cancer Biol.* **2009**, *19*, 50–56. [CrossRef]
28. Wu, H.; Yang, P.; Hu, W.; Wang, Y.; Lu, Y.; Zhang, L.; Fan, Y.; Xiao, H.; Li, Z. Overexpression of pkm2 promotes mitochondrial fusion through attenuated p53 stability. *Oncotarget* **2016**, *7*, 78069–78082. [CrossRef]
29. Fujiki, H.; Suganuma, M.; Imai, K.; Nakachi, K. Green tea: Cancer preventive beverage and/or drug. *Cancer Lett.* **2002**, *188*, 9–13. [CrossRef]
30. Tsang, W.P.; Kwok, T.T. Epigallocatechin gallate up-regulation of mir-16 and induction of apoptosis in human cancer cells. *J. Nutr. Biochem.* **2010**, *21*, 140–146. [CrossRef]
31. Zhou, H.; Chen, J.X.; Yang, C.S.; Yang, M.Q.; Deng, Y.; Wang, H. Gene regulation mediated by micrornas in response to green tea polyphenol egcg in mouse lung cancer. *BMC Genom.* **2014**, *15*, S3. [CrossRef] [PubMed]
32. Jiang, P.; Xu, C.; Chen, L.; Chen, A.; Wu, X.; Zhou, M.; Haq, I.U.; Mariyam, Z.; Feng, Q. Epigallocatechin-3-gallate inhibited cancer stem cell-like properties by targeting hsa-mir-485-5p/rxralpha in lung cancer. *J. Cell Biochem.* **2018**, *119*, 8620–8635. [CrossRef] [PubMed]
33. Yu, C.C.; Chen, P.N.; Peng, C.Y.; Yu, C.H.; Chou, M.Y. Suppression of mir-204 enables oral squamous cell carcinomas to promote cancer stemness, emt traits, and lymph node metastasis. *Oncotarget* **2016**, *7*, 20180–20192. [CrossRef] [PubMed]

34. Zhu, K.; Wang, W. Green tea polyphenol egcg suppresses osteosarcoma cell growth through upregulating mir-1. *Tumour Biol.* **2016**, *37*, 4373–4382. [CrossRef]

35. Chakrabarti, M.; Ai, W.; Banik, N.L.; Ray, S.K. Overexpression of mir-7-1 increases efficacy of green tea polyphenols for induction of apoptosis in human malignant neuroblastoma sh-sy5y and sk-n-dz cells. *Neurochem. Res.* **2013**, *38*, 420–432. [CrossRef] [PubMed]

36. Siddiqui, I.A.; Asim, M.; Hafeez, B.B.; Adhami, V.M.; Tarapore, R.S.; Mukhtar, H. Green tea polyphenol egcg blunts androgen receptor function in prostate cancer. *FASEB J.* **2011**, *25*, 1198–1207. [CrossRef]

37. Gupta, S.C.; Patchva, S.; Aggarwal, B.B. Therapeutic roles of curcumin: Lessons learned from clinical trials. *AAPS J.* **2013**, *15*, 195–218. [CrossRef] [PubMed]

38. Park, W.; Amin, A.R.; Chen, Z.G.; Shin, D.M. New perspectives of curcumin in cancer prevention. *Cancer Prev. Res.* **2013**, *6*, 387–400. [CrossRef]

39. Zhang, W.; Bai, W.; Zhang, W. Mir-21 suppresses the anticancer activities of curcumin by targeting pten gene in human non-small cell lung cancer a549 cells. *Clin. Transl. Oncol.* **2014**, *16*, 708–713. [CrossRef]

40. Liu, Z.L.; Wang, H.; Liu, J.; Wang, Z.X. Microrna-21 (mir-21) expression promotes growth, metastasis, and chemo- or radioresistance in non-small cell lung cancer cells by targeting pten. *Mol. Cell Biochem.* **2013**, *372*, 35–45. [CrossRef]

41. Gupta, A.; Verma, A.; Mishra, A.K.; Wadhwa, G.; Sharma, S.K.; Jain, C.K. The wnt pathway: Emerging anticancer strategies. *Recent Pat. Endocr. Metab. Immune. Drug Discov.* **2013**, *7*, 138–147. [CrossRef] [PubMed]

42. Prasad, C.P.; Rath, G.; Mathur, S.; Bhatnagar, D.; Ralhan, R. Potent growth suppressive activity of curcumin in human breast cancer cells: Modulation of wnt/beta-catenin signaling. *Chem. Biol. Interact.* **2009**, *181*, 263–271. [CrossRef]

43. Xiao, C.; Wang, L.; Zhu, L.; Zhang, C.; Zhou, J. Curcumin inhibits oral squamous cell carcinoma scc-9 cells proliferation by regulating mir-9 expression. *Biochem. Biophys. Res. Commun.* **2014**, *454*, 576–580. [CrossRef] [PubMed]

44. Nam, J.S.; Sharma, A.R.; Nguyen, L.T.; Chakraborty, C.; Sharma, G.; Lee, S.S. Application of bioactive quercetin in oncotherapy: From nutrition to nanomedicine. *Molecules* **2016**, *21*, 108. [CrossRef]

45. Nwaeburu, C.C.; Bauer, N.; Zhao, Z.; Abukiwan, A.; Gladkich, J.; Benner, A.; Herr, I. Up-regulation of microrna let-7c by quercetin inhibits pancreatic cancer progression by activation of numbl. *Oncotarget* **2016**, *7*, 58367–58380. [CrossRef] [PubMed]

46. Miele, L.; Miao, H.; Nickoloff, B.J. Notch signaling as a novel cancer therapeutic target. *Curr. Cancer Drug Targets* **2006**, *6*, 313–323. [CrossRef]

47. Nwaeburu, C.C.; Abukiwan, A.; Zhao, Z.; Herr, I. Quercetin-induced mir-200b-3p regulates the mode of self-renewing divisions in pancreatic cancer. *Mol. Cancer* **2017**, *16*, 23. [CrossRef]

48. Morrison, S.J.; Kimble, J. Asymmetric and symmetric stem-cell divisions in development and cancer. *Nature* **2006**, *441*, 1068–1074. [CrossRef]

49. Chen, C.; Chen, S.M.; Xu, B.; Chen, Z.; Wang, F.; Ren, J.; Xu, Y.; Wang, Y.; Xiao, B.K.; Tao, Z.Z. In vivo and in vitro study on the role of 3,3'-diindolylmethane in treatment and prevention of nasopharyngeal carcinoma. *Carcinogenesis* **2013**, *34*, 1815–1821. [CrossRef]

50. Ye, Y.; Fang, Y.; Xu, W.; Wang, Q.; Zhou, J.; Lu, R. 3,3'-diindolylmethane induces anti-human gastric cancer cells by the mir-30e-atg5 modulating autophagy. *Biochem. Pharmacol.* **2016**, *115*, 77–84. [CrossRef]

51. Fujita, N.; Itoh, T.; Omori, H.; Fukuda, M.; Noda, T.; Yoshimori, T. The atg16l complex specifies the site of lc3 lipidation for membrane biogenesis in autophagy. *Mol. Biol. Cell* **2008**, *19*, 2092–2100. [CrossRef]

52. Jin, Y. 3,3'-diindolylmethane inhibits breast cancer cell growth via mir-21-mediated cdc25a degradation. *Mol. Cell Biochem.* **2011**, *358*, 345–354. [CrossRef] [PubMed]

53. Zhang, Y.; Tang, L. Discovery and development of sulforaphane as a cancer chemopreventive phytochemical. *Acta Pharmacol. Sin.* **2007**, *28*, 1343–1354. [CrossRef] [PubMed]

54. Wang, X.; Li, Y.; Dai, Y.; Liu, Q.; Ning, S.; Liu, J.; Shen, Z.; Zhu, D.; Jiang, F.; Zhang, J.; et al. Sulforaphane improves chemotherapy efficacy by targeting cancer stem cell-like properties via the mir-124/il-6r/stat3 axis. *Sci. Rep.* **2016**, *6*, 36796. [CrossRef] [PubMed]

55. Heinrich, P.C.; Behrmann, I.; Haan, S.; Hermanns, H.M.; Muller-Newen, G.; Schaper, F. Principles of interleukin (il)-6-type cytokine signalling and its regulation. *Biochem. J.* **2003**, *374*, 1–20. [CrossRef] [PubMed]

56. Lewinska, A.; Adamczyk-Grochala, J.; Deregowska, A.; Wnuk, M. Sulforaphane-induced cell cycle arrest and senescence are accompanied by DNA hypomethylation and changes in microrna profile in breast cancer cells. *Theranostics* **2017**, *7*, 3461–3477. [CrossRef]

57. Fotsis, T.; Pepper, M.; Adlercreutz, H.; Hase, T.; Montesano, R.; Schweigerer, L. Genistein, a dietary ingested isoflavonoid, inhibits cell proliferation and in vitro angiogenesis. *J. Nutr.* **1995**, *125*, 790S–797S.

58. Hirata, H.; Hinoda, Y.; Shahryari, V.; Deng, G.; Tanaka, Y.; Tabatabai, Z.L.; Dahiya, R. Genistein downregulates onco-mir-1260b and upregulates sfrp1 and smad4 via demethylation and histone modification in prostate cancer cells. *Br. J. Cancer* **2014**, *110*, 1645–1654. [CrossRef]

59. Hirata, H.; Ueno, K.; Nakajima, K.; Tabatabai, Z.L.; Hinoda, Y.; Ishii, N.; Dahiya, R. Genistein downregulates onco-mir-1260b and inhibits wnt-signalling in renal cancer cells. *Br. J. Cancer* **2013**, *108*, 2070–2078. [CrossRef]

60. Sun, Q.; Cong, R.; Yan, H.; Gu, H.; Zeng, Y.; Liu, N.; Chen, J.; Wang, B. Genistein inhibits growth of human uveal melanoma cells and affects microrna-27a and target gene expression. *Oncol. Rep.* **2009**, *22*, 563–567.

61. Xia, J.; Cheng, L.; Mei, C.; Ma, J.; Shi, Y.; Zeng, F.; Wang, Z.; Wang, Z. Genistein inhibits cell growth and invasion through regulation of mir-27a in pancreatic cancer cells. *Curr. Pharm. Des.* **2014**, *20*, 5348–5353. [CrossRef]

62. Xu, L.; Xiang, J.; Shen, J.; Zou, X.; Zhai, S.; Yin, Y.; Li, P.; Wang, X.; Sun, Q. Oncogenic microrna-27a is a target for genistein in ovarian cancer cells. *Anticancer Agents Med. Chem.* **2013**, *13*, 1126–1132. [CrossRef]

63. Yang, Y.; Zang, A.; Jia, Y.; Shang, Y.; Zhang, Z.; Ge, K.; Zhang, J.; Fan, W.; Wang, B. Genistein inhibits a549 human lung cancer cell proliferation via mir-27a and met signaling. *Oncol. Lett.* **2016**, *12*, 2189–2193. [CrossRef]

64. Shah, B.A.; Qazi, G.N.; Taneja, S.C. Boswellic acids: A group of medicinally important compounds. *Nat. Prod. Rep.* **2009**, *26*, 72–89. [CrossRef]

65. Jones, V.; Katiyar, S.K. Emerging phytochemicals for prevention of melanoma invasion. *Cancer Lett.* **2013**, *335*, 251–258. [CrossRef]

66. Li, J.; Zhang, B.; Cui, J.; Liang, Z.; Liu, K. Mir-203 inhibits the invasion and emt of gastric cancer cells by directly targeting annexin a4. *Oncol. Res.* **2019**. [CrossRef]

67. Wang, B.; Li, X.; Zhao, G.; Yan, H.; Dong, P.; Watari, H.; Sims, M.; Li, W.; Pfeffer, L.M.; Guo, Y.; et al. Mir-203 inhibits ovarian tumor metastasis by targeting birc5 and attenuating the tgfbeta pathway. *J. Exp. Clin. Cancer Res.* **2018**, *37*, 235. [CrossRef] [PubMed]

68. Sultana, N.; Afolayan, A.J. A novel daucosterol derivative and antibacterial activity of compounds from arctotis arctotoides. *Nat. Prod. Res.* **2007**, *21*, 889–896. [CrossRef] [PubMed]

69. Xu, H.; Li, Y.; Han, B.; Li, Z.; Wang, B.; Jiang, P.; Zhang, J.; Ma, W.; Zhou, D.; Li, X.; et al. Anti-breast-cancer activity exerted by beta-sitosterol-d-glucoside from sweet potato via upregulation of microrna-10a and via the pi3k-akt signaling pathway. *J. Agric. Food Chem.* **2018**, *66*, 9704–9718. [CrossRef]

70. Awale, S.; Lu, J.; Kalauni, S.K.; Kurashima, Y.; Tezuka, Y.; Kadota, S.; Esumi, H. Identification of arctigenin as an antitumor agent having the ability to eliminate the tolerance of cancer cells to nutrient starvation. *Cancer Res.* **2006**, *66*, 1751–1757. [CrossRef] [PubMed]

71. Wang, P.; Solorzano, W.; Diaz, T.; Magyar, C.E.; Henning, S.M.; Vadgama, J.V. Arctigenin inhibits prostate tumor cell growth in vitro and in vivo. *Clin. Nutr. Exp.* **2017**, *13*, 1–11. [CrossRef]

72. Popolo, A.; Piccinelli, L.A.; Morello, S.; Cuesta-Rubio, O.; Sorrentino, R.; Rastrelli, L.; Pinto, A. Antiproliferative activity of brown cuban propolis extract on human breast cancer cells. *Nat. Prod. Commun.* **2009**, *4*, 1711–1716. [CrossRef]

73. Cheung, K.W.; Sze, D.M.; Chan, W.K.; Deng, R.X.; Tu, W.; Chan, G.C. Brazilian green propolis and its constituent, artepillin c inhibits allogeneic activated human cd4 t cells expansion and activation. *J. Ethnopharmacol.* **2011**, *138*, 463–471. [CrossRef]

74. Kumazaki, M.; Shinohara, H.; Taniguchi, K.; Yamada, N.; Ohta, S.; Ichihara, K.; Akao, Y. Propolis cinnamic acid derivatives induce apoptosis through both extrinsic and intrinsic apoptosis signaling pathways and modulate of mirna expression. *Phytomedicine* **2014**, *21*, 1070–1077. [CrossRef]

75. Cha, H.J.; Lee, G.T.; Lee, K.S.; Lee, K.K.; Hong, J.T.; Lee, N.K.; Kim, S.Y.; Lee, B.M.; An, I.S.; Hahn, H.J.; et al. Photoprotective effect of arctiin against ultraviolet b-induced damage in hacat keratinocytes is mediated by microrna expression changes. *Mol. Med. Rep.* **2014**, *10*, 1363–1370. [CrossRef] [PubMed]

76. Lee, K.S.; Cha, H.J.; Lee, G.T.; Lee, K.K.; Hong, J.T.; Ahn, K.J.; An, I.S.; An, S.; Bae, S. Troxerutin induces protective effects against ultraviolet b radiation through the alteration of microrna expression in human hacat keratinocyte cells. *Int. J. Mol. Med.* **2014**, *33*, 934–942. [CrossRef] [PubMed]

77. Lee, S.; Shin, S.; Kim, H.; Han, S.; Kim, K.; Kwon, J.; Kwak, J.H.; Lee, C.K.; Ha, N.J.; Yim, D.; et al. Anti-inflammatory function of arctiin by inhibiting cox-2 expression via nf-kappab pathways. *J. Inflamm.* **2011**, *8*, 16. [CrossRef]

78. Matsuzaki, Y.; Koyama, M.; Hitomi, T.; Yokota, T.; Kawanaka, M.; Nishikawa, A.; Germain, D.; Sakai, T. Arctiin induces cell growth inhibition through the down-regulation of cyclin d1 expression. *Oncol. Rep.* **2008**, *19*, 721–727. [CrossRef]

79. Fan, S.H.; Zhang, Z.F.; Zheng, Y.L.; Lu, J.; Wu, D.M.; Shan, Q.; Hu, B.; Wang, Y.Y. Troxerutin protects the mouse kidney from d-galactose-caused injury through anti-inflammation and anti-oxidation. *Int. Immunopharmacol.* **2009**, *9*, 91–96. [CrossRef]

80. Zhang, Z.F.; Fan, S.H.; Zheng, Y.L.; Lu, J.; Wu, D.M.; Shan, Q.; Hu, B. Troxerutin protects the mouse liver against oxidative stress-mediated injury induced by d-galactose. *J. Agric. Food Chem.* **2009**, *57*, 7731–7736. [CrossRef]

81. Rottiers, V.; Naar, A.M. Micrornas in metabolism and metabolic disorders. *Nat. Rev. Mol. Cell Biol.* **2012**, *13*, 239–250. [CrossRef] [PubMed]

82. Jeon, T.I.; Park, J.W.; Ahn, J.; Jung, C.H.; Ha, T.Y. Fisetin protects against hepatosteatosis in mice by inhibiting mir-378. *Mol. Nutr. Food Res.* **2013**, *57*, 1931–1937. [CrossRef] [PubMed]

83. Carthew, R.W.; Sontheimer, E.J. Origins and mechanisms of mirnas and sirnas. *Cell* **2009**, *136*, 642–655. [CrossRef] [PubMed]

84. Kumar, M.S.; Lu, J.; Mercer, K.L.; Golub, T.R.; Jacks, T. Impaired microrna processing enhances cellular transformation and tumorigenesis. *Nat. Genet.* **2007**, *39*, 673–677. [CrossRef]

International Journal of
Molecular Sciences

Article

In Vitro Study of the Anticancer Effects of Biotechnological Extracts of the Endangered Plant Species *Satureja Khuzistanica*

Abbas Khojasteh [1], Isidoro Metón [2], Sergio Camino [2], Rosa M. Cusido [1], Regine Eibl [3] and Javier Palazon [1,*]

[1] Secció de Fisiologia i Biotecnologia Vegetal, Departament de Biologia, Sanitat i Medi Ambient, Facultat de Farmàcia i Ciències de l'Alimentació, Universitat de Barcelona, Joan XXIII 27-31, 08028 Barcelona, Spain; abbaskhojasteh@aol.com (A.K.); rcusido@ub.edu (R.M.C.)

[2] Secció de Bioquímica i Biologia Molecular, Departament de Bioquímica i Fisiologia, Facultat de Farmàcia i Ciències de l'Alimentació, Universitat de Barcelona, Joan XXIII 27-31, 08028 Barcelona, Spain; imeton@ub.edu (I.M.); scamino93@gmail.com (S.C.)

[3] Institute of Chemistry and Biotechnology, Biochemical Engineering and Cell Cultivation Techniques, Campus Grüental, Zurich University of Applied Sciences, 8820 Wädenswill, Switzerland; eibs@zhaw.ch

* Correspondence: javierpalazon@ub.edu

Received: 25 April 2019; Accepted: 13 May 2019; Published: 15 May 2019

Abstract: Many medicinal plant species are currently threatened in their natural habitats because of the growing demand for phytochemicals worldwide. A sustainable alternative for the production of bioactive plant compounds are plant biofactories based on cell cultures and organs. In addition, plant extracts from biofactories have significant advantages over those obtained from plants, since they are free of contamination by microorganisms, herbicides and pesticides, and they provide more stable levels of active ingredients. In this context, we report the establishment of *Satureja khuzistanica* cell cultures able to produce high amounts of rosmarinic acid (RA). The production of this phytopharmaceutical was increased when the cultures were elicited with coronatine and scaled up to a benchtop bioreactor. *S. khuzistanica* extracts enriched in RA were found to reduce the viability of cancer cell lines, increasing the sub-G0/G1 cell population and the activity of caspase-8 in MCF-7 cells, which suggest that *S. khuzistanica* extracts can induce apoptosis of MCF-7 cells through activation of the extrinsic pathway. In addition, our findings indicate that other compounds in *S. khuzistanica* extracts may act synergistically to potentiate the anticancer activity of RA.

Keywords: *Satureja khuzistanica*; rosmarinic acid; plant cell cultures; coronatine; MCF-7 human breast adenocarcinoma cells; HepG2 human hepatoma cells

1. Introduction

Rosmarinic acid (RA), an ester of caffeic acid and 3,4-dihydroxyphenyl lactic acid, is widely distributed in the plant kingdom, including the genera *Ajuga, Agastache, Calamintha, Cedronella, Coleus, Collimsonia, Dracocephalum, Elsholtzia, Glechoma, Hornium, Lavandula, Lycopus, Melissa, Mentha, Micromeria, Monarda, Origanum, Perilla, Perovskia, Plectranthus*, and *Salvia* from the Lamiaceae family, *Cerinthe, Echium, Heliotropium, Lindefolia, Lithospermum, Nonea, Symphytum, Hydrophyllum, Nemophila* and *Phacelia* from the Boraginaceae family, *Chlorantus* from the Chloranteaceae family and *Blechnum* from the Blechnaceae subfamily [1,2].

Interest in RA has grown due to increasing awareness of its potential benefits for human health as a pharmaceutical or dietary supplement. Among its promising biological activities are cognitive-enhancing and cardioprotective effects, cancer chemoprevention properties, and a potential use in the treatment of Alzheimer's disease [3].

Among diseases affecting our society, cancer is one of the most widespread and has the highest mortality rate. Thus, the search for new antineoplastic agents and the confirmation of recently discovered action mechanisms is a major challenge for medicine. In a review of the biological effects of RA, Moore et al. [4] concluded that it could be used as a phytochemical to induce apoptosis. RA can reduce survival of cancer cell lines such as HT-28 (colorectal adenocarcinoma), MCF-7 (breast carcinoma), DU145 (prostate carcinoma) or MKN45 (gastric carcinoma), among others. In the same pipeline, RA, which is currently exploited as an antioxidant and food additive, could also function as a nutraceutical to enhance the effects of other chemotherapeutics.

As a result of overexploitation, many species of medicinal plants are now under threat of extinction in their original habitats. This is the case of *Satureja khuzistanica*, an endangered native Iranian plant that accumulates up to 1.81% of RA [5]. When the natural source of a phytochemical cannot meet the market demand, or is becoming increasingly limited because of over-harvesting or habitat deterioration, plant biotechnology can provide an alternative production system. Plant cell cultures producing phytochemicals have several advantages over field cultivation: (a) The target product can be harvested anywhere in the world, while maintaining strict production and quality control; (b) herbicides and pesticides are not required; (c) problems related with climate and ecology are avoided; and (d) growth cycles are significantly reduced compared to the intact plant, taking weeks rather than years [6].

In this scenario, our group has recently been developing a biotechnological platform for the production of RA based on plant cell cultures of *S. khuzistanica* [7]. In optimal conditions, plant cells cultured in shake flasks and elicited with 100 µM methyl jasmonate reached an RA production of 245 mg·g·DW^{-1} after 16 days of culture. Moreover, when the system was scaled up to a wave-mixed bag of the BIOSTAT CultiBag RM (working volume of 1 L) running in batch mode for a growth period of 21 days, the RA productivity was 3601 mg·L^{-1}, which demonstrated the suitability of the process for the commercial production of RA.

In the current work, with the aim of further improving the RA productivity of this culture system, we studied the action of the new elicitor coronatine (COR), which acts as a molecular mimic of the isoleucine-conjugated form of jasmonic acid [8]. We also tested the anticancer activity of biotechnologically produced *S. khuzistanica* extracts by analyzing their effects on the viability of MCF-7 and HepG2 cancer-derived cell lines, as well as the cell cycle and caspase activity of MCF-7 cells.

2. Results and Discussion

2.1. Effect of Coronatine on a Small Scale

Elicitation with COR has been shown to have a positive impact on intracellular taxane accumulation in in vitro cultures of Taxus spp. [9]. However, the eliciting effects of COR on metabolite production depend on the concentration used and time of exposure, the plant species and cell line.

To optimize the growth capacity of the *S. khuzistanica* cell suspension cultures, a starting packed cell volume (PCV) of 10% was prepared under previously established optimal conditions [7]. As shown in Figure 1A, after more than a week in the lag phase (until day 11), growth measured as cell fresh weight (CFW) shifted to the exponential phase, which lasted until day 16. Thereafter, a slight increase of CFW was observed and at day 18 the cultures entered the stationary phase, reaching a final biomass of 339.5 g·L^{-1} in control conditions.

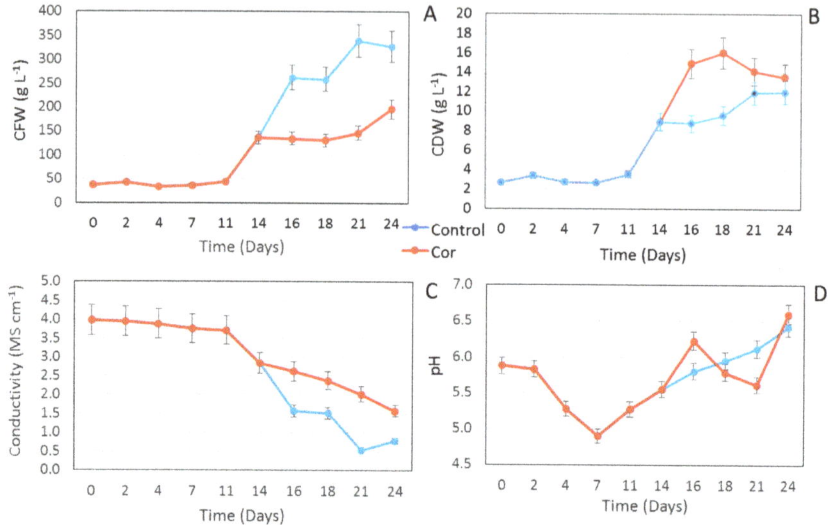

Figure 1. Time course of changes in cell fresh weight (CFW) (**A**), cell dry weight (CDW) (**B**) expressed as g·L^{-1}, conductivity (**C**) and pH (**D**), over a growth period of 24 days. Each result is the average of 3 replicates ± SD.

The growth capacity of the suspension cultures was significantly reduced ($p < 0.05$) by the addition of COR on day 14. A similar effect was reported for methyl jasmonate (MeJA) in the same cell line [7], but the reduction in biomass was considerably less (1%) compared to COR (nearly 10%), indicating the latter is more toxic. These results differ from those obtained with Taxus cell cultures, in which COR was more effective than MeJA as an inducer of taxol production without significantly affecting the growth capacity of the cultures [10].

The time course of growth measured as cell dry weight (CDW) was very similar to that of the CFW. The lag phase also lasted until day 11 and the CDW significantly decreased in the COR-treated cultures compared to the control during the exponential growth phase, being significantly lower at the end of the culture period (Figure 1B). Thus, whether measured by CFW or CDW, the growth capacity of *S. khuzistanica* cell suspension cultures was significantly reduced by the elicitor.

A comparison of the time course of growth measured as CFW and CDW (Figure 1A,B) with previous growth curves obtained by Khojasteh et al. [7] with the same *S. khuzistanica* cell culture revealed that the cell line growth capacity had changed over time and after subculturing, even in control conditions. Differences in growth between different subcultures could be attributed to small variations in the culture conditions and the inherent changes associated with the culture age [11–13]. In this context, our group has recently demonstrated that epigenetic modifications appear in cell lines with successive subcultures, which can affect not only the growth capacity but also secondary metabolite production [14]. This could also have occurred in our *S. khuzistanica* cell line, although an epigenetic study of the degree of DNA methylation is required to prove this hypothesis.

Cells absorb nutrients from the medium, including the macro- and microelements that are added as ions/salts. Thus, when cells grow, the concentration of ions decreases, which in turn reduces the medium conductivity [15]. In our study, the conductivity decreased continuously throughout the experiment. Under the control conditions it dropped significantly at day 16 (exponential growth phase), whereas in the COR-treated suspension cultures the decline on this day and thereafter continued to be slight (Figure 1C). This difference was probably due to the higher consumption of medium salts in control cultures and the lower growth capacities of the treated cultures. In other words, the increase in

CFW and CDW in control cultures inversely correlated with nitrate consumption and the decrease in conductivity.

The cell suspension cultures were initiated at a pH of 5.8, which dropped to 4.9 at day 7, possibly as a result of the high uptake of ammonium. When the exponential growth phase began, the pH started to increase again, probably due to the nitrate uptake. COR had a clear effect on the pH, which increased after the day of elicitation (day 14). The pH then decreased gradually until day 21, before increasing again when the cells entered the death phase (Figure 1C).

The specific production of RA in *S. khuzistanica* suspension cultures was determined every two days throughout the experiment and expressed as $mg \cdot g \cdot CDW^{-1}$ and in $mg \cdot L^{-1}$. The presence of RA was confirmed in all the *S. khuzistanica* methanolic extracts, as shown in Supplementary Figure S1, which depicts the corresponding UV chromatogram (330 nm) with a main peak for RA at a retention time of 16 min. In control conditions (untreated cells), the specific production of RA increased throughout the culture period, reaching a final content of 164 $mg \cdot g \cdot CDW^{-1}$ (Table 1). The addition of COR significantly enhanced the accumulation of RA (Table 1). In the treated cells, the RA production reached a maximum of nearly 221 $mg \cdot g \cdot CDW^{-1}$ between days 4 and 10 after elicitation. At day 18 (96 h after elicitation), the RA content was about 1.5-fold higher than in the control cells.

The RA production of the biotechnological system was also expressed as $mg \cdot L^{-1}$ of culture volume (Table 1), which takes into consideration the biomass-producing capacity of the system, unlike when expressed as $mg \cdot g \cdot CDW^{-1}$. As COR significantly reduced the CDW, the productivity of RA also decreased significantly after the elicitation (Table 1). However, at day 21 (168 h after elicitation), the RA production levels of COR-treated cells started to overtake those of the control cells, reaching more than 2600 $mg \cdot L^{-1}$ after 240 h, which was significantly ($p > 0.05$) higher than the control.

In previous experiments [7], MeJA elicitation of cells resulted in a maximum RA production of about 3858 $mg \cdot L^{-1}$ at day 16, whereas the maximum production with COR was reached at day 24 (Table 1). In the study by Khojasteh et al. [7], production in control conditions peaked at day 14 (1436 $mg \cdot L^{-1}$), compared to day 24 (2221 $mg \cdot L^{-1}$) in the current work, indicating that age affects the production rate and growth course of the cell line. In contrast with studies on other plant species, which lost the capacity to produce secondary metabolites with age, production in our *S. khuzistanica* cell line did not decrease with time [13,14,16].

Table 1. Time courses of rosmarinic acid (RA) production expressed as $mg \cdot g \cdot CDW^{-1}$ and $mg \cdot L^{-1}$ in *S. khuzistanica* suspension cultures elicited with pre-optimized coronatine (COR) (1 μM), up to 10 days after inoculation. Each result is the average of 3 replicates ± SD.

Day	Specific Production ($mg \cdot g \cdot CDW^{-1}$)		Production ($mg \cdot L^{-1}$)	
	Control	COR	Control	COR
0	67.1 ± 2.1		179.4 ± 3.7	
2	94.6 ± 1.9		319.3 ± 3.8	
4	96.5 ± 2.3		264.7 ± 4.6	
7	117.3 ± 2.4		312.0 ± 4.0	
11	128.4 ± 2.9		456.0 ± 4.9	
14	132.5 ± 2.7		1184.2 ± 8.6	
16	127.4 ± 3.1	147.3 ± 2.6	1901.8 ± 11.0	1288.5 ± 13.1
18	132.0 ± 2.1	195.5 ± 3.2	2121.3 ± 11.7	1873.8 ± 7.6
21	150.9 ± 3.2	206.2 ± 3.6	2134.0 ± 9.0	2460.8 ± 12.1
24	163.9 ± 3.5	221.6 ± 4.0	2221.8 ± 7.0	2665.6 ± 12.6

2.2. Scaling Up the Process to a Benchtop Bioreactor

The elicitation assay was scaled up to a 2 L culture bag (CellBag). This type of power input and bioreactor are suitable for the culture of mammalian and plant cells with a low oxygen demand [17]. The time course of the biomass production measured as CFW showed a typical growth curve with a lag phase of 2–4 days, followed by an exponential growth phase that finished at day 14, when the culture entered the stationary phase. Thereafter, the CFW did not increase for the remainder of the culture period (up to day 21) (Figure 2A). A similar curve was obtained when the growth was measured as CDW, although in this case the lag phase of the culture was not apparent (Figure 3B). In control conditions (untreated cells), the lag phase in the bioreactor was clearly shorter than in the shake flask experiment (Figures 1 and 2). In contrast, under elicitation, the negative effect of COR on growth capacity was lower in the bioreactor cultures (Figures 1 and 2).

As in the shake flasks (Table 1), COR significantly increased ($p > 0.05$) the RA production capacity of the bioreactor system, achieving a specific production of 338.2 mg·g·CDW^{-1} at day 16, which was 1.7 times higher than in control conditions (untreated cells) (Figure 3). When measured as mg·L^{-1} of culture medium, the RA production pattern was similar, and from day 14 the yield was higher in the COR-treated cultures (Figure 2). Notably, the cultures grew better in the orbitally shaken bag than in shake flasks and the negative effects of COR on biomass production were much less apparent. This fact, together with a high specific RA production (mg·g·CDW^{-1}) in bioreactor conditions, resulted in yields that were 2.3-fold higher at day 16 than in the small-scale system.

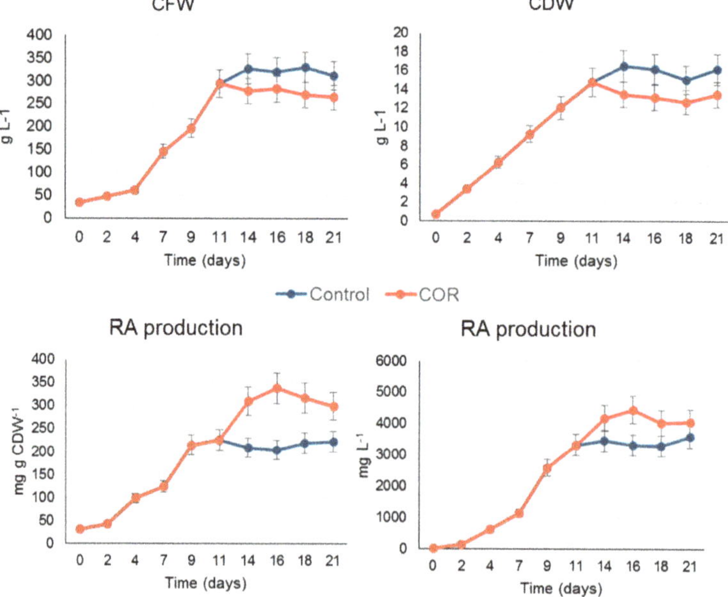

Figure 2. Biomass measured as cell fresh weight (CFW) and dry cell weight (CRW), and rosmarinic acid (RA) production in *S. khuzistanica* control and COR-elicited cultures in a benchtop bioreactor during a growth period of 21 days. The data are the average of two replicates ± SD.

Taken as a whole, these results demonstrate the suitability of the orbitally shaken CellBag for scaling up the suspension cultures of *S. khuzistanica*. Similar results have been achieved with *Centella asiatica* cell cultures, where growth and centelloside production also improved at bioreactor level [18]. Comparison of biomass and RA production of the *S. khuzistanica* cell line obtained in the wave-mixed and the orbitally shaken bag [7] revealed that both bioreactor systems are effective. In all cases, the growth rate was higher at bioreactor level than in shake flasks, as was the case for maximum RA production (Figure 2). However, if we compare the final yield achieved in control conditions (un-elicited cells) in both types of bioreactor, it was higher (3600 mg·L^{-1}) in the orbitally shaken bag than in the wave-mixed bag (3102 mg·L^{-1}). This points to different degrees of adaptation of the *S. khuzistanica* cell cultures to the two systems, but as mentioned, the two experiments were not developed simultaneously, and as reported in other studies, the age of a cell culture can have a dramatic effect on the system productivity [13,16].

2.3. Reduction of MCF-7 Cell Viability by S. Khuzistanica Extracts

The predominance of RA in the methanolic *S. khuzistanica* extract (SKE), obtained as described in Section 3.5 (see Supplementary Figure S1), was confirmed by electrospray ionization (ESI-MS) in positive and negative mode (Figure 3). The effect of SKE on cell viability was studied in MCF-7 and HepG2 cells, either non-treated or incubated for 48 h with different amounts of SKE, RA (positive control known to reduce cell viability) or vehicle (dimethyl sulfoxide, DMSO).

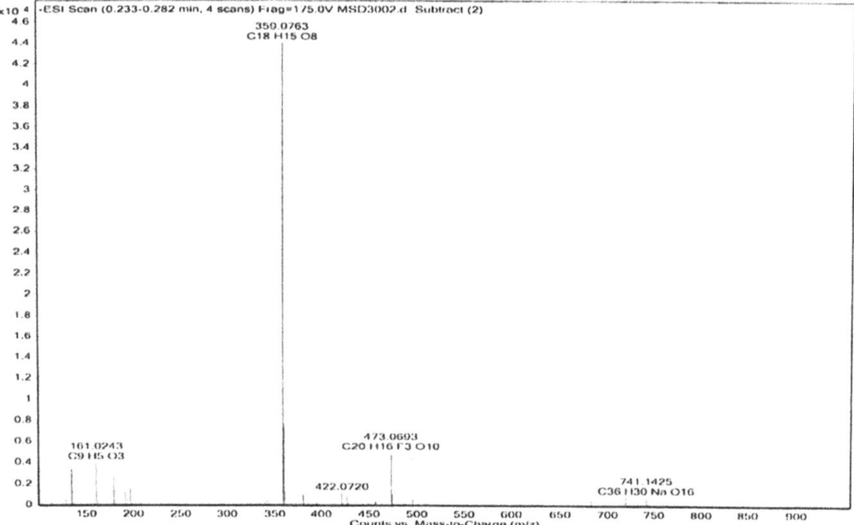

Figure 3. Electrospray ionization (ESI-MS) base peak chromatogram of a rosmarinic acid (RA)-enriched methanolic extract of *S. khuzistanica* cell cultures (SKE) in negative ion mode. Mass spectra were detected in the range of m/z 100–1000. A molecular weight of 359 was assigned to [M−H]$^-$, and 741 to [2M−2H+Na]$^-$ of RA.

As shown in Figure 4, the use of up to 0.4% (*v/v*) of DMSO as a vehicle did not significantly affect MCF-7 and HepG2 cell viability. Therefore, DMSO was selected as the solvent for subsequent experiments at concentrations of up to 0.4%. The highest concentration of SKE assayed (0.6 mg·L^{-1}) significantly reduced the viability of HepG2 cells up to about 25% of the values observed in control cells, and decreased MCF-7 cell viability to barely detectable levels. A similar cytotoxic effect on MCF-7 and HepG2 cells was observed with the highest concentration of RA used (0.2 mM). Remarkably, even though RA content in 0.6 mg·L^{-1} SKE was estimated to be 0.064 mM, the biotechnologically produced extract had a stronger effect on MCF-7 and HepG2 cell viability than 0.2 mM RA.

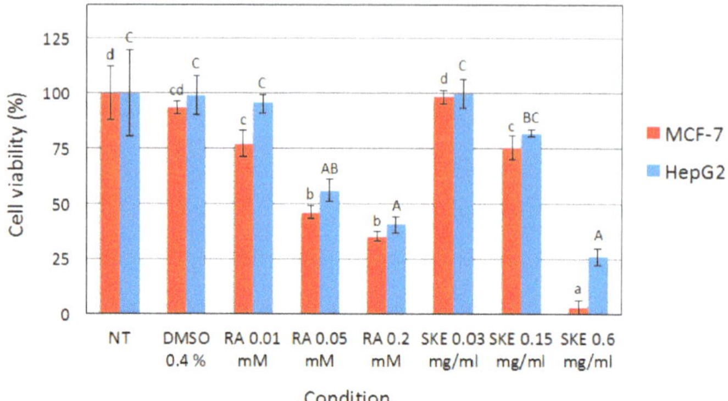

Figure 4. Effect of *S. khuzistanica* extracts on viability of MCF-7 and HepG2 cells. Cell viability was assayed 48 h following treatment with 0.03 mg·mL^{-1}, 0.15 mg·mL^{-1} and 0.6 mg·mL^{-1} *S. khuzistanica* extract (SKE), 0.01 mM, 0.05 mM and 0.2 mM rosmarinic acid (RA) or 0.4% DMSO (vehicle; D). Control cells were non-treated (NT). Each bar represents the mean ± SD of three replicates. Different letters (lower case for MCF-7 cells and upper case for HepG2 cells) indicate significant differences between treatments ($p < 0.01$).

Yousefzadi et al. [19] demonstrated the *in vitro* cytotoxicity of the essential oil of *S. khuzistanica*, which reduced the cell viability of human colon adenocarcinoma (SW480), MCF7 and choriocarcinoma (JET3) cells. This suggests that other parts of *S. khuzistanica* plants, in this case essential oil obtained from leaves, which contain carvacrol as the main component but no RA, could have a similar anticancer activity to our biotechnological extract based on *S. khuzistanica* cell cultures, containing RA as the major component. More recently, Esmaeili-Mahani et al. [20] also reported cytotoxic activity of an *S. khuzistanica* ethanolic extract obtained from dry leaves against the MCF-7 cell line. Unfortunately, the authors did not determine the composition of the leaf extract, but they attributed the biological activity to carvacrol, being the main compound of *S. khuzistanica* aerial parts. Both RA and carvacrol have anti-proliferative activity, which may explain their similar biological effects, despite having a different phytochemical composition [4,21].

2.4. Induction of MCF-7 Cell Cycle Arrest by S. Khuzistanica Extracts

Given that SKE had a stronger impact on the viability of MCF-7 cells than HepG2 cells, and based on reports showing that RA induces cell cycle arrest in human-derived cell lines such as MCF-7 [22], the effect of SKE on the cell cycle was studied in the latter, using flow cytometry. To this end, MCF-7 cells were incubated for up to 48 h with SKE, RA or DMSO (vehicle). As previously reported, RA significantly increased the sub-G0/G1 cell population at 48 h post-treatment. Flow cytometry analysis revealed that 0.6 mg·mL^{-1} of SKE also significantly increased the percentage of sub-G0/G1 cells both at 24 h (Figure 5A) and 48 h (Figure 5B) post-treatment, and a trend to reduce the cell fractions in the G0/G1 and G2/M phases was found. Similarly as for cell viability, 0.6 mg·mL^{-1} SKE had a stronger effect on MCF-7 cell cycle than the maximum RA concentration used (0.4 mM).

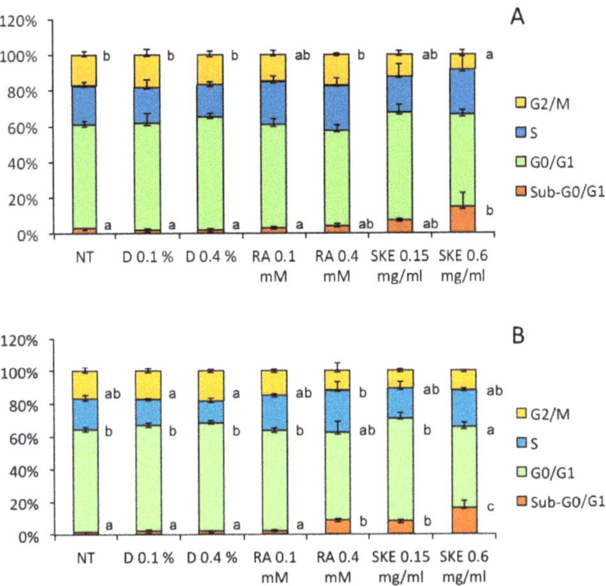

Figure 5. Effect of *S. khuzistanica* extracts on the percentage of MCF-7 cells in each phase of the cell cycle. Cell cycle distribution of MCF-7 cells treated for 24 h (**A**) and 48 h (**B**) with 0.15 mg·mL^{-1} and 0.6 mg·mL^{-1} *S. khuzistanica* extract (SKE), 0.1 mM and 0.4 mM rosmarinic acid (RA) or 0.1% and 0.4% DMSO (vehicle; D). Control cells were non-treated (NT). Each bar represents the mean ± SD of three replicates. For each phase of the cell cycle, different letters indicate significant differences between treatments ($p < 0.05$).

2.5. Effect of S. Khuzistanica Extracts on Caspase Activity of MCF-7 Cells

The fact that induction of the sub-G0/G1 cell population is associated with cell apoptosis and that RA is known to induce apoptosis in human cancer-derived cell lines [23,24] prompted us to analyze the impact of SKE on the activity of inflammatory caspase-1 and -5, initiator caspase-2, -8 and 9, as well as executioner caspase-6 (Figure 6).

No effect on the activity of inflammatory caspases was observed at 48 h after treatment. However, compared to non-treated cells, SKE significantly enhanced caspase-8 initiator activity, which is known to mediate activation of the extrinsic apoptosis pathway [25]. Although not significant, the same trend was observed in MCF-7 cells incubated with RA. No significant effects were found on the activity of caspase-2 and -9. Regarding executioner caspases, SKE, and to a lesser extend RA, showed a tendency to increase the activity of caspase-6. Previous studies have reported that involvement of caspases in apoptosis induction by RA is cell type-specific [22]. Taken together, our findings suggest that SKE and RA may induce apoptosis of MCF-7 cells by activating the extrinsic pathway. Conceivably, ligand binding to death receptors in the cell membrane would trigger caspase-8 dimerization and activation, which in turn may lead to direct cleavage and activation of executioner caspases. However, SKE and RA failed to significantly modify the activity of caspase-9, which is the initiator caspase responsible for the intrinsic apoptosis pathway. Given that 0.6 mg·mL^{-1} SKE (containing 0.064 mM RA) had a stronger effect on MCF-7 cells than 0.4 mM RA, it may be due to the anticancer activity of other minority components of the extract, although this hypothesis would need to be confirmed by their chemical characterization.

The activation of caspase-3 by extracts of *S. khuzistanica* leaves has been previously demonstrated [20]. In the current work, this particular caspase was not studied, so we cannot affirm if our *S. khuzistanica* extract has a similar effect. As mentioned above, the plant extract is rich in carvacrol, whereas the predominant compound in our biotechnological extract was RA.

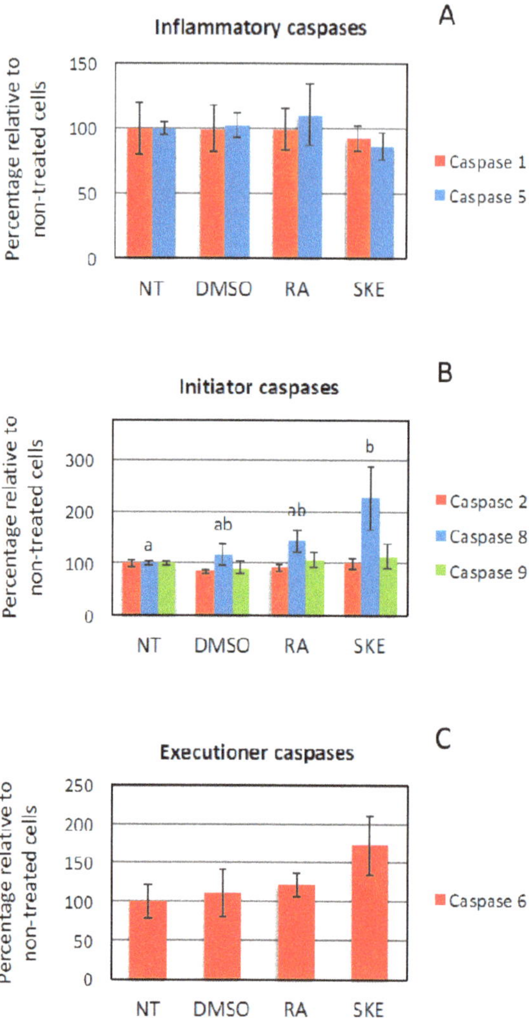

Figure 6. Assay of caspase activity in MCF-7 cells treated with *S. khuzistanica* extracts. The activity of inflammatory (**A**), initiator (**B**) and executioner (**C**) caspases was assayed in MCF-7 cells treated for 48 h with 0.6 mg·mL^{-1} *S. khuzistanica* extract (SKE), 0.4 mM rosmarinic acid (RA) or 0.4% DMSO (vehicle). Control cells were non-treated (NT). Each bar represents the mean ± SD of three replicates. Different letters indicate significant differences between treatments ($p < 0.05$).

3. Materials and Methods

3.1. S. Khuzistanica Plant Cell Cultures

The experiment was performed using 125 mL flasks with a working volume (WV) of 20 mL and an inoculum PCV of 10% from a week-old *S. khuzistanica* cell suspension obtained as previously described by Khojasteh et al. [7]. As in the previous experiments, the culture conditions were as follows: T of 25 °C, 110 rpm shaking frequency and darkness. After a growth period of 14 days (near the end of the exponential phase), COR at a final concentration of 1 µM was added as an elicitor to the flasks. The culture period consisted of 21 days and samples were harvested at 0, 2, 4, 7, 11, 14, 16, 18, 21, and 24 days of growth in order to determine in-process control measurements CFW, CDW, pH, conductivity, as well as RA production. Samples were taken in triplicate and the results compared with control conditions (without elicitation).

3.2. Benchtop Bioreactor Scale

In order to scale up the process, a 2 L CellBag (GE Healthcare Bio-Sciences AB, Uppsala, Sweden) with a working volume of 1 L, shaken in a Khuhner orbital shaker (Birsfelden, Basel, Switzerland) in the dark at 25 °C and 35–38 rpm, with a shaking diameter of 50 mm, was used. The culture was initiated at 35 rpm and gradually the shaking was increased up to 38 rpm to obtain a good distribution of cell biomass and oxygen transfer without foaming. The sterile airflow was 0.2 L·min^{-1} and the inoculum size was 10% (*w/v*). In previous experiments, these conditions were determined as optimal for the growth of the cell line and to avoid out-of-phase phenomena [7]. After 11 days of culture (near the end of the exponential phase in these conditions), 1 µM COR was added. Samples were taken at days 0, 4, 7, 9, 11, 14, 16, 18 and 21. The experiment was run in duplicate.

3.3. In-Process Controls

PCV, CFW, CDW, pH and conductivity were determined as previously described by Khojasteh et al. [7]. In brief, for the PCV, 10 mL of cell suspension was transferred to 15 mL Falcon tubes, then centrifuged at 4000 rpm for 10 min and the PCV was read in the tube. For CFW, 10 mL of the cell suspension was centrifuged at 4000 rpm, the pellet was transferred to a tube connected to a vacuum pump for 3 min and the CFW was measured. For the CDW, frozen biomass was freeze-dried for 24 h and the CDW of each sample was registered.

3.4. Rosmarinic Acid Extraction and Quantification

Extraction and quantification of RA from the samples of lyophilized cells and supernatants were performed according to Georgiev et al. [26] with some modifications as described in Khojasteh et al. [7]. In brief, 20 mg of the freeze-dried cells were suspended with 9 mL methanol. The extracts were vortexed for 2 min and incubated for 20 min in an ultrasonic bath, and then centrifuged (4000 rpm). The supernatants were evaporated under reduced pressure (vacuum evaporator, BUCHI Corporation, New Castle, Delaware, USA) at 40–45 °C, and the residue was dissolved in 1.5 mL of methanol. For cell culture extraction, 10 mL of filtered culture medium was frozen and lyophilized. The extract was dissolved in 5 mL of methanol, incubated for 24 h at 4 °C, and then filtered (0.45 µm) to remove sugars. The methanolic extracts were passed through a 0.2 µm filter. An aliquot of 20 µL of the filtrate was injected into the HPLC for RA analysis following the method previously described and validated by Sahraroo et al. [27] with some modifications [7]. The HPLC column was a Spherisorb ODS-2 (5 µm) reverse phase 4.6 mm × 250 mm connected to an HPLC-UV system ((Agilent 1100, Santa Clara, California, USA). The mobile phase A was 0.1% (*v/v*) formic acid solution in water and acetonitrile was the mobile phase B. A gradient system with A and B was used as follows: (0 min) 88% A and 12% B, (30 min) 80% A and 20% B, (45 min) 70% A and 30% B up to 60 min. Throughout the chromatography the flow rate was 0.1 mL·min^{-1} and the injection volume 40 µL. For calibration, an RA standard was used in various concentrations ranging from 0.5 to 200.0 µg·mL^{-1}. The retention time for RA was 16 min (Supplementary Figure S1). Standard graphs were prepared by plotting concentration versus area. Quantification was carried out from integrated peak areas of the samples using the corresponding standard graph.

3.5. Preparation of the S. Khuzistanica Extract for Biological Assays

In order to prepare a *S. khuzistanica* methanolic extract enriched in RA (SKE), 10 g of CDW from a 24-old cell suspension treated with COR was distributed in Falcon tubes, each containing 100 mg of

CDW supplemented with 40 mL of methanol. RA was extracted as described above from the tubes and the extracts were dried in a rotary evaporator. Due to the toxicity of methanol for the biological analyses, we utilized the minimum quantity of DMSO to dissolve the dry extract and obtain the *S. khuzistanica* RA-enriched extract. An aliquot of the extract was evaporated and re-dissolved in methanol for HPLC analyses. The final concentration of RA in the *S. khuzistanica* extract was 5.8 ± 0.2 mg·RA·mL^{-1}.

3.6. Electrospray (ESI-MS) Analyses

The instrument LC/MSD-TOF (2006) (Agilent Technologies, Santa Clara, California, USA) was utilized to detect the mass of the components of the SKE. The electrospray ionization method (ESI-MS) was used in both positive and negative mode, at a fragmentation voltage of 215 V for the positive and 175 V for the negative mode, with a drying gas temperature of 300 °C, and drying gas (N_2). Flow was 7.01 L·min^{-1}, with a nebulizer pressure of 15 psi, and capillary voltages of 3.5 KV (negative) and 4 KV (positive). The mass spectra were detected in the range of m/z 100–1000. Samples were introduced into the source with an HPLC system (Agilent 1100, Santa Clara, California, USA), using the mixture of H_2O/CH_3CN 1:1 with a flow of 200 µL·min^{-1}.

3.7. Human Cancer-Derived Cell Lines

Human breast adenocarcinoma-derived MCF-7 (ATCC no. HTB-22) and human hepatoma-derived HepG2 (ATCC no. HB-8065) cells were grown at 37 °C and 5% CO_2 in Dulbecco's Modified Eagle's medium (DMEM) supplemented with 2 mM glutamine, 110 mg·L^{-1} sodium pyruvate, 10% fetal bovine serum, 100 IU·mL^{-1} penicillin and 100 µg·mL^{-1} streptomycin.

3.8. Cell Viability Assay

MCF-7 and HepG2 cells were seeded in 24-well plates at a density of 2×10^4 cells/well. Twenty-four hours later, different amounts of SKE, RA or vehicle (DMSO) were added to the cell cultures. Forty-eight hours after the addition of the compounds, cell viability was determined by means of the 3-(4,5-dimethylthiazol-2-yl)-2,5-diphenyltetrazolium bromide (MTT) assay as previously described [28]. The cells were incubated in the presence of 0.63 mM of MTT and 18.4 mM of sodium succinate for 3 h at 37 °C. Following removal of the medium, formazan was re-suspended in DMSO supplemented with 0.57% CH_3COOH and 10% sodium dodecyl sulphate. Spectrophotometric determinations were performed at 600 nm in a Cobas Mira S analyzer (Hoffman-La Roche, Basel, Switzerland). The results are expressed as a percentage of cell survival relative to non-treated control cells.

3.9. Cell Cycle Analysis

Analysis of the cell cycle was performed using the PI/Cell Cycle Analysis Kit (Canvax Biotech, Córdoba, Spain). MCF-7 cells were seeded in 6-well plates at a density of 5×10^5 cells/well and left to grow for 24 h. Afterwards, the cells were incubated in the presence of 0.15 mg·mL^{-1} or 0.6 mg·mL^{-1} of SKE, 0.1 mM or 0.4 mM of RA or DMSO (vehicle) for 48 h. The cells were harvested, sedimented by centrifugation (1000 rpm, 5 min), washed with ice cold PBS, centrifuged again in the same conditions and fixed in ice cold 70% ethanol for 45 min. Cells were recovered by centrifugation, washed with PBS, centrifuged again, re-suspended with 200 µL of staining solution containing propidium iodide and RNase A, and incubated at 37 °C for 30 min in the dark. The cells were analyzed by flow cytometry using a Gallios analyzer and Kaluza software (Beckman Coulter, Brea, CA, USA). The percentage of cells in each phase of the cell cycle was calculated and the apoptotic cells were considered to constitute the sub-G0/G1 cell population.

3.10. Caspase Activity Assays

Activity assays for caspase-1, -2, -5, -6, -8 and -9 were performed using the Caspase-Family Colorimetric Substrate Set Plus kit (BioVision, Milpitas, CA, USA). Following treatment with

0.6 mg·mL^{-1} of SKE, 0.4 mM RA or DMSO (vehicle) for 48 h, 5 × 10^6 MCF-7 cells were harvested, sedimented by centrifugation, re-suspended in 50 μL of chilled cell lysis buffer and incubated on ice for 10 min. After centrifugation at 10,000 g for 1 min, the supernatant was recovered to assay the protein concentration and caspase activities. Total protein in cell extracts was assayed with the Bradford method [29] using BSA as a standard. For each caspase activity, the reaction mixture contained 100 μg of cell extract protein, 4.8 mM dithiothreitol and 190 μM of the corresponding caspase p-nitroaniline conjugated substrate in a total volume of 52.5 μL. The reaction proceeded at 37 °C for 2 h. Spectrophotometric determinations were performed at 30 °C in a Cobas Mira S analyzer (Hoffman-La Roche, Basel, Switzerland) at a wavelength of 405 nm (caspase activity assays) and 600 nm (total protein). The results are presented as a percentage of caspase activity relative to the control (non-treated cells).

3.11. Statistical Analysis

The statistical analysis was performed with SPSS version 24 (IBM, Armonk, NY, USA). Data were submitted to one-way ANOVA. Significant differences among treatments were determined with the Scheffé post hoc test.

4. Conclusions

Taken as a whole, our results demonstrate that plant cell cultures of *S. khuzistanica* could constitute a biosustainable source of RA and that COR is an effective elicitor for increasing its production, despite negatively affecting growth in the small-scale system. We also demonstrated that when the system was scaled up to the benchtop scale (orbitally shaken 2 L culture bag with working volume of 1 L) the negative effect of COR on biomass production was reduced and the productivity of the biotechnological system increased significantly. RA and even more so SKE reduced the viability of HepG2 and MCF-7 cancer cell lines. The fact that SKE increased the sub-G0/G1 cell population and enhanced caspase-8 initiator activity supports the notion that it activated the extrinsic apoptosis pathway in the MCF-7 cancer cell line.

Supplementary Materials: Supplementary materials can be found at http://www.mdpi.com/1422-0067/20/10/2400/s1.

Author Contributions: Conceptualization, R.E., J.P., I.M. and R.M.C; formal analysis, A.K. and S.C; writing, review and editing, J.P., I.M., R.M.C. and R.E; funding acquisition, J.P. and I.M.

Funding: The study was financially supported by the project 2017 SGR 242 from the Generalitat de Catalunya and BIO2017-82374-R from the Spanish Ministerio de Ciencia, Innovación y Universidades and AEI/FEDER, EU.

Conflicts of Interest: The authors declare no conflict of interest.

References

1. Petersen, M.; Abdullah, Y.; Benner, J.; Eberle, D.; Gehlen, K.; Hücherig, S.; Janiak, V.; Kim, K.H.; Sander, M.; Weitzel, C.; et al. Evolution of rosmarinic acid biosynthesis. *Phytochemistry* **2009**, *70*, 1663–1679. [CrossRef] [PubMed]
2. Petersen, M. Rosmarinic acid: New aspects. *Phytochem. Rev.* **2013**, *12*, 207–227. [CrossRef]
3. Bulgakov, V.P.; Inyushkina, Y.V.; Fedoreyev, S.A. Rosmarinic acid and its derivatives: Biotechnology and applications. *Crit. Rev. Biotechnol.* **2012**, *32*, 203–217. [CrossRef]
4. Moore, J.; Yousef, M.; Tsiani, E. Anticancer effects of rosemary (*Rosmarinus officinalis* L.) extract and rosemary extract polyphenols. *Nutrients* **2016**, *8*, 731. [CrossRef]
5. Hadian, J.; Mirjalili, M.H.; Kanani, M.R.; Salehnia, A.; Ganjipoor, P. Phytochemical and morphological characterization of *Satureja khuzistanica* Jamzad populations from Iran. *Chem. Biodiv.* **2011**, *8*, 902–914. [CrossRef] [PubMed]
6. Rao, S.R.; Ravishankar, G.A. Plant cell cultures: Chemical factories of secondary metabolites. *Biotechnol. Adv.* **2002**, *20*, 101–153. [PubMed]

7. Khojasteh, A.; Mirjalili, M.H.; Palazon, J.; Eibl, R.; Cusido, R.M. Methy-jasmonate enhanced production of rosmarinic acid in cell cultures of *Satureja khuzistanica* in a bioreactor. *Eng. Life Sci.* **2016**, *16*, 740–749. [CrossRef]

8. Ramirez-Estrada, K.; Vidal-Limon, H.; Hidalgo, D.; Moyano, E.; Golenioswki, M.; Cusidó, R.M.; Palazon, J. Elicitation, an effective strategy for the biotechnological production of bioactive high-added value compounds in plant cell factories. *Molecules.* **2016**, *21*, 182. [PubMed]

9. Onrubia, M.; Moyano, E.; Bonfill, M.; Cusidó, R.M.; Goossens, A.; Palazón, J. Coronatine, a more powerful elicitor for inducing taxane biosynthesis in *Taxus media* cell cultures than methyl jasmonate. *J. Plant Physiol.* **2013**, *170*, 211–219. [CrossRef]

10. Ramirez-Estrada, K.; Osuna, L.; Moyano, E.; Bonfill, M.; Tapia, N.; Cusido, R.M.; Palazon, J. Changes in gene transcription and taxane production in elicited cell cultures of *Taxus x media* and *Taxus globosa*. *Phytochemistry* **2015**, *117*, 174–184. [CrossRef]

11. Deus-Neumann, B.; Zenk, M.H. Instability of indole alkaloid production in *Catharanthus roseus* cell suspension cultures. *Planta Med.* **1984**, *50*, 427–431. [CrossRef]

12. Sierra, M.I.; van der Heijden, R.; van der Leer, T.; Verpoorte, R. Stability of alkaloid production in cell suspension cultures of *Tabernaemontana divaricata* during long- term subculture. *Plant Cell Tiss Org Cult.* **1992**, *28*, 59–68. [CrossRef]

13. Kim, K.H.; Janiak, V.; Petersen, M. Purification, cloning and functional expression of hydroxyphenylpyruvate reductase involved in rosmarinic acid biosynthesis in cell cultures of *Coleus blumei*. *Plant Mol. Biol.* **2004**, *54*, 311–332. [CrossRef]

14. Sanchez-Muñoz, R.; Bonfill, M.; Cusidó, R.M.; Palazon, J.; Moyano, E. Advances in the regulation of in vitro paclitaxel production: Methylation of a Y-Patch promoter region alters BAPT gene expression in *Taxus* cell cultures. *Plant Cell Physiol.* **2018**, *59*, 2255–2267.

15. Mustafa, N.R.; de Winter, W.; van Iren, F.; Verpoorte, R. Initiation, growth and cryopreservation of plant cell suspension cultures. *Nat. Protoc.* **2011**, *6*, 715–742. [CrossRef]

16. Fu, C.; Li, L.; Wu, W.; Li, M.; Yu, X.; Yu, L. Assessment of genetic and epigenetic variation during long-term *Taxus* cell culture. *Plant Cell Rep.* **2012**, *31*, 1321–1331. [CrossRef]

17. Klöckner, W.; Diederichs, S.; Büchs, J. Orbitally shaken single-use bioreactors. *Adv. Biochem. Eng. Biotechnol.* **2014**, *138*, 45–60. [PubMed]

18. Hidalgo, D.; Steinmetz, V.; Brossat, M.; Tournier-Couturier, L.; Cusido, R.M.; Corchete, P.; Palazon, J. An optimized biotechnological system for the production of centellosides based on elicitation and bioconversion of *Centella asiatica* cell cultures. *Eng. Life Sci.* **2017**, *17*, 413–419. [CrossRef]

19. Yousefzadi, M.; Riahi-Madvar, A.; Hadian, J.; Rezaee, F.; Rafiee, R.; Biniaz, M. Toxicity of essential oil of *Satureja khuzistanica*: In vitro cytotoxicity and anti-microbial activity. *J. Immunotoxicol.* **2014**, *11*, 50–55. [CrossRef]

20. Esmaeili-Mahani, S.; Samandari-Bahraseman, M.R.; Yaghoobi, M.M. In-vitro anti-proliferative and pro-apoptotic proterties of *Sutureja Khuzestanica* on human breast cancer (MCF-7) and its synergic effects with anticancer drug vincristine. *Iran J Pharm Res.* **2018**, *17*, 343–352.

21. Yin, Q.H.; Yan, F.X.; Zu, X.Y.; Wu, Y.H.; Wu, X.P.; Liao, M.C.; Deng, S.W.; Yin, L.L.; Zhuang, Y.Z. Anti-proliferative and pro-apoptotic effect of carvacrol on human hepotocelular carcinoma cell line HepG-2. *Cytotechnology.* **2012**, *64*, 43–51. [CrossRef] [PubMed]

22. Wu, C.F.; Hong, C.; Klauck, S.M.; Lin, Y.L.; Efferth, T. Molecular mechanisms of rosmarinic acid from *Salvia miltiorrhiza* in acute lymphoblastic leukemia cells. *J Ethnopharmacol.* **2015**, *176*, 55–68. [CrossRef] [PubMed]

23. Jang, Y.G.; Hwang, K.A.; Choi, K.C. Rosmarinic Acid, a component of rosemary tea, induced the cell cycle arrest and apoptosis through modulation of HDAC2 expression in prostate cancer cell lines. *Nutrients* **2018**, *10*, 1784. [CrossRef]

24. Huang, Y.; Cai, Y.; Huang, R.; Zheng, X. Rosmarinic acid combined with adriamycin induces apoptosis by triggering mitochondria-mediated signaling pathway in HepG2 and Bel-7402 Cells. *Med. Sci. Monit.* **2018**, *24*, 7898–7908. [CrossRef] [PubMed]

25. McIlwain, D.R.; Berger, T.; Mak, T.W. Caspase functions in cell death and disease. *Cold Spring Harb. Perspect. Biol.* **2013**, *5*, a008656. [CrossRef] [PubMed]

26. Georgiev, M.; Kuzeva, S.; Pavlov, A.; Kovacheva, E.; Ilieva, M. Enhanced rosmarinic acid production by *Lavandula vera* MM cell suspension culture through elicitation with vanadyl sulfate. *Z. Naturforsch. C.* **2006**, *61*, 241–244. [CrossRef]

27. Sahraroo, A.; Babalar, M.; Mirjalili, M.H.; Fattahi Moghaddam, M.R.; Nejad Ebrahimi, S. *In-vitro* callus induction and rosmarinic acid quantification in callus culture of *Satureja khuzistanica* Jamzad (Lamiaceae). *Iran J. Pharm. Res.* **2014**, *13*, 1447–1456.

28. Gallego, A.; Metón, I.; Baanante, I.V.; Ouazzani, J.; Adelin, E.; Palazon, J.; Bonfill, M.; Moyano, E. Viability-reducing activity of *Coryllus avellana* L. extracts against human cancer cell lines. *Biomed. Pharmacother.* **2017**, *89*, 565–572. [CrossRef]

29. Bradford, M.M. A rapid and sensitive method for the quantitation of microgram quantities of protein utilizing the principle of protein-dye binding. *Anal Biochem.* **1976**, *72*, 248–254. [CrossRef]

International Journal of
Molecular Sciences

Article

Curcumin Treatment in Combination with Glucose Restriction Inhibits Intracellular Alkalinization and Tumor Growth in Hepatoma Cells

So Won Kim [1,†], Min-Ji Cha [2,†], Seul-Ki Lee [3], Byeong-Wook Song [4], Xinghai Jin [3], Jae Myun Lee [5,6], Jeon Han Park [6,*] and Jong Doo Lee [2,7,*]

[1] Department of Pharmacology, Catholic Kwandong University College of Medicine, Gangneung 25601, Korea; kswlab2015@gmail.com

[2] Institute for Translational and Clinical Research, Catholic Kwandong University International St. Mary's Hospital, Incheon 22711, Korea; minjicha619@gmail.com

[3] KANT Science Research Institute, Incheon 22711, Korea; seulki1011@nate.com (S.-K.L.); jxh630@gmail.com (X.J.)

[4] Department of Medical Science, Catholic Kwandong University College of Medicine, Gangneung 25601, Korea; songbw@gmail.com

[5] Brain Korea 21 PLUS Project for Medical Sciences, Yonsei University College of Medicine, Seoul 03722, Korea; jaemyun@yuhs.ac

[6] Department of Microbiology and Immunology, Institute for Immunology and Immunological Diseases, Yonsei University College of Medicine, Seoul 03722, Korea

[7] Department of Nuclear Medicine, Catholic Kwandong University International St. Mary's Hospital, Incheon 22711, Korea

*   Correspondence: jhpark5277@yuhs.ac (J.H.P.); jdlee@yuhs.ac (J.D.L.); Tel.: +82-2-2228-1815 (J.H.P.); +82-10-2829-1159 (J.D.L.)

†   These authors contributed equally to this work.

Received: 18 April 2019; Accepted: 11 May 2019; Published: 14 May 2019

Abstract: Dysregulation of cellular energy metabolism is closely linked to cancer development and progression. Calorie or glucose restriction (CR or GR) inhibits energy-dependent pathways, including IGF-1/PI3K/Akt/mTOR, in cancer cells. However, alterations in proton dynamics and reversal of the pH gradient across the cell membrane, which results in intracellular alkalinization and extracellular acidification in cancer tissues, have emerged as important etiopathogenic factors. We measured glucose, lactate, and ATP production after GR, plant-derived CR-mimetic curcumin treatment, and curcumin plus GR in human hepatoma cells. Intracellular pH regulatory effects, in particular, protein–protein interactions within mTOR complex-1 and its structural change, were investigated. Curcumin treatment or GR mildly inhibited Na+/H+ exchanger-1 (NHE1). vATPase, monocarboxylate transporter (MCT)-1, and MCT4 level. Combination treatment with curcumin and GR further enhanced the inhibitory effects on these transporters and proton-extruding enzymes, with intracellular pH reduction. ATP and lactate production decreased according to pH change. Modeling of mTOR protein revealed structural changes upon treatments, and curcumin plus GR decreased binding of Raptor and GβL to mTOR, as well as of Rag A and Rag B to Raptor. Consequently, 4EBP1 phosphorylation was decreased and cell migration and proliferation were inhibited in a pH-dependent manner. Autophagy was increased by curcumin plus GR. In conclusion, curcumin treatment combined with GR may be a useful supportive approach for preventing intracellular alkalinization and cancer progression.

Keywords: hepatoma; intracellular pH; curcumin; glucose restriction; tumor suppression

1. Introduction

Dysregulation of cellular energy metabolism, known as the Warburg effect, is closely linked to cancer development and progression. Inhibition of enhanced glycolytic activity in cancer cells via calorie restriction (CR) or glucose restriction (GR) is under clinical investigation as a supportive anticancer therapy. CR inhibits energy-dependent signaling pathways, including the insulin-like growth factor 1 (IGF-1)/phosphoinositide 3-kinase (PI3K)/Akt/mammalian target of rapamycin (mTOR) pathway, and activates AMP-dependent protein kinase (AMPK) activity [1].

Recently, alterations in proton dynamics have emerged as an important factor contributing to the etiopathogenesis of cancer cells, and energy dysregulation is thought to be attributable to increased intracellular pH (pHi) [2]. In fact, intracellular alkalinization and extracellular acidification are commonly observed in malignant tumor tissues. This occurs as a result of the export of excess intracellular lactate and protons into the extracellular space by the transmembrane monocarboxylate transporter-4 (MCT4), coupled with the secretion of protons by Na^+/H^+ exchanger-1 (NHE1) or proton-extruding enzymes, including vacuolar H^+-ATPase (v-ATPase) and carbonic anhydrases. Altered activities of transporters or enzymes in cancer cells, and reversal of the pH gradient across the cancer cell membrane, play pivotal roles in cancer progression and metastasis [3,4]. A decrease in the extracellular pH (pHe) stimulates angiogenesis, activates stromal cells for tumor invasion, and impairs the immune reaction as a result of decreased cytolytic functions of T cells [5]. Resistance to chemotherapeutics is also related to increased extracellular acidity [6]. Intracellular alkalinization has important biological effects, and even small changes in the pHi significantly affect protein activities. Therefore, inhibition of intracellular alkalinization is a potential anticancer approach, and various NHE1 inhibitors, such as cariporide, are under investigation [7].

GR decreases lactate production in cancer cells [8], which suggests that energy restriction might be an alternative approach for preventing intracellular alkalinization. GR lowers intracellular ATP concentrations and decreases the pHi [9]. Curcumin prevents the extrusion of intracellular protons by functioning as an NHE1 inhibitor [10]. Furthermore, the pHi affects the conformation and binding affinity of various proteins [11].

As we have previously demonstrated that GR combined with CR-mimetic plant-derived polyphenols has synergistic anticancer effects in malignant tumors [12], we investigated the anti-cancer effects of GR or curcumin treatments, and especially combination treatment in terms of intracellular pH regulation in human hepatoma cell lines because glucose metabolism is closely linked to cytosolic pH regulation. Furthermore, protein–protein interactions within mTOR complex-1 (mTORC1) following curcumin and/or GR treatment were studied as mTORC1 plays critical roles in cell size regulation, cell growth, and nutrient sensing.

2. Results

2.1. Curcumin and GR Inhibit Intracellular Alkalinization and Function as NHE1 Inhibitors

Enhanced aerobic glycolysis (Warburg effect) is one of the hallmarks of cancer cells. Because glycolysis is closely linked to the cytosolic pH and export of H^+ ions, whether curcumin and/or GR affect the pHi was investigated in the hepatoma cell lines HepG2, Hep3B, and SNU449. The pHi in these cell lines was decreased when cells were cultured with standard RPMI-1640 medium (11 mM glucose) with curcumin, low glucose concentration (5.5 mM, GR), and more decreased under GR plus curcumin as compared to standard medium (11 mM glucose) without curcumin. Treatment effects were more prominent in HepG2 cells (Supplementary Figure S1). The pHi of HepG2 cells grown in the standard medium was 8.15 ± 0.16, and it was significantly decreased after curcumin administration (7.19 ± 0.05, $p < 0.05$), and mildly decreased under GR condition (7.73 ± 0.04, $p > 0.05$). Curcumin administration under GR condition decreased the pHi to a lower normal limit (6.91 ± 0.16, $p < 0.01$). Curcumin inhibited intracellular alkalinization as effectively as the NHE1 inhibitor, cariporide

(7.25 ± 0.11, $p < 0.05$). However, the NHE1 activator PMA did not significantly increase the pHi (7.89 ± 0.08, $p > 0.05$) (Figure 1A).

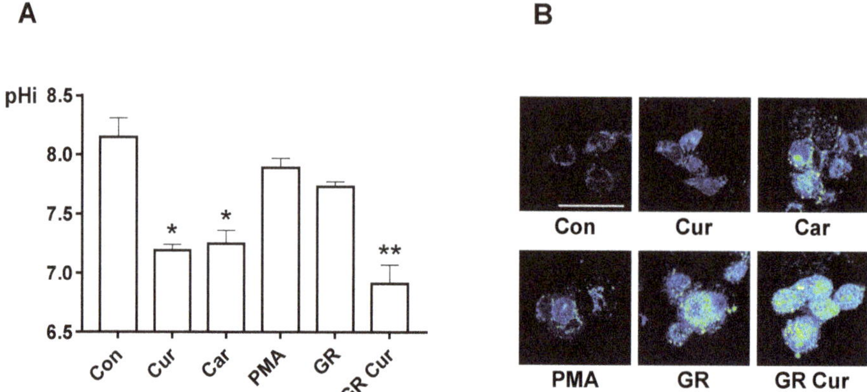

Figure 1. pHi-lowering effect of curcumin and glucose restriction. (**A**) HepG2 cells were cultivated with standard medium, standard medium containing 20 nM curcumin, 100 nM cariporide, or 100 nM PMA, GR (5.5 mM), or GR containing 20 nM curcumin, then pHi was measured. The experiment was conducted five times independently. (**B**) pHi imaging was performed using confocal microscopy (400×). Bright green color and dark blue color indicate acidic and alkaline condition, respectively. The scale bar is 50 μm. Con, standard RPMI-1640 medium; Cur, curcumin; Car, cariporide; PMA, phorbol-12-myristate-13-acetate; GR, glucose restriction, 5.5 mM glucose medium; GR Cur, glucose restriction plus curcumin. * $p < 0.05$ vs. control; ** $p < 0.01$ vs. control.

Fluorescence visualization of the pHi by BCECF-AM confirmed that curcumin decreased the pHi similar to cariporide, and combination of curcumin with GR resulted in more effective pHi suppression on fluorescent imaging (Figure 1B). However, the pHi of human dermal fibroblast cells was within the normal range after GR plus curcumin (Supplementary Figure S2).

2.2. Curcumin and GR Inhibit Level of Proton-Extruding Proteins

To elucidate the pHi regulatory mechanisms of curcumin and GR, the effect of curcumin and GR on the level of the proton-extruding proteins NHE1, MCTs, and v-ATPase was investigated in HepG2 cells by immunoblotting. Protein level of NHE1 was decreased in HepG2 cells grown in standard medium with curcumin, or GR alone, and these effects were more prominent in the GR plus curcumin group (Figure 2). Protein level of MCT1 and MCT4 was also significantly decreased under the treatment conditions. ATP synthase (ATP subunit alpha, ATP5A) and v-ATPase were decreased under the same treatment conditions (Figure 2). These findings indicated that the level changes of these proteins by curcumin and GR were correlated with pHi changes. Thus, curcumin and GR might in part regulate pHi by modulating the level of proton-extruding proteins. Curcumin suppressed NHE1 mRNA to the same level as cariporide. Upon treatment with PMA, the mRNA level of NHE1 was slightly increased. Combination of GR and curcumin reduced the mRNA level of NHE1 the most significantly (Supplementary Figure S3). In contrast, AMPK and p-AMPK were markedly increased under GR conditions (>3-fold increases, $p < 0.01$) (Figure 2).

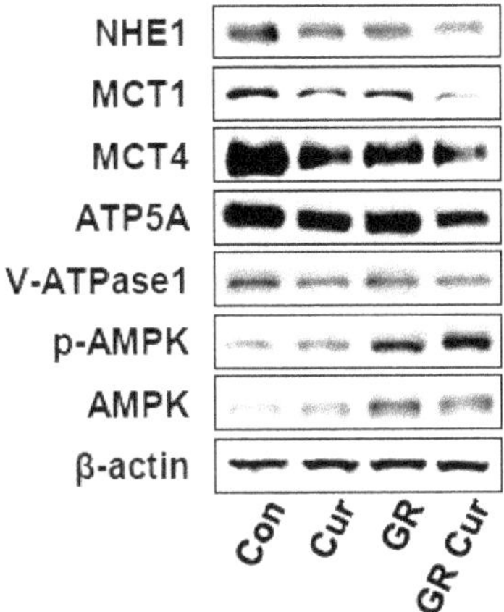

Figure 2. Effect of curcumin and glucose restriction on the protein level of transporters, enzymes regulating pHi, and the energy regulator AMPK. HepG2 cells were cultivated under the conditions indicated in the legend of Figure 1, and immunoblotting was performed using appropriate antibodies to NHE1, MCT1, MCT4, ATP5A, v-ATPase1, p-AMPK, and AMPK, respectively. β-Actin was used as a loading control. The experiment was conducted three times independently. Con, standard RPMI-1640 medium; Cur, curcumin; GR, glucose restriction, 5.5 mM glucose medium; GR Cur, glucose restriction plus curcumin.

2.3. Glucose Uptake and Lactate Production are Affected by pHi, and Inhibited by Curcumin and GR

Because enhanced glucose uptake and enhanced lactate formation are key features of cancer cells, whether changes of pHi by curcumin and GR affect glucose uptake and lactate formation was investigated in HepG2 cells. Glucose uptake was significantly decreased after treatment with curcumin, and/or GR as NHE-1 inhibitor cariporide when compared with the control (Figure 3A). Lactate production was mildly decreased after treatment with curcumin, cariporide, and/or GR (Figure 3B). Therefore, glucose uptake and lactate production appear to be associated with pHi changes by curcumin and GR, although curcumin did not produce synergic effect with GR.

2.4. Intracellular ATP is Linked to pHi Change

Next, ATP production in HepG2 cells treated with curcumin and GR was assessed. ATP production was mildly decreased after curcumin treatment or GR alone. It was significantly further decreased after curcumin treatment in a GR condition (Figure 4A). Confocal imaging of intracellular ATP confirmed these results with a marked decrease in ATP, and smaller cell size after curcumin treatment in a GR compared to cells grown in the standard medium (Figure 4B). ATP concentrations were also decreased in Hep3B and SNU449 cells after curcumin treatment in a GR condition (Supplementary Figure S4A,B). These findings suggested that ATP production is controlled by pHi changes due to curcumin and GR. No significant treatment effects on ATP production were observed in human melanocytes and dermal fibroblasts (Supplementary Figure S4C,D)

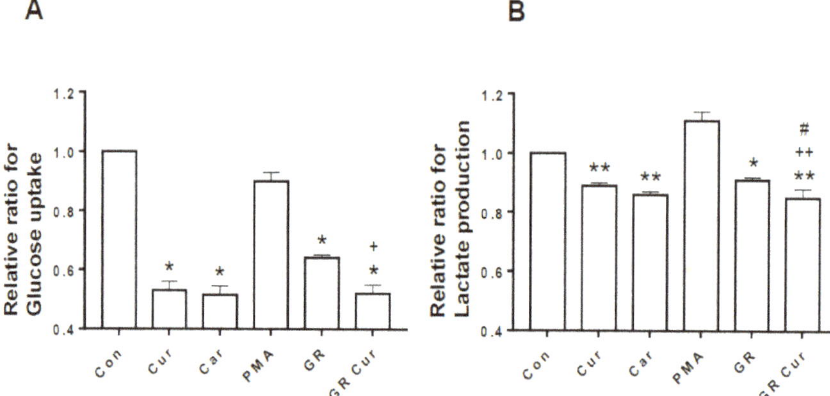

Figure 3. Effect of curcumin and glucose restriction on glucose uptake and lactate production. HepG2 cells were cultivated under the conditions indicated in the legend of Figure 1, and (**A**) glucose uptake and (**B**) lactate production were quantified. Con, standard RPMI-1640 medium; Cur, curcumin; Car, cariporide; PMA, phorbol-12-myristate-13-acetate; GR, glucose restriction, 5.5 mM glucose medium; GR Cur, glucose restriction plus curcumin. Experiment was conducted five times independently. * $p < 0.05$ vs. control; ** $p < 0.01$ vs. control; + $p < 0.05$ vs. GR; ++ $p < 0.01$ vs. GR; # $p < 0.05$ vs. Cur.

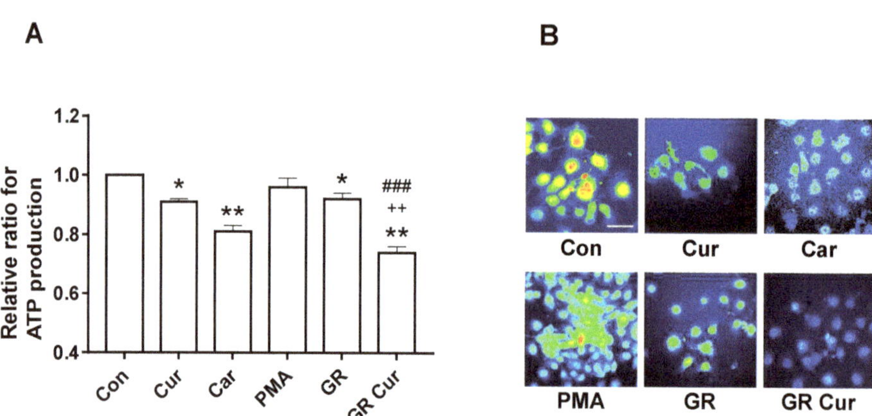

Figure 4. Effect of curcumin and glucose restriction on ATP production. HepG2 cells were cultivated under the conditions indicated in the legend of Figure 1, and (**A**) ATP production was measured. Experiment was conducted five times independently. (**B**) Confocal imaging of ATP production in HepG2 cells. High and low ATP concentrations are indicated by bright red and dark blue color, respectively. The scale bar is 50 μm. Con, standard RPMI-1640 medium; Cur, curcumin; Car, cariporide; PMA, phorbol-12-myristate-13-acetate; GR, glucose restriction, 5.5 mM glucose medium; GR Cur, glucose restriction plus curcumin. * $p < 0.05$ vs control; ** $p < 0.01$ vs. control; ++ $p < 0.01$ vs. GR; ### $p < 0.005$ vs. Cur.

2.5. Curcumin and GR Induce Structural Changes in the mTOR Protein and Changes in the Binding of mTORC1 Interacting Proteins by Affecting pHi

To compare the original structure of mTOR protein crystallized at pH 8, ionization of the mTOR-GβL protein-binding site at pH 7 was evaluated by computational analysis using Pipeline Pilot 8.5. The pattern and total numbers of hydrogen bonds between ATP and mTOR protein were changed, and the active binding cavity size increased from 7.08 A to 8.09 A (Figure 5A).

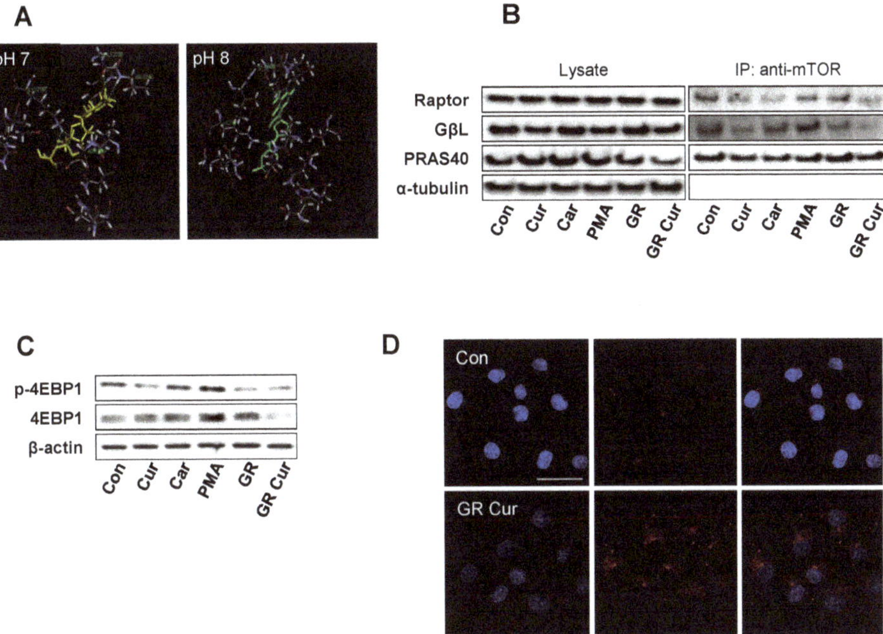

Figure 5. pH-dependent structural changes and alteration of mTORC1 interaction degree according to pHi changes induced by curcumin and glucose restriction. (**A**) Structural changes of the mTOR-GβL complex at the ATP-binding site (PDB code: 4JSV) were analyzed by computational modeling using the Pipeline Pilot 8.5. In the original structure at pH 8, the activity-site cavity size is 7.08 A, whereas at pH 7, the activity-site cavity size is 8.09 A. The ATP-binding site is indicated by yellow (pH 7) and green (pH 8) solid lines, respectively. Green dashed lines indicate hydrogen bonds. The names of residues involved in hydrogen bonding are indicated. (**B**) Using HepG2 cells, co-immunoprecipitation was used to detect mTOR interacting proteins. The experiment was conducted three times independently. (**C**) For the determination of mTOR activity, the mTOR downstream signal phosphor-4EBP1 was detected by immunoblotting. β-Actin was used as a loading control. The experiment was conducted three times independently. (**D**) Confocal imaging of autophagosomes. Cells were immunostained with an LC3 antibody. The scale bar is 50 μm. Con, standard RPMI-1640 medium; Cur, curcumin; Car, cariporide; PMA, phorbol-12-myristate-13-acetate; GR, glucose restriction, 5.5 mM glucose medium; GR Cur, glucose restriction plus curcumin. * $p < 0.05$ vs. control.

Because GR alters energy-dependent signaling pathways, such as the IGF-1-PI3K-Akt-mTOR pathway [1], and pHi affects the conformation and binding affinities of various proteins [11], mTOR co-IP and Raptor co-IP were carried out. Co-IP of the mTOR protein showed a decrease in binding of both GβL and Raptor to mTOR after curcumin and/or GR treatment as cariporide, especially GβL binding was more significantly decreased by GR plus curcumin. However, PRAS40 binding to mTOR was not significantly reduced (Figure 5B). Consequently, phosphorylation of eukaryotic translation initiation factor 4E-binding protein 1 (p-4EBP1), one of the downstream targets of mTOR [13], was decreased upon treatment with curcumin, cariporide, or GR. In accordance with the pHi pattern, phosphorylation of 4EBP1 was the most strongly decreased by GR plus curcumin treatment (Figure 5C). Consequently, autophagy was remarkably induced after curcumin treatment under GR condition in both HepG2 and Hep3B cells (Figure 5D and Supplementary Figure S5A). In human dermal fibroblasts, the treatments had no significant effects (Supplementary Figure S5B).

2.6. GR Plus Curcumin Treatment Results in Diminished Cell Migration, Growth Inhibition, and Apoptosis

Migration and proliferation are malignant phenotypes of cancer cells. Therefore, the migration and proliferation potential of HepG2 cells treated with curcumin and/or GR were assessed. Significantly attenuated migration potential was observed in HepG2 cells treated with curcumin and/or GR when compared to cells cultivated in standard medium (Figure 6A). The cell migration rate was found to be related with pHi changes induced by curcumin and/or GR. GR plus curcumin led to a significant reduction in cell proliferation potential of HepG2 cells. Glutamine addition in the culture media by 2 mM did not increase cell proliferation (Figure 6B). IP experiments conducted using Raptor antibody showed that the RagA and RagB proteins did not achieve good complex with raptor under CR+Cur condition (Figure 6C).

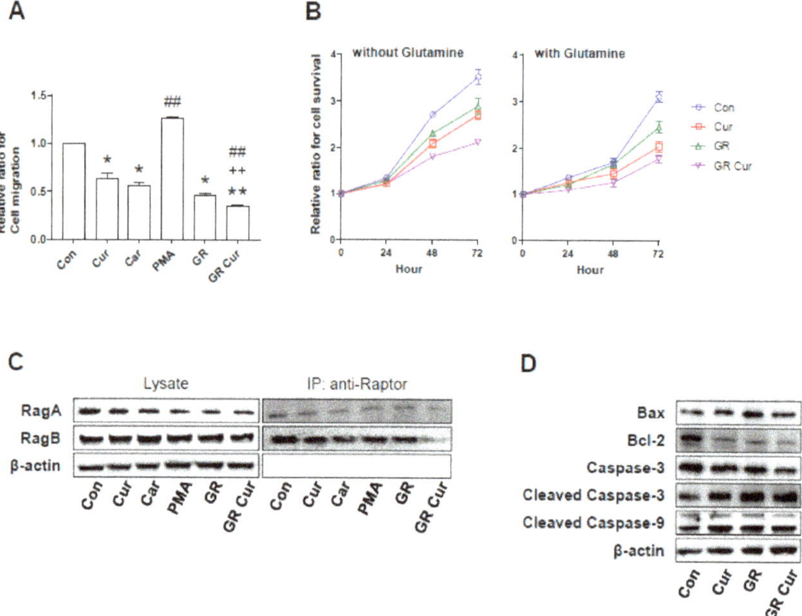

Figure 6. Effect of curcumin and glucose restriction on cell migration, proliferation, and death in HepG2 cells. (**A**) Transwell migration of HepG2 cells. Results were obtained from three independent experiments, and bar graphs represent the cell number per image field (mean ± SE). The experiment was conducted five times independently. (**B**) Cell proliferation rates of HepG2 cells were measured using a Cell Counting Kit-8 (CCK-8). The presence or absence of glutamine was observed. The experiment was conducted five times independently. (**C**) Co-immunoprecipitation was used to detect RagA and RagB. The experiment was conducted three times independently. (**D**) Bax, Bcl-2, Caspase-3 and -9, cleaved Caspase-3 and -9 proteins were expressed by immunoblotting. Beta actin was used as an internal control. The experiment was conducted three times independently. Con, standard RPMI-1640 medium; Cur, curcumin; Car, cariporide; PMA, phorbol-12-myristate-13-acetate; GR, glucose restriction, 5.5 mM glucose medium; GR Cur, glucose restriction plus curcumin. * $p < 0.05$ vs. control; ** $p < 0.01$ vs. control; ++ $p < 0.01$ vs. GR; ## $p < 0.01$ vs. Cur.

Bax protein level was increased, and protein level of Bcl-2, a negative regulator of Bax, was decreased with curcumin and/or GR. Caspase-3 protein level was suppressed by curcumin and/or GR treatment, whereas cleaved caspase-3 was increased by curcumin and/or GR treatment. Cleaved caspase-9 protein level was similar to that of cleaved caspase-3 (Figure 6D). Thus, curcumin and/or GR suppressed Bcl-2 protein level and induced Bax protein level and cleavage of Caspase-3 and -9, which indicates activation of apoptotic pathways.

3. Discussion

Reversal of the pH gradient across the cell membrane with intracellular alkalinization and extracellular acidification is a hallmark of cancer cells. In healthy cells, the pHi is approximately 7.2 and the pHe is 7.4. In tumor cells, the pHi is >7.2, whereas the pHe decreases to 6.7–7.1 [3]. Reversal of the pH gradient has important effects on cell physiology. Protonation or deprotonation significantly affect the charge status of many amino acid side chains, resulting in post-translational modifications of various proteins via phosphorylation, acetylation, and ubiquitination. It may disrupt protein–protein binding affinities, localization of proteins on the cell membrane, and the assembly of macromolecular subunits [14,15]. Limited increases in pHi, especially alkaline pHi, have important advantages for cell growth [3], inhibit apoptosis [16], and enhance glycolytic enzyme activity [17]. However, acidification of the extracellular microenvironment potentiates tumor invasion and metastasis.

Previous studies have demonstrated that intracellular alkalinization is the primary transforming event in cancer cells. Transfection of a yeast proton-pumping ATPase into mouse NIH3T3 cells or monkey Vero fibroblasts resulted in intracellular alkalinization with malignant transformation. Thus, yeast ATPase gene behaves as an oncogene [18]. Another study revealed that NIH3T3 cells infected with recombinant retroviruses expressing E7 oncogene of the human papilloma virus type 16 displayed cytoplasmic alkalinization driven by NHE1 activation. Subsequently, glycolytic activity was increased [2]. Conversely, NHE1 inhibition diminishes tumorigenic potential [19], and acidification of the cytoplasm leads to apoptosis [20]. Therefore, targeting proton dynamics associated with the intracellular pH gradient has been proposed as a potential cancer prevention strategy and therapeutic approach.

This study revealed that curcumin plus GR lowered the pHi in several hepatoma cell lines as effectively as the NHE1 inhibitor cariporide. Among the hepatoma cell lines, HepG2 cells were the most significantly affected, although GR alone could mildly inhibit NHE1. As NHE1 activation is energy- and Akt-dependent [21,22]. GR appears to potentiate the ability of curcumin to inhibit NHE1 synergistically, and vice versa.

The exact mechanism by which curcumin inhibits NHE1 needs to be elucidated. However, studies have demonstrated that curcumin affects various ion channels and transporters [23] and intracellular proton extrusion [10]. In this study, both mRNA and protein levels of NHE1 were decreased by curcumin treatment regardless of GR. This suggests that pHi alterations might affect both transcription and translation of various proteins, as previously reported [24]. In addition, proton-extruding enzymes, including v-ATPase and ATP synthase, were also decreased in our study. This is consistent with a previous report that v-ATPase can be inhibited by GR through disruption of v-ATPase assembly [9].

MCT4 regulates pHi by exporting excess lactate and protons from the cytoplasm of cancer cells. Lactate in the extracellular space affects cancer development and progression by acidifying the cellular microenvironment and by inducing the secretion of cytokines and growth factors needed for tumor growth. When lactate is extruded from hypoxic cancer cells by MCT4, it can be imported into less hypoxic cancer cells, stromal cells, or vascular endothelial cells by MCT1 as an energy metabolite [25]. Imported lactate activates hypoxia inducible factor-1 (HIF-1) through stabilization of the HIF-1α subunit, independently of hypoxia [26]. Therefore, MCT4 and MCT1 form a lactate shuttle between cancer cells and stromal cells (reverse Warburg effect), hypoxic cancer cells, and oxidative cancer cells (metabolic symbiosis), as well as between hypoxic cancer cells and vascular endothelial cells [27,28]. This provides the energy required for tumor growth [29]. In this study, the protein level of both MCT1 and MCT4 was decreased by curcumin treatment in a low-glucose condition. This suggests that the curcumin plus GR might restrain not only hypoxic cancer cells, but also oxidative cancer cells, stromal, and vascular endothelial cells.

In this study, glucose uptake and lactate formation were decreased by curcumin plus GR. These findings suggest that low glucose uptake and low lactate production were associated with the pHi changes induced by curcumin and GR. The intracellular ATP concentration was decreased by 26% in HepG2 cells upon curcumin treatment under low-glucose condition ($p < 0.01$). This was a more profound reduction than that induced by cariporide treatment. In contrast, curcumin treatment or low-glucose condition alone induced only a mild reduction. The dynamics of intracellular ATP concentrations appear to correlate with changes in pHi and lactate production. Although curcumin treatment alone can inhibit ATP synthase [30], the low-glucose condition restricted the energy supply needed for intracellular ATP synthesis (Figure 4). Finally, growth inhibition and cell death of cancer cells was induced by treatment with curcumin under low-glucose condition, likely through a decline in ATP synthesis and a decrease in pHi [31,32]. Therefore, curcumin plus GR might be an effective anticancer approach by reducing the pHi. Additionally, this treatment strategy might prevent chemoresistance because high intracellular ATP levels are associated with chemoresistance, particularly in colon cancer cells [33].

Intriguingly, the protein levels of AMPK and p-AMPK significantly increased in the low-glucose condition, possibly as a result of diminished glucose availability and low intracellular ATP levels. However, it is also possible that curcumin might mildly activate AMPK [34–36]. Therefore, GR would synergistically enhance the anticancer activity of pHi-lowering agents because GR-induced AMPK activation enhances NHE1 inhibitor activity [37].

Next, structural changes in mTOR protein and pH-dependent protein–protein interactions required for mTORC1 assembly were investigated using computer modeling and co-IP, respectively. It is well known that mTOR forms a complex, especially with Raptor [38] and GβL [39]. Computer modeling of mTOR protein showed that the mTOR structure is affected by pHi, and mTOR binding with GβL and Raptor was significantly diminished by curcumin treatment under low-glucose condition. This was correlated with the reduction in pHi. Consequently, downstream 4EBP1 phosphorylation was decreased, because GβL is a positive regulator of mTOR kinase activity in conjunction with Raptor [39]. In addition, hepatoma cells were smaller, with intensely stained autophagosomes, especially under curcumin treatment plus GR. mTOR is well known as a regulator of autophagy. mTOR basically is an inhibitor to autophagosome formation. PI3K and AKT control mTOR, and under nutrient starvation, mTOR induces autophagy [40,41]. Accordingly, in this study, autophagy was observed in the starvation condition, and the addition of curcumin further enhances autophagy induction. GR and curcumin treatment reduced the intracellular pH, which might have altered the structure of mTOR and consequently weakened interactions with interactors such as Raptor, GβL, which together form mTORC1.

In cancer cell metabolism, amino acids are as important as glucose as an energy source. mTORC1 plays a pivotal role in amino acid-dependent cell proliferation. Therefore, it needs to be elucidated whether high concentrations of amino acids attenuate the anticancer effects of GR. During amino acid-dependent mTORC1 activation, Ras-related small guanosine triphosphate (GTP)-binding proteins, RagA or RagB, bind to RagC or RagD to form heterodimers. These heterodimers actively bind to Raptor [42], which recruits mTORC1 to the surface of the lysosomal membrane [43]. In addition, v-ATPase is necessary for lysosome-associated mTORC1 activation [44]. However, our data demonstrated that Raptor binding to mTOR, RagA, and RagB, and v-ATPase protein level were all diminished after curcumin treatment under GR. Furthermore, amino acids such as leucine (data not shown) or glutamine supplementation did not affect cell growth. These findings imply that in clinical practice, protein restriction—e.g., through ketogenic diet—may not be necessary.

Reducing cytoplasmic alkalinization might significantly affect the conformation of various proteins [9], which in turn affects post-translational modifications of other proteins in a chain reaction, leading to tumor cell death or delayed cell growth. Additionally, polymerization and remodeling of the actin filaments that regulate cancer cell migration [4,11,45] can be inhibited. Likewise, HepG2 cell migration was significantly inhibited by curcumin or GR, and most effectively by curcumin plus GR. These effects were observed to be pHi-dependent.

Curcumin has long been reported to have a problem with bioavailability such as low serum levels and limited tissue distribution due to its poor absorption and rapid metabolism [46]. For these reasons, various strategies are used to enhance bioavailability through innovative drug delivery systems such as liposome, nanoparticle, phospholipid encapsulation, and developing new curcumin analogues [47–49]. Currently, products that enhance bioavailability are being sold, and it is thought curcumin products will continue to be released in the future [49].

4. Materials and Methods

4.1. Chemicals and Antibodies

Cariporide, curcumin, and phorbol-12-myristate-13-acetate (PMA) were purchased from Sigma-Aldrich (St. Louis, MO, USA). Antibodies against NHE1, MCT1, MCT4, ATP5A, v-ATPase, β-actin, α-tubulin, Rag-A, Rag-B, Raptor, Caspase 3, Caspase 9, cleaved Caspase 3, and cleaved Caspase 9 were purchased from Santa Cruz Biotechnology (Dallas, TX, USA). Antibodies against AMPK, p-AMPK, GβL, PRAS40, and p-4EBP1 were purchased from Cell Signaling Technology (Danvers, MA, USA).

4.2. Cell Lines and Cell Culture

HepG2 and Hep3B human hepatocellular carcinoma cell lines were obtained from the American Type Culture Collection (ATCC, Manassas, VA, USA). The HDFa normal human dermal fibroblast line and HEMa normal human epidermal melanocyte line were obtained from the ATCC. The human hepatoma cell line SNU449 was obtained from the Korean Cell Line Bank (Seoul, Korea). All cells were maintained in RPMI-1640 medium (Welgene, Gyeongsan, Korea) supplemented with 10% fetal bovine serum (FBS) (HyClon, Tauranga, New Zealand), 100 U/mL penicillin, and 100 μg/mL streptomycin (HyClone). The cells were maintained at 37 °C in a 5% CO_2 humidified incubator. For growing cells at high glucose concentration, standard RPMI-1640 medium was used. For growing cells at low glucose concentration, a mixture (1:1, *v/v*) of standard and glucose-free RPMI-1640 medium was used.

4.3. Reverse Transcriptase Polymerase Chain Reaction (RT-PCR)

Total RNA was isolated from HepG2 cells using TRIZOL Reagent (Takara, Kusatsu, Shiga, Japan) and the RNA concentration was measured spectrophotometrically (Gen5TM) (BioTek, Winooski, VT, USA). RNA (2 μg) was reverse-transcribed using 5× PrimeScript RT Master Mix (Takara, Japan). The cDNA was used as a template for PCR amplification using the following thermal cycles: 95 °C for 5 min, 30 cycles of 95 °C for 30 s, 58 °C for 30 s, and 72 °C for 60 s. PCR products were analyzed on a 1% agarose gel. PCR primer sequences were as follows: NHE1, 5′-TCT TCA CCG TCT TTG TGC AG-3′ and 5′-CTT GTC CTT CCA GTG GTG GT-3′; GAPDH, 5′-CTC ATG ACC ACA GTC CAT GCC ATC-3′ and 5′-CTG CTT CAC CAC CTT CTT GAT GTC-3′.

4.4. Real-Time (q)RT-PCR

RNA was extracted from cells using TRIZOL reagent (Invitrogen, Carlsbad, CA, USA) and reverse-transcribed using random hexamers and SuperScript IV Reverse Transcriptase (Invitrogen). cDNA was used for qPCR with TaqMan Universal Master Mix II (Applied Biosystems, Foster City, CA, USA). Taqman primers used were: NHE1 (SLC9A1): Hs00300047_m1, GAPDH: Hs03929097_g1. Thermal cycles were: 95 °C for 20 s, 40 cycles of 95 °C for 3 s and 60 °C for 30 s, and finally, 95 °C for 15 s, 60 °C for 1 min, and 95 °C for 15 s. Relative level was analyzed according to the comparative cycle threshold (Ct) method and was normalized to the mRNA level of *GAPDH* in each sample [50]. *NHE1* mRNA level was examined in triplicate and the experiment was repeated at least three times.

4.5. Immunoblotting

After harvesting cells by centrifugation, pellets were resuspended in 100 μL of lysis buffer (20 mM HEPES, (pH 7.5), 1.5 mM $MgCl_2$, 10 mM KCl, 1 mM EDTA, 1 mM EGTA, 250 mM sucrose, 0.1 mM

PMSF, 1 mM dithiothreitol, 4 µg/mL pepstatin, 4 µg/mL leupeptin, and 5 µg/mL aprotinin). After incubation on ice for 10 min, the cells were centrifuged at 750× g at 4 °C for 10 min. The supernatant was collected and centrifuged at 10,000× g for 10 min at 4 °C. Protein concentrations were determined using bicinchoninic acid (Thermo Fisher Scientific, Waltham, MA, USA). Proteins were separated by sodium dodecyl sulfate–polyacrylamide gel electrophoresis (SDS-PAGE) using 8–15% polyacrylamide gels and then electrotransferred to methanol-treated polyvinylidene difluoride membranes. The membranes were blocked with Tris-buffered saline with 0.1% Tween 20 (TBS-T) containing 5% fat-free powdered milk at room temperature for 1 h. Then, they were washed twice with TBS-T and incubated with primary antibodies at room temperature for 1 h or overnight. The membranes were washed three times with TBS-T for 10 min and then incubated with horseradish peroxidase-conjugated secondary antibodies at room temperature for 1 h. After thorough washing, protein bands were detected using chemiluminogenic reagent (GE Healthcare Life Sciences, Marlborough, MA, USA).

4.6. pHi Measurement

Cells were cultured on coverslips with 10% standard RPMI-1640 medium containing 11 mM glucose (control), RPMI-1640 containing 5.5 mM glucose (GR), standard RPMI-1640 medium with 20 nM curcumin dissolved in dimethylsulfoxide, or low-glucose medium with curcumin. The concentration of curcumin may vary depending on the cells and experimental conditions. Experiments were conducted at various concentrations and changed in physiological phenomenon were observed at 20 nM. In addition, HepG2 cells were incubated with 100 nM cariporide (an NHE1 inhibitor) or 100 nM PMA (an NHE1 activator) for 30 min. Then, the cells were washed with cold phosphate-buffered saline (PBS; 137 mM NaCl, 2.7 mM KCl, 10 mM Na_2HPO_4, 2 mM $KH_2 PO_4$). The cells were incubated with 1.5 mg/L of the fluorescent pH probe 2′,7′-bis-(2-carboxyethyl)-5-(and-6)-carboxyfluorescein acetoxymethyl ester (BCECF-AM; Thermo Fisher Scientific) for 10 min. The cells were resuspended in gradient pH calibration solution (4 mM KOH, 2 mM H3PO2, 135 mM KCl, 20 mM HEPES, 1.2 mM CaCl2, 0.8 mM MgSO4). Read fluorescence intensity was determined by fluorescence-assisted cell sorting using the FL-1 channel (Accuri C6 plus; BD Biosciences, Franklin Lakes, NJ, USA).

4.7. Glucose Uptake Measurement

Uptake of 2-deoxyglucose was estimated by an enzymatic NADPH-amplifying system. Briefly, HepG2 cells were seeded into 4-well plates and incubated overnight. Then, the cells were incubated with each culture medium at 37 °C for 12 h. The cells were washed three times with PBS, and lyzed to prepare a 50-µL reaction system. Cell lysates were diluted 1:1, and HepG2 cells were diluted 1:10. After a series of reactions, samples were analyzed spectrophotometrically at excitation and emission wavelengths of 535 nm and 587 nm, respectively. A standard curve was generated by placing 2-deoxyglucose-6-phosphate standard solutions in the wells of a culture plate that had been prepared without cells. Cells were allowed to rest for at least 5 min before measurements.

4.8. Lactate Assay

From each well of the 24-well plates, 25 µL of medium was collected and mixed with 100 µL of NADH solution (0.03% β-NAD in the reduced form of disodium salt in phosphate buffer) and 25 µL of pyruvate solution (22.7 mM pyruvic acid in phosphate buffer) at room temperature. NADH consumption was quantified for 2 min by measuring the absorbance at 340 nm. The mean absorbance change (ΔA/min) was calculated for each treatment condition and was expressed as a percentage of the level from cells (100%) induced by 500 µM NMDA. Each treatment was tested in four replicate wells, and the experiment was repeated five to six times.

4.9. ATP Assay

Intracellular ATP was analyzed using an ATP Fluorometric Assay Kit (Abcam, Cambridge, MA, USA), according to the manufacturer's instructions. HepG2 cells were cultured on coverslips with

10% standard RPMI-1640 medium containing 11 mM glucose (control), RPMI-1640 containing 5.5 mM glucose (GR), standard RPMI-1640 medium with 20 nM curcumin dissolved in dimethylsulfoxide, or low-glucose medium with curcumin. In addition, HepG2 cells were incubated with 100 nM cariporide, or 100 nM PMA for 12 h. In total, 1×10^6 cells were washed twice with cold PBS and resuspended in 100 µL of ATP assay buffer. A 50-µL volume of cell suspension (0.5×10^6 cells) was mixed with 2 µL of ATP probe. After the mixture was gently vortexed, the cells were incubated in the dark at room temperature for 30 min and then subjected to immunofluorescence analysis using a microplate reader (Varioskan Flash; Thermo Fisher Scientific, Waltham, MA, USA). The ATP assay protocol relies on the phosphorylation of glycerol to generate a product that is easily quantified by fluorometry (Ex/Em = 535/587 nm).

4.10. In Silico Analysis of mTOR Protein Structure

Structural changes of mTOR-GβL complex at the ATP-binding site (PDB code: 4JSV) [51] were analyzed using Pipeline Pilot 8.5 (Biovia, San Diego, CA, USA). Chemistry at Harvard Macromolecular Mechanics (CHARMm) force-field was assigned to the structure, and the Momany–Rone method [52] was used for partial charge estimation. Parameters were set to default settings.

4.11. Co-Immunoprecipitation (-IP)

Cells were lyzed in IP lysis buffer (Thermo Fisher Scientific) on ice for 30 min. Cell lysates were centrifuged at 10,000× *g* for 20 min, and the supernatant was collected. After protein quantification, the lysates were incubated with antibodies against mTOR, Raptor, GβL, PRAS40, Rag-A, or Rag-B. The tubes were rotated at 4 °C for 1 h. Then, Protein G Dynabeads (Thermo Fisher Scientific) were added to the lysates and incubated overnight at 4 °C. The Dynabeads-antibody-mTOR complexes and antibody-Raptor complexes were centrifuged at 2500× *g* at 4 °C for 30 min, and the antibody-mTOR and antibody-Raptor complexes were eluted with elution buffer. The eluted proteins were heated at 99 °C for 10 min and subjected to SDS-PAGE.

4.12. Immunohistochemistry for Autophagy

Cells were cultured in 4-well slide chambers, washed twice with PBS, and fixed in a 1% paraformaldehyde solution for 10 min. The cells were washed twice with PBS prior to permeabilization in 0.1% Triton X-100 for 10 min. Next, the cells were blocked in blocking solution (2% BSA and 10% horse serum in PBS) for 1 h. They were then incubated with LC3 primary antibody (1:100 dilution, Santa Cruz Biotechnology), followed by incubation with FITC-conjugated mouse or Texas Red-conjugated rabbit secondary antibody (Jackson ImmunoResearch Laboratories, West Grove, PA, USA). Nuclei were counterstained with DAPI (Thermo Fisher Scientific) for 30 min. The cells were rinsed six times with PBS to remove excess DAPI. Immunofluorescence was detected by confocal microscopy (LSM780; Carl Zeiss, Oberkochen, Germany).

4.13. Cell Viability Assay

Cell viability was determined using a WST-8 (2-[2-methoxy-4-nitrophenyl]-3-[4-nitrophenyl]-5-[2,4-disulfophenyl]-2H-tetrazolium) assay kit (CCK-8 assay kit; Dojindo, Kumamoto, Japan). HepG2 cells were seeded into 96-well plates (Corning Inc., Corning, NY, USA) at 5×10^3 cells/well. After incubation for 24 h, the cells were treated with curcumin (20 nM) in 10% RPMI-1640 containing glucose at different concentrations (11, 5.5, or 2 mM). The cells were washed twice with culture medium, and 100 µL of CCK-8 reagent was added to each well. The plates were incubated at 37 °C for 2 h. Absorbance in each well was measured at 450 nm using a microplate reader (Bio-Rad, Hercules, CA, USA) and corrected for background. L-glutamine concentration was 2 mM.

4.14. Cell Migration Assay

Cell migration was assayed using a modified version of the Boyden chamber method [53], which employs microchemotaxis chambers and polycarbonate filters with a pore size of 8.0 µm. Cells were trypsinized and suspended at 5×10^5 cells/mL in culture medium supplemented with 10% FBS. Cell suspension (100 µL) was placed in the upper chamber, and the experimental medium (600 µL/well) was added to the lower chamber. The chamber was incubated at 37 °C under 5% CO_2 for 48 h. Then, the filter was removed, and cells on the upper side of the filter were scraped off using a cotton tip. Cells migrated to the lower side of the filter were fixed in methanol and stained with hematoxylin. Cells in three randomly selected fields were counted at a magnification of 200× under an inverted microscope. Each treatment was analyzed in triplicate and the experiment was repeated at least three times.

4.15. Statistical Analysis

Data are expressed as the means ± standard errors (SEs). Data for two groups were compared using Student's *t*-test. Comparisons among more than two groups were conducted by one-way analysis of variance (ANOVA) followed by Bonferroni tests. $p < 0.05$ was considered significant.

5. Conclusions

In conclusion, curcumin treatment in combination with GR without calorie or protein restriction could be a useful anticancer approach to prevent intracellular alkalinization and inhibit enhanced glycolysis. In clinical practice, this might be a more useful chemoprevention strategy than CR or GR diets, such as a ketogenic diet.

Supplementary Materials: Supplementary materials can be found at http://www.mdpi.com/1422-0067/20/10/2375/s1. **Supplementary Figure S1.** pHi-lowering effect of curcumin and glucose restriction in the human hepatoma cell lines HepG2, Hep3B, and SNU449. The experiment was conducted five times independently. Con, standard RPMI-1640 medium; Cur, curcumin; GR, glucose restriction, 5.5 mM glucose medium; GR Cur, glucose restriction plus curcumin. * $p < 0.05$ vs. control; ** $p < 0.01$ vs. control. **Supplementary Figure S2.** Effect of curcumin and glucose restriction on the pHi in HDFa human dermal fibroblasts. The pHi is within the normal range, even after curcumin treatment under glucose restriction condition. experiment was conducted five times independently. The experiment was conducted five times independently. Con, standard RPMI-1640 medium; Cur, curcumin; Car, cariporide; PMA, phorbol-12-myristate-13-acetate; GR, glucose restriction, 5.5 mM glucose medium; GR Cur, glucose restriction plus curcumin **Supplementary Figure S3**. NHE1 mRNA level affected by curcumin, cariporide, PMA, and glucose restriction conditions. NHE1 mRNA level was detected by (**A,B**) RT-PCR and (**C**) qRT-PCR. The experiment was conducted three times independently. Con, standard RPMI-1640 medium; Cur, curcumin; Car, cariporide; PMA, phorbol-12-myristate-13-acetate; GR, glucose restriction, 5.5 mM glucose medium; GR Cur, glucose restriction plus curcumin. * $p < 0.05$ vs. control; ***** $p < 0.0005$ vs. control; + $p < 0.05$ vs. GR. **Supplementary Figure S4.** Effect of curcumin and glucose restriction on ATP production in (**A**) Hep3B, (**B**) SNU449, (**C**) HEMa, and (**D**) HDFa cells. Confocal ATP images were taken after 30-min treatment of human melanocytes with curcumin under low-glucose condition (magnification, 400 µL). Con, standard RPMI-1640 medium; GR Cur, glucose restriction plus curcumin. **Supplementary Figure S5.** Confocal imaging of autophagosomes of (**A**) HepG3B and (**B**) HDFa cells. Cells were immunostained with an LC3 antibody. Con, standard RPMI-1640 medium; GR Cur, glucose restriction plus curcumin

Author Contributions: Conceptualization, J.D.L.; Acquisition of data, M.-J.C., S.-K.L., B.-W.S., X.J., and J.M.L.; Data analysis and interpretation, J.D.L., J.H.P. and S.W.K.; Manuscript preparation, S.W.K.

Funding: This study was carried out in part at the Yonsei-Carl Zeiss Advanced Imaging Center, Yonsei University College of Medicine. This research was supported by the Basic Science Research Program through the National Research Foundation of Korea (NRF) funded by the Ministry of Education, Science and Technology (2013-31-0342).

Conflicts of Interest: The authors have no financial conflicts of interest.

Abbreviations

CR	calorie restriction
GR	glucose restriction
IGF-1	insulin-like growth factor 1
PI3K	phosphoinositide 3-kinase
mTOR	mammalian target of rapamycin
AMPK	AMP-dependent protein kinase
pHi	intracellular pH
MCT	monocarboxylate transporter
NHE1	Na+/H+ exchanger-1
v-ATPase	vacuolar H+-ATPase
pHe	extracellular pH
mTORC1	mTOR complex-1
4EBP1	4E-binding protein 1
HIF-1	hypoxia inducible factor-1
GTP	guanosine triphosphate
PMA	phorbol-12-myristate-13-acetate
FBS	fetal bovine serum
SDS-PAGE	sodium dodecyl sulfate–polyacrylamide gel electrophoresis
TBS-T	tris-buffered saline with 0.1% Tween 20
PBS	phosphate-buffered saline
BCECF	2′,7′bis-(2-carboxyethyl)-5-(and-6)-carboxyfluorescein acetoxymethyl ester
SEs	standard errors
ANOVA	analysis of variance

References

1. Meynet, O.; Ricci, J.E. Caloric restriction and cancer: Molecular mechanisms and clinical implications. *Trends Mol. Med.* **2014**, *20*, 419–427. [CrossRef]
2. Reshkin, S.J.; Bellizzi, A.; Caldeira, S.; Albarani, V.; Malanchi, I.; Poignee, M.; Alunni-Fabbroni, M.; Casavola, V.; Tommasino, M. Na+/H+ exchanger-dependent intracellular alkalinization is an early event in malignant transformation and plays an essential role in the development of subsequent transformation-associated phenotypes. *FASEB J.* **2000**, *14*, 2185–2197. [CrossRef]
3. Webb, B.A.; Chimenti, M.; Jacobson, M.P.; Barber, D.L. Dysregulated pH: A perfect storm for cancer progression. *Nat. Rev. Cancer* **2011**, *11*, 671–677. [CrossRef] [PubMed]
4. Damaghi, M.; Wojtkowiak, J.W.; Gillies, R.J. pH sensing and regulation in cancer. *Front. Physiol.* **2013**, *4*, 370. [CrossRef] [PubMed]
5. Fischer, K.; Hoffmann, P.; Voelkl, S.; Meidenbauer, N.; Ammer, J.; Edinger, M.; Gottfried, E.; Schwarz, S.; Rothe, G.; Hoves, S.; et al. Inhibitory effect of tumor cell-derived lactic acid on human T cells. *Blood* **2007**, *109*, 3812–3819. [CrossRef]
6. Thews, O.; Gassner, B.; Kelleher, D.K.; Schwerdt, G.; Gekle, M. Impact of extracellular acidity on the activity of P-glycoprotein and the cytotoxicity of chemotherapeutic drugs. *Neoplasia* **2006**, *8*, 143–152. [CrossRef] [PubMed]
7. Harguindey, S.; Arranz, J.L.; Polo Orozco, J.D.; Rauch, C.; Fais, S.; Cardone, R.A.; Reshkin, S.J. Cariporide and other new and powerful NHE1 inhibitors as potentially selective anticancer drugs—an integral molecular/biochemical/metabolic/clinical approach after one hundred years of cancer research. *J. Transl. Med.* **2013**, *11*, 282. [CrossRef]
8. Schroeder, U.; Himpe, B.; Pries, R.; Vonthein, R.; Nitsch, S.; Wollenberg, B. Decline of lactate in tumor tissue after ketogenic diet: In vivo microdialysis study in patients with head and neck cancer. *Nutr. Cancer* **2013**, *65*, 843–849. [CrossRef] [PubMed]
9. Dechant, R.; Binda, M.; Lee, S.S.; Pelet, S.; Winderickx, J.; Peter, M. Cytosolic pH is a second messenger for glucose and regulates the PKA pathway through V-ATPase. *EMBO J.* **2010**, *29*, 2515–2526. [CrossRef] [PubMed]

10. Naz, R.K. The effect of curcumin on intracellular pH (pHi), membrane hyperpolarization and sperm motility. *J. Reprod. Infertil.* **2014**, *15*, 62–70.

11. Srivastava, J.; Barber, D.L.; Jacobson, M.P. Intracellular pH sensors: Design principles and functional significance. *Physiology (Bethesda)* **2007**, *22*, 30–39. [CrossRef]

12. Lee, J.D.; Choi, M.A.; Ro, S.W.; Yang, W.I.; Cho, A.E.; Ju, H.L.; Baek, S.; Chung, S.I.; Kang, W.J.; Yun, M.; et al. Synergic chemoprevention with dietary carbohydrate restriction and supplementation of AMPK-activating phytochemicals: The role of SIRT1. *Eur. J. Cancer Prev.* **2016**, *25*, 54–64. [CrossRef] [PubMed]

13. Showkat, M.; Beigh, M.A.; Andrabi, K.I. mTOR signaling in protein translation regulation: Implications in cancer genesis and therapeutic interventions. *Mol. Biol. Int.* **2014**, *2014*, 686984. [CrossRef] [PubMed]

14. Schonichen, A.; Webb, B.A.; Jacobson, M.P.; Barber, D.L. Considering protonation as a posttranslational modification regulating protein structure and function. *Annu. Rev. Biophys.* **2013**, *42*, 289–314. [CrossRef] [PubMed]

15. Petukh, M.; Stefl, S.; Alexov, E. The role of protonation states in ligand-receptor recognition and binding. *Curr. Pharm. Des.* **2013**, *19*, 4182–4190. [CrossRef]

16. Matsuyama, S.; Llopis, J.; Deveraux, Q.L.; Tsien, R.Y.; Reed, J.C. Changes in intramitochondrial and cytosolic pH: Early events that modulate caspase activation during apoptosis. *Nat. Cell. Biol.* **2000**, *2*, 318–325. [CrossRef]

17. Trivedi, B.; Danforth, W.H. Effect of pH on the kinetics of frog muscle phosphofructokinase. *J. Biol. Chem.* **1966**, *241*, 4110–4112.

18. Perona, R.; Serrano, R. Increased pH and tumorigenicity of fibroblasts expressing a yeast proton pump. *Nature* **1988**, *334*, 438–440. [CrossRef]

19. Lagarde, A.E.; Franchi, A.J.; Paris, S.; Pouyssegur, J.M. Effect of mutations affecting Na+: H+ antiport activity on tumorigenic potential of hamster lung fibroblasts. *J. Cell Biochem.* **1988**, *36*, 249–260. [CrossRef]

20. Matsuyama, S.; Reed, J.C. Mitochondria-dependent apoptosis and cellular pH regulation. *Cell Death Differ.* **2000**, *7*, 1155–1165. [CrossRef]

21. Goss, G.G.; Woodside, M.; Wakabayashi, S.; Pouyssegur, J.; Waddell, T.; Downey, G.P.; Grinstein, S. ATP dependence of NHE-1, the ubiquitous isoform of the Na+/H+ antiporter. Analysis of phosphorylation and subcellular localization. *J. Biol. Chem.* **1994**, *269*, 8741–8748. [PubMed]

22. Meima, M.E.; Webb, B.A.; Witkowska, H.E.; Barber, D.L. The sodium-hydrogen exchanger NHE1 is an Akt substrate necessary for actin filament reorganization by growth factors. *J. Biol. Chem.* **2009**, *284*, 26666–26675. [CrossRef]

23. Zhang, X.; Chen, Q.; Wang, Y.; Peng, W.; Cai, H. Effects of curcumin on ion channels and transporters. *Front. Physiol.* **2014**, *5*, 94. [CrossRef]

24. Torigoe, T.; Izumi, H.; Yoshida, Y.; Ishiguchi, H.; Okamoto, T.; Itoh, H.; Kohno, K. Low pH enhances Sp1 DNA binding activity and interaction with TBP. *Nucleic Acids Res* **2003**, *31*, 4523–4530. [CrossRef]

25. Doherty, J.R.; Cleveland, J.L. Targeting lactate metabolism for cancer therapeutics. *J. Clin. Investig.* **2013**, *123*, 3685–3692. [CrossRef] [PubMed]

26. Spugnini, E.P.; Sonveaux, P.; Stock, C.; Perez-Sayans, M.; De Milito, A.; Avnet, S.; Garcìa, A.G.; Harguindey, S.; Fais, S. Proton channels and exchangers in cancer. *Biochim. Biophys. Acta* **2015**, *1848*, 2715–2726. [CrossRef]

27. San-Millán, I.; Brooks, G.A. Reexamining cancer metabolism: Lactate production for carcinogenesis could be the purpose and explanation of the Warburg Effect. *Carcinogenesis* **2017**, *38*, 119–133. [CrossRef]

28. Pérez-Escuredo, J.; Van Hée, V.F.; Sboarina, M.; Falces, J.; Payen, V.L.; Pellerin, L.; Sonveaux, P. Monocarboxylate transporters in the brain and in cancer. *Biochim. Biophys. Acta* **2016**, *1863*, 2481–2497. [CrossRef]

29. Benjamin, D.; Robay, D.; Hindupur, S.K.; Pohlmann, J.; Colombi, M.; El-Shemerly, M.Y.; Maira, S.M.; Moroni, C.; Lane, H.A.; Hall, M.N. Dual inhibition of the lactate transporters MCT1 and MCT4 is synthetic lethal with metformin due to NAD+ depletion in cancer cells. *Cell Rep.* **2018**, *25*, 3047–3058. [CrossRef]

30. Zheng, J.; Ramirez, V.D. Inhibition of mitochondrial proton F0F1-ATPase/ATP synthase by polyphenolic phytochemicals. *Br. J. Pharmacol.* **2000**, *130*, 1115–1123. [CrossRef]

31. Comelli, M.; Di Pancrazio, F.; Mavelli, I. Apoptosis is induced by decline of mitochondrial ATP synthesis in erythroleukemia cells. *Free Radic. Biol. Med.* **2003**, *34*, 1190–1199. [CrossRef]

32. Nagata, H.; Che, X.F.; Miyazawa, K.; Tomoda, A.; Konishi, M.; Ubukata, H.; Tabuchi, T. Rapid decrease of intracellular pH associated with inhibition of Na+/H+ exchanger precedes apoptotic events in the MNK45 and MNK74 gastric cancer cell lines treated with 2-aminophenoxazine-3-one. *Oncol. Rep.* **2011**, *25*, 341–346.

33. Zhou, Y.; Tozzi, F.; Chen, J.; Fan, F.; Xia, L.; Wang, J.; Gao, G.; Zhang, A.; Xia, X.; Brasher, H.; et al. Intracellular ATP levels are a pivotal determinant of chemoresistance in colon cancer cells. *Cancer Res.* **2012**, *72*, 304–314. [CrossRef]

34. Kim, T.; Davis, J.; Zhang, A.J.; He, X.; Mathews, S.T. Curcumin activates AMPK and suppresses gluconeogenic gene expression in hepatoma cells. *Biochem. Biophys. Res. Commun.* **2009**, *388*, 377–382. [CrossRef] [PubMed]

35. Xiao, K.; Jiang, J.; Guan, C.; Dong, C.; Wang, G.; Bai, L.; Sun, J.; Hu, C.; Bai, C. Curcumin induces autophagy via activating the AMPK signaling pathway in lung adenocarcinoma cells. *J. Pharmacol. Sci.* **2013**, *123*, 102–109. [CrossRef] [PubMed]

36. Kim, J.; Yang, G.; Kim, Y.; Kim, J.; Ha, J. AMPK activators: Mechanisms of action and physiological activities. *Exp. Mol. Med.* **2016**, *48*, e224. [CrossRef] [PubMed]

37. Hallows, K.R.; Alzamora, R.; Li, H.; Gong, F.; Smolak, C.; Neumann, D.; Pastor-Soler, N.M. AMP-activated protein kinase inhibits alkaline pH- and PKA-induced apical vacuolar H+-ATPase accumulation in epididymal clear cells. *Am. J. Physiol. Cell. Physiol.* **2009**, *296*, C672–C681. [CrossRef]

38. Dunlop, E.A.; Dodd, K.M.; Seymour, L.A.; Tee, A.R. Mammalian target of rapamycin complex 1-mediated phosphorylation of eukaryotic initiation factor 4E-binding protein 1 requires multiple protein–protein interactions for substrate recognition. *Cell Signal.* **2009**, *21*, 1073–1084. [CrossRef]

39. Kim, D.H.; Sarbassov, D.D.; Ali, S.M.; Latek, R.R.; Guntur, K.V.; Erdjument-Bromage, H.; Tempst, P.; Sabatini, D.M. GbetaL, a positive regulator of the rapamycin-sensitive pathway required for the nutrient-sensitive interaction between raptor and mTOR. *Mol. Cell* **2003**, *11*, 895–904. [CrossRef]

40. Jung, C.H.; Ro, S.H.; Cao, J.; Otto, N.M.; Kim, D.H. mTOR regulation of autophagy. *FEBS Lett.* **2010**, *584*, 1287–1295. [CrossRef]

41. Heras-Sandoval, D.; Pérez-Rojas, J.M.; Hernández-Damián, J.; Pedraza-Chaverri, J. The role of PI3K/AKT/mTOR pathway in the modulation of autophagy and the clearance of protein aggregates in neurodegeneration. *Cell Signal.* **2014**, *26*, 2694–2701. [CrossRef] [PubMed]

42. Sancak, Y.; Peterson, T.R.; Shaul, Y.D.; Lindquist, R.A.; Thoreen, C.C.; Bar-Peled, L.; Sabatini, D.M. The Rag GTPases bind raptor and mediate amino acid signaling to mTORC1. *Science* **2008**, *320*, 1496–1501. [CrossRef]

43. Shimobayashi, M.; Hall, M.N. Multiple amino acid sensing inputs to mTORC1. *Cell Res.* **2016**, *26*, 7–20. [CrossRef]

44. Zoncu, R.; Bar-Peled, L.; Efeyan, A.; Wang, S.; Sancak, Y.; Sabatini, D.M. mTORC1 senses lysosomal amino acids through an inside-out mechanism that requires the vacuolar H (+)-ATPase. *Science* **2011**, *334*, 678–683. [CrossRef]

45. Bravo-Cordero, J.J.; Magalhaes, M.A.; Eddy, R.J.; Hodgson, L.; Condeelis, J. Functions of cofilin in cell locomotion and invasion. *Nat. Rev. Mol. Cell. Biol.* **2013**, *14*, 405–415. [CrossRef]

46. Anand, P.; Kunnumakkara, A.B.; Newman, R.A.; Aggarwal, B.B. Bioavailability of curcumin: Problems and promises. *Mol. Pharm.* **2007**, *4*, 807–818. [CrossRef]

47. Kanai, M.; Otsuka, Y.; Otsuka, K.; Sato, M.; Nishimura, T.; Mori, Y.; Kawaguchi, M.; Hatano, E.; Kodama, Y.; Matsumoto, S.; et al. A phase I study investigating the safety and pharmacokinetics of highly bioavailable curcumin (Theracurmin) in cancer patients. *Cancer Chemother. Pharmacol.* **2013**, *71*, 1521–1530. [CrossRef]

48. De Leo, V.; Milano, F.; Mancini, E.; Comparelli, R.; Giotta, L.; Nacci, A.; Longobardi, F.; Garbetta, A.; Agostiano, A.; Catucci, L. Encapsulation of Curcumin-Loaded Liposomes for Colonic Drug Delivery in a pH-Responsive Polymer Cluster Using a pH-Driven and Organic Solvent-Free Process. *Molecules* **2018**, *23*, 739. [CrossRef] [PubMed]

49. Martí Coma-Cros, E.; Biosca, A.; Lantero, E.; Manca, M.L.; Caddeo, C.; Gutiérrez, L.; Ramírez, M.; Borgheti-Cardoso, L.N.; Manconi, M.; Fernàndez-Busquets, X. Antimalarial Activity of Orally Administered Curcumin Incorporated in Eudragit®-Containing Liposomes. *Int. J. Mol. Sci.* **2018**, *19*, 1361. [CrossRef]

50. Pfaffl, M.W. A new mathematical model for relative quantification in real-time RT-PCR. *Nucleic Acids Res.* **2001**, *29*, e45. [CrossRef]

51. Yang, H.; Rudge, D.G.; Koos, J.D.; Vaidialingam, B.; Yang, H.J.; Pavletich, N.P. mTOR kinase structure, mechanism and regulation. *Nature* **2013**, *497*, 217–223. [CrossRef] [PubMed]

52. Momany, F.A.; Rone, R. Validation of the general-purpose QUANTA 3.2/CHARMm force field. *J. Comput. Chem.* **1992**, *13*, 888–900. [CrossRef]
53. Senger, D.R.; Ledbetter, S.R.; Claffey, K.P.; Papadopoulos-Sergiou, A.; Peruzzi, C.A.; Detmar, M. Stimulation of endothelial cell migration by vascular permeability factor/vascular endothelial growth factor through cooperative mechanisms involving the alphavbeta3 integrin, osteopontin, and thrombin. *Am. J. Pathol.* **1996**, *149*, 293–305. [PubMed]

International Journal of
Molecular Sciences

Article

β3-Adrenoreceptor Activity Limits Apigenin Efficacy in Ewing Sarcoma Cells: A Dual Approach to Prevent Cell Survival

Amada Pasha [1,2], Marina Vignoli [1,2], Angela Subbiani [1,2], Alessio Nocentini [3,4],
Silvia Selleri [3], Paola Gratteri [4], Annalisa Dabraio [1,2], Tommaso Casini [1], Luca Filippi [5],
Ilaria Fotzi [1], Claudio Favre [1] and Maura Calvani [1,*]

[1] Division of Pediatric Oncology/Hematology, Meyer University Children's Hospital, 50139 Florence, Italy;
    amanda.pasha@yahoo.it (A.P.); marina.vignoli@unifi.it (M.V.); angela.subbiani@gmail.com (A.S.);
    annalisa.d.92@gmail.com (A.D.); tommaso.casini@meyer.it (T.C.); ilaria.fotzi@meyer.it (I.F.);
    claudio.favre@meyer.it (C.F.)
[2] Department of Health Sciences, University of Florence, 50139 Florence, Italy
[3] NEUROFARBA Department, Pharmaceutical and nutraceutical section, University of Florence,
    50019 Sesto Fiorentino, Italy; alessio.nocentini@unifi.it (A.N.); silvia.selleri@unifi.it (S.S.)
[4] Laboratory of Molecular Modeling Cheminformatics & QSAR, NEUROFARBA Department, Pharmaceutical
    and nutraceutical section, University of Florence, 50019 Sesto Fiorentino, Italy; paola.gratteri@unifi.it
[5] Neonatal Intensive Care Unit, Medical Surgical Fetal-Neonatal Department, Meyer "University Children's
    Hospital, 50139 Florence, Italy; luca.filippi@meyer.it
* Correspondence: maura.calvani@meyer.it; Tel.: +39-055-7944-573

Received: 29 March 2019; Accepted: 27 April 2019; Published: 30 April 2019

Abstract: Ewing Sarcoma (ES) is an aggressive paediatric tumour where oxidative stress and antioxidants play a central role in cancer therapy response. Inhibiting antioxidants expression, while at the same time elevating intracellular reactive oxygen species (ROS) levels, have been proposed as a valid strategy to overcome ES cancer progression. Flavonoid intake can affect free radical and nutritional status in children receiving cancer treatment, but it is not clear if it can arrest cancer progression. In particular, apigenin may enhance the effect of cytotoxic chemotherapy by inducing cell growth arrest, apoptosis, and by altering the redox state of the cells. Little is known about the use of apigenin in paediatric cancer. Recently, β3-adrenergic receptor (β3-AR) antagonism has been proposed as a possible strategy in cancer therapy for its ability to induce apoptosis by increasing intracellular levels of ROS. In this study we show that apigenin induces cell death in ES cells by modulating apoptosis, but not increasing ROS content. Since ES cells are susceptible to an increased oxidative stress to reduce cell viability, here we demonstrate that administration of β3-ARs antagonist, SR59230A, improves the apigenin effect on cell death, identifying β3-AR as a potential discriminating factor that could address the use of apigenin in ES.

Keywords: apigenin; β3-adrenoreceptor; Ewing Sarcoma

---

1. Introduction

Ewing Sarcoma (ES) is one of the most common paediatric malignant tumours, accounting for 2% of all childhood cancers [1]. In recent decades, the therapeutic choice, consisting of a multi-drug chemotherapy regimen combined with radiotherapy and surgery, has significantly improved the survival to 70% in localised disease, but the outcome remains poor for patients with metastatic disease at diagnosis, occurring in approximately 25–30% of ES patients [2,3]. These cases often show resistance to multi chemotherapeutic agents [4]. It has been observed that response to the induction of cell death carried out by chemotherapeutic agents is different between ES patients and cell lines, reflecting the different levels of intracellular antioxidants and the variable ability to neutralize the action of reactive oxygen species (ROS) [5]. ES cells show a dysfunction in oxidative phosphorylation's activity and a redox state imbalance due to their abnormal metabolic activity with an increased glucose uptake and activation of accelerated glycolysis that provide the energy required by cancer cells. These cells also show a high sensitivity to rapid changes in the intracellular redox environment. In this context a fine regulation of ROS production and detoxification is critical for the growth or reduction of an ES tumour. Therapeutic drugs act on the reduction of cellular antioxidant activity to promote oxidative stress increase and induction of cell death in ES. Cellular antioxidants are differentially expressed in ES cell lines and their levels can be associated with poor prognosis [6]. Glutathione (GSH) is one of the cellular antioxidants whose function is important in ES cell lines. It was observed that depletion of intracellular GSH decreases cell viability and enhances the efficacy of ROS-generating anticancer agents such as fenretinide [5].

Nutraceutical antioxidant supplement may improve tumour response to therapy and patient survival, leading to a long-term outcome or interference with chemo- and radiation- therapy by reducing their effectiveness [7,8]. Among nutraceutical antioxidants, flavonoid supplementation during cancer therapy is a controversial and questionable subject. Flavonoids have been implicated in tumour regression by either antioxidant or prooxidant activity. A plant flavonoid naturally abundant in vegetables and fruits is apigenin (5,7,4′-trihydroxyflavone) [9]. It is a bioactive flavonoid that shows anti-inflammatory, antioxidant, and anticancer properties against several human-cancer cell lines, including prostate carcinoma, colon carcinoma, breast cancer, leukemia cells, cervical carcinoma, lung cancer, and hepatoma [10–19]. The cellular effects of apigenin are associated with cell-cycle arrest and apoptosis induction mediated by the increase of p21 levels, regardless of the Rb status and p53 involvement [16,18,20,21], alteration of the Bax/Bcl-2 ratio, release of cytochrome C, and induction of Apaf-1, leading to caspase activation and PARP-cleavage [15–17,21,22].

Recently, β3 adrenergic receptors (β3-ARs) became incredibly attractive in cancer biology because of their role in reducing tumour growth and metastasis. Overexpression of β3-ARs has been associated with cancer growth, recruitment of circulating stromal cell precursors to the tumour sites, and enhancement of stem cell traits [23]. It has also been reported that β3-ARs stimulation exhibits dual antioxidant properties: it directly inhibits NADPHoxidase (NADPHox) activity, which regulates ROS production, and induces the expression of Catalase, which has a role as an endogenous antioxidant. Moreover, it has been observed that the stimulation of β3-ARs by noradrenaline increased the intracellular GSH levels [24].

In this study we investigated the effect of apigenin treatment on human Ewing Sarcoma cell lines, showing the partial decrease of cell viability and induction of cell apoptosis. Interestingly, the impact of apigenin in these cancer cells results in an improvement of the contemporaneous treatment with β3-ARs antagonist in a synergistic effect revealing a new possible approach for ES therapy.

2. Results

2.1. Apigenin Induces Cell Death and Inhibition of ROS and Antioxidant Activity in ES Cells

In order to investigate the effect of apigenin on the human ES cells, A673 cells were exposed to different concentrations of apigenin and the cellular viability was determined with four different methodologies. MTT analysis, Viobility™ Fixable Dyes and Annexin V/PI staining showed that apigenin 50 µM decreased partially ES cell viability of about 35–39% (Figure 1A–C) and showed that the treatment affected early apoptosis to a higher extent than late apoptosis and necrocrotic cell death (Tables 1 and 2). Moreover, investigation of the expression and cleavage of PARP-1 enzyme confirmed induction of the apoptotic process (Figure 1D). Furthermore, to exclude a massive toxic effect of high dose apigenin treatment, we performed cell viability assays on healthy human peripheral lymphocytes:

no evidence of any toxic effect was observed after apigenin treatment (Figure 1E). Altogether, these results demonstrated that apigenin 50 µM reduced ES cell viability by activating the apoptotic pathway.

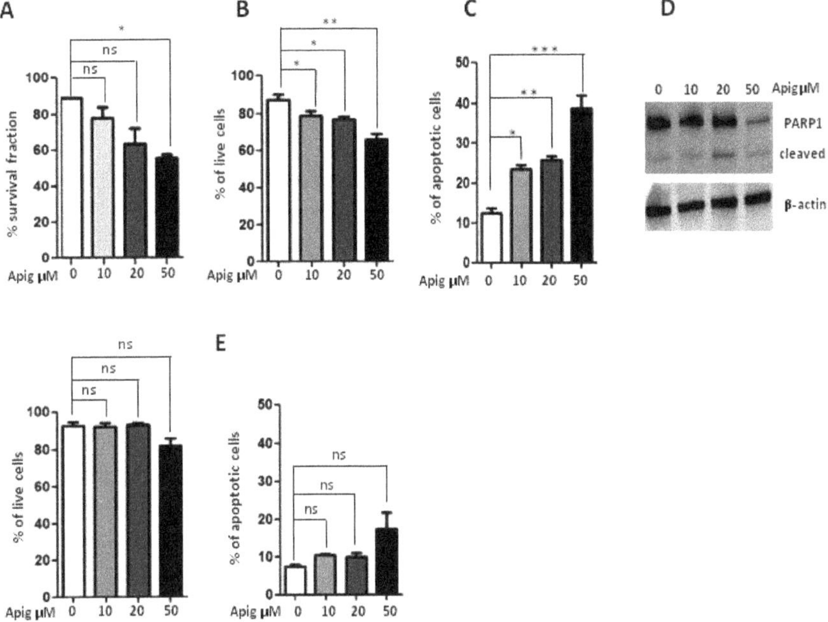

Figure 1. (**A**) Analysis of cellular viability with MTT survival experiment in human A673 ES cells after 24 h of treatment with apigenin (10-20-50 µM); (**B**) Percentage of live cells after Viobility™ Fixable Dyes assay after 24 h of treatment with apigenin (10-20-50 µM); (**C**) The apoptotic effect of apigenin analysed after 24 h of treatment (10-20-50 µM); (**D**) Western Blot analysis of PARP1 enzyme after treatment with apigenin (10-20-50 µM) with β-actin as loading control; (**E**) Percentage of live cells after Viobility™ Fixable Dyes assay and apoptotic effect on healthy lymphocytes after 24 h of treatment with apigenin (10-20-50 µM). Apig: apigenin, ns: not significant. p values for treatments: * $p < 0.05$, ** $p < 0.01$ and *** $p < 0.001$.

Table 1. Percentage of early apoptotic, late apoptotic and dead cells expressed by the annexin V assay in A673 cells and normal lymphocytes. APIG: apigenin.

A673 Cells	% Dead Cells			
	Early	Late	Necrotic	Total
0	2.59	4.95	5.36	12.09
APIG 10 µM	6.51	8.36	8.56	23.43
APIG 20 µM	7.89	9.70	8.11	25.70
APIG 50 µM	19.00	16.3	3.68	38.68
Lymphocytes				
0	4.26	1.35	2.02	7.63
APIG 10 µM	5.66	2.42	2.28	10.36
APIG 20 µM	5.74	2.23	2.05	10.02
APIG 50 µM	7.04	3.22	7.01	17.27

Table 2. Percentage of live cells expressed by the Viobility^{TM} Fixable Dyes assay in A673 cells and normal lymphocytes. APIG: apigenin.

A673 Cells	% Live Cells
0	87.20
APIG 10 µM	78.88
APIG 20 µM	76.90
APIG 50 µM	65.90
Lymphocytes	
0	92.70
APIG 10 µM	92.14
APIG 20 µM	93.41
APIG 50 µM	81.95

The expression levels of antioxidants were examined upon treatment with apigenin at different concentrations 10-20-50 µM after 24 h. Apigenin treatment in A673 cells displayed a significantly lower amount of protein levels of all antioxidants examined, superoxide dismutase 2 (SOD2), Catalase, Thioredoxin, sirtuin-1 (SIRT1), thioredoxin interacting protein TXNIP (VDUP-1), glutathione S-transferase Mu4 (GSTM4), and nuclear factor erythroid 2-related factor 2 (Nrf2) (Figure 2A). The analysis of various ROS species levels measured at different times showed that the amount of most of ROS species decreased at the highest dose of apigenin (50 µM) after 24 h of treatment when it inhibits the expression of the most antioxidant proteins examined. Also, the amount of peroxide levels decreased after 6h of treatment (Figure 2B–D). Results indicated that apigenin partially reduced ES cell viability principally by inducing cell apoptosis.

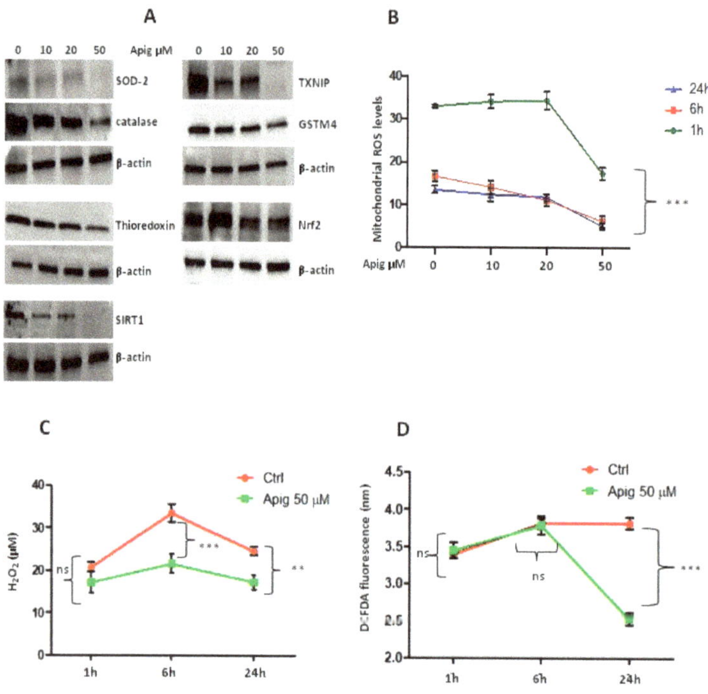

Figure 2. (**A**) WB analysis of apigenin effect on antioxidants expression: SOD2, Catalase, Thioredoxin, SIRT1, TXNIP, GSTM4, Nrf2 with β-actin as loading control; (**B**) Effect of apigenin after different times

of treatment (1-6-24 h) on oxygen reactive species production (ROS) with MitoSOX^{TM} Red assay. (**C**) Effect of apigenin 50 μM after different times of treatment (1-6-24 h) on peroxide levels with Amplex® Red Hydrogen Peroxidase Assay Kit (**D**) Effect of apigenin 50 μM after different times of treatment (1-6-24 h) on oxygen reactive species production (ROS) with DCFDA assay. ns: not significant. P values for treatments: ** $p < 0.01$, and *** $p < 0.001$.

2.2. Apigenin Controls ROS Levels by Activation of UCP2 and GSH Accumulation

Recently, it has been reported that a role of β3-adrenoreceptors in ROS balancing both in melanoma cells and in glioma cells, respectively controlling Uncoupling Protein 2 (UCP2) and glutathione levels [25]. In order to elucidate the mechanism by which apigenin reduced ROS levels, expression of UCP2 and GSH contents were analysed upon different time and doses of apigenin treatment. Results indicated that apigenin induced UCP2 protein expression and increased GSH levels after 24 h of treatment (Figure 3A,B), thus causing ROS levels decrease. Moreover, here we demonstrated the expression of β3-ARs in mitochondria of ES cells as it has been previously reported in melanoma cells [25] (Figure 3C). To address the involvement of β3-AR receptor in controlling ROS levels in ES cells, we used the selective antagonist of β3-AR, SR59230A. We showed an increase of mitochondrial ROS levels and an inhibition of GSH amount after 24h of treatment with SR59230A (Figure 3D). Interestingly, SR59230A inhibited the UCP2 expression in accord with previous data reported in melanoma cells (Figure 3E) [25]. These results indicate that the treatment with SR59230A could improve the effects of apigenin action by increasing ROS mitochondrial levels. Therefore, we tested the impact of the administration of apigenin and/or SR59230A (10 μM) on the survival of A673 cells (Figure 3F). Results clearly indicate that double treatment reduced cell viability with a higher extent respect to single treatments confirming the synergistic effect of both drug usage.

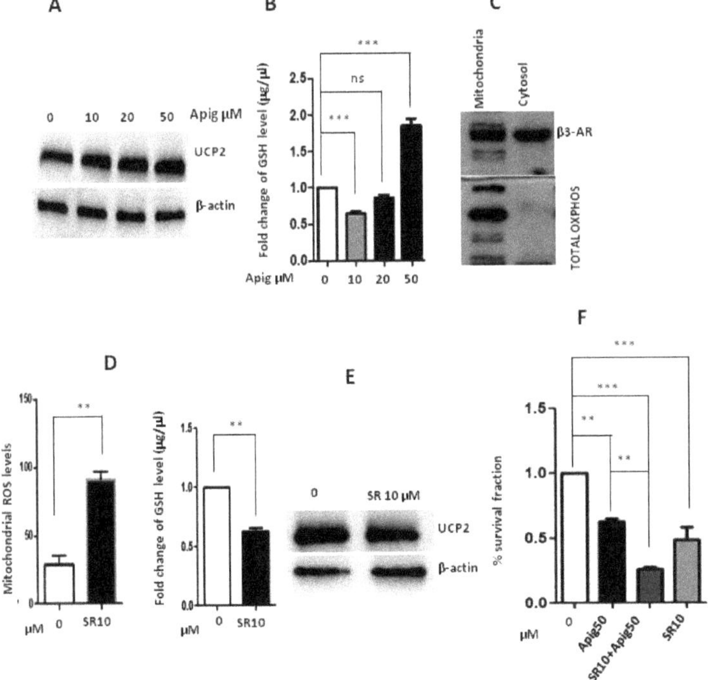

Figure 3. (**A**) Western Blot analysis of apigenin (10-20-50 μM) effect on UCP2 expression, with β-actin as loading control; (**B**) Measurement of reduced glutathione levels (GSH) after 24 h of treatment with

apigenin; (**C**) WB analysis of β3-AR on mitochondria proteins; (**D**) Mitochondria mtROS measurement after treatment with β3-AR antagonist, SR59230A, at the concentration of 10 μM and measurement of GSH levels at the same time and concentration of SR59230A; (**E**) WB analysis of UCP2 expression after treatment with β3-AR antagonist SR59230A with β-actin as loading control; (**F**) MTT survival experiment with double treatment with SR59230A (10 μM) and apigenin (50 μM). SR10: SR59230A 10 μM, Apig50: apigenin 50 μM, ns: not significant. P values for treatments: ** $p < 0.01$ and *** $p < 0.001$.

2.3. The Agonism of β3-AR Reproduces the Effect of Apigenin

Even if β3-AR antagonism increased the levels of ROS, apigenin treatment did not increase β3-AR expression in A673 cells (Figure 4A), and so therefore we hypothesised that apigenin could work as β3-AR agonist. To address this question, we analysed the expression of UCP2 and the GSH production under the agonism of β3-AR with BRL37344 (10 μM), and we observed an increased expression of the protein and production of GSH comparable to the treatment with apigenin 50 μM (Figure 4B,C). Moreover, we observed that the expression of antioxidant levels was decreased after 24 h of treatment with BRL37344, and the same reduction was observed with apigenin treatment (Figure 4D). In addition, results clearly indicated that the agonism of β3-AR dramatically decreased ROS levels after 24 h of treatment in the same way as the apigenin treatment (Figure 4E).

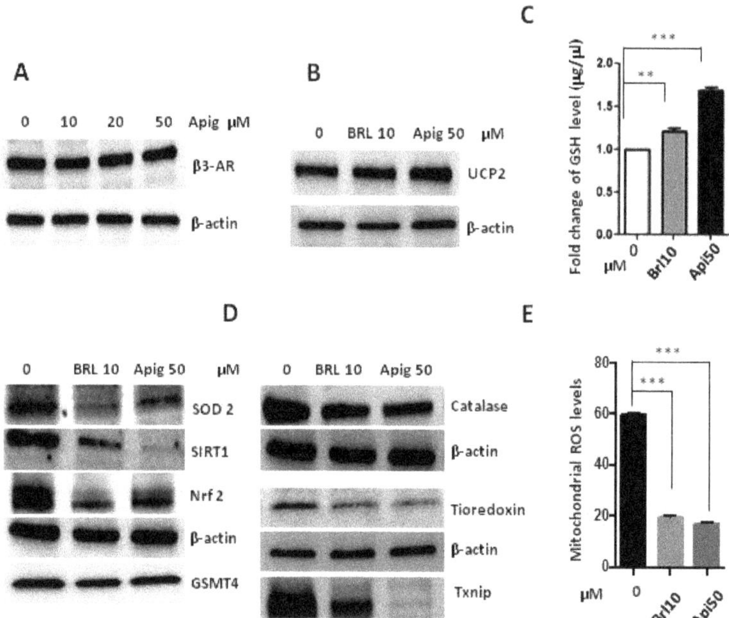

Figure 4. (**A**) Western Blot analysis of β3-AR expression after 24 h of treatment with apigenin (10-20-50 μM) with β-actin as loading control; (**B**) WB analysis of UCP2's expression under treatment with β3-AR's agonist, BRL37344 (10 μM) and apigenin (50 μM); (**C**)Measurement of GSH levels after 24 h of treatment with BRL37344 (10 μM) and apigenin (50 μM); (**D**) WB analysis of antioxidants SOD2, SIRT1, Nrf2, GSTM4, Catalase, Thioredoxin, TXNIP after 24 h of treatment with BRL37344 (10 μM) and apigenin (50 μM) with β-actin as loading control; (**F**) Mitochondrial mtROS measurement after treatment with BRL37344 (10 μM) and apigenin (50 μM) for 24 h. Brl10: BRL37344 10 μM, Apig50: apigenin 50 μM. *p* values for treatments: ** $p < 0.01$, and *** $p < 0.001$.

2.4. Apigenin Could Be a β3-AR Agonist

To support the hypothesis on the β3-AR agonist profile of apigenin, in silico studies were performed and the ability of the ligand to bind the receptor was evaluated using two homology-built models (HM1 and HM2) based on the crystal structures of turkey β1-AR (pdb code 2Y03) [26] and human β2-AR (pdb code 3PDS) [27] as single template. The templates were chosen according to their sequence homology (55% and 42% for 2Y03 and 3PDS, respectively) as well as to their activation state. Indeed, agonist-bound templates were chosen basing on the putative agonist activity shown by apigenin in the biological assays.

The conformational changes within the HMs induced upon ligand (apigenin) binding were evaluated by means of the induced fit docking procedure in order to allow both the receptor and the ligand to freely move during docking. The best protein-ligand complex (poses, one for each homology HM1 and HM2 models) resulting from this procedure were selected basing on the IFD scored values and binding energies estimated by applying the MM-GBSA method [28].

Insights into the binding mode of apigenin with the two modelled targets revealed a key role played by the residues involved in the binding of β-ARs agonists/antagonists with the receptor, i.e., D117, F309, and N332. In particular, the 7-OH moiety of apigenin established two H-bond interactions with the side chains of D117 and N331, acting as a donor and acceptor, respectively. Moreover, the chromone core (i.e., the moiety formed by A+B in Figure 4A) of the ligand is sandwiched by F309 and V118, forming π-π and π-alkyl interactions. Other common interactions are a T-shaped π-π stacking engaged by the phenyl ring of apigenin (C in Figure 5A) and the side chain of F198 as well as a H-bond formed by the hydroxyl group in position 4′ and the guanidinium group of the R315 side chain. The outcomes of the in silico study substantiate the binding of apigenin to the β3-AR and support the findings of the biological assays.

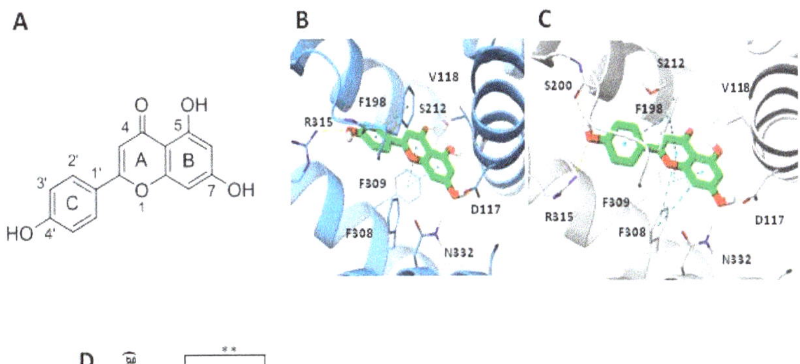

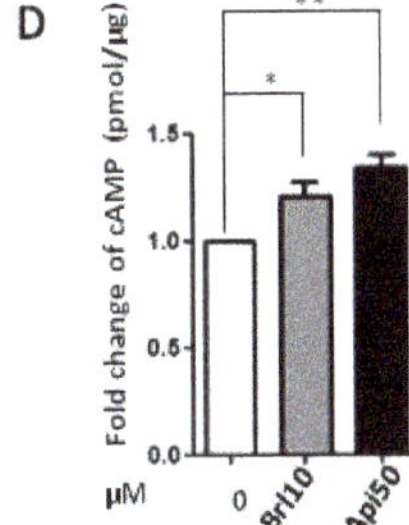

Figure 5. (**A**) Structure of apigenin. (**B,C**) Predicted binding mode of apigenin in the β3-AR binding pocket of (**B**) HM1 and (**C**) HM2; (**D**) Measurement of cAMP concentration after 30 min of treatment with BRL37344 (10 μM) and apigenin (50 μM). Brl10: BRL37344 10 μM, Apig50: apigenin 50 μM. *p* values for treatments: * $p < 0.05$ and ** $p < 0.01$.

As further proof, an analysis of the second messenger cyclic AMP (cAMP) levels was performed showing increased cAMP levels in the presence of BRL37344 and in the same way with apigenin 50 μM after 30 min of exposition, confirming that apigenin could act as β3-AR agonist (Figure 5B).

3. Discussion

The fine regulation of ROS production and detoxification is fundamental for the growth or reduction of a Ewing Sarcoma tumour. Therapeutic drugs that act on the reduction of cellular antioxidant activity, promote oxidative stress increase, and induction of cell death in ES. In this context, an antioxidant-inhibiting strategy has been evaluated in order to enhance the efficacy of some chemotherapeutics used in the standard treatment of ES such as doxorubicin and etoposide, which increase generation of ROS and oxidative stress in mitochondria [29]. In this study we demonstrated that apigenin induces partial cell death by activating the apoptotic pathway without increasing mitochondrial ROS production, which conversely is observed by administration of β3-AR antagonist.

Therapeutic strategies that are aimed to disrupt the redox homeostasis with bioactive nutrients of malignant cells of ES is constantly under debate. Unfortunately, despite the great interest, there is no consistent data regarding the use of apigenin and other flavonoids in humans as an anticancer therapy. Inconsistent reporting and study design for the investigation of flavonoids in both epidemiologic and intervention trials have significantly prevented development of clear recommendations about the intake of flavonoids to support or promote human health [30]. Apigenin has been reported to induce apoptosis in HepG2 cells by inhibiting Catalase activity and enhancing the expression of high levels of ROS, as well as preventing hepatocellular carcinogenesis via decreasing oxidative stress [31]. Flavonoids used in clinical practice may not confirm data of preclinical efficacy due to low/moderate anti-cancer activity when used alone at human physiological dosages [32–34]. Furthermore, usage at higher dosages has no known safety profile, with the possibility of unacceptable toxicities. Transition ions, as $Cu2^+$ and $Fe2^+$ present in biological systems, can affect the pro-oxidant activity of flavonoids, leading to inhibition of mitochondrial breathing. This activity of natural antioxidant plays an important role in their selective cytotoxicity toward cancer cells that contain more copper than normal cells [35–38].

Here we demonstrated that apigenin inhibits the expression of antioxidant proteins such as SOD2, Catalase, SIRT1, GSTM4, TXNIP, Thioredon1, and Nrf2, but increased the level of UCP2 and GSH which conversely are strongly inhibited by β3-AR antagonism. UCPs have been related to ROS production for the first time in 1997 in experiments where GDP, an inhibitor of UCP1, caused an increase of ROS production. Subsequent studies demonstrated that superoxide directly activates UCPs, leading to negative feedback controlling both ROS production and UCPs levels [39,40].

The β3-ARs antioxidant activity could be mediated by UCP2 protein expression that could work as a guardian for the redox homeostasis in ES cells. The redox homeostasis of cells is balanced by ROS generation and ROS quenching capacity. Undoubtedly, an imbalance in favor of increased ROS production within the cellular microenvironment by disruption of UCP2 signalling, and by inhibiting β3-ARs, can lead to excessive oxidative stress resulting in massive cell death.

The link between β3-ARs and UCP2 has been well described in white and brown adipocytes. Selective pharmacological β3-ARs stimulation has been shown to affect adipose tissue morphology and metabolism. The activity of CL-316,243, a potent and highly selective β3-ARs agonist [41] leads to improved thermogenesis in brown adipose tissue (BAT), lipolysis in white adipose tissue (WAT), and an acute decrease in food consumption [42,43]. Thermogenesis in BAT is mediated by activation by UCP1. β3-ARs agonists reduce fat stores, improved obesity-induced insulin resistance and increased brown adipocytes content in WAT tissue [44,45]. It is well known that β3-ARs control thermogenesis through activation of UCPs, in particular UCP1. UCPs maintain redox state of the cells in the respiratory chain transport. Interestingly, UCP2 has been shown to control GSH/GSSH in beta pancreatic cells [24,46].

More recently, data reported the β3-AR antioxidant activity by showing dual antioxidant properties: the reduction of NADPHox activity and induction of the expression of Catalase [46]. β3-ARs are expressed and functional in the human macrophages where their antioxidant effects lead to a potent anti-inflammatory response and play an important role in PPARγ activation through the Erk1/2 pathway [46]. In the second paper, authors show an increased intracellular concentration of GSH induced by GLCc protein in both U-251 MG cells and mouse astrocytes through noradrenaline-mediated β3-adrenoreceptor activation. This study revealed the importance of β3-ARs in the maintenance of GSH homeostasis in glioma cells by Gi/0-protein but not by Gs-protein in U-251 cells. These results indicated that the activation of intracellular signalling in response to β3-ARs stimulation may have been required for the induction of GSH by noradrenaline [24]. Even if apigenin inhibits antioxidant proteins, it works as β3-AR agonist, by avoiding the elevation of mitochondrial ROS useful to reach a massive cell death in ES cells.

In this work we confirm that the inhibition of antioxidants may be strategically useful in Ewing sarcoma therapy, and that the use of β3-ARs antagonist could be the limiting factor to reach a large amount of cell death.

4. Materials and Methods

4.1. Materials

Human Ewing Sarcoma (ES) cells A673 were purchased from the American Type Culture Collection (ATCC® CRL-1598 ™, Manassas, VA, USA). Dulbecco's modified Eagle's medium (DMEM) high-glucose, fetal bovine serum (FBS), penicillin-streptomycin, L-glutamine, and trypsin-enzyme were obtained from Euroclone Group (Pero, MI, Italy). Phosphate buffered saline (PBS) was purchased from Gibco (Gaithersburg, MD, USA). Dimethylsulfoxide (DMSO), apigenin (>97%), MTT (3-[4,5-dymethilthiazol-2-yl-]-2,5-diphenyltetrazolium bromide; thiazol blue) assay, BRL37344 and SR59230A were obtained from Sigma-Aldrich (St. Louis, MO, USA). Precast gels, PVDF membranes, Milk Blotting-Grade Blocker, and ECL, HRP Chemiluminescent Substrate Reagent kit were obtained from Biorad® (Hercules, CA, USA); Tween®20 was obtained from Sigma-Aldrich. The primary antibodies used include the following: β3-adrenergic receptor (ab-77588), Catalase (ab-16731) and glutathione S-transferase Mu4 (GSTM4) from Abcam (Cambridge, MA, USA), superoxide dismutase-2 (SOD2) (sc-137254), β-actin (sc-1615), uncoupling protein 2 (UCP2) (sc-390189), erythroid 2-related factor 2 (Nrf2) (sc-305948) from Santa Cruz Biotechnology (Dallas, TX, USA), Thioredoxin 1 (MA5-14941) from Invitrogen by Thermo Fisher® (Waltham, MA, USA); thioredoxin interacting protein TXNIP (VDUP-1) from Life Technologies (Waltham, MA, USA), Sirtuin 1 from Millipore (Darmstadt, Germany); Poly (ADP ribose) polymerase 1 (PARP-1) from Cell Signaling Technology (Beverly, MA, USA). The specific secondary antibodies, conjugated with horseradish peroxidase (HRP), anti-rabbit, anti-goat, and anti-mouse were purchased from Santa Cruz Biotechnology. Glutathione Fluorometric Assay Kit and cAMP Direct ImmunoAssay Kit (Colorimetric), were purchased from Biovision® (Milpitas, CA, USA). Mitochondria isolation kit, and Annexin V-FITC kit were obtained from Miltenyi Biotec® (Bergisch Gladbach, Germany). For ROS detection, MitoSOX™ Red mitochondrial superoxide indicator from Invitrogen by Thermo Fisher® was used. Viobility™ Fixable Dyes were obtained from Miltenyi Biotec®. H2DCFDA Assay Kit and Amplex® Red Hydrogen Peroxidase Assay Kit was obtained from Invitrogen by Thermo Fisher®. Separation Media Lymphosep for lymphocytes separation was obtained from Biowest (Nuaillè, France).

4.2. Cell Cultures

Human Ewing Sarcoma (ES) cells A673 were cultured in 100mm plates in DMEM high glucose medium supplemented with 10% fetal bovine serum (FBS), 1% of L-glutamine, 1% of penicillin-streptomycin and were maintained at 37 °C in a 5% CO_2 humidified atmosphere incubator. Cells were usually stored in liquid nitrogen in a freezing solution, containing 95% complete DMEM medium and 5% DMSO and then plated in petri p100. For defrosting, the vials were rapidly brought to 37 °C by immersion in the thermostat bath, then centrifuged to remove the toxic DMSO from the cells, re-suspended in DMEM high glucose FBS 10, and appropriately plated.

Sub-confluent cells were detached from the plate with trypsin-enzyme after aspirating the medium and one wash with PBS to eliminate medium and serum residues. Then DMEM high glucose was added and the cell suspension obtained was counted and plated in fresh DMEM high glucose with appropriate dilutions. Human lymphocytes were isolated from peripheral blood with Separation Media Lymphosep to test the effect of apigenin in healthy control cells.

4.3. Cell Treatments

A673 ES cells were plated to reach 70% confluence in complete high-glucose DMEM medium. In order to promote cell entry into a G0 phase and better evaluate cells responsiveness to exogenous treatments, after 24 h the medium was removed, the cells were washed in PBS solution, and finally starved overnight with starvation medium (DMEM high glucose without FBS). The consequent morning, cells were treated with a single dose of apigenin at the concentrations of 10 µM, 20 µM, 50 µM [47–50] and subsequently left in an incubator for 24 h, then collected for the experiments. Apigenin was dissolved in DMSO 2.7 mg/mL to obtain a final concentration stock of 10mM, then appropriate dilutions from the stock solution were made for treatments.

4.4. MTT Assay

Viability of tumour cells after treatment with apigenin was detected by MTT (3-[4,5-dymethilthiazol-2-yl-]-2,5-diphenyltetrazolium bromide; thiazol blue) assay. A673 cells were transferred into a 96-well plate at a density of 10×10^4 cells/well in 150 µL DMEM complete and were incubated with apigenin at different concentrations (10 µM, 20 µM, 50 µM) for 24 h. A total of 10µL of MTT was added to each well and incubated under darkness for 1h at 37 °C. Then, the culture medium was removed and 150 µL of DMSO was added to each well. The intensity of absorbance was detected at 570 nm using a spectrophotometer (PerkinElmer, Waltham, MA, USA).

4.5. Western Blot Analysis

After homogenization and protein quantification, samples (15–20 µg of total proteins) were loaded on SDS-PAGE and subjected to Western Blot analysis. Subsequently PVDF membranes were incubated for 1 h in slow agitation at room temperature in a blocking solution of non-fat dry milk 5% and Tween PBS 0.1% in order to avoid the formation of unspecific ties. Membranes were then incubated with the following primary antibodies: β3-adrenergic receptor, Catalase, Superoxide dismutase-2, TXNIP, Thioredoxin 1, Sirtuin 1, β-actin, UCP2, Nrf2, GSTM4. The primary antibody was added generally in a concentration of 1:1000 and incubated, in shaking, over-night at 4 °C. The next day membranes were washed three times with a washing solution containing Tween PBS 0.1% in order to remove unbound primary antibody in excess. Then, the specific secondary antibody, which was conjugated with horseradish peroxidase (HPR), was added, in a dilution of 1:5000 in Tween PBS 0.1% and incubated for 1 h. Chemiluminescent protein's revelation was carried out with ECL reagent and developing of blots was carried out by Chemidoc Imaging System (Biorad®). To verify the application of equal amounts of protein, the intensity of the corresponding protein bands of interest was normalised based on that of the the the β-actin band for each sample.

4.6. Glutathione Fluorometric Assay Kit

For the detection of reduced glutathione (GSH) the Glutathione Fluorometric Assay Kit was used, following the manufacturer instructions and intensity of fluorescence being analysed with spectrophotometer (PerkinElmer).

4.7. Cell Viability Analysis

For the detection and discrimination of live and dead cells (apoptotic, necrotic) in A673 line and lymphocytes, the Viobility™ Fixable Dyes and Annexin V-FITC Kit were used after 24 h of treatment with apigenin at different concentrations, following the manufacturer's instructions. Results were analysed by flow cytometry MACSQuant FACS (Miltenyi Biotec®).

4.8. cAMP Direct ImmunoAssay Kit

The Adenosine 3′,5′-cyclic monophosphate (cyclic AMP, cAMP) was measured by the cAMP Direct ImmunoAssay Kit (Colorimetric) after 30 min of treatment with BRL37344 (10 μM) and apigenin (50 μM) following the manufacturer's instructions and absorbance at 450 nm detected using spectrophotometer (PerkinElmer).

4.9. ROS Analysis

For the intracellular ROS measurements, A673 cells plated into 24-wells plates at a density of 10×10^4 cells in 1 mL complete high glucose DMEM medium, were treated with apigenin as described above. After 1-6-24 h of treatment, cells were stained with 1 μL of MitoSOX™ Red mitochondrial superoxide indicator reagent at a concentration of 2.5 μM. After 15 min of incubation at room temperature under darkness, cells were washed with PBS, detached with 250 μL of trypsine-enzyme, spun at 1300 rpm for 5 min, and pellet was resuspended in 300 μL of PBS + 0.2% of FBS, and then evaluated by flow cytometry MACSQuant FACS (Miltenyi Biotec®). For peroxide measurements Amplex® Red Hydrogen Peroxidase Assay Kit was used, following the manufacturer's instructions. H2DCFDA Assay Kit was performed according to manufacturer's instructions to measure levels of different species of ROS.

4.10. Mitochondria Isolation

To isolate the mitochondria from A673 cells the Mitochondria Isolation Kit was used. At the end of the kit procedure after the centrifugation at $13,000 \times g$ for 2 min at 4 °C, the supernatant was aspirated and the mitochondria pellet was resuspended in an adequate buffer for further analysis. We resuspended in lysis buffer for Western Blot analysis, and in PBS for analysis by flow cytometry MACSQuant FACS.

4.11. In Silico Methods

The primary sequence of the human β3-AR was retrieved from UniProt. The crystal structures of turkey β1-AR (2Y03) and human β2-AR (3PDS) were downloaded from the Protein Data Bank and used as a template in the homology modelling procedure. Prime module of the Schrödinger suite was used in the sequence alignment and model building procedures [51]. Then, the models were submitted to loop refinements and the quality checked by the analysis of the Ramachandran plots and evaluation of the QMEAN value.

The two HMs were prepared with Maestro (a of [51]) applying an energy minimization with RMSD value of 0.30 using the OPLS-3 force field. Apigenin structure was prepared by Maestro (a of [51]) and Macromodel (e of [51]) with the OPLS-3 force field was used for energy minimization.

The Induced Fit Docking protocol implemented in the Schrödinger package was used. The procedure consists of a Glide SP docking, (f of [51]) followed by a Prime refinement of the residue side chains within 5 Å and then by a final Glide XP docking of the ligand into the receptor in the refined conformation. The docking grid were centred on the centre of mass of the bound ligands. In the initial Glide SP docking, the vdW scaling was set to 0.7 for non-polar atoms of receptor and 0.5 those of the ligand.

The best poses for each model were submitted to MM-GBSA calculations (d of [51]) in VSGB solvent model enabling residue flexibility 5 A around the ligand to compute the binding free energies [52,53].

4.12. Statistical Analysis

In vitro data are presented as means ± Standard Deviation (SD) from at least three experiments. Results were normalised versus control expression levels.

Statistical analysis was performed using Graph Pad Prism software (GraphPad, San Diego, CA, USA) by one-way analysis of variance (ANOVA) and two-way analysis for all experiments except for the analysis of GSH levels after SR59230A treatment, followed by Bonferroni post hoc analysis.

Author Contributions: M.C. conceived and designed the project; A.P., M.V., A.S. and A.D. performed cell biology experiments. A.N., S.S. and P.G. performed chemical analysis; I.F. and T.C. collected the literatures. M.C. and A.P. wrote the manuscript; L.F. and C.F. revised the manuscript. C.F. supervised all the project. All authors have read the final manuscript and approved it.

Funding: This research received no external funding.

Conflicts of Interest: The authors declare no conflict of interest.

References

1. Kovar, H. Ewing's Sarcoma and peripheral primitive neuroectodermal tumors after their genetic union. *Curr. Opin. Oncol.* **1998**, *10*, 334–342. [CrossRef] [PubMed]
2. Riggi, N.; Stamenkovic, I. The Biology of Ewing sarcoma. *Cancer Lett.* **2007**, *254*, 1–10. [CrossRef] [PubMed]
3. Balamuth, N.J.; Womer, R.B. Ewing's sarcoma. *Lancet Oncol.* **2010**, *11*, 184–192. [CrossRef]
4. Jain, S.; Kapoor, G. Chemotherapy in Ewing's sarcoma. *Indian J. Orthop.* **2010**, *44*, 369–377. [PubMed]
5. Magwere, T.; Myatt, S.S.; Burchill, S.A. Manipulation of oxidative stress to induce cell death in Ewing's sarcoma family of tumors. *Eur. J. Cancer* **2008**, *44*, 2276–2287. [CrossRef]
6. Scotlandi, K.; Remondini, D.; Castellani, G.; Manara, M.C.; Nardi, F.; Cantiani, L.; Francesconi, M.; Mercuri, M.; Caccuri, A.M.; Serra, M.; et al. Overcoming resistance to conventional drugs in Ewing sarcoma and identification of molecular predictors of outcome. *J. Clin. Oncol.* **2009**, *27*, 2209–2216. [CrossRef]
7. Conklin, K. Dietary antioxidants during cancer chemotherapy: Impact on chemotherapeutic effectiveness and development of side effects. *Nutr. Cancer* **2000**, *37*, 1–18. [CrossRef]
8. Ladas, E.; Kelly, K.M. The antioxidant debate. *Explore (NY)* **2010**, *6*, 75–85. [CrossRef]
9. Peterson, J.; Dweyer, J. Flavonoids: Dietary occurrence and biochemical activity. *Nutr. Res.* **1998**, *18*, 1995–2018. [CrossRef]
10. Shukla, S.; Gupta, S. Apigenin: A promising molecule for cancer prevention. *Pharm. Res.* **2010**, *27*, 962–978. [CrossRef] [PubMed]
11. Gupta, S.; Afaq, F.; Mukhtar, H. Selective growth inhibitory, cell-cycle deregulatory and apoptotic response of apigenin in normal versus human prostate carcinoma cells. *Biochem. Biophys. Res. Commun.* **2001**, *287*, 914–920. [CrossRef]
12. Morrissey, C.; O'Neill, B.; Spengler, B.; Christoffel, V.; Fitzpatrick, J.M.; Watson, R.W. Apigenin drives the production of reactive species and initiates a mitochondrial mediated cell death pathway in prostate epithelial cells. *Prostate* **2005**, *63*, 131–142. [CrossRef] [PubMed]
13. Wang, W.; Heideman, L.; Chung, C.S.; Pelling, K.J.; Koehler, D.F.; Birt, D.F. Cell cycle arrest at G2/M and growth inhibition by apigenin in human colon carcinoma cell lines. *Mol. Carcinog.* **2000**, *28*, 102–110. [CrossRef]
14. Way, T.D.; Kao, M.C.; Lin, J.K. Apigenin induces apoptosis through proteasomal degradation of HER2/neu in HER/neu-overexpressing breast cancer cells via the phosphatidylinositol 3-kinase/Akt-dependent pathway. *J. Biol. Chem.* **2004**, *279*, 4479–4489. [CrossRef]
15. Wang, I.K.; Lin-Shiau, J.K.; Lin, J.K. Induction of apoptosis by apigenin and related flavonoids through cytochrome c release and activation of caspase-9 and caspase-3 in leukemia HL-60 cells. *Eur. J. Cancer* **1999**, *35*, 1517–1525. [CrossRef]
16. Zheng, P.W.; Chiang, L.C.; Lin, L.L. Apigenin induced apoptosis through p53-dependent pathway in human cervical carcinoma cells. *Life Sci.* **2005**, *75*, 1367–1379. [CrossRef] [PubMed]
17. Lu, H.F.; Chie, Y.S.; Tan, T.W.; Wu, S.H.; Ma, Y.S.; Ip, S.W.; Chung, J.W. Apigenin induces caspase-dependent apoptosis in human lung cancer A549 cells through Bax- and Bcl-2-triggered mitochondrial pathway. *Int. J. Oncol.* **2010**, *36*, 1477–1484.

18. Chiang, L.C.; Ng, N.T.; Lin, I.C.; Kuo, P.I.; Lin, C.C. Anti-proliferative effect of apigenin and its apoptotic induction in human Hep G2 cells. *Cancer Lett.* **2006**, *237*, 207–214. [CrossRef] [PubMed]

19. Choi, S.I.; Jeong, C.S.; Cho, S.Y.; Lee, Y.S. Mechanism of apoptosis induced by apigenin in HepG2 human hepatoma cells: Involvement of reactive oxygen species generated by NADPH oxidase. *Arch. Pharm. Res.* **2007**, *30*, 1328–1335. [CrossRef]

20. Gupta, S.; Afaq, F.; Mukhtar, H. Involvement of nuclear factor-kappa B, Bax and Bcl-2 in induction of cell cycle arrest and apoptosis by apigenin in human prostate carcinoma cells. *Oncogene* **2002**, *21*, 3727–3738. [CrossRef]

21. Shukla, S.; Gupta, S. Molecular mechanisms for apigenin-induced cell-cycle arrest and apoptosis of hormone refractory human prostate carcinoma DU145 cells. *Mol. Carcinog.* **2004**, *39*, 114–126. [CrossRef] [PubMed]

22. Shukla, S.; Gupta, S. Apigenin-induced prostate cancer cell death is initiated by reactive oxygen species and p53 activation. *Free Radic. Biol. Med.* **2008**, *44*, 1833–1845. [CrossRef] [PubMed]

23. Calvani, M.; Pelon, F.; Comito, G.; Taddei, M.L.; Moretti, S.; Innocenti, S.; Nassini, R.; Gerlini, G.; Borgognoni, L.; Bambi, F.; et al. Norepinephrine promotes tumor microenvironment reactivity through β3-adrenoreceptors during melanoma progression. *Oncotarget* **2014**, *6*, 4615–4632.

24. Yoshioka, Y.; Kadoi, H.; Yamamuro, A.; Ishimaru, Y.; Maeda, S. Noradrenaline increases intracellular glutathione in human astrocytoma U-251 MG cells by inducing glutamate-cysteine ligase protein via β3-adrenoceptor stimulation. *Eur. J. Pharm.* **2015**, *772*, 51–61. [CrossRef] [PubMed]

25. Calvani, M.; Cavallini, L.; Tondo, A.; Spinelli, V.; Ricci, L.; Pasha, A.; Bruno, G.; Buonvicino, D.; Bigagli, E.; Vignoli, M.; et al. β3-Adrenoreceptors Control Mitochondrial Dormancy in Melanoma and Embryionic Stem Cells. *Oxid. Med. Cell Longev.* **2018**. [CrossRef]

26. Warne, T.; Moukhametzianov, R.; Baker, J.G.; Nehmé, R.; Edwards, P.C.; Leslie, A.G.; Schertler, G.F.; Tate, C.G. The structural basis for agonist and partial agonist action on a β(1)-adrenergic receptor. *Nature* **2011**, *469*, 241–244. [CrossRef] [PubMed]

27. Rosenbaum, D.M.; Zhang, C.; Lyons, J.A.; Holl, R.; Aragao, D.; Arlow, D.H.; Rasmussen, S.G.; Choi, H.J.; Devree, B.T.; Sunahara, R.K.; et al. Structure and function of an irreversible agonist-β(2) adrenoceptor complex. *Nature* **2011**, *469*, 236–240. [CrossRef]

28. Genheden, S.; Ryde, U. The MM/PBSA and MM/GBSA methods to estimate ligand-binding affinities. *Expert Opin. Drug Discov.* **2015**, *10*, 449–461. [CrossRef]

29. Zhou, S.; Starkov, A.; Froberg, M.K.; Leino, R.L.; Wallace, K.B. Cumulative and irreversible cardiac mitochondrial dysfunction induced by doxorubicin. *Cancer Res.* **2001**, *61*, 771–777.

30. Balentine, D.A.; Dwyer, J.T.; Erdman, J.W.; Ferruzzi, M.G.; Gaine, P.C.; Harnly, J.M.; Kwik Uribe, C.L. Recommendations on reporting requirements for flavonoids in research. *Am. J. Clin. Nutr.* **2015**, *101*, 1113–1125. [CrossRef]

31. Jeyabal, P.V.; Syed, M.B.; Venkataraman, M.; Sambandham, J.K.; Sakthisekaran, D. Apigenin inhibits oxidative stress-induced macromolecular damage in N-nitrosodiethylamine (NDEA)-induced hepatocellular carcinogenesis in Wistar albino rats. *Mol. Carcinog.* **2005**, *44*, 11–20. [CrossRef]

32. Shao, H.; Jing, K.; Mahmoud, E.; Huang, H.; Fang, X.; Yu, C. Apigenin sensitizes colon cancer cells to antitumor activity of ABT-263. *Mol. Cancer Ther.* **2013**, *12*, 2640–2650. [CrossRef]

33. Salmani, J.M.M.; Zhang, X.P.; Jacob, J.A.; Chen, B.A. Apigenin's anticancer properties and molecular mechanisms of action: Recent advances and future prospectives. *Chin. J. Nat. Med.* **2017**, *15*, 321–329. [CrossRef]

34. Ganai, S.A. Plant-derived flavone apigenin: The small-molecule with promising activity against therapeutically resistant prostate cancer. *Biomed. Pharmacother* **2017**, *85*, 47–56. [CrossRef]

35. Eghbaliferiz, S.; Iranshahi, M. Prooxidant Activity of Polyphenols, Flavonoids, Anthocyanins and Carotenoids: Updated Review of Mechanisms and Catalyzing Metals. *Phytother. Res.* **2016**, *30*, 1379–1391. [CrossRef]

36. Cheng, Z.; Li, Y.; Chang, W. Kinetic deoxyribose degradation assay and its application in assessing the antioxidant activities of phenolic compounds in a Fenton-type reaction system. *Anal. Chim. Acta* **2003**, *478*, 129–137. [CrossRef]

37. Fukumoto, L.; Mazza, G. Assessing antioxidant and prooxidant activities of phenolic compounds. *J. Agric. Food Chem.* **2000**, *48*, 3597–3604. [CrossRef]

38. Perron, N.R.; Brumaghim, J.L. A Review of the Antioxidant Mechanisms of Polyphenol Compounds Related to Iron Binding. *Cell Biochem. Biophys.* **2009**, *53*, 75–100. [CrossRef]

39. Negre-Salvayre, A.; Hirtz, C.; Carrera, G.; Cazenave, R.; Troly, M.; Salvayre, R.; Pènicaud, L.; Casteilla, L. A role for uncoupling protein-2 as a regulator of mitochondrial hydrogen peroxide generation. *FASEB J.* **1997**, *11*, 809–815. [CrossRef]

40. Echtay, K.S.; Roussel, D.; St-Pierre, J.; Jekabsons, M.B.; Cadenas, S.; Stuart, J.A.; Harper, J.A.; Roebuck, S.J.; Morrison, A.; Pickering, S. Superoxide activates mitochondrial uncoupling proteins. *Nature* **2002**, *415*, 96–99. [CrossRef]

41. Klaus, S.; Seivert, A.; Boeuf, S. Effect of the β3-adrenergic agonist Cl316,243 on functional differentiation of white and brown adipocytes in primary cell culture. *Biochim. Byophis. Acta* **2001**, *1539*, 85–92. [CrossRef]

42. Susulic, V.C.; Frederich, R.C.; Lawitts, J.; Tozzo, E.; Kahn, B.B.; Harper, M.E.; Himms-Hagen, J.; Flier, J.S.; Lowell, B.B. Targeted disruption of the beta 3-adrenergic receptor gene. *J. Biol. Chem.* **1995**, *270*, 29483–29492. [CrossRef] [PubMed]

43. Atgiè, C.; D'Allaire, F.; Bukowiecki, L.J. Role of β1- and β3-adrenoceptors in the regulation of lipolysis and thermogenesis in rat brown adipocytes. *Am. J. Physiol.* **1997**, *273*, C1136–C1142. [CrossRef]

44. Himms-Hagen, J.; Cui, J.; Danforth, E., Jr.; Taatjes, D.J.; Lang, S.S.; Waters, B.L.; Claus, T.H. Effect of CL-316,243, a thermogenic beta 3-agonist, on energy balance and brown and white adipose tissues in rats. *Am. J. Physiol.* **1994**, *266*, R1371–R1382.

45. Ghorbani, M.; Claus, T.H.; Himms-Hagen, J. Hypertrophy of brown adipocytes in brown and white adipose tissues and reversal of diet-induced obesity in rats treated with a β3-adrenoceptor agonist. *Biochem. Pharmacol.* **1997**, *54*, 121–131. [CrossRef]

46. Hadi, T.; Douhard, R.; Dias, A.M.M.; Wendremaire, M.; Pezzè, M. Beta3 adrenergic receptor stimulation in human macrophages inhibits NADPHoxidase activity and induces catalase expression via PPARγ activation. *Biochim. Biophys. Acta* **2017**, *1854*, 1769–1784. [CrossRef]

47. Scherbakov, A.M.; Andreeva, O.E. Apigenin Inhibits Growth of Breast Cancer Cells: The Role of ERα and HER2/neu. *Acta Naturae* **2015**, *7*, 133–139. [PubMed]

48. Zhao, G.; Han, X.; Cheng, W.; Ni, J.; Zhang, Y.; Lin, J.; Song, Z. Apigenin inhibits proliferation and invasion, and induces apoptosis and cell cycle arrest in human melanoma cells. *Oncol. Rep.* **2017**, *37*, 2277–2285. [CrossRef]

49. Ujiki, M.B.; Ding, X.Z.; Salabat, M.R.; Bentrem, D.J.; Golkar, L.; Milam, B.; Talamonti, M.S.; Bell, R.H.; Iwamura, T.; Adrian, T.E. Apigenin inhibits pancreatic cancer cell proliferation trought G2/M cell cycle arrest. *Mol. Cancer* **2006**, *5*, 76. [CrossRef]

50. Meyer, H.; Bolarinwa, A.; Wolfram, G.; Linseisen, J. Bioavailability of Apigenin from Apiin-Rich Parsley in Humans. *Ann. Nutr. Metab.* **2006**, *50*, 167–172. [CrossRef]

51. *Schrödinger Suite Release 2018-2 ((a) Maestro v.11.6; (b) Epik, v.4.4; (c) Impact, v.7.9; (d) Prime, v.5.2; (e) Macromodel v.12.0. (f) Glide, v.7.9; (g) Jaguar, v.10.0.)*; Schrödinger, L.L.C.: New York, NY, USA, 2018.

52. Nocentini, A.; Bonardi, A.; Gratteri, P.; Cerra, B.; Gioiello, A.; Supuran, C.T. Steroids interfere with human carbonic anhydrase activity by using alternative binding mechanisms. *J. Enzyme Inhib. Med. Chem.* **2018**, *33*, 1453–1459. [CrossRef]

53. Ibrahim, H.S.; Allam, H.A.; Mahmoud, W.R.; Bonardi, A.; Nocentini, A.; Gratteri, P.; Ibrahim, E.S.; Abdel-Aziz, H.A.; Supuran, C.T. Dual-tail arylsulfone-based benzenesulfonamides differently match the hydrophobic and hydrophilic halves of human carbonic anhydrases active sites: Selective inhibitors for the tumor-associated hCA IX isoform. *Eur. J. Med. Chem.* **2018**, *152*, 1–9. [CrossRef] [PubMed]

International Journal of
Molecular Sciences

Article

Theaflavin-Enriched Fraction Stimulates Adipogenesis in Human Subcutaneous Fat Cells

Phil June Park [1,*,†], Chan-Su Rha [2,†] and Sung Tae Kim [3,*]

[1] Basic Research & Innovation Research Institute, AmorePacific Corporation R&D Unit., 1920, Yonggu-daero, Giheung-gu, Yongin-si, Gyeonggi-do 17074, Korea

[2] Vital Beautie Research Institute, AmorePacific Corporation R&D Unit, 1920, Yonggu-daero, Giheung-gu, Yongin-si, Gyeonggi-do 17074, Korea; teaman@amorepacific.com

[3] Department of Pharmaceutical Engineering, Inje University, Gimhae-si 50834, Korea

* Correspondence: philjunep@gmail.com (P.J.P.); stkim@inje.ac.kr (S.T.K.);
   Tel.: +82-31-280-5639 (P.J.P.); +82-55-320-4038 (S.T.K.)

† These authors contributed equally to this work.

Received: 26 March 2019; Accepted: 19 April 2019; Published: 25 April 2019

Abstract: Skin provides the first defense line against the environment while preserving physiological homeostasis. Subcutaneous tissues including fat depots that are important for maintaining skin structure and alleviating senescence are altered during aging. This study investigated whether theaflavin (TF) in green tea (GT) has skin rejuvenation effects. Specifically, we examined whether high ratio of TF contents can induce the subcutaneous adipogenesis supporting skin structure by modulating lipid metabolism. The co-fermented GT (CoF-GT) fraction containing a high level of TF was obtained by co-fermentation with garland chrysanthemum (*Chrysanthemum coronarium*) and the conventionally fermented GT (F-GT) fraction was also obtained. The effects of the CoF- or F-GT fractions on adipogenesis were assessed using primary human subcutaneous fat cells (hSCF). Adipogenesis was evaluated based on lipid droplet (LD) formation, as visualized by Oil Red O staining; by analyzing of adipogenesis-related factors by real-time quantitative polyperase chain reaction (RT-qPCR); and by measuring the concentration of adiponectin released into the culture medium by enzyme-linked immunosorbent assay. TF-enriched CoF-GT fraction did not adversely affect hSCF cell viability but induced their adipogenic differentiation, as evidenced by LD formation, upregulation of adipogenesis-related genes, and adiponectin secretion. TF and TF-enriched CoF-GT fraction promoted differentiation of hSCFs and can therefore be used as an ingredient in rejuvenating agents.

Keywords: anti-aging; adipogenesis; green tea; human subcutaneous fat cells; theaflavin

---

1. Introduction

Skin senescence is caused by intrinsic and extrinsic factors and leads to a loss of integrity and physiological functions of skin [1]. A decline in skin stiffness is a characteristic of aging [2]. Intrinsic factors related to skin aging include ethnicity (e.g., pigmentation), anatomical variations, and hormonal changes; extrinsic factors include environment (e.g., temperature and humidity), lifestyle (e.g., smoking/nicotine intake), and exposure to sunlight (e.g., ultraviolet radiation, UVR) [3,4]. Healthy skin maintains homeostasis and metabolic functions through communication among dermal cells such as keratinocytes, fibroblasts, melanocytes, and subcutaneous fat cells [5].

Subcutaneous adipose tissue plays an important role in skin rejuvenation [6]; subcutaneous adipocytes interact with fibroblasts and associate with elastic fibers in the dermal layer, thereby influencing the mechanical and structural properties of skin layers [7]. However, these fat-storing cells become thinner with aging and show a reduction in thermogenic capacity and structural stability including dermal elasticity, leading to skin wrinkling [8,9]. For example, UVR-induced photo-aging can modulate lipid metabolism, leading to reduced free fatty acid and triglyceride (TG) levels in adipocytes [10]. Therefore, the decreased function of adipocytes influences lipid metabolism in skin and cellular uptake of circulating free fatty acids, which can cause adverse health outcomes such as dyslipidemia [11], metabolic syndrome [12], and insulin resistance [13].

Green tea (GT) made from *Camellia sinensis* leaves is a widely consumed beverage that contains polyphenolic compounds with various health benefits [14] such as catechins, theaflavins (TFs), and thearubigins [15], whose abundance may be altered during the fermentation process [16]. TFs are known to influence lipid metabolism [17–19]. There are four major types of TFs—i.e., theaflavin (TF), theaflavin-3-gallate (TF3G), theaflavin-3′-gallate (TF3′G), and theaflavin-3,3′-digallate (TFDG)—that are produced in vitro from fresh tea leaves through oxidation by polyphenol oxidase; additionally, tea catechins in black tea leaves are generated by horseradish peroxidase [20,21]. Black tea polyphenols including TFs were found to have an anti-obesity effect in a mouse model [19]. Consumption of black tea prevents fat storage in liver, lowers lipid as well as glucose levels, increases fecal excretion of TGs, and diminishes adipose tissue [22]. TFs were also shown to block UV-induced skin cancer by suppressing UVB-induced activator protein-1 activity via inhibition of extracellular signal-regulated protein kinase and c-Jun NH2-terminal kinase [23]. TFDG has anti-melanogenic effects that are exerted via downregulation of tyrosinase protein and mRNA levels [24]. Black tea extract containing TFs may be used as a skin-lightening agent in the cosmetic industry owing to its anti-melanogenic effect [25]. Therefore, TFs have both anti-obesity and hypopigmentation-inducing properties. However, although TFs have been linked to lipid metabolism, their effects on subcutaneous tissue—and particularly adipocytes—are not known.

To address this issue, the present study was conducted to investigate the potential anti-aging effects of TFs. TF-enriched co-fermented GT extract (CoF-GT) was obtained from natural GT by co-fermentation with garland chrysanthemum (GC; *Chrysanthemum coronarium*) to increase TF concentration, and its effects were evaluated in cultured human subcutaneous fat cells (hSCF).

2. Results

2.1. Preparation of Theaflavin (TF)-Enriched Fraction by Co-Fermentation with Chrysanthemum Coronarium (GC)

Polyphenols in tea leaves are altered during the transformation from GT to black tea by fermentation via enzymatic oxidation [21,26]. Secondary polyphenols are generated through this process, with some changing the color of tea to brown [15,27]. To achieve high concentrations of TF, we prepared a TF-enriched fraction by co-fermentation with GC, and changes in the components were assessed by high-performance liquid chromatography (HPLC) (Figure 1). It was first confirmed that the four major TFs were well-isolated and showed different retention times after injection of TFs as standard chemicals (Figure 1A), comparing to the GT extract (Figure 1B). Then, it was verified whether CoF- or F-GT had different composition of polyphenols. As result, the content ratio of TF was higher than that of other components in CoF-GT (Figure 1C). F-GT fermented in a conventional manner without GC showed similarly increases for the four major TFs, without a significant change in overall TF content (Figure 1D and Table S1). CoF- and F-GT fractions obtained in the preceding fermentation step were used for subsequent experiments. In addition, representative polyphenols such as TF and TFDG were also used as a positive control for a better understanding of CoF- and F-GT effects.

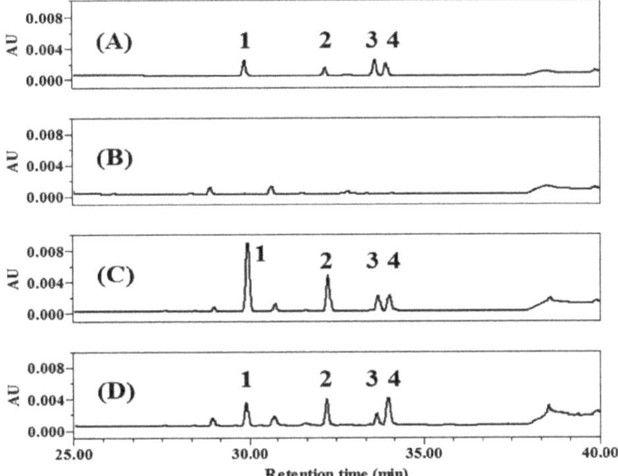

Figure 1. High-performance liquid chromatography (HPLC) analysis of theaflavins (TFs). (**A**) Standard chemicals; (**B**) Green tea extract; (**C**) co-fermented green tea (CoF-GT) fraction; (**D**) fermented green tea (F-GT) fraction. Data were monitored at 270 nm and compounds were assigned as follows: Peak 1, theaflavin (TF); Peak 2, theaflavin-3-gallate (TF3G); Peak 3, theaflavin-3′-gallate (TF3′G); and Peak 4, theaflavin-3,3′-digallate (TFDG).

2.2. TF-Enriched Fraction Does Not Adversely Affect Human Subcutaneous Fat Cells (hSCF) Cell Viability

We next investigated the effects of TFs on cell proliferation and viability by treating undifferentiated hSCF cells with CoF-GT, F-GT, TF, or TFDG at concentrations ranging from 1 to 5 µg/mL for 24 and 72 h (Figure 2). After 24 h, cell proliferation was increased in the CoF- and F-GT groups relative to the TF and TFDG groups (Figure 2A). After 72 h, there was no evidence of cytotoxicity in any group and in some cases cell viability was slightly increased (Figure 2B). Therefore, polyphenols in CoF- and F-GT extracts promote hSCF cell growth and are safe for human cells.

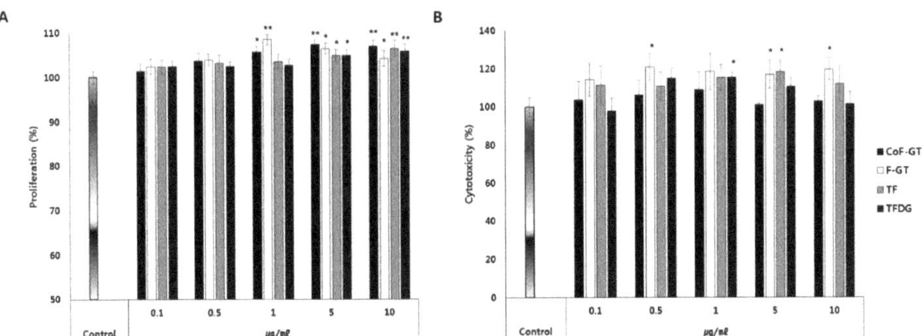

Figure 2. Effect of TF on human subcutaneous fat cells (hSCFs) viability. (**A,B**) Proliferation (**A**) and cytotoxicity (**B**) were evaluated 24 and 72 h after treatment, respectively. Data are presented as mean ± standard deviation (SD). * $p < 0.05$, ** $p < 0.01$.

2.3. TF-Enriched Fraction Induces Lipogenesis in hSCF Cells

Pre-adipocytes and mature adipocytes were prepared from hSCF cells. The differentiation of hSCF cells was confirmed by the formation of intracellular lipid droplets after 21 days of culture (Figure 3A, lower left panel) because lipid droplet formation is closely related to adipocyte differentiation and its observation could be useful for monitoring the differentiation. The amount of intracellular lipid in mature hSCF cells was evaluated by Oil Red O staining (Figure 3A,B) and by analysis of triglyceride (TG) content (Figure 3C). We examined the effect of TF and TFDG on hSCF cell differentiation and found that TF at concentrations ranging from 0.5 to 2 µg/mL increased the abundance of intracellular lipid droplets (Figure 3A, the right upper panel). Similar results were obtained for TG content. TFDG had the opposite effects in the same concentration range. Therefore, TF induces whereas TFDG inhibits lipogenesis in hSCF cells.

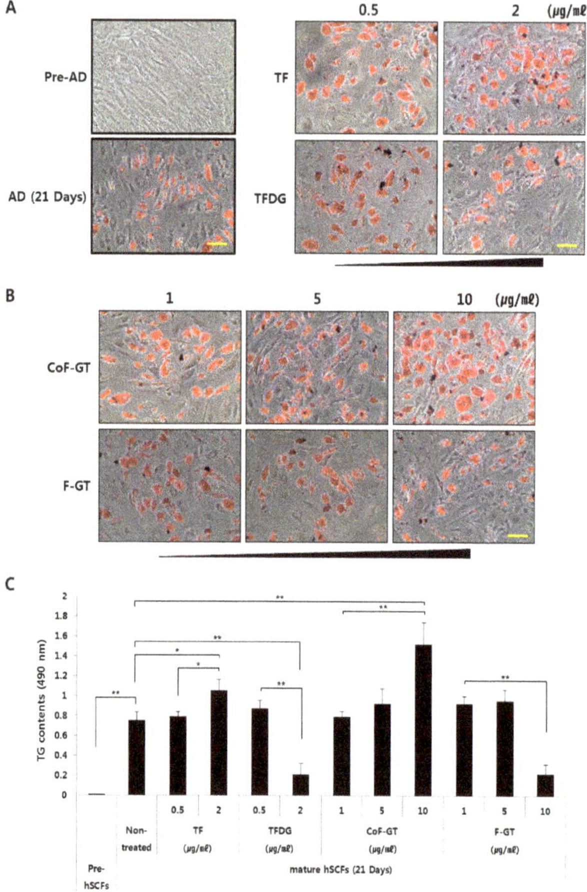

Figure 3. Effect of TF on hSCF differentiation. Pre- and mature adipocyte states of hSCFs fixed and stained with Oil Red O (ORO) at indicated time points and examined by light microscopy. (**A**) **Left**, untreated group; **right**, TF- and TFDG-treated groups used as a positive control; (**B**) CoF- and F-GT treated groups. Scale bar, 200 µm; (**C**) TG content was determined by measuring the absorbance at 490 nm. Data are presented as mean ± SD. * $p < 0.05$, ** $p < 0.01$.

To assess the effects of CoF- and F-GT on hSCF cell differentiation, cells were treated with the test substances at concentrations ranging from 1 to 10 µg/mL for 21 days (Figure 3B). Treatment with 10 µg/mL CoF-GT induced lipogenesis (Figure 3B, upper panel), while the same concentration of F-GT had the opposite effect. We analyzed the content of TGs, a major component of lipid droplets, and found that TF (2 µg/mL) and CoF-GT (10 µg/mL) increased whereas TFDG (2 µg/mL) and F-GT

(10 µg/mL) decreased TG levels (Figure 3C). The latter effect was accompanied by inhibition of hSCF cell differentiation. These results indicate that TF and TF-enriched fractions (e.g., CoF-GT) stimulate the lipogenic differentiation of hSCF cells.

2.4. TF-Enriched Fraction Increases Expression of Lipogenesis-Related Genes in hSCF Cells

We measured the mRNA expression of lipogenesis-related genes including *peroxisome proliferator activated receptor gamma* (PPARγ) and *adiponectin* (ADIPOQ) by real-time quantitative polyperase chain reaction (RT-qPCR) (Figure 4A). The transcript level of PPARγ, a master regulator of adipogenesis, was about six times higher in mature as compared to undifferentiated hSCFs after 21 days of differentiation (Figure 4A, black bar). TF treatment increased PPARγ expression in a dose-dependent manner; however, the level was decreased and reached a minimum value upon treatment with 2 µg/mL TFDG. In the CoF-GT group, PPARγ gene expression showed a dose-dependent increase, which is in accordance with the ORO staining results (Figure 3B). Furthermore, PPARγ expression was higher in cells treated with 10 µg/mL CoF-GT than in those treated with 2 µg/mL TF, and was downregulated in a dose-dependent manner in the F-GT treatment group.

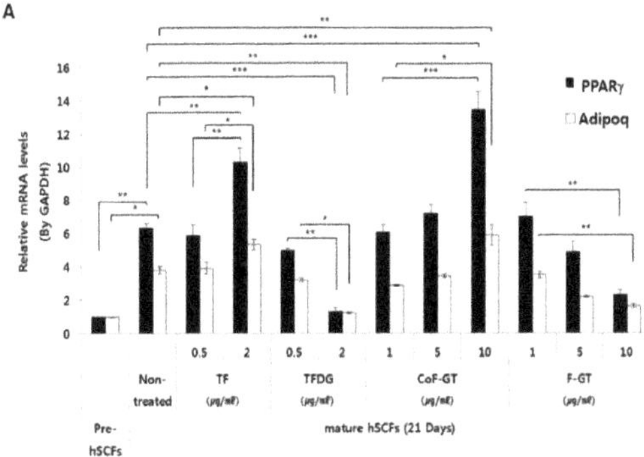

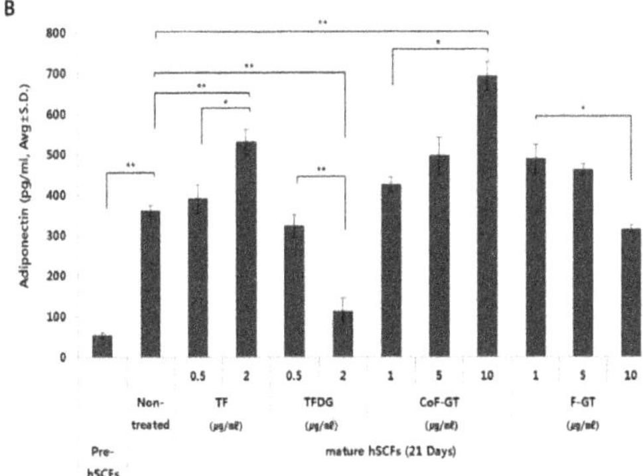

Figure 4. Changes in the expression of adipogenesis-related genes and quantitative analysis of adiponectin release into culture medium. (**A**) After 21 days of hSCF differentiation, *peroxisome proliferator activated receptor gamma* (PPARγ) and *adiponectin* (ADIPOQ) mRNA expression levels were evaluated by real-time quantitative polyperase chain reaction (RT-qPCR); (**B**) Adiponectin concentration in cell culture supernatant after 21 days of differentiation was measured by enzyme linked immunosorbent assays (ELISA). Data are presented as mean ± SD. * $p < 0.05$, ** $p < 0.01$, *** $p < 0.001$.

The expression of ADIPOQ, a mature adipocyte marker [28], was four times higher in mature as compared to undifferentiated hSCF cells (Figure 4A, white bar). ADIPOQ level was increased by 2 µg/mL TF and decreased by 2 µg/mL TFDG treatment. Overall, the changes in ADIPOQ expression in the presence of TF and TFDG were similar to those observed for PPARγ, and were in agreement with the histological findings.

2.5. TF-Enriched Fraction Stimulates Adiponectin Secretion in hSCFs

Adiponectin (APN) is one of the adipocyte-secreted cytokine (adipokine) that is synthesized in LDs. The amount of APN in the culture media was higher for differentiated cells than for pre-adipocytes (Figure 4B). APN level was increased by treatment with TF (2 µg/mL) and CoF-GT (10 µg/mL) and reduced by TFDG (2 µg/mL) and F-GT (10 µg/mL), exhibiting a trend comparable to that of ADIPOQ mRNA expression in the RT-qPCR analysis. These results confirm that TF and CoF-GT promote lipogenesis in hSCF cells.

Based on the above findings, we propose the following model for the role of TFs in skin aging. TF and TF-enriched CoF-GT fraction promotes the differentiation of hSCF cells whereas conventionally fermented F-GT fraction has the opposite effect, as evidenced by up- and downregulation, respectively, of *PPARγ* and *ADIPOQ* expression. The increase in adipocyte marker gene expression was accompanied by intracellular LD formation. Therefore, TF enhances whereas TFDG inhibits hSCF differentiation. Additionally, although TF-enriched CoF-GT stimulated adipogenic differentiation, there were no changes in polyphenol content. Therefore, polyphenol content ratio varies according to the fermentation conditions and is the main factor regulating hSCF differentiation. These results suggest that TF and TF-enriched extracts stabilize skin structure by inducing subcutaneous fat production (Figure 5).

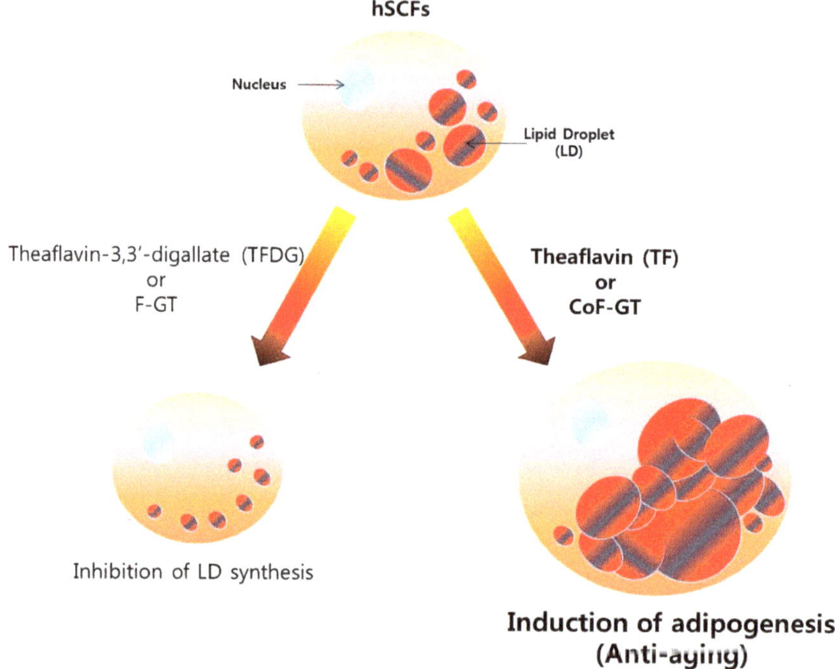

Figure 5. Function model of CoF- and F-GT fraction in hSCFs differentiation. Conventionally fermented F-GT fraction decreases the number of LDs while TF-enriched CoF-GT fraction induces their production in hSCFs.

3. Discussion

Tea is a widely consumed drink around the world that is a dietary source of bioactive compounds with numerous health benefits [29]. Various types of tea can be produced via different manufacturing processes, such as non-fermented (e.g., GT), semi-fermented (e.g., oolong tea), and fermented (e.g., black tea) [30]. Fermentation or extraction modifies chemical components in tea, which can alter their medicinal effects [30,31]. Polyphenolic compounds in tea are of particular interest due to their antioxidant and anti-obesity activities [32,33]; indeed, many previous studies have demonstrated the health benefits of polyphenol-enriched tea extracts [34–36]. However, few studies to date have focused on the individual chemical components such as TF, TF3G, TF3′G, and TFDG (Figure S1), especially in the context of lipid metabolism. Possible reasons for this limited research are as follows: (1) it is easier to evaluate the effects of the whole extract rather than individual components; (2) effective separation of individual components is technically challenging; and (3) fat cells are derived from visceral tissues that benefit from reduction. In this study, we investigated the effects of TF on subcutaneous adipocytes since these cells play an important role in lipid metabolism beneath the dermal layer and are a potential target for anti-aging products.

Adipocytes of murine origin (e.g., 3T3-L1 and 3T3-F442A cells) have been widely used as a model system since they can form LDs within 14 days after differentiation. However, in this study we used hSCF cells as a model of subcutaneous fat cells despite their long period of differentiation, since they are derived from human tissue. hSCF cells required approximately three weeks for differentiation, which was observed at a rate of 60–80%. LDs formed in the differentiated hSCF cells, accompanied by upregulation of lipogenesis-related gene expression (Figure 3A, left panel and Figure 4A).

The effect of CoF-GT on hSCFs differed markedly from that of conventional polyphenols. Unlike F-GT, CoF-GT stimulated lipogenesis in hSCF cells; this effect is presumably distinct from its anti-obesity effects such as lipid disruption or suppression of differentiation (Figure 3). The treatment of F-GT, which increased overall polyphenol content, indicates that TF content ratio in the extract is a critical determinant of LD formation in hSCF cells (Figure 3B, lower panel and Figure 3C).

The hSCFs used in this experiment are responsible for the structure of human skin despite being fat-producing adipocytes, and function as brown adipose tissue to maintain thermogenesis [37–39]. As such, hSCFs are more suitable for anti-skin aging experiments than cells derived from white adipose tissue in visceral fat, which increases with age and is a site of inflammation [40,41]. Cellular changes caused by TF or TFDG treatment were accompanied by altered expression of lipogenesis-related genes and adipokine release (Figure 4). In particular, increased level of PPARγ could stimulate adiponectin with anti-aging properties through inhibiting destruction of extracellular matrix (e.g., type 1 collagen and elastin) in skin [41,42]. The opposing effects of CoF- and F-GT on hSCF cell differentiation are due to the ratio of TF contents. CoF-GT, which has an exceptionally high TF content, stimulates lipogenesis and the formation of LDs (Figure 5).

4. Materials and Methods

4.1. Reagents and Materials

Fresh C. sinensis leaves were harvested between August and September 2017 in Jeju, Korea. *C. coronarium* L. was purchased from a market in Kyungdong, Korea. TFs were from Wako Pure Chemical Industries (Osaka, Japan). Acetonitrile and methanol for chromatography were from Thermo Fisher Scientific (Waltham, MA, USA). Water for analytical high-performance liquid chromatography (HPLC) was from Burdick and Jackson (Morris Plains, NJ, USA). Ultra-pure deionized water for preparative HPLC (18.2 MΩ·cm) was prepared from a Direct-Q system (Merck, Darmstadt, Germany). Dimethylsulfoxide (DMSO) and formic acid were from Sigma-Aldrich (St. Louis, MO, USA). All other chemicals were of analytical grade or higrer.

4.2. Production of TF-Enriched CoF-GT and Conventionally Fermented F-GT Fractions

Fresh GT and GC leaves were washed twice with deionized water and excess water was removed by light tapping. The leaves were soaked in liquid nitrogen and then crushed into a fine powder that was stored at −80 °C. CoF- and F-GT were prepared as follows. CoF-GT was fermented from a mixture of fresh GT powder (100 g) and frozen fresh GC (50 mg). F-GT was also fermented from frozen fresh GT powder (100 g) in a thermal jacket z-blade mixer (IKA, Staufen im Breisgau, Germany) at 37.5 °C. After 3 h incubation, the mixture was extracted with 70% ethyl alcohol (*v/v*) for 2 h. Solid particles/debrises were removed by passage through a 90-mesh sieve followed by a 0.22 μm pore filter (Dow Corning, Corning, NY, USA). Filtered samples were evaporated with a Hei-VAP Rotary Evaporator (Heidolph Instruments, Schwabach, Germany) and then powdered using a FreeZone freeze dryer (Labconco, Kansas City, MO, USA).

4.3. HPLC Analysis of CoF- and F-GT Fractions

Each powered extract (8 g) was dissolved in a mixture of 100 mL DMSO/methanol/ethanol (5:45:50, *v/v*) with 30 min sonication and then centrifuged, followed by filtration through a 0.45 μm polytetrafluoroethylene syringe filter (Pall Corp., Port Washington, NY, USA). An 82 mL volume of filtered sample was injected into ÄKTApurifier 10 (GE Healthcare, Stockholm, Sweden) equipped with a 50 mL sample loop and a photo diode array detector at 275 and 365 nm. Preparative separation was performed with an AQ-HG octadecylsilyl column (120 Å, 10 μm, 20 × 250 mm, column volume = 78.5 mL) (YMC Co., Kyoto, Japan). Gradient elution was carried out with pure water (solvent A) and acetonitrile (solvent B). The flow rate of the mobile phase was 10 mL/min with an injection volume of 8.2 mL. All solvents were filtered, degassed, and maintained under pressure. Fractions (110 mL) were collected after sample injection. Four fractions were prepared for each cycle and collected in separate bottles over 10 injection cycles (Figure S2). Every fourth fraction was evaporated in a Hei-VAP Rotary Evaporator (Heidolph Instruments), freeze dried, and stored at −20 °C until analysis.

4.4. Cell Culture and Differentiation

hSCF cells and subcutaneous pre-adipocyte medium were purchased from ZenBio (Research Triangle Park, NC, USA). The cells were cultured in a humidified 5% CO_2 incubator. To induce differentiation, the cells were cultured in Dulbecco's modified Eagle's medium (Lonza, Walkersville, MD, USA) containing 10% fetal bovine serum (PAA, Pasching, Austria) along with 10 μg/mL insulin, 0.5 mM 3-isobutyl-1-methylxanthine, 1 μM dexamethasone, and 1 μM troglitazone (all from Sigma-Aldrich) for 21 days. The medium was replaced every other day.

4.5. Cell Viability Assay

The viability of hSCF cells was measured with the EZ-Cytox assay kit (Daeil Lab Service, Seoul, Korea) according to the manufacturer's instructions. Briefly, cells were cultured for 7 days and treated with various concentrations of test substance for 24 and 72 h. EZ-Cytox solution (10 μL) was added to each well followed by incubation at 37 °C for 2 h. Absorbance at 450 nm was measured with a spectrophotometer (Synergy H2; BioTek, Winooski, VT, USA). The experiment was performed in triplicate and data are presented as the absorbance value.

4.6. Analysis of Triglyceride (TG) Content by Oil Red O (ORO) Staining

Differentiated hSCF cells were washed twice with cold phosphate-buffered saline (PBS) and fixed with 3.7% formaldehyde (Sigma-Aldrich) for 1 h. The fixed cells were washed with 60% propylene glycol (Sigma-Aldrich) in PBS and then stained with a working solution of Oil Red O (0.7% ORO stock in 60% propylene glycol; Sigma-Aldrich) for 30 min. The cells were washed three times with 85% propylene glycol and rinsed with tap water. Lipid droplets (LDs) stained with ORO dye were visualized with an IX71 microscope (Olympus, Tokyo, Japan).

To quantify LDs, stained cells were washed with 70% ethyl alcohol (Sigma-Aldrich) and the triglyceride (TG)-bound ORO was extracted with 4% Nonidet P-40 (Sigma-Aldrich) in isopropyl alcohol for 20 min. The absorbance of the extract at 490 nm was measured with a spectrophotometer.

4.7. Real-Time Quantitative Plymerase Chain Reaction (RT-qPCR)

Total RNA was extracted using TRIzol reagent (Life Technologies, Carlsbad, CA, USA) according to the manufacturer's instructions, and cDNA was synthesized from approximately 1 μg total RNA using a reverse transcription kit (Promega, Madison, WI, USA). The cDNA was used as a template for RT-qPCR amplification on a 7500 Fast Real-Time PCR System (Life Technologies) using the following TaqMan probes: peroxisome proliferator activated receptor gamma (PPARγ) (#Hs01115513_m1), adiponectin (ADIPOQ; #Hs00605917_m1), and glyceraldehyde 3-phosphate dehydrogenase (GAPDH; #4352339E). Data were obtained from three independent experiments and are presented as fold change relative to the GAPDH level in the sample.

4.8. Enzyme-Linked Immunosorbent Assay (ELISA) for Secreted Adiponectin

hSCFs were treated with various concentrations of TG, TFDG, CoF-GT, and F-GT and differentiated for 21 days. The culture medium was collected and centrifuged at 13,000 rpm for 15 min to remove any debris. Secreted adiponectin level was measured by ELISA using a commercial kit (Enzo Life Sciences, Farmingdale, NY, USA) according to the manufacturer's instructions.

4.9. Statistical Analysis

Data are presented as mean ± SD. One- or two-way ANOVA analysis of variance were used to analyze differences between two and multiple groups, respectively. The threshold for statistical significance was set at $p < 0.05$.

5. Conclusions

The results of this study demonstrate that TF and TF-enriched CoF-GT plays an important role in hSCF cell differentiation due to different biological effects depending on cell types whereas TFDG and conventionally fermented GT inhibits LD synthesis. Our findings suggest that with rapid fermentation and effective trans-dermal delivery, TF and TF-enriched CoF-GT can be an effective agent for preventing human skin aging.

Supplementary Materials: Supplementary materials can be found at http://www.mdpi.com/1422-0067/20/8/2034/s1.

Author Contributions: P.J.P. and C.-S.R. conceived the study and designed the experiments. P.J.P. and C.-S.R. performed the experiments. P.J.P., C.-S.R., and S.T.K. analyzed the data and wrote the paper.

Funding: This research received no external funding.

Conflicts of Interest: The authors declare no conflicts of interest.

Abbreviations

ADIPOQ	Adiponectin of human gene
APN	Adiponectin, secreted form mature adipocytes cytokine
CoF-GT	Co-fermented green tea fraction with high TF contents
F-GT	Conventionally fermented green tea
GC	Garland chrysanthemum
GT	Green tea
PPARγ	Peroxisome proliferator-activator receptor γ
TF	Theaflavin

References

1. Friedman, O. Changes associated with the aging face. *Facial Plast. Surg. Clin. N. Am.* **2005**, *13*, 371–380. [CrossRef] [PubMed]
2. Wollina, U.; Wetzker, R.; Abdel-Naser, M.B.; Kruglikov, I.L. Role of adipose tissue in facial aging. *Clin. Interv. Aging* **2017**, *12*, 2069–2076. [CrossRef]
3. Farage, M.A.; Miller, K.W.; Elsner, P.; Maibach, H.I. Intrinsic and extrinsic factors in skin ageing: A review. *Int. J. Cosmet. Sci.* **2008**, *30*, 87–95. [CrossRef]
4. Südel, K.M.; Venzke, K.; Mielke, H.; Breitenbach, U.; Mundt, C.; Jaspers, S.; Koop, U.; Sauermann, K.; Knussman-Hartig, E.; Moll, I.; et al. Novel aspects of intrinsic and extrinsic aging of human skin: Beneficial effects of soy extract. *Photochem. Photobiol.* **2005**, 581–587. [CrossRef]
5. Sotiropoulou, P.A.; Blanpain, C. Development and homeostasis of the skin epidermis. *Cold Spring Harb. Perspect. Biol.* **2012**, *4*, a008383. [CrossRef]
6. Kruglikov, I.L.; Scherer, P.E. Skin aging: Are adipocytes the next target? *Aging (Albany NY)* **2016**, *8*, 1457–1469. [CrossRef]
7. Ezure, T.; Amano, S. Increment of subcutaneous adipose tissue is associated with decrease of elastic fibres in the dermal layer. *Exp. Dermatol.* **2015**, *24*, 924–929. [CrossRef] [PubMed]
8. Kruglikov, I.L.; Scherer, P.E. General theory of skin reinforcement. *PLoS ONE* **2017**, *12*, e0182865. [CrossRef]
9. Pawlaczyk, M.; Lelonkiewicz, M.; Wieczorowski, M. Age-dependent biomechanical properties of the skin. *Postepy Dermatol. Alergol.* **2013**, *30*, 302–306. [CrossRef]
10. Kim, E.J.; Kim, Y.K.; Kim, J.E.; Kim, S.; Kim, M.K.; Park, C.H.; Chung, J.H. UV modulation of subcutaneous fat metabolism. *J. Investig. Dermatol.* **2011**, *131*, 1720–1726. [CrossRef] [PubMed]
11. Lemieux, I. Energy partitioning in gluteal-femoral fat: does the metabolic fate of triglycerides affect coronary heart disease risk? *Arterioscler. Thromb. Vasc. Biol.* **2004**, *24*, 795–797. [CrossRef]
12. Miranda, P.J.; De Fronzo, R.A.; Califf, R.M.; Guyton, J.R. Metabolic syndrome: Definition, pathophysiology, and mechanisms. *Am. Heart J.* **2005**, *1*, 33–45. [CrossRef]
13. Petrofsky, J.S.; Prowse, M.; Lohman, E. The influence of ageing and diabetes on skin and subcutaneous fat thickness in different regions of the body. *J. Appl. Res.* **2008**, *8*, 55–62.
14. Mukhtar, H.; Ahmad, N. Tea polyphenols: Prevention of cancer and optimizing health. *Am. J. Clin. Nutr.* **2000**, *71*, 1698S–1702S. [CrossRef]
15. Tanaka, T.; Matsuo, Y.; Kouno, I. Chemistry of secondary polyphenols produced during processing of tea and selected foods. *Int. J. Mol. Sci.* **2010**, *11*, 14–40. [CrossRef]
16. Lee, L.S.; Kim, Y.C.; Park, J.D.; Kim, Y.B.; Kim, S.H. Changes in major polyphenolic compounds of tea (*Camellia sinensis*) leaves during the production of black tea. *Food Sci. Biotechnol.* **2016**, *25*, 1523–1527. [CrossRef]
17. Lin, C.L.; Huang, H.C.; Lin, J.K. Theaflavins attenuate hepatic lipid accumulation through activating AMPK in human HepG2 cells. *J. Lipid Res.* **2007**, *48*, 2334–2343. [CrossRef]
18. Pan, H.; Gao, Y.; Tu, Y. Mechanisms of body weight reduction by black tea polyphenols. *Molecules* **2016**, *21*, 1659. [CrossRef]
19. Yamashita, Y.; Wang, L.; Wang, L.; Tanaka, Y.; Zhang, T.; Ashida, H. Oolong, black and pu-erh tea suppresses adiposity in mice via activation of AMP-activated protein kinas. *Food Funct.* **2014**, *5*, 2420–2429. [CrossRef]
20. Subramanian, N.; Venkatesh, P.; Ganguli, S.; Sinkar, V.P. Role of polyphenol oxidase and peroxidase in the generation of black tea theaflavins. *J. Agric. Food Chem.* **1999**, *47*, 2571–2578. [CrossRef]
21. Li, S.; Lo, C.Y.; Pan, M.H.; Lai, C.S.; Ho, C.T. Black tea: Chemical analysis and stability. *Food Funct.* **2013**, *4*, 10–18. [CrossRef]
22. Hamdaoui, M.H.; Snoussi, C.; Dhaouadi, K.; Fattouch, S.; Ducroc, R.; Le Gall, M.; Bado, A. Tea decoctions prevent body weight gain in rats fed high-fat diet; black tea being more efficient than green tea. *J. Nutr. Intermed. Metab.* **2016**, *6*, 33–40. [CrossRef]
23. Nomura, M.; Ma, W.Y.; Huang, C.; Yang, C.S.; Bowden, G.T.; Miyamoto, K.I.; Dong, Z. Inhibition of ultraviolet B-induced AP-1 activation by theaflavins from black tea. *Mol. Carcinog.* **2000**, *28*, 148–155. [CrossRef]
24. Yamaoka, Y.; Ohguchi, K.; Itoh, T.; Nozawa, Y.; Akao, Y. Effects of theaflavins on melanin biosynthesis in mouse B16 melanoma cells. *Biosci. Biotechnol. Biochem.* **2009**, *73*, 1429–1431. [CrossRef]

25. Kim, Y.C.; Choi, S.Y.; Park, E.Y. Anti-melanogenic effects of black, green, and white tea extracts on immortalized melanocytes. *J. Vet. Sci.* **2015**, *16*, 135–143. [CrossRef]
26. Pou, K.J. Fermentation: The key step in the processing of black tea. *J. Biosyst. Eng.* **2016**, *41*, 85–92. [CrossRef]
27. Obanda, M.; Owuor, P.O.; Mang'oka, R. Changes in the chemical and sensory quality parameters of black tea due to variations of fermentation time and temperature. *Food Chem.* **2001**, *75*, 395–404. [CrossRef]
28. Torre-Villalvazo, I.; Bunt, A.E.; Alemán, G.; Marquez-Mota, C.C.; Diaz-Villaseñor, A.; Noriega, L.G.; Estrada, I.; Figueroa-Juárez, E.; Tovar-Palacio, C.; Rodriguez-López, L.A.; et al. Adiponectin synthesis and secretion by subcutaneous adipose tissue is impaired during obesity by endoplasmic reticulum stress. *J. Cell. Biochem.* **2018**, *119*, 5970–5984. [CrossRef]
29. Khan, N.; Mukhtar, H. Tea polyphenols for health promotion. *Life Sci.* **2007**, *81*, 519–533. [CrossRef]
30. Chacko, S.M.; Thambi, P.T.; Kuttan, R.; Nishigaki, I. Beneficial effects of green tea: A literature review. *Chin. Med.* **2010**, *5*, 13. [CrossRef]
31. Khan, N.; Mukhtar, H. Tea and health: Studies in humans. *Curr. Pharm. Des.* **2013**, *19*, 6141–6147. [CrossRef]
32. Meydani, M.; Hasan, S.T. Dietary polyphenols and obesity. *Nutrients* **2010**, *2*, 737–751. [CrossRef]
33. Wang, S.; Moustaid-Moussa, N.; Chen, L.; Mo, H.; Shastri, A.; Su, R.; Bapat, P.; Kwun, I.; Shen, C.L. Novel insights of dietary polyphenols and obesity. *J. Nutr. Biochem.* **2014**, *25*, 1–8. [CrossRef]
34. Sajilata, M.G.; Bajaj, P.R.; Singhal, R.S. Tea polyphenols as nutraceuticals. *Compr. Rev. Food Sci. Food Saf.* **2008**, *7*, 229–254. [CrossRef]
35. Higdon, J.V.; Frei, B. Tea catechins and polyphenols: Health effects, metabolism, and antioxidant functions. *Crit. Rev. Food Sci. Nutr.* **2003**, *43*, 89–143. [CrossRef]
36. Lin, J.K.; Lin-Shiau, S.Y. Mechanisms of hypolipidemic and anti-obesity effects of tea and tea polyphenols. *Mol. Nutr. Food Res.* **2006**, *50*, 211–217. [CrossRef]
37. Cannon, B.; Nedergaard, J. Brown adipose tissue: Function and physiological significance. *Physiol. Rev.* **2004**, *84*, 277–359. [CrossRef]
38. Wu, J.; Boström, P.; Sparks, L.M.; Ye, L.; Choi, J.H.; Giang, A.H.; Khandekar, M.; Virtanen, K.A.; Nuutila, P.; Schaart, G.; et al. Beige adipocytes are a distinct type of thermogenic fat cell in mouse and human. *Cell* **2012**, *150*, 366–376. [CrossRef]
39. Kraemer, F.B.; Shen, W.J. Hormone-sensitive lipase: Control of intracellular tri-(di-)acylglycerol and cholesteryl ester hydrolysis. *J. Lipid Res.* **2002**, *43*, 1585–1594. [CrossRef]
40. Oikarinen, A. Connective tissue and aging. *Int. J. Cosmet. Sci.* **2004**, *26*, 107. [CrossRef]
41. Madison, K.C. Barrier function of the skin: "la raison d'être" of the epidermis. *J. Investig. Dermatol.* **2003**, *121*, 231–241. [CrossRef]
42. Kim, S.O.; Han, Y.; Ahn, S.; An, S.; Shin, J.C.; Choi, H.; Kim, H.J.; Park, N.H.; Kim, Y.J.; Jin, S.H.; et al. Kojyl cinnamate esters are peroxisome proliferator-activated receptor α/γ dual agonists. *Bioorg. Med. Chem.* **2018**, *26*, 5654–5663. [CrossRef] [PubMed]

International Journal of
Molecular Sciences

Review

Natural Product Mediated Regulation of Death Receptors and Intracellular Machinery: Fresh from the Pipeline about TRAIL-Mediated Signaling and Natural TRAIL Sensitizers

Durray Shahwar [1], Muhammad Javed Iqbal [2,3], Mehr-un Nisa [4], Milica Todorovska [5], Rukset Attar [6], Uteuliyev Yerzhan Sabitaliyevich [7], Ammad Ahmad Farooqi [1], Aamir Ahmad [8] and Baojun Xu [9,*]

[1] Laboratory for Translational Oncology and Personalized Medicine, Rashid Latif Medical College, Lahore 54840, Pakistan; durrayshahwar.kh@gmail.com (D.S.); ammadfarooqi@rlmclahore.com (A.A.F.)

[2] Department of Biochemistry and Biotechnology, University of Gujrat, Gujrat 34120, Pakistan; m.javediqbal@uog.edu.pk

[3] Department of Biotechnology, University of Sialkot, Sialkot 51310, Pakistan

[4] National Institute of Food Science and Technology, UAF, Faisalabad 38000, Pakistan; mehru912@gmail.com

[5] Department of Pharmacy, Medical Faculty, University of Niš, Dr Zorana Djindjica Boulevard 81, 18000 Niš, Serbia; mimatod@gmail.com

[6] Department of Obstetrics and Gynecology, Yeditepe University, Ataşehir/İstanbul 34755, Turkey; ruksetattar@hotmail.com

[7] Kazakhstan Medical University "KSPH", Almaty 050060, Kazakhstan; e.uteuliyev@ksph.kz

[8] Department of Oncologic Sciences, Mitchell Cancer Institute, University of South Alabama, Mobile, AL 36604, USA; aahmad@health.southalabama.edu

[9] Food Science and Technology Program, Beijing Normal University-Hong Kong Baptist University United International College, Zhuhai 519087, Guangdong, China

* Correspondence: baojunxu@uic.edu.hk; Tel.: +86-0756-3620636

Received: 26 March 2019; Accepted: 18 April 2019; Published: 24 April 2019

Abstract: Rapidly developing resistance against different therapeutics is a major stumbling block in the standardization of therapy. Tumor necrosis factor (TNF)-related apoptosis-inducing ligand (TRAIL)-mediated signaling has emerged as one of the most highly and extensively studied signal transduction cascade that induces apoptosis in cancer cells. Rapidly emerging cutting-edge research has helped us to develop a better understanding of the signaling machinery involved in inducing apoptotic cell death. However, excitingly, cancer cells develop resistance against TRAIL-induced apoptosis through different modes. Loss of cell surface expression of TRAIL receptors and imbalance of stoichiometric ratios of pro- and anti-apoptotic proteins play instrumental roles in rewiring the machinery of cancer cells to develop resistance against TRAIL-based therapeutics. Natural products have shown excellent potential to restore apoptosis in TRAIL-resistant cancer cell lines and in mice xenografted with TRAIL-resistant cancer cells. Significantly refined information has previously been added and continues to enrich the existing pool of knowledge related to the natural-product-mediated upregulation of death receptors, rebalancing of pro- and anti-apoptotic proteins in different cancers. In this mini review, we will set spotlight on the most recently published high-impact research related to underlying mechanisms of TRAIL resistance and how these deregulations can be targeted by natural products to restore TRAIL-mediated apoptosis in different cancers.

Keywords: apoptosis; cancer; death receptors; natural products; TRAIL

1. Introduction

Cancer is a multifaceted and therapeutically challenging disease. Groundbreaking discoveries in the past few decades have enabled us to develop a sharper understanding of intra- and inter-tumor heterogeneity, loss of apoptosis, oncogenic overexpression, inactivation of tumor suppressors, and deregulation of spatio-temporally controlled signal transduction cascades which play a central role in cancer onset and progression [1–4]. Loss of apoptosis has been, and still remains, a subject of intense discussion for basic and clinical scientists. Cutting-edge research has sequentially revealed that the inhibition of cell death in combination with mitogenic oncogenes promoted cancer in animal models [5]. There were some exciting developments which highlighted that many oncogenic pathways inhibited apoptosis, whereas tumor suppressors (e.g., p53) played an instrumental role in the induction of apoptosis [6]. More importantly, approval of BH3 (BCL-2 homology domain-3) mimetics by the Food and Drug Administration (FDA) for the treatment of 17p-deleted CLL is a milestone in the field of molecular therapy. Therefore, researchers have focused on the identification of pathways and proteins that can efficiently induce apoptosis and simultaneously induce regression in xenografted mice. There has always been a quest to identify molecules that can cause maximum damage to cancer cells while leaving normal cells intact. In accordance with this concept, TRAIL (TNF-related apoptosis-inducing ligand) has emerged as a scientifically approved protein that can induce apoptosis, specifically in cancer cells. Initial findings reported by researchers were tremendously encouraging, and urged cotemporary scientists to further dissect this intriguing and therapeutically important pathway. Consequently, substantial excitement encompasses the premium potential of natural products to effectively restore apoptosis in TRAIL-resistant cancers.

There are some detailed reviews about TRAIL-mediated signaling in different cancers, but we have exclusively focused on the most recent evidence related to positive and negative regulators of TRAIL-mediated signaling in this review. We have also summarized how different natural products effectively restored apoptosis in TRAIL-resistant cancers. Before providing an overview of the natural product-mediated regulation of the TRAIL-driven pathway, we will discuss some of the most important advancements in the TRAIL-mediated signaling pathway.

2. Molecular Insights of TNF-Related Apoptosis-Inducing Ligand (TRAIL)-Mediated Signaling

Overwhelmingly, increasing and continuously upgrading discoveries in the field of apoptotic cell death have enabled us to develop a better knowledge of this area. Researchers have extensively characterized two primary death-signaling cascades, the extrinsic and intrinsic pathways. When the signaling is "switched on", it results in the activation of downstream effector molecules for both pathways that mediate apoptosis. Caspases belong to a group of cysteine proteases which proteolytically process a variety of cytoplasmic and nuclear substrates [7]. The extrinsic, or death -receptor-mediated, pathway is activated and functionalized through the binding of death ligands such as TRAIL, TNFα (tumor necrosis factor α) and Fas ligand (FasL) to specific receptors (e.g., TRAILR1/DR4, TRAIL2/DR5, TNFR, Fas). Ligand–receptor interaction results in the recruitment of the cytoplasmic adaptor protein FADD (Fas-associated protein with death domain) to death domains present in the cytoplasmic segment of the death receptor (shown in Figure 1). Death domains present in death receptors served as recruiting modules and heterodimerized with the death domains of client proteins. FADD contains death domains that can link the death receptor to procaspase-8 to form a death-inducing signaling complex (DISC).

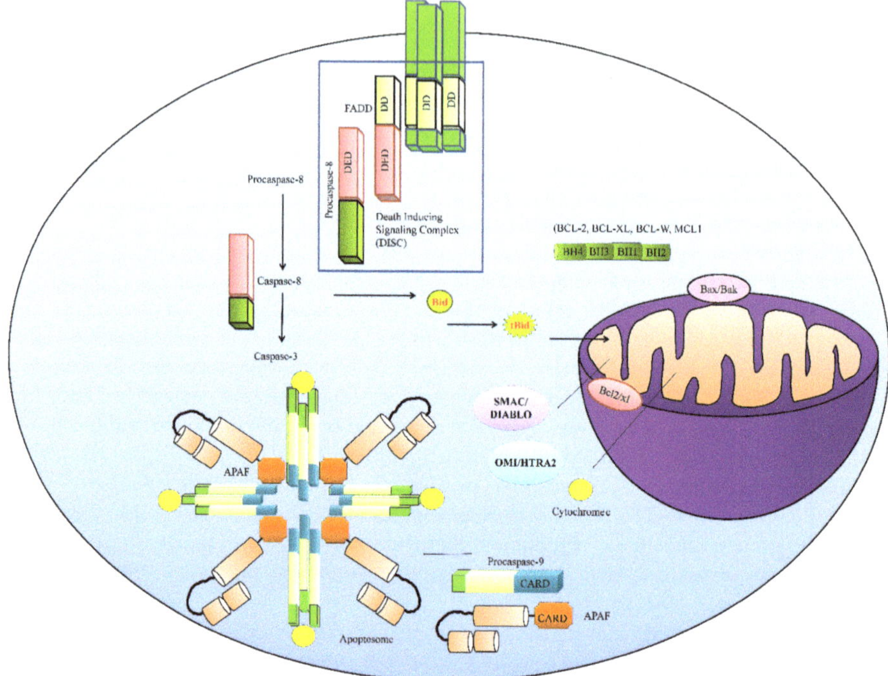

Figure 1. TRAIL (TNF-related apoptosis-inducing ligand)-mediated signaling. TRAIL transduces the signals intracellularly through death receptors. Death-inducible signaling complex (DISC) is formed by the interaction of the death receptor Fas-associated protein with death domain (FADD) and pro-caspase-8. Formation of DISC is necessary for the activation of caspase-8. Caspase-8 activates its downstream effector caspase-3. However, caspase-8 may also proteolytically process Bid to initialize the intrinsic pathway. The intrinsic pathway is triggered following entry of truncated Bid into mitochondria. Cytochrome c and SMAC are released from mitochondria and an apoptosome was formed in the cytoplasm by the assembly of apoptotic protease activating factor (APAF), cytochrome c, and pro-caspase-9. The apoptosome is necessary for the activation of caspase-9 and it can further activate caspase-3 to induce apoptosis in cancer cells. In healthy cells, APAF is present as an autoinhibitory monomer. However, mitochondrial outer membrane permeabilization (MOMP) and subsequent release of cytochrome c unlocks APAF.

Activation of the mitochondrial or intrinsic pathway is induced either through radiation or chemotherapy. Caspase-8-mediated truncation of Bid also played a dominant role in activating the mitochondrially driven pathway. Translocation of truncated Bid into mitochondria caused mitochondrial permeabilization and release of apoptogenic proteins, including cytochrome c and second mitochondrial-derived activator of caspases (SMAC) from the mitochondria into the cytosol. Cytosolic cytochrome c interacted with apoptotic protease activating factor-1 (APAF1) and formed a multimeric complex termed the apoptosome (shown in Figure 1). The apoptosome recruited, cleaved, and activated caspase-9 and -3. SMAC promoted apoptosis by binding to and degrading multiple inhibitors of apoptotic proteins (IAPs): XIAP, c-IAP1, and c-IAP2. Therefore, it seems clear that TRAIL-mediated signaling following the activation of caspase-8 is dichotomously branched. Either caspase-8 can activate its downstream effector caspases or it can proteolytically process the Bid protein to initialize the intrinsic pathway that routes through mitochondria. It appears to be important that the TRAIL-induced intracellular signaling pathway is not as simple as previously surmised. A wealth of information points towards myriad signaling pathways which crosstalk with different proteins of the TRAIL-mediated signaling pathway, and play a critical role. Therefore, we partitioned our review into negative and positive regulators of TRAIL-mediated signaling to comprehensively analyze the most recent breakthroughs made in uncovering mechanisms that inhibit or potentiate TRAIL-triggered apoptotic cell death.

3. Negative Regulators of TRAIL-Mediated Signaling

Because TRAIL-based therapies have entered into various phases of clinical trials, it is essential to drill down deep into TRAIL-mediated signaling and further unfold the mysterious aspects of this pathway. The overexpression and stabilization of anti-apoptotic proteins is necessary to induce resistance against TRAIL. Various lines of evidence have revealed that cancer cells developed resistance against TRAIL mainly through the overexpression and stabilization of anti-apoptotic proteins. We have partitioned this section into modes and strategies used by cancer cells to develop resistance against TRAIL. (1) Stabilization of anti-apoptotic proteins; (2) Noncoding RNAs also play their role in rewiring cell-signaling pathways to potentiate TRAIL resistance; (3) Ubiquitination is a very critical mechanism in the context of anti-apoptotic and pro-apoptotic protein degradations. It has been seen that pro-apoptotic proteins are degraded but anti-apoptotic proteins escape from these death-tagging molecules in TRAIL-resistant cancer cells; (4) Recent discoveries have unmasked unique locations and functionalities of death receptors. Death receptors not only activate classical pathways to induce apoptosis, but their movement has also been tracked in the nucleus to modulate miRNA biogenesis. Therefore, we will discuss these interesting topics in the upcoming section and try to critically evaluate their implications.

GADD34 (growth arrest and DNA damage-inducible protein 34) overexpression resulted in an increase in MCL-1 levels. Knockdown of GADD34 resulted in marked reduction in MCL-1 levels in both SMMC-7721 and HepG2 cells [8]. GADD34 suppressed proteasomal degradation of MCL-1 mainly through promoting extracellular signal-regulated kinase (ERK1/2)-mediated phosphorylation of proline (P), glutamic acid (E), serine (S), and threonine (T) (PEST) domains in MCL-1 (shown in Figure 2). TNF receptor associated factor 6 (TRAF6) played a central role in directing ERK1/2 phosphorylation and the enhancement of the stability of MCL-1 levels in HepG2 cells. GADD34 knockdown exerted repressive effects on the levels of MCL-1 in SMMC-7721 and HepG2 cells. Furthermore, TRAF6 knockdown also enhanced TRAIL-mediated apoptosis, as evidenced by considerably reduced levels of p-ERK1/2 and MCL-1 [8].

CIB1 (calcium and integrin-binding protein 1) played a contributory role in the development of resistance against chemotherapeutics and TRAIL [9]. Intriguingly, expression levels of DR5 were noted to be dramatically enhanced in CIB1-depleted MDA-436 breast cancer cells [9]. These findings appear to be exciting, but scientists have not provided a comprehensive pathway opted by CIB1 to inhibit DR5. There is an urgent need to put the missing pieces of information together to uncover the underlying mechanisms which inhibit, repress, or degrade death receptors.

Ovarian adenocarcinoma-amplified lncRNA (OVAAL) has been shown to play an instrumental role in the development of resistance against TRAIL [10]. OVAAL interacted with STK3 (serine/threonine-protein kinase-3), which consequently enhanced the structural association between Raf-1 and STK3. Studies have shown that the ternary complex OVAAL/STK3/Raf-1 activated rapidly accelerated fibrosarcoma/mitogen-activated protein kinase kinase (RAF/MEK/ERK signaling pathway and promoted Mcl-1-mediated survival and c-Myc-driven proliferation of the ME4405 and HCT116 cells. c-Myc has also been noted to transcriptionally upregulate OVAAL (Figure 2). However, expectedly, TRAIL-mediated apoptosis was considerably enhanced in OVAAL-silenced ME4405 cells [10]. These findings clearly indicate that non-coding RNAs stabilize anti-apoptotic proteins via the rewiring of signaling pathways.

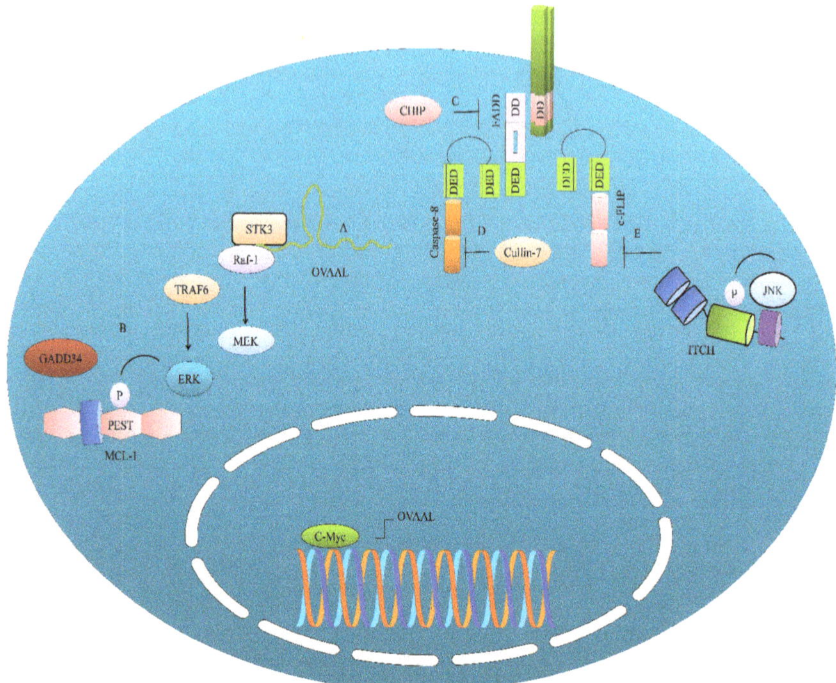

Figure 2. Negative regulation of the TRAIL-driven pathway. (**A**) OVAAL, a long non-coding RNA, has been shown to promote interaction between STK3 and Raf-1. STK3 and Raf-1 work synchronously and activate the MEK/ERK pathway; (**B**) ERK has also been shown to stabilize MCL-1. ERK is also activated by TRAF6 and GADD34 to stabilize MCL-1. OVAAL is transcriptionally upregulated by c-Myc; (**C**) Post-translational modifications have been shown to effectively regulate caspase-8 and FADD. FADD is negatively regulated by C terminus HSC70-interacting protein (CHIP); (**D**) Caspase-8 is negatively regulated by Cullin-7; (**E**) ITCH involved in the negative regulation of c-FLIP.

Detailed mechanistic insights revealed that cytoplasmic PARP-1 was recruited into the TRA-8-activated DISC and sustained Src-mediated pro-survival signals [11]. However, the knockdown of PARP-1 not only interfered with the activation of Src but also improved TRA-8-mediated apoptotic cell death in BxPc-3 and MiaPaCa-2 cells [11].

Cullin-7 has recently been shown to physically interact with caspase-8 [12]. CUL7 prevented the activation of caspase-8 by promoting post-translational modifications of caspase-8 by the addition of non-degradative polyubiquitin chains at the 215th lysine (shown in Figure 2). Knockdown of CUL7 re-sensitized cancer cells to TRAIL-triggered apoptotic cell death. Tumor growth was significantly inhibited in mice xenografted with CUL7-silenced MDA-MB-231 cells [12]. CHIP (C terminus HSC70-interacting protein) induced the K6-linked polyubiquitylation of FADD and suppressed the formation of the DISC (Figure 2) [13].

Monocyte chemotactic protein-induced protein-1 (MCPIP1), a deubiquitinating enzyme, promoted the lysosomal degradation of DR5 [14]. MCPIP1 knockdown facilitated DISC formation [14]. At a more basic level, it would be extremely interesting to see how different proteins sort death receptors as well as pro-apoptotic and anti-apoptotic proteins for degradation in different cancers.

Karyopherin β1 (KPNB1) played an instrumental role in the nuclear import of DR5. KPNB1 transported DR5 into the nucleus, while inhibition of KPNB1 restored DR5 levels on the cell surface of glioblastoma cells [15]. These findings are exciting, and it needs to be seen how DR5 behaves in the nucleus. Certain hints have emerged which have scratched the surface of regulatory role of DR5 in the biogenesis of microRNAs. Nuclear DR5 inhibited the maturation of a miRNA, let-7, in pancreatic cancer cells and increased their proliferation abilities [15]. Astonishingly, two functional nuclear localization signal (NLS) sequences have previously been identified in DR5 [16]. Additionally, it was shown that importin-β1 interacted with DR5 and shipped it to the nucleus in HeLa cells [16]. It will not be invalid if we say that shuttling of death receptors in the nucleus is the "tip of the iceberg" and needs detailed and in-depth research. At the moment, we have a segmented view about the internalization of death receptors from the cell surface and "moonlight activity" of death receptors in the nucleus.

It has recently been convincingly revealed that circulating tumor cells (CTCs) demonstrated rapid autophagic flux, characterized by an accumulation of autophagosome organelles [17]. Notably, there was substantial evidence highlighting the presence of DR5 in the autophagosomes, followed by degradation by lysosomes [17]. Overall, these findings clearly suggest that CTCs escape from TRAIL-mediated killing activities by reducing the cell surface expression of DR5.

4. Positive Regulators of TRAIL-Mediated Signaling

RUNX3 (RUNT-related transcription factor-3) played a central role in stimulating the expression of DR5 [18]. DR5 was markedly increased in RUNX3-overexpressing HT29 cells. RUNX3-mediated reactive oxygen species (ROS) can lead to an enhanced ER stress in cancer cells. Therefore, the overexpression of RUNX3 induced DR5 via IRE1α-JNK-CHOP pathway. Treatment of the cells with an ROS scavenging chemical, *N*-acetyl-L-cysteine, severely compromised TRAIL-mediated apoptosis in RUNX3-overexpressing cancer cells [18]. Superoxide dismutases (SODs) constitute the antioxidant defense grid. RUNX3 has been shown to transcriptionally repress SOD3 to induce ROS generation and upregulate DR5. Mechanistically it has been shown that RUNX3 occupied RUNX3 binding sites present within the promoter region of SOD3 and inhibited its transcription [18]. The findings of this study are exciting, and future studies must converge on the analysis of the role of RUNX3 in different cancers. It will be informative to see if RUNX3 is functionally active in other TRAIL-resistant cancers.

Fucosylation is a post-translational modification of critical importance that plays a crucial role in improving TRAIL-mediated apoptosis [19]. Stably and transiently overexpressed FUT3 and FUT6 dramatically enhanced TRAIL-mediated apoptosis in HCT116 and DLD-1 cells. Activation and cleavage mechanisms of caspase-8 and PARP-1 were noted to be more pronounced in cells overexpressing FUT6 and FUT3. More importantly, a significantly higher fraction of signalosomes was noticed, as evidenced by highly increased DISC-associated caspase-8 complexes in FUT3-overexpressing cells [19]. Harakiri (HRK), a BH3-only protein of the Bcl-2 family, has been shown to promote TRAIL-mediated apoptosis in glioblastoma cells [20].

ITCH, a homologous to the E6AP carboxyl terminus (HECT) domain E3 ligase, is reportedly involved in the negative regulation of c-FLIP [21]. JNK phosphorylation sites have been mapped in the protein sequence of ITCH. Levels of phosphorylated ITCH (p-ITCH) were found to be enhanced in MCF-7- and T47D-derived tamoxifen-resistant and faslodex-resistant cells. Moreover, inhibition of JNK resulted in the inactivation of ITCH and reduced p-ITCH levels, and consequently c-FLIP levels were restored in cancer cells [21,22]. Therefore, in future studies it will be paramount to investigate if additional kinases are involved in the activation of ITCH and if ITCH can post-translationally modify various other negative regulators of apoptosis in different cancers.

5. Natural-Product-Mediated Restoration of TRAIL-Mediated Apoptosis in Different Cancers

Natural products have captivated tremendous attention because of their premium pharmacological properties. There has been a longstanding quest to identify products that can be combined with TRAIL to maximize the apoptosis in TRAIL-resistant cancers. Therefore, in this specific section, we provide an update about products obtained from natural sources which can restore apoptosis in resistant cancer cells.

Auriculasin, a prenylated isoflavone, was highly effective when used in combination with TRAIL [23]. Auriculasin and TRAIL combinatorially increased the expression of Bax, AIF, endo G, and cytochrome c. Auriculasin triggered the upregulation of DR5 but levels of DR4 remained unchanged [23]. Cannabidiol, pharmacologically potent cannabinoid isolated from the *Cannabis* plant, induced the upregulation of the protein and cell surface expression of TRAIL-R2/DR5 [24].

Andrographolide isolated from *Andrographis paniculata* enhanced the expression of DR4 and DR5 mainly through increasing the levels of p53 [25]. As expected, andrographolide-mediated upregulation of DR4 and DR5 was not observed in p53 knockdown T24 cells [25].

Periplocin obtained from Cortex Periplocae led to a dose-dependent enhancement of the expression levels of DR4 and DR5 on the surface of MGC-803 and SGC-7901 cells [26]. Periplocin and TRAIL combinatorially enhanced the levels of p-ERK1/2 and EGR1 (early growth response-1). However, as expected, treatment with an inhibitor of MEK (PD98059) severely interfered with the upregulation of DR4 and DR5 in cancer cells. EGR1 overexpression induced the stimulation of DR4 and DR5, while EGR1 knockdown exerted repressive effects on the expression levels of DR4 and DR5 [26]. Weights and volumes of tumors from mice treated combinatorially with TRAIL and periplocin were found to be significantly reduced as compared to mice treated with periplocin or TRAIL alone [26].

In combination with TRAIL, the sesquiterpene coumarin galbanic acid, worked effectively against non-small-cell lung cancer cells [27]. Galbanic acid and TRAIL induced the upregulation of DR5 and simultaneously suppressed DcR1. Galbanic acid and TRAIL attenuated the expression of Bcl-2 (B-cell lymphoma-2), Bcl-xL (B-cell lymphoma-extra-large) and XIAP (X-linked inhibitor of apoptotic proteins) in H460/R cells (Figure 3) [28].

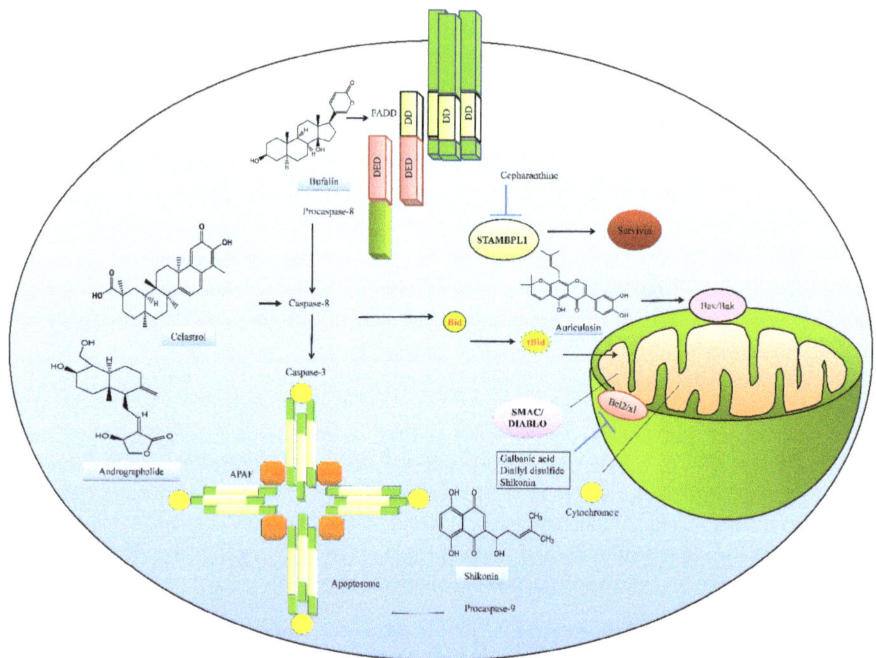

Figure 3. Natural-product-mediated multi-step regulation of the TRAIL-driven pathway.

Diallyl disulfide (DADS) and TRAIL jointly repressed Bcl-2 in colorectal cancer cells [28]. Bufalin, a cardiotonic steroid isolated from the secretion of parotid glands and skin of Chansu and black-spectacled toad has recently been shown to restore apoptosis in TRAIL-resistant cancer cells by the regulation of pro- and anti-apoptotic proteins [29]. Bufalin increased ER stress associated proteins (GRP78, caspase-4, IRE-1α, IRE-1β, ATF-6α, GADD153, and Calpain 1). Bufalin increased Bax; cytochrome c; caspase-3, -8, and -9; AIF; and Endo G, but simultaneously reduced Bcl-2 in NPC-TW 076 cells. Furthermore, bufalin elevated the expression levels of TRAIL, FADD, DR4, and DR5 [29].

Shikonin is a medicinally important product isolated from the root of *Lithospermum erythrorhizon* that was shown to work effectively with TRAIL against A549 cells [30]. Shikonin and TRAIL synergistically reduced Mcl-1, Bcl-2, Bcl-xL, XIAP, and c-FLIP and upregulated the levels of Bid (Figure 3) [30]. Celastrol, a triterpenoid isolated from *Tripterygium wilfordii* worked effectively with TRAIL and inhibited autophagic influx in A549 cells [31].

Remarkable advancements have been made in the molecular biology of autophagy, and scientists are focusing on solving questions regarding how this pathway can be harnessed to improve clinical outcomes [32,33]. In 2016 the scientific community acknowledged the true potential of autophagy in health and disease, and Yoshinori Ohsumi was awarded the Nobel Prize for Physiology or Medicine for his outstanding work, which elevated our understanding about intricate mechanisms of autophagy to the next level. Autophagy plays a central role in the maintenance of homeostasis. Autophagy begins when double-membrane autophagosomes engulf fractions of the cytoplasm, which is followed by the fusion of these vesicles with lysosomes, and autophagic contents are consequently degraded [33,34]. Phosphatidylethanolamine (PE) is conjugated to cytoplasmic LC3I to generate the lipidated form, LC3II, and consequently LC3II is incorporated into the growing membrane [34].

Celastrol treatment induced an increase in the levels of LC3II and p62. However, authors did not report the effects induced by combinatorial treatment with celastrol and TRAIL [31]. It was only suggested that celastrol and TRAIL synergistically enhanced ROS generation in cancer cells. Various other natural products, particularly 6-shogaol, were also found to increase the levels of LC3II and p62 in Huh7 cells [35].

2-Deoxy-D-glucose potentiated TRAIL-triggered apoptotic cell death, in part through suppressing JNK-mediated cytoprotective autophagic signaling in SGC7901 and MGC803 cells [36]. Chloroquine and TRAIL synergistically enhanced LC3II levels in pancreatic cancer cells [37].

Excitingly, flow cytometry analyses revealed that toosendanin induced a reduction of membrane DR5 [38]. However, these effects were prevented by inhibitors of autophagy (3-methyladenine). 3-Methyladenine increased the basal level of membrane DR5, which clearly indicated that autophagy centrally regulated the membrane distribution of DR5 [38,39].

It is important to mention that autophagy has a dualistic role in TRAIL-mediated apoptosis. There is also sufficient evidence related to the positive regulation of TRAIL-mediated apoptosis by autophagy. Different natural products have also been shown to potentiate TRAIL-mediated apoptosis through the induction of autophagy. Ginsenosides are biologically active constituents of ginseng, and potentiate TRAIL-mediated apoptosis through the induction of autophagy [40]. Juglanin also induced autophagy and consequently enhanced TRAIL-mediated apoptosis in cancer cells. Juglanin induced the regression of tumors in mice subcutaneously injected with A549 cells [41].

Ursolic acid stimulated the expression of DR4 and DR5 and simultaneously downregulated c-FLIP$_L$ and re-sensitized TRAIL-resistant triple-negative breast cancer cells to apoptosis [42].

Cepharanthine, a biscoclaurine alkaloid isolated from *Stephania cepharantha*, has been shown to be effective against renal carcinoma cells [43]. Cepharanthine time-dependently induced the downregulation of survivin protein levels. It has been mechanistically demonstrated that cepharanthine promoted c-FLIP degradation. It was observed that use of proteasome inhibitor reversed cepharanthine-induced c-FLIP degradation. STAMBPL1 is one of the JAB1/MPN/Mov34 metalloenzymes (JAMM) deubiquitin enzymes, and is reportedly involved in the regulation of different processes. Cepharanthine dose-dependently downregulated STAMBPL1 and increased USP53 expression. Ectopic expression of STAMBPL1 significantly inhibited cepharanthine-induced reduction in the levels of survivin [43]. Overall, these findings clearly suggest that cepharanthine enhances TRAIL-induced apoptosis by promoting the degradation of survivin through STAMBPL1 downregulation in renal carcinoma cells (Figure 3).

C-27-carboxylated oleanolic acid derivatives have been shown to stimulate the expression of DR5 through CHOP [44]. Treatment of glioma U251MG and LN428 cells with 3β-hydroxyolean-12-en-27-oic acid (C27OA-1) induced an increase in the expression of CHOP. C27OA-1 considerably activated p38 and ERK (extracellular signal regulated kinase). Treatment of U251MG cells with p38 inhibitors (SB203580) severely abrogated C27OA-1-mediated increase in expression levels of CHOP and DR5 [44].

Lambertianic acid is a biologically active product isolated from *Pinus koraiensis* that was efficient against non-small-cell lung cancer. Lambertianic acid and TRAIL upregulated DR4. Furthermore, lambertianic acid and TRAIL markedly reduced p-NF-κB, p-IκB, and c-FLIP in A549 and H1299 cells [45].

6. Concluding Remarks

The cancer genome atlas (TCGA) network groups have comprehensively reported the genomic landscape for over 30 different cancer types [46]. More importantly, many of these malignancies have a subset of cases which harbored genomic alterations in components of extrinsic or intrinsic pathways, including overexpression and amplification of FADD and IAP (inhibitor of apoptotic proteins), as well as the identification of mutations in caspase-encoding genes [47,48].

It seems surprising to note that although scientists have uncovered tremendous information about the TRAIL-mediated signaling pathway, we have not sufficiently investigated the natural-product-mediated regulation of the TRAIL-driven pathway. Even though we have seen that natural products triggered the upregulation of death receptors and induced the re-balancing of pro- and anti-apoptotic proteins, we still have unanswered questions and visible knowledge gaps in our understanding related to the realization of products derived from medicinally important natural sources.

Different ubiquitin ligases (e.g., MARCH8) have been shown to ubiquitinate DR4 at 273rd lysine and induce degradation [49]. CHIP (C terminus HSC70-interacting protein) induced the K6-linked polyubiquitylation of FADD and suppressed the formation of the DISC [13]. CUL7 prevented the activation of caspase-8 mainly by promoting the modification of caspase-8 with non-degradative polyubiquitin chains [12]. However, the natural-product-mediated targeting of ubiquitin ligases to restore TRAIL-mediated apoptosis is an insufficiently studied area of research. There is a need to focus on the ubiquitin-ligase-targeting abilities of natural products which can later be used effectively in TRAIL-resistant cancers.

Another important and exciting area of research that needs detailed research is the microRNA regulation of the TRAIL-driven pathway. Rapidly emerging scientific reports have started to shed light on the central role of microRNAs in the modulation of the TRAIL-mediated pathway. Certain pieces of evidence have suggested that maritoclax, isolated from marine bacteria, promoted miR-708-mediated targeting of c-FLIP and restored apoptosis [50]. Interestingly, α-mangostin, a xanthone isolated from the mangosteen fruit, restored apoptotic death in TRAIL-resistant colon cancer DLD-1 cells [51]. A-Mangostin effectively promoted DR5 oligomerization. A-Mangostin exerted repressive effects on miR-133b and stimulated the expression of DR5 [51].

However, we have not yet witnessed considerable experimental work related to the natural-product-mediated regulation of miRNAs to restore apoptosis in TRAIL-resistant cancers. Likewise, different xenografted mice model studies are necessary for an effective evaluation of the potential of natural products in inducing tumor regression.

Conflicts of Interest: The authors declare no conflict of interest.

Abbreviations

APAF1	apoptotic protease activating factor-1
Bcl-2	B-cell lymphoma-2
Bcl-xL	B-cell lymphoma-extra-large
CHIP	C terminus HSC70-interacting protein
CIB1	calcium and integrin-binding protein 1
CTCs	circulating tumor cells
DISC	death-inducing signaling complex
EGR1	early growth response-1
FADD	Fas-associated protein with death domain
GADD	growth arrest and DNA damage-inducible protein
IAP	inhibitor of apoptotic proteins
OVAAL	ovarian adenocarcinoma-amplified lncRNA
RUNX3	RUNT-related transcription factor-3
SMAC	second mitochondrial-derived activator of caspases
SOD	superoxide dismutase
TNFα	tumor necrosis factor α
TRAIL	TNF-related apoptosis-inducing ligand
XIAP	X-linked inhibitor of apoptotic proteins

References

1. Maman, S.; Witz, I.P. A history of exploring cancer in context. *Nat. Rev. Cancer* **2018**, *18*, 359–376. [CrossRef]
2. Sud, A.; Kinnersley, B.; Houlston, R.S. Genome-wide association studies of cancer: Current insights and future perspectives. *Nat. Rev. Cancer* **2017**, *17*, 692–704. [CrossRef] [PubMed]
3. Winters, I.P.; Murray, C.W.; Winslow, M.M. Towards quantitative and multiplexed in vivo functional cancer genomics. *Nat. Rev. Genet.* **2018**, *19*, 741–755. [CrossRef]
4. Dongre, A.; Weinberg, R.A. New insights into the mechanisms of epithelial-mesenchymal transition and implications for cancer. *Nat. Rev. Mol. Cell Biol.* **2019**, *20*, 69–84. [CrossRef]
5. Finch, A.; Prescott, J.; Shchors, K.; Hunt, A.; Soucek, L.; Dansen, T.B.; Swigart, L.B.; Evan, G.I. Bcl-xL gain of function and p19 ARF loss of function cooperate oncogenically with Myc in vivo by distinct mechanisms. *Cancer Cell.* **2006**, *10*, 113–120. [CrossRef] [PubMed]
6. Yonish-Rouach, E.; Resnitzky, D.; Lotem, J.; Sachs, L.; Kimchi, A.; Oren, M. Wild-type p53 induces apoptosis of myeloid leukaemic cells that is inhibited by interleukin-6. *Nature* **1991**, *25*, 345–347. [CrossRef] [PubMed]
7. Degterev, A.; Boyce, M.; Yuan, J. A decade of caspases. *Oncogene* **2003**, *53*, 8543–8567. [CrossRef] [PubMed]
8. Song, P.; Yang, S.; Hua, H.; Zhang, H.; Kong, Q.; Wang, J.; Luo, T.; Jiang, Y. The regulatory protein GADD34 inhibits TRAIL-induced apoptosis via TRAF6/ERK-dependent stabilization of myeloid cell leukemia 1 in liver cancer cells. *J. Biol. Chem.* **2019**, *19*. [CrossRef]
9. Chung, A.H.; Leisner, T.M.; Dardis, G.J.; Bivins, M.M.; Keller, A.L.; Parise, L.V. CIB1 depletion with docetaxel or TRAIL enhances triple-negative breast cancer cell death. *Cancer Cell Int.* **2019**, *19*, 26. [CrossRef]
10. Sang, B.; Zhang, Y.Y.; Guo, S.T.; Kong, L.F.; Cheng, Q.; Liu, G.Z.; Thorne, R.F.; Zhang, X.D.; Jin, L.; Wu, M. Dual functions for OVAAL in initiation of RAF/MEK/ERK prosurvival signals and evasion of p27-mediated cellular senescence. *Proc. Natl. Acad. Sci. USA* **2018**, *11*, 115. [CrossRef]
11. Xu, F.; Sun, Y.; Yang, S.Z.; Zhou, T.; Jhala, N.; McDonald, J.; Chen, Y. Cytoplasmic PARP-1 promotes pancreatic cancer tumorigenesis and resistance. *Int. J. Cancer* **2019**, *7*. [CrossRef] [PubMed]
12. Kong, Y.; Wang, Z.; Huang, M.; Zhou, Z.; Li, Y.; Miao, H.; Wan, X.; Huang, J.; Mao, X.; Chen, C. CUL7 promotes cancer cell survival through promoting caspase-8 ubiquitination. *Int. J. Cancer* **2019**, *26*. [CrossRef] [PubMed]
13. Seo, J.; Lee, E.W.; Shin, J.; Seong, D.; Nam, Y.W.; Jeong, M.; Lee, S.H.; Lee, C.; Song, J. K6 linked polyubiquitylation of FADD by CHIP prevents death inducing signaling complex formation suppressing cell death. *Oncogene* **2018**, *37*, 4994–5006. [CrossRef] [PubMed]

14. Oh, Y.T.; Qian, G.; Deng, J.; Sun, S.Y. Monocyte chemotactic protein-induced protein-1 enhances DR5 degradation and negatively regulates DR5 activation-induced apoptosis through its deubiquitinase function. *Oncogene* **2018**, *37*, 3415–3425. [CrossRef]

15. Haselmann, V.; Kurz, A.; Bertsch, U.; Hübner, S.; Olempska-Müller, M.; Fritsch, J.; Häsler, R.; Pickl, A.; Fritsche, H.; Annewanter, F.; et al. Nuclear death receptor TRAIL-R2 inhibits maturation of let-7 and promotes proliferation of pancreatic and other tumor cells. *Gastroenterology* **2014**, *146*, 278–290. [CrossRef] [PubMed]

16. Kojima, Y.; Nakayama, M.; Nishina, T.; Nakano, H.; Koyanagi, M.; Takeda, K.; Okumura, K.; Yagita, H. Importin β1 protein-mediated nuclear localization of death receptor 5 (DR5) limits DR5/tumor necrosis factor (TNF)-related apoptosis-inducing ligand (TRAIL)-induced cell death of human tumor cells. *J. Biol. Chem.* **2011**, *16*, 43383–43393. [CrossRef] [PubMed]

17. Twomey, J.D.; Zhang, B. Circulating tumor cells develop resistance to TRAIL-induced apoptosis through autophagic removal of death receptor 5: Evidence from an in vitro model. *Cancers* **2019**, *15*, 94. [CrossRef] [PubMed]

18. Kim, Y.H.; Shin, E.A.; Jung, J.H.; Park, J.E.; Koo, J.; Koo, J.I.; Shim, B.S.; Kim, S.H. Galbanic acid potentiates TRAIL induced apoptosis in resistant non-small cell lung cancer cells via inhibition of MDR1 and activation of caspases and DR5. *Eur. J. Pharmacol.* **2019**, *15*, 91–96. [CrossRef]

19. Zhang, B.; van Roosmalen, I.A.M.; Reis, C.R.; Setroikromo, R.; Quax, W.J. Death receptor 5 is activated by fucosylation in colon cancer cells. *FEBS J.* **2019**, *286*, 555–571. [CrossRef]

20. Kaya-Aksoy, E.; Cingoz, A.; Senbabaoglu, F.; Seker, F.; Sur-Erdem, I.; Kayabolen, A.; Lokumcu, T.; Sahin, G.N.; Karahuseyinoglu, S.; Bagci-Onder, T. The pro-apoptotic Bcl-2 family member Harakiri (HRK) induces cell death in glioblastoma multiforme. *Cell Death Discov.* **2019**, *8*, 64. [CrossRef]

21. Zhu, Z.C.; Liu, J.W.; Yang, C.; Li, M.J.; Wu, R.J.; Xiong, Z.Q. Targeting KPNB1 overcomes TRAIL resistance by regulating DR5, Mcl-1 and FLIP in glioblastoma cells. *Cell Death Dis.* **2019**, *10*, 118. [CrossRef]

22. Piggott, L.; Silva, A.; Robinson, T.; Santiago-Gómez, A.; Simões, B.M.; Becker, M.; Fichtner, I.; Andera, L.; Young, P.; Morris, C.; et al. Acquired resistance of ER-positive breast cancer to endocrine treatment confers an adaptive sensitivity to TRAIL through posttranslational downregulation of c-FLIP. *Clin. Cancer Res.* **2018**, *24*, 2452–2463. [CrossRef]

23. Cho, H.D.; Gu, I.A.; Won, Y.S.; Moon, K.D.; Park, K.H.; Seo, K.I. Auriculasin sensitizes primary prostate cancer cells to TRAIL-mediated apoptosis through up-regulation of the DR5-dependent pathway. *Food Chem. Toxicol.* **2019**, *25*, 223–232. [CrossRef]

24. Ivanov, V.N.; Wu, J.; Wang, T.J.C.; Hei, T.K. Inhibition of ATM kinase upregulates levels of cell death induced by cannabidiol and γ-irradiation in human glioblastoma cells. *Oncotarget* **2019**, *25*, 825–846. [CrossRef]

25. Deng, Y.; Bi, R.; Guo, H.; Yang, J.; Du, Y.; Wang, C.; Wei, W. Andrographolide enhances TRAIL-induced apoptosis via *p53*-mediated death receptors up-regulation and suppression of the NF-κB pathway in bladder cancer cells. *Int. J. Biol. Sci.* **2019**, *24*, 688–700. [CrossRef]

26. Zhao, L.M.; Li, L.; Huang, Y.; Han, L.J.; Li, D.; Huo, B.J.; Dai, S.L.; Xu, L.Y.; Zhan, Q.; Shan, B.E. Antitumor effect of periplocin in TRAIL-resistant gastric cancer cells via upregulation of death receptor through activating ERK1/2-EGR1 pathway. *Mol. Carcinog.* **2019**, *9*. [CrossRef]

27. Kim, B.R.; Park, S.H.; Jeong, Y.A.; Na, Y.J.; Kim, J.L.; Jo, M.J.; Jeong, S.; Yun, H.K.; Oh, S.C.; Lee, D.H. RUNX3 enhances TRAIL-induced apoptosis by upregulating DR5 in colorectal cancer. *Oncogene* **2019**, *28*, 1. [CrossRef]

28. Kim, H.J.; Kang, S.; Kim, D.Y.; You, S.; Park, D.; Oh, S.C.; Lee, D.H. Diallyl disulfide (DADS) boosts TRAIL-Mediated apoptosis in colorectal cancer cells by inhibiting Bcl-2. *Food Chem. Toxicol.* **2019**, *125*, 354–360. [CrossRef]

29. Su, E.Y.; Chu, Y.L.; Chueh, F.S.; Ma, Y.S.; Peng, S.F.; Huang, W.W.; Liao, C.L.; Huang, A.C.; Chung, J.G. Bufalin induces apoptotic cell death in human nasopharyngeal carcinoma cells through mitochondrial ROS and TRAIL pathways. *Am. J. Chin. Med.* **2019**, *1*, 237–257. [CrossRef]

30. Guo, Z.L.; Li, J.Z.; Ma, Y.Y.; Qian, D.; Zhong, J.Y.; Jin, M.M.; Huang, P.; Che, L.Y.; Pan, B.; Wang, Y.; et al. Shikonin sensitizes A549 cells to TRAIL-induced apoptosis through the JNK, STAT3 and AKT pathways. *BMC Cell Biol.* **2018**, *29*, 29. [CrossRef]

31. Nazim, U.M.; Yin, H.; Park, S.Y. Autophagy flux inhibition mediated by celastrol sensitized lung cancer cells to TRAIL-induced apoptosis via regulation of mitochondrial transmembrane potential and reactive oxygen species. *Mol. Med. Rep.* **2019**, *19*, 984–993. [CrossRef]

32. Rubinsztein, D.C.; Codogno, P.; Levine, B. Autophagy modulation as a potential therapeutic target for diverse diseases. *Nat. Rev. Drug Discov.* **2012**, *11*, 709–730. [CrossRef] [PubMed]

33. Galluzzi, L.; Bravo-San Pedro, J.M.; Levine, B.; Green, D.R.; Kroemer, G. Pharmacological modulation of autophagy: Therapeutic potential and persisting obstacles. *Nat. Rev. Drug Discov.* **2017**, *16*, 487–511. [CrossRef]

34. Levy, J.M.M.; Towers, C.G.; Thorburn, A. Targeting autophagy in cancer. *Nat. Rev. Cancer* **2017**, *17*, 528–542. [CrossRef]

35. Nazim, U.M.; Park, S.Y. Attenuation of autophagy flux by 6-shogaol sensitizes human liver cancer cells to TRAIL-induced apoptosis via p53 and ROS. *Int. J. Mol. Med.* **2019**, *43*, 701–708. [CrossRef]

36. Xu, Y.; Wang, Q.; Zhang, L.; Zheng, M. 2-Deoxy-D-glucose enhances TRAIL-induced apoptosis in human gastric cancer cells through downregulating JNK-mediated cytoprotective autophagy. *Cancer Chemother Pharmacol.* **2018**, *81*, 555–564. [CrossRef]

37. Monma, H.; Iida, Y.; Moritani, T.; Okimoto, T.; Tanino, R.; Tajima, Y.; Harada, M. Chloroquine augments TRAIL-induced apoptosis and induces G2/M phase arrest in human pancreatic cancer cells. *PLoS ONE* **2018**, *13*, e0193990. [CrossRef] [PubMed]

38. Li, X.; You, M.; Liu, Y.J.; Ma, L.; Jin, P.P.; Zhou, R.; Zhang, Z.X.; Hua, B.; Ji, X.J.; Cheng, X.Y.; et al. Reversal of the apoptotic resistance of non-small-cell lung carcinoma towards TRAIL by natural product toosendanin. *Sci. Rep.* **2017**, *7*, 42748. [CrossRef]

39. Di, X.; Zhang, G.; Zhang, Y.; Takeda, K.; Rivera Rosado, L.A.; Zhang, B. Accumulation of autophagosomes in breast cancer cells induces TRAIL resistance through downregulation of surface expression of death receptors 4 and 5. *Oncotarget* **2013**, *4*, 1349–1364. [CrossRef]

40. Chen, L.; Meng, Y.; Sun, Q.; Zhang, Z.; Guo, X.; Sheng, X.; Tai, G.; Cheng, H.; Zhou, Y. Ginsenoside compound K sensitizes human colon cancer cells to TRAIL-induced apoptosis via autophagy-dependent and -independent DR5 upregulation. *Cell Death Dis.* **2016**, *7*, e2334. [CrossRef] [PubMed]

41. Chen, L.; Xiong, Y.Q.; Xu, J.; Wang, J.P.; Meng, Z.L.; Hong, Y.Q. Juglanin inhibits lung cancer by regulation of apoptosis, ROS and autophagy induction. *Oncotarget* **2017**, *8*, 93878–93898. [CrossRef]

42. Manouchehri, J.M.; Kalafatis, M. Ursolic acid promotes the sensitization of rhTRAIL-resistant triple-negative breast cancer. *Anticancer Res.* **2018**, *38*, 6789–6795. [CrossRef]

43. Shahriyar, S.A.; Woo, S.M.; Seo, S.U.; Min, K.J.; Kwon, T.K. Cepharanthine enhances TRAIL-mediated apoptosis through STAMBPL1-mediated downregulation of survivin expression in renal carcinoma cells. *Int. J. Mol. Sci.* **2018**, *19*, 3280. [CrossRef]

44. Byun, H.S.; Zhou, W.; Park, I.; Kang, K.; Lee, S.R.; Piao, X.; Park, J.B.; Kwon, T.K.; Na, M.; Hur, G.M. C-27-carboxylated oleanane triterpenoids up-regulate TRAIL DISC assembly via p38 MAPK and CHOP-mediated DR5 expression in human glioblastoma cells. *Biochem. Pharmacol.* **2018**, *158*, 243–260. [CrossRef]

45. Ahn, D.S.; Lee, H.J.; Hwang, J.; Han, H.; Kim, B.; Shim, B.; Kim, S.H. Lambertianic acid sensitizes non-small cell lung cancers to TRAIL-induced apoptosis via inhibition of XIAP/NF-κB and activation of caspases and death receptor 4. *Int. J. Mol. Sci.* **2018**, *19*, 1476. [CrossRef]

46. Hoadley, K.A.; Yau, C.; Wolf, D.M.; Cherniack, A.D.; Tamborero, D.; Ng, S.; Leiserson, M.D.M.; Niu, B.; McLellan, M.D.; Uzunangelov, V.; et al. Multiplatform analysis of 12 cancer types reveals molecular classification within and across tissues of origin. *Cell* **2014**, *14*, 929–944. [CrossRef]

47. Hanahan, D.; Weinberg, R.A. Hallmarks of cancer: The next generation. *Cell* **2011**, *4*, 646–674. [CrossRef]

48. Hengartner, M.O. The biochemistry of apoptosis. *Nature* **2000**, *12*, 770–776. [CrossRef]

49. Wang, Q.; Chen, Q.; Zhu, L.; Chen, M.; Xu, W.; Panday, S.; Wang, Z.; Li, A.; Røe, O.D.; Chen, R.; et al. JWA regulates TRAIL-induced apoptosis via MARCH8-mediated DR4 ubiquitination in cisplatin-resistant gastric cancer cells. *Oncogenesis* **2017**, *7*, 53. [CrossRef]

50. Jeon, M.Y.; Min, K.J.; Woo, S.M.; Seo, S.U.; Choi, Y.H.; Kim, S.H.; Kim, D.E.; Lee, T.J.; Kim, S.; Park, J.W.; et al. Maritoclax enhances TRAIL-induced apoptosis via CHOP-mediated upregulation of DR5 and miR-708-mediated downregulation of cFLIP. *Molecules* **2018**, *23*, 3030. [CrossRef]
51. Kumazaki, M.; Shinohara, H.; Taniguchi, K.; Ueda, H.; Nishi, M.; Ryo, A.; Akao, Y. Understanding of tolerance in TRAIL-induced apoptosis and cancelation of its machinery by α-mangostin, a xanthone derivative. *Oncotarget* **2015**, *22*, 25828–25842. [CrossRef]

International Journal of
Molecular Sciences

Article

Ginsenoside Rh2 Ameliorates Doxorubicin-Induced Senescence Bystander Effect in Breast Carcinoma Cell MDA-MB-231 and Normal Epithelial Cell MCF-10A

Jin-Gang Hou [1,†] , Byeong-Min Jeon [2,†], Yee-Jin Yun [2], Chang-Hao Cui [1] and Sun-Chang Kim [1,2,*]

[1] Intelligent Synthetic Biology Center, Daejeon 34141, Korea; houjg2014@kaist.ac.kr (J.-G.H.); chcui@kaist.ac.kr (C.-H.C.)
[2] Department of Biological Sciences, KAIST, Daejeon 34141, Korea; jbm0901@kaist.ac.kr (B.-M.J.); yyj07@kaist.ac.kr (Y.-J.Y.)
* Correspondence: sunkim@kaist.ac.kr; Tel.: +82-042-350-2619
† These authors contributed equally to this work.

Received: 22 February 2019; Accepted: 9 March 2019; Published: 12 March 2019

Abstract: The anthracycline antibiotic doxorubicin is commonly used antineoplastic drug in breast cancer treatment. Like most chemotherapy, doxorubicin does not selectively target tumorigenic cells with high proliferation rate and often causes serve side effects. In the present study, we demonstrated the cellular senescence and senescence associated secretory phenotype (SASP) of both breast tumor cell MDA-MB-231 and normal epithelial cell MCF-10A induced by clinical dose of doxorubicin (100 nM). Senescence was confirmed by flattened morphology, increased level of beta galactose, accumulating contents of lysosome and mitochondrial, and elevated expression of p16 and p21 proteins. Similarly, SASP was identified by highly secreted proteins IL-6, IL-8, GRO, GM-CSF, MCP-1, and MMP1 by antibody array assay. Reciprocal experiments, determined by cell proliferation and apoptosis assays and cell migration and cell invasion, indicated that SASP of MDA-MB-231 cell induces growth arrest of MCF-10A, whereas SASP of MCF-10A significantly stimulates the proliferation of MDA-MB-231. Interestingly, SASP from both cells powerfully promotes the cell migration and cell invasion of MDA-MB-231 cells. Treatment with the natural product ginsenoside Rh2 does not prevent cellular senescence or exert senolytic. However, SASP from senescent cells treated with Rh2 greatly attenuated the above-mentioned bystander effect. Altogether, Rh2 is a potential candidate to ameliorate this unwanted chemotherapy-induced senescence bystander effect.

Keywords: cellular senescence; doxorubicin; breast cancer cell; breast epithelial cell; ginsenoside Rh2

1. Introduction

Chemotherapeutic drugs are designed to eliminate tumor cells with high proliferation rates in treatment of malignancies [1]. Generally, tumor cells are deprived of reproductive potential and undergo apoptosis [2]. However, there is a growing body of literature that recognizes apoptosis may not be the confined mechanism whereby cancer cells lose their ability of self-renewal after exposure to chemotherapy treatment, especially in solid tumors [3]. Among these alternative pathways, the theory of cellular senescence provides a useful account of how thermotherapy prevents tumor growth both in vitro and in vivo [4]. Cellular senescence is characterized by an irreversible cellular growth arrest in response to DNA damage. Additionally, cellular senescence was proposed to be regulated by two main tumor suppressor pathways of cell, the ARF/53 and INK4a/RB pathways [5]. Since most clinical or under-investigation chemotherapeutic drugs induce severe DNA damage to tumor cells, they are conceivably engaged and, correspondingly, trigger cellular senescence. However, the tumor suppression of senescence is quickly challenged by massive emerging envidence. Senescent cells still metabolically active preserving the potential to secrete paracrine acting factors [6]. Manipulation of p53 and p16 activation can reverse cellular senescence in human cells, suggesting that chemotherapy resistance may be partly driven by the subsets of tumor cells emerging from a senescence state [7]. In addition, DNA damage-induced cellular senescence develops a typical senescence-associated secretory phenotype (SASP), involving a spectrum of secreted growth factors, proteases, and cytokines, many of which are attributive to the microenvironment promoting resistance [8]. On the other hand, most cytotoxic chemotherapy agents target cancer cell by sensing tumor cell characteristics, such as high proliferation rates. However, this manner also targets normal counterparts and senescent cell disrupts the normal function of tissues and organs [9]. Evidence suggests that a regimen with anthracycline and alkylating agents on patients with breast cancer durably induces cellular senescence and SASP [10]. Interestingly, senescent cells, especially SASP, are often recognized and cleared through an antigen-specific immune response termed senescence surveillance [11], which might be a potential strategy to treat cancer patients. However, SASP also can promote the proliferation and invasion of neighboring cancer cells that escape the senolytic and immune system, leading to cancer relapse [12]. Importantly, the accumulation of senescent cells in tissue with age gives rise to chronic maintenance of senescent cells since the immune system declines in function during aging [13]. This indicates that targeting senescent cells and SASP could be a strategy for cancer therapy and aging-related cancer relapse.

Ginsenoside is of interest because it has been receiving considerable attention as a general tonic on longevity and body enhancement [14,15]. Recently, there has been renewed interest in synergistic effects of ginsenoside on chemotherapy treatment. Ginsenoside Rh2 treatment combined with selected chemotherapy agents sensitizes toxicity on prostate cancer cells [16] and hepatoma cells [17], and inhibits angiogenesis and growth of lung cancer [18] and ovarian cancer [19]. Rh2 enhances efficacy of paclitaxel or mitoxantrone in prostate cancer cells [20]. Accordingly, these above synergistic effects are mainly due to regulation of drug efflux, enhancement of subcellular distribution, and induction of apoptosis. Importantly, the interaction between ginsensoides and chemotherapy agents allows lower doses in clinical application. Furthermore, this chemotherapy regimen is reported to induce unwanted impairment and a proposed protection of ginsenoside Rh2 reverses this side effect in mice with lung tumor [21]. Collectively, ginsenoside Rh2 possesses great potential in sensing chemotherapy agents and ameliorating unwanted side effects.

In the present study, we determined the effects of ginsenoside Rh2 on the following circumstances. (1) Senolytic effect on Doxorubicin-induced cellular senescence of breast cells. (2) Doxorubicin induces cellular senescence of breast epithelial cells. (3) Senescent breast cancer cells induces bystander effects on normal counterparts. (4) Senescent normal counterparts stimulate proliferation, migration, and invasion of breast cancer cells. The present work suggested that Rh2 extenuated the bystander effect induced by chemotherapy in breast cells.

2. Results

2.1. Low Dose Doxorubicin Induces Senescence of Human Breast Cells

To screen out the concentration of doxorubicin to induce senescence in human breast cells, MDA-MB-231, and MCF-10A cells were exposed to various concentrations of doxorubicin for 72 h and cell viability was determined by WST-1 method. Doxorubicin concentration-dependently reduced the cell viability of both cell lines (Figure 1A). We observed that concentrations initiating from 0.1 to 5 μM significantly decreased the cell viability than vehicle (distilled water) control. However, concentrations higher than 0.1 μM induced evidenced cellular apoptosis with a large amount of debris. Hence we used 0.1 μM doxorubicin for subsequent experiments. Concomitantly, immunostaining for Ki-67, a proliferation marker, corroborated the profound cell growth arrest at 100 nM doxorubicin.

To further identify whether cells with inhibited growth turned senescent, we evaluated typical markers for senescence. One biomarker of senescence is the accumulating lysosomal contents. Non-treated and treated (100 nM doxorubicin) cells were labeled with Lysotracker Red (Figure 1B). Notably, treated cells displayed a marked redistribution of lysosome with diffused perinuclear pattern. Apart from enhanced lysosomal content, an increased percentage of canonical marker SA-β-gal in treated cells was correspondingly observed (Figure 1C). Another biomarker is increased mitochondrial biomass. We therefore labeled the non-treated and treated (100 nM doxorubicin) cells with Mitotracker Red (Figure 1D). A remarkable mitochondrial signal was detected in treated cells. Senescent cells showed nuclear foci termed DNA-SCARs, requiring for SASP development. Treated cells significantly altered the number of 53BP1 foci compared with Nontreated con (Figure 1E). Senescence was further confirmed by elevated levels of proteins p16 and p21 in treated cells using Western blot analysis (Figure 1F). Importantly, the above evaluations indicated that 100 nM doxorubicin induces typical cellular senescence in human breast cell lines.

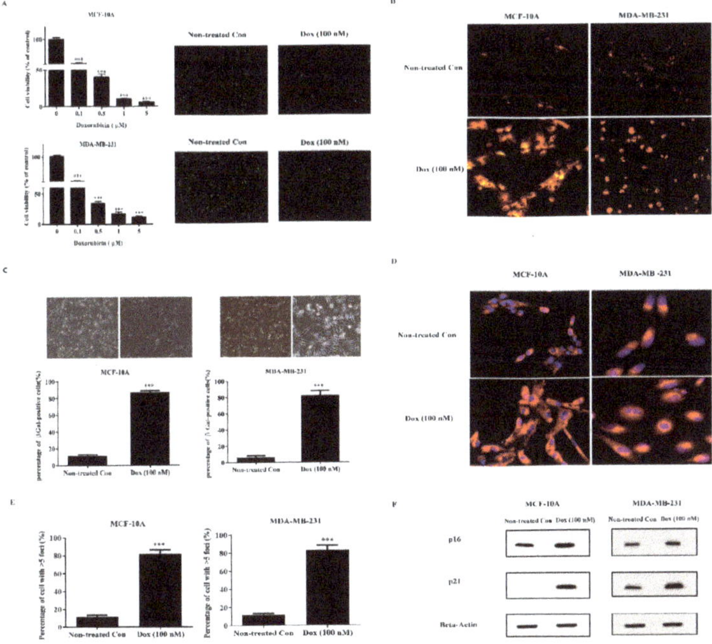

Figure 1. Doxorubicin induces senescence of breast normal and cancer cells. (**A**) MDA-MB-231 cells and MCF-10A cells were treated with indicated doses of doxorubicin for 72 h. Cells were either assayed for viability (left panel) or fixed and stained for Ki-67 (right panel, green). Cell viability was determined by WST-1 method and normalized with nontreated con cells. (**B–D**) Cells were treated with 100 nM doxorubicin for 72 h to induce senescence. Then, cells (**B**) were incubated with Lysotracker Red (200 nM) for 1 h. Cells (**C**) were fixed and stained for SA-β-gal. The upper panel shows the bright-field images. The lower panel was the percentage of positive cells (>200 cells scored). Cells (**D**) were incubated with Mitotracker Red (100 nM) for 30 min. Blue, DAPI stained nuclear. Representative images were captured by a Fluorescence Microscope (100× for **A**,**C**; 200× for (**B**,**D**). (**E**) Then days after senescence induction, nontreated con and senescent cells were immunostained with 53BP1, a DNA-SCAR marker. The number of the foci was determined by CellProfiler. Shown was the percentage of cells with >5 foci. (**F**) Extracts from nontreated con and senescent cells were measured for the indicated proteins by western blotting. Beta-actin was used as the loading control. *** indicates *p* < 0.001 versus nontreated con.

2.2. Doxorubicin-Induced SASP in Human Breast Cell Lines

To determine whether senescent cells developed SASP, a conditioned medium from senescent MDA-MB-231 and MCF-10A cells was applied to a human cytokine array assay with 120 secreted proteins. In contrast to nontreated con cells, for senescent human breast cancer MDA-MB-231 cells, the factors detected by arrays and secreted at a significant level are FGF-6, GM-CSF, IGFBP-1, MCP-1, IL-6, IL-1α, GRO a/b/g, GRO α, IL-8, MIP, MIP-1α, uPAR, ICAM-1, and MMP-1(Figure 2). In senescent nontumorigenic MCF-10A cells, proteins secreted at substantial level are FGF-6, MCP-1, GRO a/b/g, GRO α, IL-8, uPAR, IGFBP-6, OPG, TNFR1, IP10, CD14, and MMP-13 (Figure 2). Additionally, we observed in certain proteins (PDGF-AA, PDGF-BB, ANGPT2, IGFBP-2, and ALCAM) that secretion was downregulated in senescent MCF-10A cells. Intriguingly, although a similar secretion pattern of major SASP factors such IL-6 and IL-8 was observed in both cell lines, two cell lines displayed differed secretory phenotype. We postulated that these differences may lead to various paracrine effects.

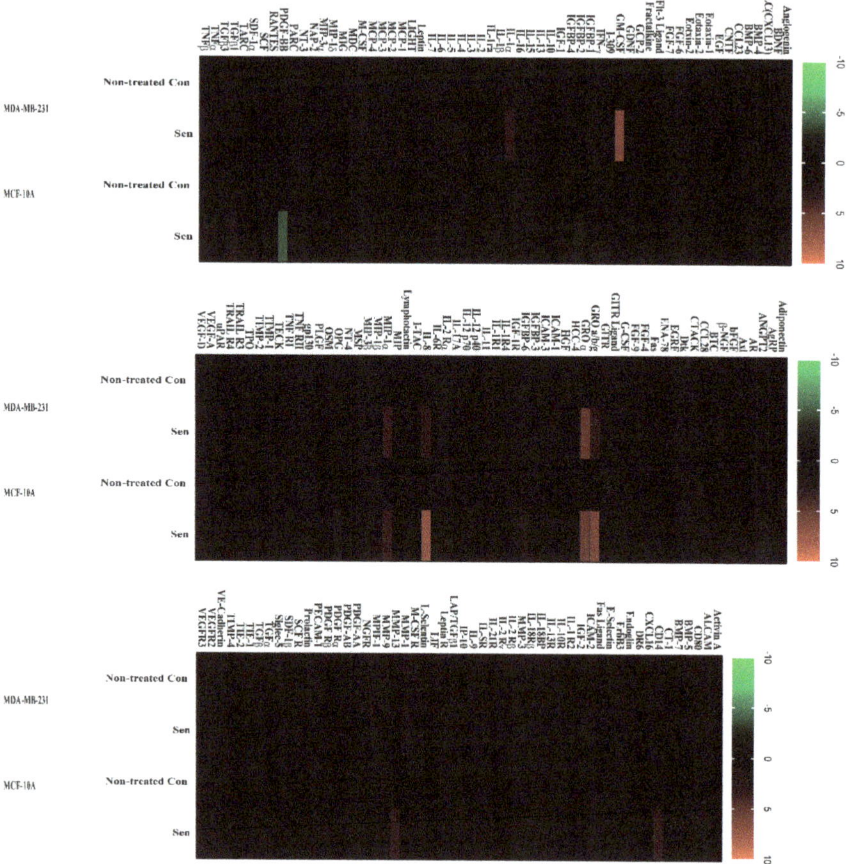

Figure 2. Senescent human breast cancer and normal cells developed SASP. Conditioned medium from nonsenescent (nontreated Con) or senescent (100 nM of doxorubicin exposure, Sen) MDA-MB-231 (A) and MCF-10A (D) cells were analyzed with human cytokine antibody arrays. Levels of each cytokine factor in untreated cells were arbitrary set to zero. Data shown represent log2-fold change in expression relative to untreated cells. Signals higher than the untreated control are shown in red; signals lower than the untreated control are shown in green.

2.3. SASP Stimulates Migration and Invasion of Breast Cancer Cells

To address the possibility that SASP (high secretions of IL-6 and IL-8) from senescent cells affects carcinoma cells migration, we examined the consequences of treatments with conditioned medium (CM) on the motogenic response of human breast cancers. Monolayers of MDA-MB-231 cells were scraped to create a cell-free area, and cell migrations were evaluated 48h later. Conditioned medium from senescent cells produced a marked increase in breast cancer migration (Figure 3A). As expected, quantitative assay showed that CM of MDA-MB-231 induced significant migration than that of non-treated con ($p < 0.01$). Importantly, similar to CM of MDA-MB-231, CM of MCF-10A also strongly stimulated breast cancer migration ($p < 0.01$). As expected, Rh2 treatment notably inhibited these elevated migrations. For invasion assay, CM of both senescent cell lines vigorously stimulated the cancer cell invasion by over 10-fold, which were noticeably mitigated by Rh2 treatment (Figure 3B). Since epithelial–mesenchymal transition (EMT) gives rise to caner invasion, we measured several hallmarks by western blot and immunofluorescence analysis. Intriguingly, CM of MDA-MB-231 elevated the expression levels of beta-catenin and snail while reduced the level of ZO-1 when in comparison to nontreated control cells. Rh2 exposure exerted decreased levels of beta-catenin and snail, while no changes of ZO-1 were noted. Importantly, Rh2 treatment powerfully abated the expression of vimentin though no noticeable increase was observed in CM treated cells. CM of MCF-10A induced higher expression level of slug, but lower than that of ZO-1 in comparison to normal cells; these alterations were ameliorated by Rh2 treatment (Figure 3C).

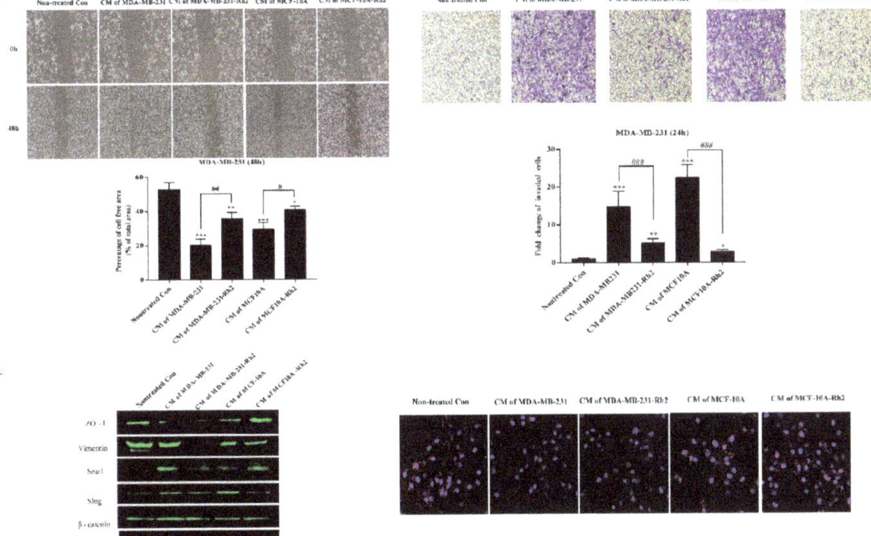

Figure 3. Rh2 inhibited SASPs-induced migration and invasion of breast cell lines. (**A**) MDA-MB-231 cells were cultured with conditioned medium from senescent MDA-MB-231 and MCF-10A cells treated with or without Rh2 for 48 h. The cell-free areas were imaged with microscope at 0 h and 48 h, respectively. The changes in cell-free area were calculated using CellProfiler Software (2.2.0). At least three wound scratches were analyzed per experiment. (**B**) Invasion of MDA-MB-231 cells (5×10^4/well) as determined by 24-well plate transwell system. Cells on the lower side of membranes were stained and quantified. (**C**) Western blot analysis of representative epithelial–mesenchymal transition (EMT) markers of MDA-MB-231 cells after CM treatment. Immunostaining for the tight junction protein ZO-1(Red) and the nuclear regions were counterstained with DAPI (blue). * indicates $p < 0.05$ versus Nontreated con, ** indicates $p < 0.01$ versus Non-treated con; *** indicates $p < 0.001$ versus nontreated con. ## indicates $p < 0.01$ versus CM alone group. ### indicates $p < 0.001$ versus CM alone group.

2.4. Ginsenoside Rh2 Does Not Exert Senolytic but Suppress the Paracrine Effects of Sasp in Human Breast Cell Lines

To check whether Rh2 exerts senolytic effects on both senescent cell lines, senescent MDA-MB-231 and MCF-10A cells were exposed to Rh2 (20 µg/mL) and caspase 3/7 activity was determined 48h later. Interestingly, Rh2 did not induce statistical apoptosis as compared to nontreated control group. Additionally, senescent cells in both cell lines showed apoptosis resistance as elaborated by reduced caspase 3/7 activity with staurosporin (Figure 4A).

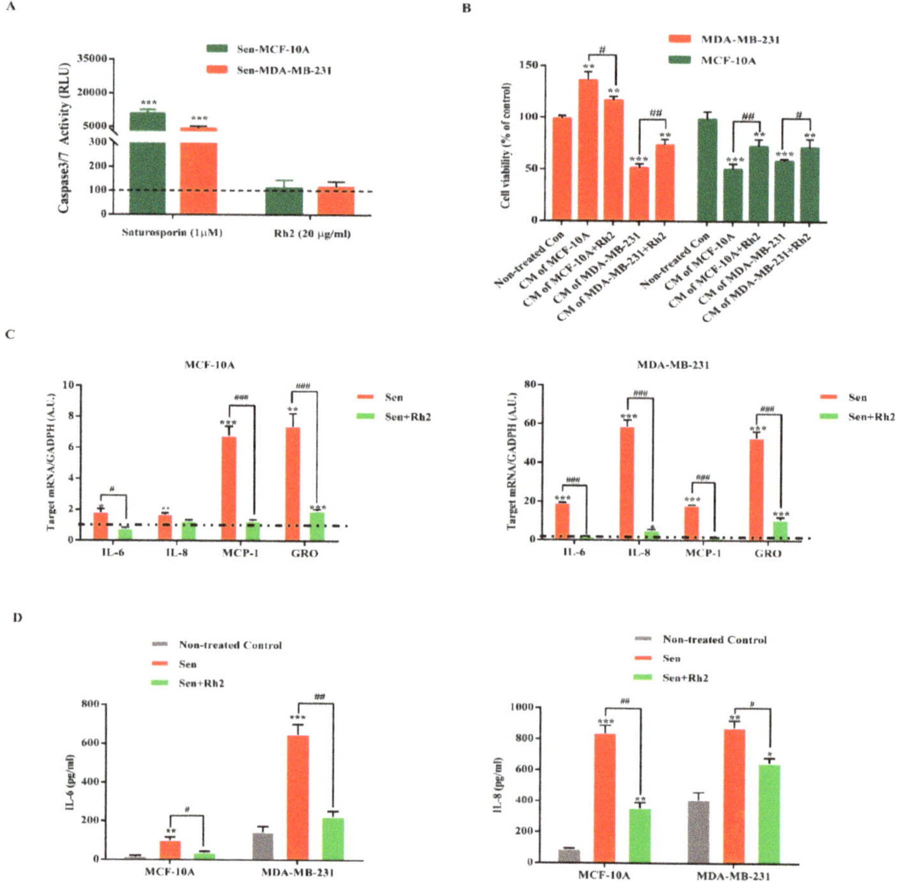

Figure 4. Rh2 suppressed paracrine effects of SASP in breast cell lines. (**A**) Senescent MDA-MB-231 and MCF-10A were exposed to Rh2 for 48 h, after which caspase 3/7 activity was quantified. The dotted line indicates fluorescence level in nontreated senescent cells. (**B**) MDA-MB-231 and MCF-10A cells were cultured with indicated CM for 24 h. The cell viability was assayed by WST-1 method. (**C**) Fold changes in mRNA levels of IL-8, IL-6, CXCL1, and MCP-1. The dotted line indicates expression level in control cells, set at 1 for each gene. $N = 3$. A.U., arbitrary units. (**D**) ELISA assay of IL-6 and IL-8. * indicates $p < 0.05$ versus Non-treated con. ** indicates $p < 0.01$ versus Non-treated con; *** indicates $p < 0.001$ versus Nontreated con. # indicates $p < 0.05$ versus CM group. ## indicates $p < 0.01$ versus CM group. ### indicates $p < 0.001$ versus CM alone group.

Given senescent cells can reprogram neighboring cells through SASP, we determined whether ginsenoside Rh2 mitigated the cancer-promoting effect on MDA-MB-231 cells and senescent paracrine effect on MCF-10A cells. As seen in Figure 4B, CM from senescent MDA-MB-231 cells significantly inhibited the cell proliferation of MCF-10A cells, which was clearly ameliorated by treatment of senescent MDA-MB-231 cells with Rh2. On the contrary, CM from senescent MCF-10A cells exerted profoundly tumorigenic effect on MDA-MB-231 cells, while CM prepared from senescent cells treated with Rh2 attenuated the stimulated effect. Intriguingly, CM from both cell lines showed obviously self-senescence paracrine effect with manifested growth arrest. Additionally, CM from Rh2 treated senescent cells slightly rescued the cell proliferative ability.

Further, we measured the effect of Rh2 on the expression of some representative SASP genes associated with inflammation, proliferation, and invasion. As shown in Figure 4C, Rh2 significantly reduced the mRNA level of MCP-1 and CXCL1 in both senescent cells lines, while it selective suppressed that of IL-6 and IL-8. Additionally, the enzyme-linked immunosorbent assay (ELISA) further confirmed the highly secreted SASP components IL-6 and IL-8, which were significantly diminished by Rh2 treatment (Figure 4D).

2.5. Ginsenoside Rh2 Suppresses Potential Signaling Pathways Inferred SASP Secretion in Human Breast Cell Lines

To address the potential signaling pathways that activated SASP, we investigated the major pathways that regulate cell growth, differentiation, response to proinflammatory cytokines, growth factor receptors, inflammation, and cellular stress in the Signaling Nodes Multitarget Sandwich ELISA Kit. Endogenous levels of Akt1, phosphor-Akt(Ser473), phosphor-MEK1(Ser217/221), phosphor-p38 MAPK(Thr180/Tyr182), phosphor-Stat3(Tyr705), and phosphor-NF-κB p65(Ser536) were determined. For MDA-MB-231 cells, senescence significantly induced the phosphorylated forms of MEK1, p38, Stat3, and NF-κB p65, while no phosphorylation of Akt1 was noted. Treatment with ginsenoside Rh2 strongly reduced those elevations to the level of nontreated control but with further statistical decrease of Stat3 (Figure 5A). For MCF-10A cells, senescence evidently induced the phosphorylated forms of p38, Stat3, and NF-κB p65, while profoundly decreased phosphorylation of Akt1 and MEK1 was noted. Treatment with ginsenoside Rh2 only significantly reduced the elevated levels of phosphorylated p38 and Stat3, with no effects on Akt1, MEK1, and NF-κB p65 (Figure 5B). Interestingly, Rh2 exerted different suppression manner between two cell lines.

Given that two signaling pathways are well-documented regulator of SASP, we herein again verified that low-dose exposure to doxorubicin induced SASP with robust activation of p38 MAPK and NF-κB pathways. Additionally, inclusion of the p38 MAPK inhibitor SB230580 (10 μM), NF-κB inhibitor Bay 11-7082 (10 μM), or Rh2 (20 μg) in the incubation medium evidently suppresses the SASP in both cell lines (Figure 6).

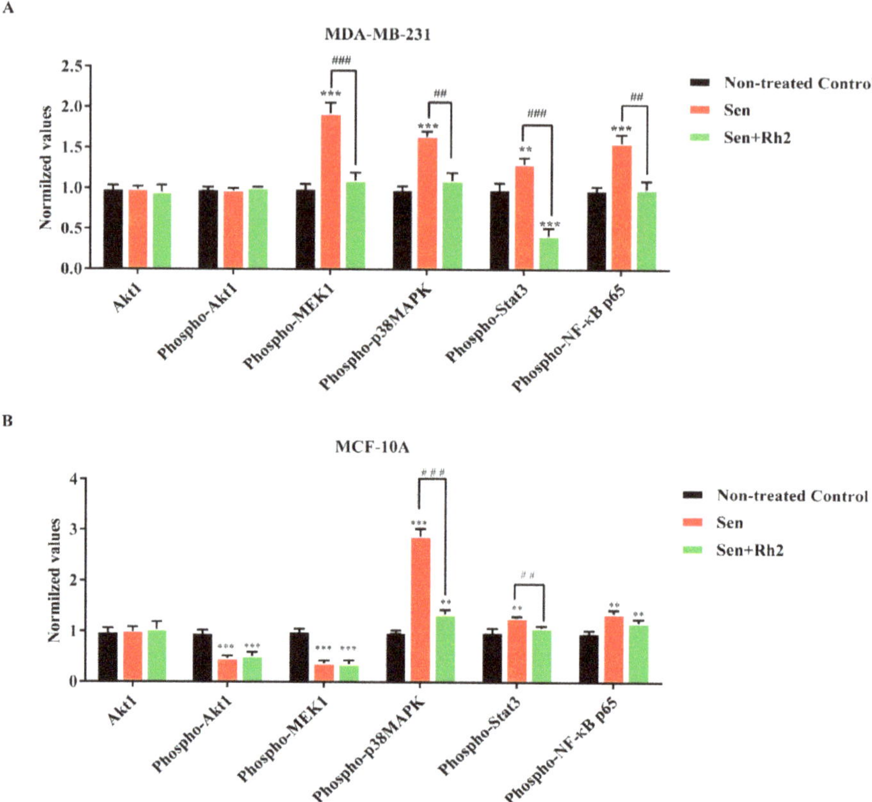

Figure 5. Rh2 suppressed potential signaling pathways regulating SASP in breast cell lines. Senescent MDA-MB-231 (**A**) and MCF-10A (**B**) cells treated with or without Rh2 were lysed and subjected to the Signaling Nodes Multitarget Sandwich ELISA assay. ** indicates $p < 0.01$ versus nontreated con; *** indicates $p < 0.001$ versus nontreated con. ## indicates $p < 0.01$ versus CM alone group. ### indicates $p < 0.001$ versus CM alone group.

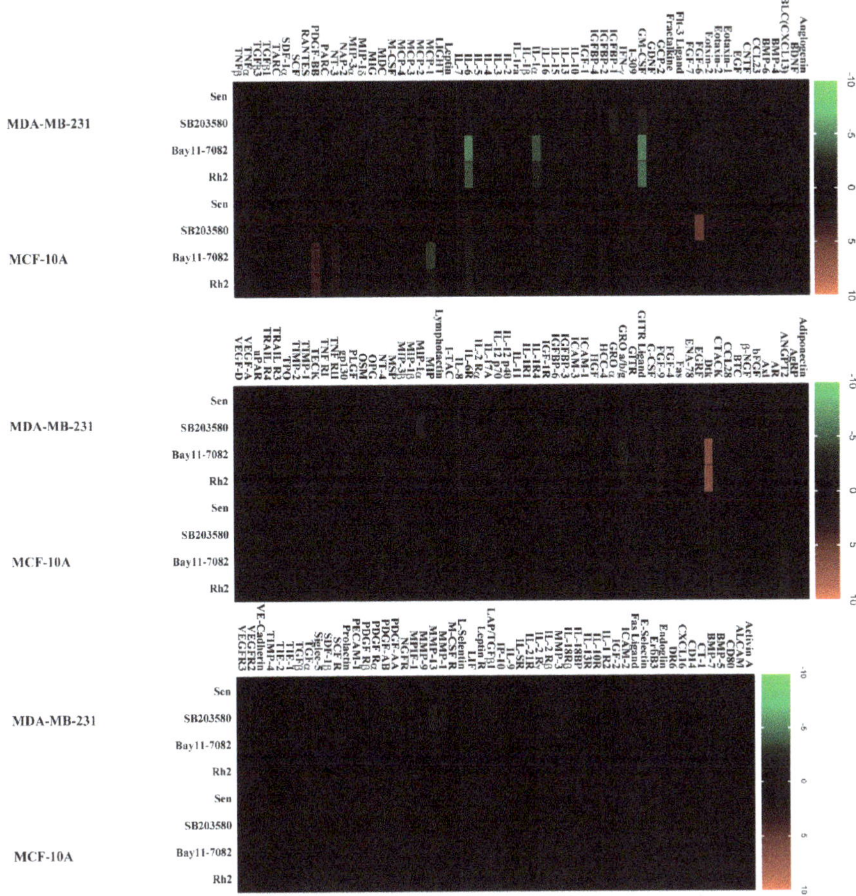

Figure 6. Rh2 suppressed SASP secretion in breast cell lines. Conditioned medium, prepared from senescent cells treated with DMSO, SB203580 (10 µM), Bay11-7082 (10 µM), and Rh2 (20 µg/mL), was analyzed with human cytokine antibody arrays. Levels of each cytokine factor in untreated cells were arbitrary set to zero. Data shown represent log2-fold change in expression relative to untreated cells. Signals higher than the senescent control are shown in red; signals lower than the senescent control are shown in green.

3. Discussion

Cellular senescence, originally described as an irreversibly proliferative arrest in normal fibroblasts after a limited number of divisions [22], is an important tumor-suppressive mechanism. However, accumulated evidence challenged that cellular senescence exerts several deleterious biological functions encompassing tumorigenesis and age-related pathologies especially involving SASP [23]. Doxorubicin is the mainstay drug in the treatment for breast cancer. Intriguingly, since the required dose of chemotherapy drug to induce cellular senescence is much lower than that necessary to kill cells, breast cancer cells exposed to doxorubicin undergo widespread senescence [24]. In this study, we provided further evidence for the reciprocal effects of senescent breast tumor and normal cells under stimulation with clinical dose of doxorubicin. Additionally, we proposed that ginsenoside Rh2 is a potential candidate for the extenuation of chemotherapy-induced senescence bystander effect.

Doxorubicin induces the formation of DNA-Dox-topoisomerase II cleavable complexes and subsequent DNA damage, by which cellular senescence was initiated in many cancer cells. Clinically, the reported steady-state plasma level of doxorubicin achieved in patients receiving a total dose of 165 kg/m^2 was 0.1 μM to avoid cardiac toxicity [25]. Accordingly, our results showed that 0.1 μM of doxorubicin successfully induced cellular senescence of breast carcinomas cell MDA-MB-231 and normal cell MCF-10A, as elaborated by typical flatten morphological changes, increased mitochondrial biomass, SA-β-gal expression, and redistribution of lysosomes. Additionally, elevated levels of senescence markers p16 and p21 were noted. Stable nuclear foci 53BP1 was observed, indicating the potential SASP development. Intriguingly, pretreatment or post-treatment of Rh2 does not affect the alterations of the senescence markers, suggesting Rh2 may acts downstream of signaling pathways involved in senescence.

To test the possibility that senescent cells develop SASP after challenge with doxorubicin, we assessed the levels of 120 secreted cytokines from CM using antibody arrays. As acknowledged, SASP presents variations relying on tissue and stimulus. A large number of cytokines consisting of SASP components secreted at significant level after doxorubicin induction, encompassing proinflammatory cytokines (e.g., IL-1α/β, IL-6, and IL-8), growth factor (e.g., PDGF and GM-CSF), chemokines (e.g., CXCL1 and MCP-1), and matrix remodeling enzymes (e.g., MMPs) were noted in the present study. Among the above typical SASP components, IL-1, IL-6, and IL-8 are reported to exert pleiotropic effects including maintenance of senescence, promotion of tumorigenesis and chemotherapy resistance [26]. GM-CSF has been reported to be overexpressed in a variety of human cancers including melanoma and hepatocellular carcinoma and was proposed to regulate tumor progression [27]. MCP-1 and CXCL1 (GRO) influence breast carcinogenesis by facilitating tumor growth and metastatic spread [28]. MMPs are another major SASP factor usually secreted by senescent fibroblasts, which can enhance the invasion of multiple epithelial cell types [29]. Apart from abovementioned SASP factors, certain cytokines were detected at remarkable levels as follows: insulin-like growth factor-binding proteins (IGFBPs), urokinase plasminogen activator (uPA), and fibroblast growth factor (FGF-6). These cytokines are closely associated with cancer invasion and metastasis [30]. Importantly, Rh2 treatment reversed these alterations of secreted cytokines in the present study. Intriguingly, nontumorigenic MCF-10A cell exhibited different pattern of SASP, with significant decreased secretion of PDGF-AA and PDGF-BB compared to nontreated normal cells, which might indirectly lead to tissue dysfunction and impaired regeneration. Similarly, Rh2 exposure restored the levels of PDGF-AA and PDGF-BB. Next, we measured the ability of SASP from both senescent cell lines to stimulate migration and invasion of MDA-MB-231 cells by a transwell system. Accordingly, significant migration and invasion were observed with CM supplement in culture medium, which were clearly suppressed by Rh2 treatment. Rh2 profoundly reduced the expression of EMT markers (beta catenin, slug, and snail) in MDA-MB-231 cells. Moreover, CM from Rh2 treated cells reduced carcinomas cell proliferation and restored normal cell doubling population when compared with CM-treated cells. Collectively, our results herein demonstrated that SASP is able to promote cancer progression and inhibit proliferation ability of normal cell in breast tissue, while Rh2 is a potential candidate for the amelioration of this deleterious effect. However, the present study did not explicitly address how Rh2 inhibited SASP-associated migration and invasion of malignant breast cancer cells. Chemotherapy-induced senescence has been proposed as a potential approach to combat cancer through induction of a persistent growth arrest state, whereas SASP is challenging, this possible strategy with reprogrammed and complicated tumor microenvironment as well as declined immune response in aging population.

How might Rh2 suppress SASP in senescent breast cell lines? Initially, DNA damage response and downstream activation of ATM are sufficient to activate certain SASP [31], which apparently can be induced by anticancer drug doxorubicin in the present study. Correspondingly, we detected DNA-SCARS via identifying 53BP1 foci in senescent cells. Currently, the NF-κB signaling pathway is proposed as the potential inducer and signaling pathway that activate NF-κB signaling subsequently triggers SASP [32]. Furthermore, the p38MAPK signaling pathway is a crucial inducer for SASP without genotoxic stress in human fibroblasts [33]. Consistent with these well-documented inducers, our results again confirmed the evident activation of p38MAPK and NF-κB signaling pathways in senescent breast cells. Additionally, we also verified noteworthy upregulation of Akt and MEK pathways, which are potential mediators for SASP development and maintenance. However, decreased expression of Akt and MEK pathway were noted in nontumorigenic cell MCF-10A. Concomitantly, inhibitors SB203580 for p38MAPK and BAY11-7082 for NF-κB significantly suppressed the SASP from both tumorigenic and nontumorigenic cells evidenced by antibody array test. Moreover, notable activation of Stat3 may potentiate the deleterious effect of doxorubicin-induced senescence, since this signaling pathway mediates immune suppression in tumor [34]. Impressively, for doxorubicin-induced senescent breast tumor cell line MDA-MB-231, Rh2 not only suppressed the accepted p38MAPK and NF-κB pathways but also prohibited Stat3 pathway. Together, Rh2 may attenuate SASP development through regulation of above mentioned multipathways.

4. Materials and Methods

4.1. Cell Culture

Human breast cancer cells MDA-MB-231 and normal breast cells MCF-10A were purchased from the American Type Culture Collection (ATCC). Breast cancer cell line MDA-MB-231 was maintained in high glucose Dulbecco's Modified Eagle's Medium (Themro Fisher Scientific, Seoul, South Korea), supplemented with 10% Fetal Bovine Serum and penicillin/streptomycin (Invitrogen, Seoul, South Korea). MCF-10A was maintained in Dulbecco's Modified Eagle's Medium/F-12 (Thermo Fisher Scientific, Seoul, South Korea), supplemented with Mammary Epithelial Cell Growth Medium SingleQuot Kit Supplement & Growth Factors (Lonza, CC-4316, Alpharetta, GA, USA) and 100 ng/mL cholera toxin (Sigma, St. Louis, MO, USA). The cells were incubated at 37 °C in a 5 % CO_2 atmosphere.

4.2. Reagents

Ginsenosides Rg1, Re, F1, Rh1, Rh1, PPT, Rb1, Rd, Gyp75, F2, Rg3, Rh2, CK, and PPD, with a purity of more than 98%, were prepared with High Performance Liquid Chromatography (HPLC, Agilent, Seoul, South Korea). Each ginsenoside was dissolved in dimethyl sulfoxide (DMSO, St. Louis, MO, USA) as 10 mg/mL solution. Bay 11-7082, SB203580 and doxorubicin were purchased from Sigma (St. Louis, MO, USA).

4.3. Senescence Induction and Assessment

Human breast cancer cells MDA-MB-231 and normal breast cells MCF-10A were induced to senescence by exposure to doxorubicin in complete culture medium. Briefly, proliferating cells were treated with indicated concentrations of doxorubicin for 72 h. Later, cells were scored for senescence markers, including senescence-associated β-galactosidase (SA-β-gal) activity and the amount of persistent DNA damage foci. SA-β-gal staining was performed using a SA-β-gal kit (#9860, Cell Signaling Technology Inc., Beverly, MA, USA) in accordance with the manufacture's manual. The cells were fixed for 15 min at room temperature, then rinsed with PBS and stained with staining solution at a final pH of 6.0 for overnight (at least 16 h). The SA-β-gal positive cells develop blue color and were counted under a phase-contrast microscope. DNA damage foci were estimated by immunostaining for 53BP1. For DNA damage foci and SA-β-gal positivity, random fields were selected. Fluorescent images were quantified using CellProfiler (2.2.0, Cambridge, MA, USA), an open source software program. SA-β-gal staining was quantified by researcher that was blind to the treatments.

For SASP development, cells were treated with 100 nM doxorubicin for 2 cycles (day 0 and day 4) for 10 days. Specific inhibitor of p38MAPK (SB203580), NF-κB inhibitor, Bay 11-7082, and ginsenoside Rh2 were added at day 6. Then 48 h later (day 8), cultures were replaced with fresh serum-free medium. Lastly, medium were collected and prepared for the subsequent tests (day 10).

4.4. Cell Proliferation Assay

Cell viability (proliferation) was evaluated by the WST-1 assay, which is based on the enzymatic cleavage of the tetrazolium salt WST-1 to formazan by cellular mitochondrial dehydrogenase present in viable cells. In brief, after 72 h treatment, 20 μL of WST-1 was added to each well and the plates were incubated at 37 °C for 2 h. The plates were then centrifuged and 100 μL of the medium was withdrawn for measuring the absorbance value at a wavelength of 450 nm using a microplate reader (Tecan, Männedorf, Switzerland).

Seventy-two hours after senescence induction, cells were also assessed by immunostaining for Ki-67, a key proliferation marker. Fluorescent images were captured by a Nikon i2 U microscope (Tokyo, Japan).

4.5. Assay of Caspase 3/7 Activation

Cells were plated in 12-well plates and subjected to senescence induction. Six days after that, DMSO or ginsenoside Rh2 was added into culture medium. Forty-eight hours after addition, live imaging was initiated by 30 min preincubation of CellEvent Caspase 3/7 green detection reagent (5 μM, Thermo Fisher Scientific, Seoul, South Korea) using a Nikon i2 U microscope (Tokyo, Japan) and quantification was measured by microplate reader (Tecan, Männedorf, Switzerland).

4.6. Cytokine Antibody Array

The conditioned medium (CM) for antibody analyses were prepared by washing approximately 6×10^6 presenescent and senescent cells 3 times with PBS, and incubating them with serum-free medium for 48 h. The conditioned medium were collected and the remained cells were counted to normalize conditioned medium volumes for cell number. Then medium were centrifuged for 20 min at 5000 rpm, filtered through 0.22 μM bottle-top filters (Sartorius Stedim Biotech, Göttingen, Germany) diluted with serum-free medium to a concentration equivalent to 1×10^6 cells per 1.5 mL, and applied to antibody array (Ray Biotech, Norcross, GA, USA). The signals were detected with Odyssey-LC chemiluminescent imaging system. Signals were averaged and expressed as described in the figure legend.

4.7. Migration and Invasion Assay

MDA-MB-231 and MCF-10A cells were plated in 24-well plates (5×10^4 cells per well) in a complete corresponded medium. Cells were scraped off from the bottom of a culture plate using a pipette tip to produce a cell-free area. Cells were washed with DMEM or DMEM: F12 (1:1) to remove the cell debris and incubated with indicated conditioned media in 3% FBS prepared from senescent MDA-MB-231 and MCF-10A cells treated with or without ginsenoside Rh2 for 24 h. The wound areas were captured at 0 h and 48 h and quantified using CellProfiler Software (2.2.0, Cambridge, MA, USA).

MDA-MB-231 cells were serum-starved overnight, trypsinized, then seeded in the upper chamber with Matrigel-coated transwells in serum-free medium, with cells migrating towards the lower chamber in response to SASP-containing CM (R&D systems, Minneapolis, MN, USA). Cells on the lower side of the membranes were stained with 0.1% crystal violet (Sigma) after 24 h and enumerated.

4.8. Quantification of SASP Major Factors

IL-6 and IL-8 levels in conditioned media were quantified using Human IL-6 and IL-8 ELISA Ray Biotech protocol respectively.

Major SASP factors were analyzed by real-time PCR. Total RNA was prepared with the RNeasy Micro Kit (Qiagen, Germantown, MD, USA). qRT-PCR reactions were performed using the QuantiNova SYBR Green RT-PCR Kit (Qiagen) according to the manufacture's protocol. Primer/probe sets for human IL-6, IL-8, MCP-1, GRO were used: IL6F: 5′-GCCCAGCTATGAACTCCTTCT-3′; IL6R: 5′-GAA GGCAGCAGGCAACAC-3′; IL-8F: 5′-AGACAGCAGAGCACACAAGC-3′; IL-8R:

5′-ATGGTTCCTTCCGGTGGT-3′; MCP-1F: 5′-AGTTCTTGCCGCCCTTCT-3′; MCP-1R: 5′-GTGACTG GGGCATTGATTG-3′; CXCL-1F: 5′-TCCTGCATCCCCCATAGTTA-3′; CXCL-1F: 5′-TCCTGCATCC CCCATAGTTA-3′; CXCL-1R: 5′-CTTCAGGAACAGCCACCAGT-3′.

The Ct-value for targets and endogenous control (GADPH) were used to calculate the relative expression of the gene of interest. Samples were determined in triplicate.

4.9. PathScan® Signaling Nodes Multitarget Sandwich ELISA Assay

Presenescent and senescent MDA-MB-231 and MCF-10A cells were collected after SASP was fully developed, and then lysed with ELISA lysis buffer. Cell lysates were applied to specific assay formulations for the indicated target proteins according to the manufacture's protocol (Ray Biotech, Norcross, GA, USA). Briefly, microwell strips were unsealed and placed at room temperature. One-hundred microliters of sample diluted with equal amount of sample diluent were applied to well and incubated overnight at 4 °C. Then the microwell strips were washed with buffer and incubated with detection antibody for 1h at 37 °C, followed by HRP-linked secondary antibody for 30 min at 37 °C. Next, TMB substrate was added and incubated for 10 min at 37 °C. After adding stop solution to each well, the microwell strips were measured using a microplate reader (Spark 10M, Tecan, Männedorf, Switzerland) at a wavelength of 450 nm within 30 min.

4.10. Western Blotting

Cell lysates were collected and prepared from indicated treatments. Lysates were subjected to 10% SDS-PAGE gels; separated proteins were transferred to polyvinylidene difluoride membrane for 1 h at 110 V. Membranes were blocked and incubated overnight at 4 °C with the following primary antibodies: p16, p21, ZO-1, Vimentin, Snail, Slug, and beta-catenin (rabbit monoclonal, 1:1000, CST, USA), with beta-actin (mouse monoclonal, 1:2000, CST, Norcross, GA, USA) as loading control. Membranes were washed and incubated with horseradish peroxidase-conjugated (1:5000; CST, Norcross, GA, USA) or IRDye 800CW or IRDye 680RD (1:10000; Li-COR, Lincoln, NE, USA) for 1 h at room temperature and washed again. Signals were detected by Odyssey-Fc imaging system (Image Studio, ver5.2, Li-COR, Lincoln, NE, USA).

4.11. Fluorescence Microscopy

Cells with various indicated treatments were washed twice with PBS, then cells were fixed with 4% paraformaldehyde for 10 min and permeabilized with 0.15% Triton X-100 in PBS for 15 min at room temperatures. Cells were then blocked with 3% BSA for 30 min and incubated with indicated primary antibody against ZO-1 (rabbit monoclonal, 1:2000, CST, USA) overnight at 4 °C, followed by incubation with Alexa fluorescein-labeled secondary antibodies (1:200, Thermo Fisher Scientific, Seoul, South Korea) for 1h and mounted with DAPI (Thermo Fisher Scientific, Seoul, South Korea). Images were captured with a Nikon i2 U microscope (Japan).

4.12. Statistical Analyses

All data that show error bars are presented as mean ±s.e.m. The significance of difference in the mean was determined using Student's *t*-test and one-way Analysis of Variance (ANOVA) unless otherwise mentioned. $p < 0.05$ was considered significant. All calculations were performed using GraphPad Prism software (7.0, San Diego, CA, USA).

5. Conclusions

In summary, our results provided Rh2 as a potential candidate for ameliorating chemotherapy-induced senescence bystander effect, which might be associated with such age-related pathology breast tumor progresses and tissue damage.

Author Contributions: Conceptualization, J.-G.H. and B.-M.J.; Writing—Original Draft, J.-G.H. and B.-M.J.; Formal Analysis, Y.-J.Y. and C.-H.C.; Funding Acquisition, S.-C.K.

Funding: This work was supported by the Intelligent Synthetic Biology Center of the Global Frontier Project, funded by the Ministry of Education, Science and Technology (2011-0031957), Republic of Korea.

Conflicts of Interest: The authors declare no conflicts of interest.

References

1. Widmer, N.; Bardin, C.; Chatelut, E.; Paci, A.; Beijnen, J.; Levêque, D.; Veal, G.; Astier, A. Review of therapeutic drug monitoring of anticancer drugs part two-targeted therapies. *Eur. J. Cancer* **2014**, *50*, 2020–2036. [CrossRef] [PubMed]
2. Hickman, J.A. Apoptosis induced by anticancer drugs. *Cancer Metastasis Rev.* **1992**, *11*, 121–139. [CrossRef] [PubMed]
3. Pietenpol, J.A.; Stewart, Z.A. Cell cycle checkpoint signaling: Cell cycle arrest versus apoptosis. *Toxicology* **2002**, *181*, 475–481. [CrossRef]
4. Chang, B.D.; Xuan, Y.Z.; Broude, E.V.; Zhu, H.M.; Schott, B.; Fang, J.; Roninson, I.B. Role of p53 and p21(waf1/cip1) in senescence-like terminal proliferation arrest induced in human tumor cells by chemotherapeutic drugs. *Oncogene* **1999**, *18*, 4808–4818. [CrossRef]
5. Campisi, J. Cellular senescence as a tumor-suppressor mechanism. *Trends Cell Biol.* **2001**, *11*, S27–S31. [CrossRef]
6. Freund, A.; Orjalo, A.V.; Desprez, P.Y.; Campisi, J. Inflammatory networks during cellular senescence: Causes and consequences. *Trends Mol. Med.* **2010**, *16*, 238–246. [CrossRef] [PubMed]
7. Beauséjour, C.M.; Krtolica, A.; Galimi, F.; Narita, M.; Lowe, S.W.; Yaswen, P.; Campisi, J. Reversal of human cellular senescence: Roles of the p53 and p16 pathways. *EMBO J.* **2003**, *22*, 4212–4222. [CrossRef] [PubMed]
8. Chien, Y.; Scuoppo, C.; Wang, X.; Fang, X.; Balgley, B.; Bolden, J.E.; Premsrirut, P.; Luo, W.; Chicas, A.; Lee, C.S. Control of the senescence-associated secretory phenotype by NF-κB promotes senescence and enhances chemosensitivity. *Genes Dev.* **2011**, *25*, 2125–2136. [CrossRef]
9. Tao, J.J.; Visvanathan, K.; Wolff, A.C. Long term side effects of adjuvant chemotherapy in patients with early breast cancer. *Breast* **2015**, *24*, S149–S153. [CrossRef]
10. Scuric, Z.; Carroll, J.E.; Bower, J.E.; Ramos-Perlberg, S.; Petersen, L.; Esquivel, S.; Hogan, M.; Chapman, A.M.; Irwin, M.R.; Breen, E.C.; et al. Biomarkers of aging associated with past treatments in breast cancer survivors. *NPJ Breast Cancer* **2017**, *3*, 50. [CrossRef]
11. Eggert, T.; Wolter, K.; Ji, J.; Ma, C.; Yevsa, T.; Klotz, S.; Medina-Echeverz, J.; Longerich, T.; Forgues, M.; Reisinger, F.; et al. Distinct Functions of Senescence-Associated Immune Responses in Liver Tumor Surveillance and Tumor Progression. *Cancer Cell* **2016**, *30*, 533–547. [CrossRef]
12. Demaria, M.; O'Leary, M.N.; Chang, J.; Shao, L.; Liu, S.; Alimirah, F.; Koenig, K.; Le, C.; Mitin, N.; Deal, A.M.; et al. Cellular Senescence Promotes Adverse Effects of Chemotherapy and Cancer Relapse. *Cancer Discov.* **2017**, *7*, 165–176. [CrossRef]
13. Weyand, C.M.; Goronzy, J.J. Aging of the immune system. Mechanisms and therapeutic targets. *Ann. Am. Thorac. Soc.* **2016**, *13*, S422–S428. [CrossRef]
14. Park, E.Y.; Kim, M.H.; Kim, E.H.; Lee, E.K.; Park, I.S.; Yang, D.C.; Jun, H.S. Efficacy comparison of Korean ginseng and American ginseng on body temperature and metabolic parameters. *Am. J. Chin. Med.* **2014**, *42*, 173–187. [CrossRef]
15. Yang, Y.; Ren, C.; Zhang, Y.; Wu, X. Ginseng: An Nonnegligible Natural Remedy for Healthy Aging. *Aging Dis.* **2017**, *8*, 708–720. [CrossRef]
16. Kim, H.S.; Lee, E.H.; Ko, S.R.; Choi, K.J.; Park, J.H.; Im, D.S. Effects of ginsenosides Rg(3) and Rh-2 on the proliferation of prostate cancer cells. *Arch. Pharm. Res.* **2004**, *27*, 429–435. [CrossRef]
17. Chen, W.; Qiu, Y. Ginsenoside Rh2 Targets EGFR by Up-Regulation of miR-491 to Enhance Anti-tumor Activity in Hepatitis B Virus-Related Hepatocellular Carcinoma. *Cell Biochem. Biophys.* **2015**, *72*, 325–331. [CrossRef]
18. Ge, G.; Yan, Y.; Cai, H. Ginsenoside Rh2 Inhibited Proliferation by Inducing ROS Mediated ER Stress Dependent Apoptosis in Lung Cancer Cells. *Biol. Pharm. Bull.* **2017**, *40*, 2117–2124. [CrossRef]

19. Kim, J.H.; Choi, J.S. Effect of ginsenoside Rh-2 via activation of caspase-3 and Bcl-2-insensitive pathway in ovarian cancer cells. *Physiol. Res.* **2016**, *65*, 1031–1037.
20. Xie, X.; Eberding, A.; Madera, C.; Fazli, L.; Jia, W.; Goldenberg, L.; Gleave, M.; Guns, E.S. Rh2 synergistically enhances paclitaxel or mitoxantrone in prostate cancer models. *J. Urol.* **2006**, *175*, 1926–1931. [CrossRef]
21. Wang, Z.; Zheng, Q.; Liu, K.; Li, G.; Zheng, R. Ginsenoside Rh2 enhances antitumour activity and decreases genotoxic effect of cyclophosphamide. *Basic Clin. Pharmacol. Toxicol.* **2006**, *98*, 411–415. [CrossRef]
22. Alcorta, D.A.; Xiong, Y.; Phelps, D.; Hannon, G.; Beach, D.; Barrett, J.C. Involvement of the cyclin-dependent kinase inhibitor p16 (INK4a) in replicative senescence of normal human fibroblasts. *Proc. Natl. Acad. Sci. USA* **1996**, *93*, 13742–13747. [CrossRef]
23. Oubaha, M.; Miloudi, K.; Dejda, A.; Guber, V.; Mawambo, G.; Germain, M.A.; Bourdel, G.; Popovic, N.; Rezende, F.A.; Kaufman, R.J.; et al. Senescence-associated secretory phenotype contributes to pathological angiogenesis in retinopathy. *Sci. Transl. Med.* **2016**, *8*, 362ra144. [CrossRef]
24. Eom, Y.W.; Kim, M.A.; Park, S.S.; Goo, M.J.; Kwon, H.J.; Sohn, S.; Kim, W.H.; Yoon, G.; Choi, K.S. Two distinct modes of cell death induced by doxorubicin: Apoptosis and cell death through mitotic catastrophe accompanied by senescence-like phenotype. *Oncogene* **2005**, *24*, 4765–4777. [CrossRef]
25. Legha, S.S.; Benjamin, R.S.; Mackay, B.; Ewer, M.; Wallace, S.; Valdivieso, M.; Rasmussen, S.L.; Blumenschein, G.R.; Freireich, E.J. Reduction of doxorubicin cardiotoxicity by prolonged continuous intravenous infusion. *Ann. Intern. Med.* **1982**, *96*, 133–139. [CrossRef]
26. Levina, V.; Marrangoni, A.M.; DeMarco, R.; Gorelik, E.; Lokshin, A.E. Drug-selected human lung cancer stem cells: Cytokine network, tumorigenic and metastatic properties. *PLoS ONE* **2008**, *3*, e3077. [CrossRef]
27. Thorn, M.; Guha, P.; Cunetta, M.; Espat, N.J.; Miller, G.; Junghans, R.P.; Katz, S.C. Tumor-associated GM-CSF overexpression induces immunoinhibitory molecules via STAT3 in myeloid-suppressor cells infiltrating liver metastases. *Cancer Gene Ther.* **2016**, *23*, 188–198. [CrossRef]
28. Takahashi, M.; Miyazaki, H.; Furihata, M.; Sakai, H.; Konakahara, T.; Watanabe, M.; Okada, T. Chemokine CCL2/MCP-1 negatively regulates metastasis in a highly bone marrow-metastatic mouse breast cancer model. *Clin. Exp. Metastasis* **2009**, *26*, 817–828. [CrossRef]
29. Pellikainen, J.M.; Ropponen, K.M.; Kataja, V.V.; Kellokoski, J.K.; Eskelinen, M.J.; Kosma, V.M. Expression of matrix metalloproteinase (MMP)-2 and MMP-9 in breast cancer with a special reference to activator protein-2, HER2, and prognosis. *Clin. Cancer Res.* **2004**, *10*, 7621–7628. [CrossRef]
30. Sieuwerts, A.M.; Martens, J.W.; Dorssers, L.C.; Klijn, J.G.; Foekens, J.A. Differential effects of fibroblast growth factors on expression of genes of the plasminogen activator and insulin-like growth factor systems by human breast fibroblasts. *Thromb. Haemost.* **2002**, *87*, 674–683.
31. Rodier, F.; Coppe, J.P.; Patil, C.K.; Hoeijmakers, W.A.; Munoz, D.P.; Raza, S.R.; Freund, A.; Campeau, E.; Davalos, A.R.; Campisi, J. Persistent DNA damage signalling triggers senescence-associated inflammatory cytokine secretion. *Nat. Cell Biol.* **2009**, *11*, 973–979. [CrossRef]
32. Salminen, A.; Kauppinen, A.; Kaarniranta, K. Emerging role of NF-κB signaling in the induction of senescence-associated secretory phenotype (SASP). *Cell. Signal.* **2012**, *24*, 835–845. [CrossRef]
33. Freund, A.; Patil, C.K.; Campisi, J. p38MAPK is a novel DNA damage response-independent regulator of the senescence-associated secretory phenotype. *EMBO J.* **2011**, *30*, 1536–1548. [CrossRef]
34. Tkach, M.; Coria, L.; Rosemblit, C.; Rivas, M.A.; Proietti, C.J.; Diaz Flaque, M.C.; Beguelin, W.; Frahm, I.; Charreau, E.H.; Cassataro, J.; et al. Targeting Stat3 induces senescence in tumor cells and elicits prophylactic and therapeutic immune responses against breast cancer growth mediated by NK cells and CD4+ T cells. *J. Immunol.* **2012**, *189*, 1162–1172. [CrossRef]

International Journal of
Molecular Sciences

Article

Peanut Sprout Extracts Attenuate Triglyceride Accumulation by Promoting Mitochondrial Fatty Acid Oxidation in Adipocytes

Seok Hee Seo [1,†], Sang-Mi Jo [2,†], Jiyoung Kim [3], Myoungsook Lee [4], Yunkyoung Lee [2] and Inhae Kang [2,*]

1   Department of Marine Life Science, Jeju National University, Jeju 63243, Korea; bossni3@naver.com
2   Department of Food Science and Nutrition, Jeju National University, Jeju 63243, Korea; iris1603@hanmail.net (S.-M.J.); lyk1230@jejunu.ac.kr (Y.L.)
3   Department of Food and Nutrition, Kyungnam College of Information & Technology, Pusan 47011, Korea; kjy1@eagle.kit.ac.kr
4   Department of Food and Nutrition, Sungshin Women's University, Seoul 01133, Korea; mlee@sungshin.ac.kr
*   Correspondence: inhaek@jejunu.ac.kr; Tel.: +82-64-754-3554
†   These authors contributed equally to this work.

Received: 13 February 2019; Accepted: 6 March 2019; Published: 11 March 2019

Abstract: Peanut sprouts (PS), which are germinated peanut seeds, have recently been reported to have anti-oxidant, anti-inflammatory, and anti-obesity effects. However, the underlying mechanisms by which PS modulates lipid metabolism are largely unknown. To address this question, serial doses of PS extract (PSE) were added to 3T3-L1 cells during adipocyte differentiation. PSE (25 µg/mL) significantly attenuated adipogenesis by inhibiting lipid accumulation in addition to reducing the level of adipogenic protein and gene expression with the activation of AMP-activated protein kinase (AMPK). Other adipocyte cell models such as mouse embryonic fibroblasts C3H10T1/2 and primary adipocytes also confirmed the anti-adipogenic properties of PSE. Next, we investigated whether PSE attenuated lipid accumulation in mature adipocytes. We found that PSE significantly suppressed lipogenic gene expression, while fatty acid (FA) oxidation genes were upregulated. Augmentation of FA oxidation by PSE in mature 3T3-L1 adipocytes was confirmed via a radiolabeled-FA oxidation rate experiment by measuring the conversion of $[^3H]$-oleic acid (OA) to $[^3H]$-H_2O. Furthermore, PSE enhanced the mitochondrial oxygen consumption rate (OCR), especially maximal respiration, and beige adipocyte formation in adipocytes. In summary, PSE was effective in reducing lipid accumulation in 3T3-L1 adipocytes through mitochondrial fatty acid oxidation involved in AMPK and mitochondrial activation.

Keywords: peanut sprouts; resveratrol; adipogenesis; fatty acid oxidation; mitochondrial respiration

---

1. Introduction

The prevalence of obesity has reached epidemic proportions in the United States and all over the world [1,2]. Obesity is characterized by the abnormal expansion of white adipose tissue, either due to an increase in the number of adipocytes from pre-adipocytes or an increase in cell size [3,4]. Obesity increases the risk of metabolic disorders, such as type 2 diabetes, heart disease, hypertension, and cancer [5]. Therefore, understanding the molecular mechanisms by which hyperplastic and hypertrophic obesity is driven by adipocytes and identifying the novel molecules for regulating lipid metabolism is necessary for the prevention of obesity.

The peanut (*Arachis hypogaea* L.) is a naturally occurring, common, nutritious food, which contains high levels of protein, unsaturated fatty acid, fiber, potassium, magnesium, copper niacin, arginine, fiber, α-tocopherol, and folates. Bioactive compounds, such as phytosterols, and flavonoids are also rich in the peanut [6]. Peanut seeds can be germinated to create a form called peanut sprouts (PS). PS have a high total polyphenolic content, including a high content of resveratrol, protocatechuic acid, gallic acid, and caffeic acid [7]. Among them, resveratrol (3,5,4′-*trihydroxystilbene*) is a well-known major component of PS. It has been previously reported that the amount of resveratrol significantly increases from 4.5 μg/g on day 0 of germination to around 30 μg/g on day 9 of germination [7]. The health-promoting effects of resveratrol are mainly due to the activation of AMP-activated protein kinase (AMPK)/sirtuin 1 (SIRT1), which regulates energy metabolism in the whole body [8,9]. In particular, resveratrol supplementations have been shown to increase mitochondrial content/activity in skeletal muscle, brown adipose tissue, and the liver. This enhances body basal energy expenditure, which protects against diet-induced obesity and metabolic complications such as insulin resistance and fatty liver disease [8]. Accumulating evidence has reported that PS has anti-oxidant [10], anti-inflammatory [11], and lipid-lowering effects in vivo [12,13] and in vitro [14]. However, most studies were descriptive reports and few studies have explored the role of PS in lipid accumulation and its effect on adipocytes.

In the present study, we first report that PS extract (PSE) is a potent negative regulator of adipogenesis in 3T3-L1 cells adipocytes. We also confirm the anti-adipogenic properties of PSE by using another adipocyte model that does not need clonal expansion such as C3H10T1/2 and primary adipocytes from ear mesenchymal stem cells (EMSCs) [15]. To investigate the regulatory effects of PSE on lipid metabolism, 3T3-L1 adipocytes were used to examine the induction of fatty acid oxidation, energy metabolism-related genes and mitochondrial bioenergetics.

2. Results

2.1. Total Polyphenol, Flavonoid, and Resveratrol Contents of PS Extracts

The total polyphenol content of the PSE was 10.87 mg gallic acid/g extract, while the total flavonoid content was 3.79 mg catechin/g extract (Table 1). The resveratrol content of PSE, determined by LC/MS, was 18 μg/g.

Table 1. Total polyphenol, flavonoid, and resveratrol contents of peanut sprout extract.

Total polyphenols, Flavonoids, and Resveratrol Contents	Peanut Sprout Extract
Total polyphenols (mg Gallic acid/extract g)	10.87 ± 0.11
Total flavonoids (mg Catechin/extract g)	3.79 ± 0.14 [1]
Resveratrol (μg/g)	18

[1] Values are presented as the mean ± SEM of three independent experiments.

2.2. Effects of PSE on Cell Viability 3T3-L1 and C3H10T1/2 Preadipocytes

To determine whether PSE affects the cell viability of pre-adipocytes, the cytotoxic effects of PSE (10–100 μg/mL) were determined in both 3T3-L1 and C3H10T1/2 pre-adipocytes. PSE was incubated for 24 h before the 2,3-Bis-(2-methoxy-4-nitro-5-sulfophenyl)-2H-tetrazolium-5-carboxanilide salt (XTT) assay. As shown in Figure 1, there was no significant reduction in cell viability after different doses of PSE were applied.

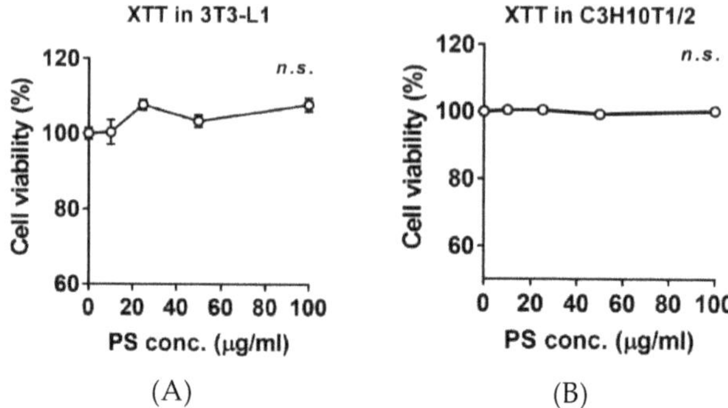

(A)   (B)

Figure 1. Effects of PSE on cell viability in 3T3-L1 and C3H10T1/2 pre-adipocytes. The culture of 3T3-L1 (**A**) and C3H1051/2 cells (**B**) were treated with 10-100 µg/mL of PSE for 24 h. XTT reagent was added 3 h before measurement of OD 450 nm. Data are expressed as a percentage of the vehicle control (dimethyl sulfoxide (DMSO)). *n.s.* represents no significance. Data are represented as the mean ± SEM of three independent experiments. Values that do not share the same superscript are significantly different, as determined by one-way ANOVA ($p < 0.05$).

2.3. PSE Inhibits Adipogenesis in 3T3-L1 Adipocytes

To determine whether PSE was able to inhibit adipogenesis, PSE was added to 3T3-L1 cells during differentiation; this was maintained for 10 days. The presence (50, 100, 200 µg/mL) of PSE caused a significant reduction in triglyceride (TG) accumulation, as measured by Oil-red-O (ORO) staining (Figure 2A–C), which is compatible with 40 µM of resveratrol treatment (resveratrol was used as a positive control). Next, we investigated whether a low dose of PSE (5–25 µg/mL) has anti-adipogenic effects. Adipogenic gene and protein expressions were determined by quantitative PCR (qPCR) and Western blot. PSE treatment, especially 25 µg/mL, significantly suppressed adipogenic gene expression, including peroxisome proliferator-activated receptor gamma (PPARγ), fatty acid binding protein 4 (FABP4, aP2), CCAAT/enhancer binding protein α (C/EBPα), and fatty acid synthase (Fas) (Figure 2D). Based on these results, we used the 25 µg/mL concentration of PSE for the rest of the experiments to ensure we did not cause cellular damage. Adipogenic protein expression, including PPARγ and aP2, was also reduced in murine cultures treated with 25 µg/mL of PSE (Figure 2E).

AMP kinase (AMPK) activation is a well-known mechanism behind the anti-adipogenic properties of several phytochemicals [16–19]. In our study, treatment with PSE (25 µg/mL) increased AMPK phosphorylation (Figure 2E).

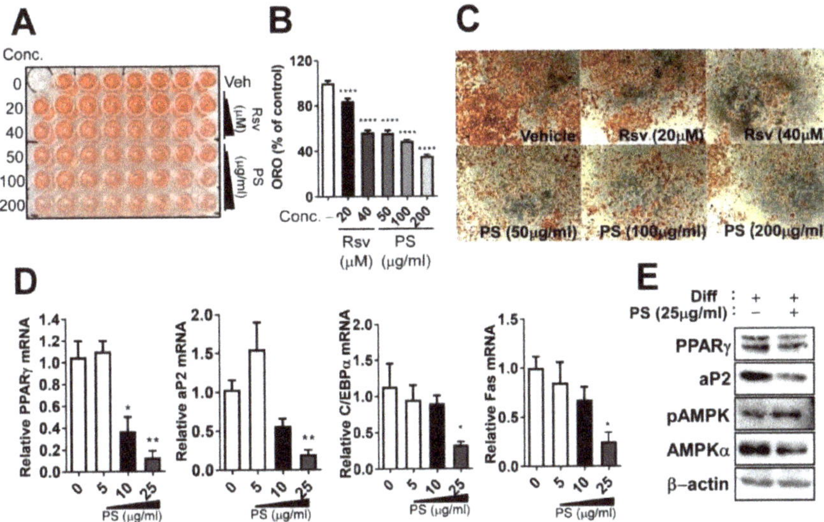

Figure 2. PSE inhibits adipogenesis in 3T3-L1 adipocytes. 3T3-L1 cells were seeded and induced to differentiation in the presence of DMSO (vehicle control), resveratrol (20–40 µM) or PSE (50–200 µg/mL) for 10 days: (**A**) TG accumulation in 96-well culture plates was visualized by ORO staining; (**B**) extracted ORO staining was quantified (OD 500 nm); (**C**) representative images from three separate experiments (magnified 4×); (**D**) adipogenic gene expression of PPARγ, aP2, C/EBP α, and Fas by qPCR; (**E**) adipogenic protein expressions of PPARγ, aP2, phosphor-specific, or total antibodies targeting AMPK and β-actin by Western blot analysis. All values are presented as the mean ±S.E.M. * $p < 0.05$; ** $p < 0.01$; *** $p < 0.001$; **** $p < 0.0001$ compared with the vehicle control (DMSO treated cells) by one-way ANOVA with Bonferroni's comparison test. +; treatment, -; non-treatment.

2.4. PSE Inhibits Adipogenesis in C3H10T1/2 Mouse Embryonic Fibroblasts Adipocytes and EMSCs

We also utilized a complementary approach by using the models do not required clonal expansion to confirm the anti-adipogenic effects of PSE, which were C3H10T1/2 mouse embryonic fibroblasts mimicking primary mouse embryo-fibroblasts [15] and mouse primary adipocytes from EMSCs. PSE (50 and 100 µg/mL) caused a significant reduction in TG accumulation, as measured by ORO staining in C3H10T1/2 cells (Figure 3A). To further confirm the anti-adipogenic effects of PSE, we prepared primary adipocytes derived from EMSCs. The bright-field picture represents the differentiated primary adipocytes and PSE (25 µg/mL) significantly inhibited lipid droplet formation (Figure 3B). Consistent with reduced TG accumulation, the adipogenic gene PPARγ, aP2, C/EBP α, and Fas expression were significantly suppressed by PSE (Figure 3C).

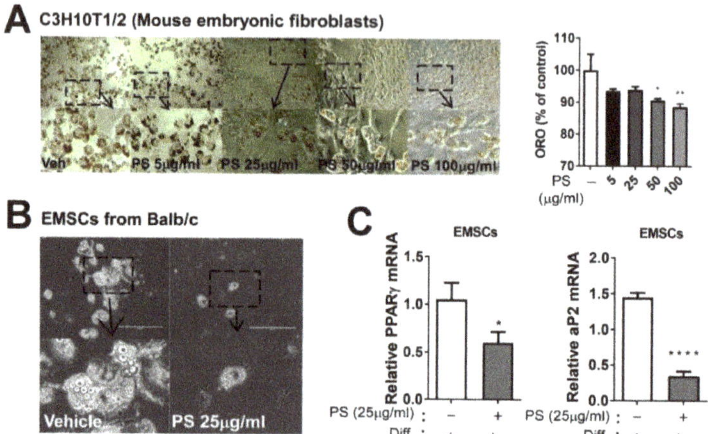

Figure 3. PSE inhibits adipogenesis in C3H10T1/2 mouse embryonic fibroblasts, adipocytes, and EMSCs. C3H10T1/2 cells were seeded and induced to differentiation in the presence of either DMSO (vehicle control) and PSE (5–100 μg/mL) for four days: (**A**) Triglyceride accumulation was visualized by Oil red-O staining and representative images from three separate experiments are shown in (**left**) (magnified 4×); extracted ORO staining was quantified (OD 500 nm) (**right**). Primary adipocytes were prepared from EMSC of Balb/c mice: (**B**) Phase contrast images of primary adipocytes were differentiated with or without PSE (25 μg/mL) for seven days (magnified 4×, scale bar = 400 μm); (**C**) adipogenic gene expression of PPARγ and aP2 by qPCR. * $p < 0.05$; ** $p < 0.01$; **** $p < 0.0001$ compared with the vehicle control (DMSO treated cells) by one-way ANOVA with Bonferroni's comparison test or Student's t-test. +; treatment, -; non-treatment.

2.5. PSE Attenuates Lipid Accumulation in Cultures of Adipocytes by Upregulating Fatty Acid Oxidation and Mitochondrial Oxygen Consumption

We next postulated that PSE would antagonize adipocyte hypertrophy. To address this hypothesis, fully differentiated cultures of 3T3-L1 adipocytes were exposed to PSE (25 μg/mL) for three days based on the experimental design (Figure 4A). Exposure to PSE (25 μg/mL) caused a significant reduction of lipogenic activation as measured by mRNA expression while there was upregulation of fatty acid oxidation-related gene expression (peroxisome proliferator-activated receptor gamma coactivator 1-alpha (PGC1α), carnitine palmitoyltransferase I (CPT1), PPARα) (Figure 4B).

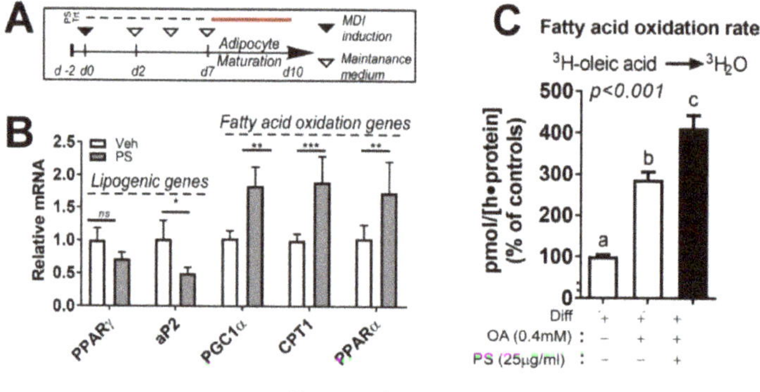

Figure 4. *Cont.*

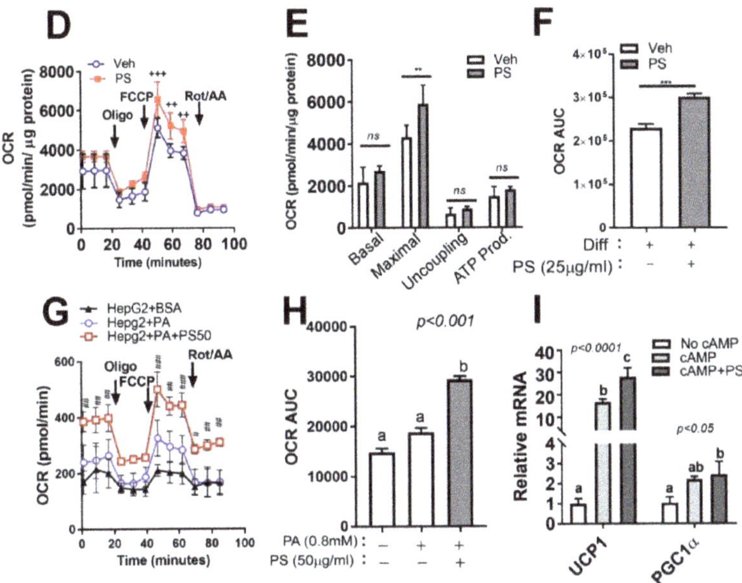

Figure 4. PSE attenuates lipid accumulation in cultures of adipocytes by upregulating fatty acid oxidation and mitochondrial oxygen consumption. (**A**) Experimental scheme. 3T3-L1 were seeded on the second day before differentiation (*d-2*) and induced to differentiation (*d0*, MDI: methyl isobutyl-xanthine, dexamethasone, and insulin). Keep 3T3-L1 cells differentiated into fully differentiated adipocytes until d7. Fully differentiated adipocytes (*d7*) were incubated with PSE (25 μg/mL) for three days. (**B**) Lipogenic and fatty acid oxidation-related gene expression of PPARγ, aP2, PGC1α, CPT1, and PPARα as determined by qPCR. (**C**) Conversion of [^{3}H]-OA into [^{3}H]-H$_2$O. (**D–F**) Oxygen consumption rate (OCR) in 3T3-L1 adipocytes treated with Veh (blue) and PSE (red) as determined by Seahorse extracellular analyzer. 3T3-L1 cells differentiated into fully differentiated adipocytes. Fully differentiated adipocytes (*d7*) were incubated with PSE (25 μg/mL) for one day. Arrow indicates the addition of respiratory inhibitors of oligomycin (Oligo), carbonyl cyanide 4-trifluoromethoxy phenylhydrazone (FCCP) and a combination of antimycin A and rotenone (Rot/AA). (**G–H**) OCR in HepG2 cells treated with BSA (black), PA (blue), and PA + PSE (red) as determined by Seahorse extracellular analyzer. HepG2 cells were pre-incubated with PSE (50 μg/mL)for 48 h. BSA or 0.8 mM BSA-PA complex was loaded for 3 h. (**I**) Relative expressions of UCP1 and PGC1α by qPCR. Pre-treatment of the 3T3-L1 cell with PSE for 7 d during adipogenesis, followed by Bt2-cAMP stimulation for 6 h. All values are presented as the mean ±SEM. *n.s.* represents no significance. * $p < 0.05$; ** $p < 0.01$; *** $p < 0.001$ compared with the vehicle control (DMSO-treated cells) by Student's *t*-test or one-way ANOVA with Bonferroni's comparison test. Means that do not share a common superscript are significantly different as determined by one-way ANOVA with Bonferroni's comparison test. ++ $p < 0.01$; +++ $p < 0.001$ compared with the vehicle control (DMSO treated cells) # $p < 0.05$; ## $p < 0.01$; ### $p < 0.001$ compared with PA-treated HepG2 cells by two-way ANOVA with Bonferroni's comparison test. +; treatment, -; non-treatment.

Next, we investigated the enhancement of FA oxidation in PSE-treated cells by determination of the actual FA oxidation rate using a radiolabeled precursor. [^{3}H]-OA was added to fully differentiated adipocytes for 2 h, then the conversion of [^{3}H]-H$_2$O released from [^{3}H]-OA was measured by a liquid scintillation counter. BSA-OA complex treatment significantly increased the FA oxidation rate, which confirmed the accuracy of this method. PSE (25 μg/mL) markedly increased the conversion of [^{3}H]-OA to [^{3}H]-H$_2$O (Figure 4C).

To determine the impact of PSE on mitochondrial activities, OCR was determined by the Seahorse system, which is a standard method for exhibiting the key parameters of the mitochondrial function. PSE significantly increased maximal respiration compared to the vehicle control. Moreover, we also found an increase in the area under the curve (AUC) of OCR (Figure 4D–F). Mitochondrial bioenergetics properties of PSE were also confirmed in hepatoma HepG2 cells by measuring OCR. Palmitic acid (PA) treatment conjugated with BSA has a marginal increase of OCR compared to the BSA-treated group. Pretreatment 50 µg/mL of PSE enhanced OCR even more than PA-treated HepG2 cells (Figure 4G,H).

After this, we investigated whether the upregulation of FA oxidation and alterations in white adipocyte mitochondrial respiratory function of PSE are involved in the augmentation of beige adipocyte formation. Differentiated 3T3-L1 adipocytes were treated with dibutyryl cyclic AMP (Bt2-cAMP), an analog of cAMP, for mimicking beige adipocytes in the presence or absence of PSE (25 µg/mL). The incubation of mature adipocytes with Bt2-cAMP was correlated with the induction of uncoupling protein 1 (UCP1) gene expression, which confirmed the accuracy of this analog. Interestingly, PSE significantly enhanced UCP1 levels at an even higher level compared to Bt2-cAMP-treated adipocytes. PSE also significantly enhanced the PGC1α gene expression (Figure 4I).

3. Discussion

Peanuts (*Arachis hydrogaea* L.) have nutritional characteristics that can promote human health [6,20–22]. Recent reports have suggested that peanut sprouts prepared from the germination of peanut kernels are rich in phytochemicals. It is suggested that resveratrol (*3,5,4'-trihydroxystilbene*), which is a naturally occurring polyphenol present in grapes, berries, and other vegetables, is the major phytochemical in peanut sprouts. Resveratrol is synthesized by several plants as a defense mechanism against stress, such as UV irradiation and/or microbial infection [23,24]. In rodents, resveratrol improves mitochondrial and metabolic health by modulating energy metabolism via activation of the AMPK/Sirtuin 1 (SIRT1)/PGC-1α axis [8,9]. Some clinical studies have confirmed the metabolic effects of resveratrol [25,26]. Since several recent reports showed the beneficial health effects of PSE, such as anti-inflammatory, anti-oxidant, and anti-obesity effects [10–13], it is plausible to assume that PSE may regulate AMPK/SIRT activation in a similar way to resveratrol to control lipid metabolism in adipocytes. However, the effect of PSE on lipid metabolism through stimulating changes in adipocytes and uncovering its detailed molecular mechanisms have not been sufficiently studied. Here, we demonstrate that PSE is effective at attenuating TG accumulation and increasing mitochondrial FA oxidation involved in AMPK mechanisms, which may ultimately modulate energy metabolism. These results may provide insight into PSE as a unique therapeutic method for controlling adiposity.

In this study, we noticed that PSE from PS that germinated for nine days had a high total polyphenols content (TPC) and total flavonoid content (TFC); moreover, the resveratrol contents were consistent with other reports [7] (Table 1), which indicates that PS may be a potentially functionally beneficial food. Therefore, we decided that it is appropriate to investigate the lipid modulating properties of PSE in adipocytes. By using the high potency of PSE, which has high TPC, TFC, and resveratrol contents, we first assessed the cytotoxicity of PSE in 3T3-L1 and C3H10T1/2 pre-adipocyte cells. Consistently, the accumulating data suggested that PSE treatment does not seem to affect cell viability in several non-carcinogenic cells [27,28]. In a study by Kim et al. [14], there was no significant difference in the cell cytotoxicity of 3T3-L1 cells within two days of PS ethanol treatment (up to 40 µg/mL). In our data, up to 100 µg/mL concentrations of PSE showed no sign of cytotoxicity in the tested time frames for both 3T3-L1 and C3H10T1/2 cells (Figure 1). Based on ORO data, adipogenic gene and protein data, and a literature review of routine PSE concentrations for cellular studies (25–100 µg/mL), we chose the 25 µg/mL concentration of PSE as a non-toxic concentration to assess the potency of PSE in regulating lipid metabolism in adipocytes.

In our study, PSE treatment in adipocytes was associated with at least three metabolic consequences: (i) inhibition of adipogenesis in 3T3-L1 adipocytes and other adipocyte models,

(ii) enhancement of fatty acid oxidation and mitochondrial respiration, which was correlated to AMPK activation, and (iii) increase in beige adipogenic conversion pathways of white adipocytes.

The first metabolic outcome that we immediately noticed was that PSE impacted adipogenesis in several adipocyte cell models (3T3-L1, C3H10T1/2 cells and primary adipocytes from EMSCs) (Figures 1 and 2). In our data, neutral TG accumulation was reduced significantly from 50–200 µg/mL, which is similar to 40 µM of resveratrol treatment (Figure 1A–C). We wondered whether the lower concentration of PSE was also able to inhibit adipogenic and protein expression during adipogenesis and 25 µg/mL of PSE concentration effectively inhibited adipogenic conversion in 3T3-L1 cells. In agreement with our notion, Kim et al. reported that PS ethanol extract (40 µg/mL) significantly attenuated adipogenesis in 3T3-L1 cells [14]. We extracted PSE components with the pressurized hot water extraction method, not ethanol extraction. It is unclear what concentrations and/or types of bioactive components are created by PS extraction (pressurized hot water extraction vs. ethanol extraction), so it would be interesting to compare these two methods regarding the biochemical and functional aspects of PSE. One of the possible mechanisms that involve inhibiting adipogenesis of PSE is the inhibition of matrix metalloproteinases (MMP), which are differentially expressed in adipose tissue during obesity and modulate adipocyte differentiation [29]. Extensive extracellular matrix occurs during adipose tissue growth and MMP, especially MMP2 and MMP9, which play a pivotal role in regulating adipose tissue remodeling. Interestingly, a recent report showed that PSE attenuates MMP2 and MMP9 activity during adipogenesis in 3T3-L1 adipocytes [14]. Apart from PSE, several other functional food components, such as ginseng [30], have been found to regulate MMP during adipogenesis. Further studies are necessary to validate the aforementioned MMP inhibition activity by PSE in adipogenic differentiation. In vivo experiments confirmed the lipid-lowering effects of PSE by using high-fat-diet-fed Sprague–Dawley rats [12,13]. Although we confirmed the in vitro anti-adipogenic properties of PSE in our study, we next need to confirm whether it is translated into an animal and/or human.

Subsequently, we examined the anti-lipogenic properties of PSE by using mature adipocytes. PSE directly acts on adipocytes and triggers changes in gene expression and related biochemical parameters, which are consistent with reduced lipogenesis and enhanced substrate oxidation (Figure 4B). This is confirmed by the FA oxidation rate using a radiolabeled precursor (Figure 4C) and maximal OCR (Figure 4D–F). Since AMPK is a major energy sensor that triggers a variety of catabolic processes and suppresses anabolic pathways simultaneously, it is plausible to connect the lipid-lowering effects of PSE and increase mitochondrial fatty acid oxidation with AMPK activation (Figure 2E). Until now, little information has been available regarding the AMPK activation by PSE, although a considerable number of reports have shown an increase in AMPK/SIRT1 due to resveratrol. Resveratrol activates the AMPK/SIRT1 axis, which inhibits adipogenesis and enhances brown adipocyte formation in vivo and in vitro [9]. Resveratrol 20 µM is required to inhibit adipogenesis ([19,31] and Figure 2A–C), and our PS extract contains 18 µg/g resveratrol (Table 1). Although 18 µg/g resveratrol is a significant amount in PSE, it is possible to assume that the upregulation of mitochondrial FA oxidation and AMPK activation by PSE was not solely due to resveratrol. We should examine whether the beneficial effects of PSE in our study were due to resveratrol or the synergistic effects of other bioactive components. Several plant-derived nutrients, such as Salvianolic acid B [32], anthocyanins [33], and ecliptal, a natural compound isolated from the herb *Eclipta alba*, were recently reported to promote mitochondrial respirations in 3T3-L1 adipocytes. In our study, maximal respiration from OCR, which is regarded as an index of energetic reserve capacity, increased after PSE treatment. This indicated that PSE has the potential to have mitochondrial bioenergetic properties. Although HepG2 cells require a higher concentration of PSE (50 µg/mL) than adipocytes, OCR was also upregulated in a non-adipocyte model in our study (Figure 4G–H). We guess that the low sensitivity of PSE was due to the characteristics of HepG2 cells, which are hepatocyte carcinoma cells, not primary hepatocytes [34]. We are currently undertaking a dose- and time-dependent experiment using HepG2 cells to unravel these issues. Since the liver and adipose tissue are the two major organs regulating systemic lipid metabolism, there needs to be an investigation into the lipid-lowering properties of PSE in the liver. Gomez-Zorita et al. also showed that resveratrol-treated obese Zucker rats had reduced fat accumulation and increased fatty acid oxidation in the liver [35]. Our data showed possible mitochondrial bioenergetic properties in the liver, but it is still unclear whether PSE has the potential to modulate hepatic steatosis. We are currently undertaking several experiments related to the hepatic lipid modulation of PSE by mitochondrial activations.

Finally, we have found the impact of PSE on the beigeing of white adipocytes. It has been suggested that resveratrol could trigger an enhancement in mitochondrial function and metabolic homeostasis for beigeing and brown adipogenesis [31]. In agreement with this notion, our data showed that PSE treatment increased the level of Bt2-cAMP-mediated UCP1 expression to higher than the expression triggered by Bt2-cAMP treatment alone (Figure 4I). This is also involved in the induction of maximal respiration confirmed by seahorse oxygen consumption rate (Figure 4D–F). Since 3T3-L1 cells are a classical white adipocyte model, not a 'good' beige adipocyte model, we are currently conducting mechanistic experiments on PSE for the beigeing of white adipocytes in other brite adipocyte models and also examining the protective effects of PSE on *Lipopolysaccharides* (LPS), which are known as lipoglycans/endotoxin-mediated downregulation of UCP1 levels. Further studies are warranted to investigate these possibilities by measuring metabolic activities in the presence or absence of PSE.

In conclusion, the present study determined that PSE impeded lipid accumulation in adipocytes by the augmentation of mitochondrial fatty acid oxidation, which may mimic resveratrol. A major limitation of our study is that it is still vague whether a 25 µg/mL PSE concentration is feasible in the plasma or adipose depots after nutritional intervention. Thus, determining the clinical relevance and efficacy of PSE supplementation should be conducted with caution. We are currently planning to perform in vivo experiments to confirm PSE's lipid-modulating effects. Nevertheless, we believe that these discoveries support the potential benefit of PSE as a novel anti-adipogenic and anti-lipogenic agent in future clinical studies.

4. Materials and Methods

4.1. Materials

All cell cultures were purchased from SPL (Seoul, Korea). Fetal bovine serum (FBS) and penicillin-streptomycin were purchased from Cellgro Mediatech, Inc. (Herndon, VA, USA). Rosiglitazone (BRL49653) was purchased from Cayman Chemical (Ann Arbor, MI, USA). All other chemicals and reagents were purchased from Sigma Chemical Co (St. Louis, MO, USA) unless otherwise stated.

4.2. Sample Preparation

PS, germinated for nine days, were kindly provided by WooYoung E&T (Jeju, South Korea). PS were dried, freeze-dried, powdered, and extracted using the pressurized hot water extraction method (modified from [36]). A 10-g sample (dry weight) of PS was mixed with 100 mL of Milli-Q water. The extracted solutions were combined and centrifuged at 3000 rpm for 3 min. Next, the obtained extracts were filtered using Whatman®filter paper and the filtrate was lyophilized to obtain the powdered extract. Finally, the obtained sample was dissolved in dimethyl sulfoxide (DMSO, Sigma, St. Louis, MO, USA) at a concentration of 100 mg/mL with several aliquots and utilized freshly in the in vitro experiments.

4.3. Total Polyphenol and Flavonoid Contents of PS Extracts

The total polyphenols content (TPC) of the PS extract was determined using the modified Folin–Denis method [37]. The same volume of PS extracts (50 µL) and 1 M Folin–Ciocalteu's phenol reagent (FMD Millipore Corporation, Darmstadt, Germany) was added to each well of a 96-well plate and left at room temperature for 5 min. A total of 100 µL of 4% Na_2CO_3 solution was added to the reaction, which was incubated for another 60 min at room temperature and protected from light using aluminum foil. The absorbance was obtained at 725 nm at room temperature by a spectrophotometer (Molecular Devices, San Jose, CA, USA) and the results were expressed as gallic acid concentration equivalents.

The total flavonoid content (TFC) was determined by the method of Moreno et al. [38]. We mixed 0.1 mL of the sample with 500 μL of water before 30 μL of a solution containing 5% sodium nitrate was added and the mixture was incubated for 6 min at room temperature. A total of 60 μL of aluminum nitrate (10%, w/v) was added and the mixture was incubated at 25 °C for 5 min. A total of 60 μL of sodium hydroxide was added and the reactant absorbance was measured at 510 nm. The calibration curve was calculated using catechin and the results were expressed in mg of catechin equivalent.

4.4. LC/MS Analyses of PS Extracts

Resveratrol content was determined by LC/MS. Extracts analyzed by HPLC tandem mass spectrometry (LC-MS) were prepared in 0.1% formic acid. A Shimadzu LC system (Kyoto, Japan) was equipped with a Poroshell 120 EC-C18 (3.0 × 50 mm) column. The column temperature was fixed at 40 °C. Mobile phases were 0.1% formic acid in water (solvent A) and 0.1% formic acid in Acetonitrile (solvent B). The gradient was 0 min (60% A), 5 min (30% A), 8 min (5% A), 10 min (5% A), 10.01 min (60% A), and 15 min (60% A). This was followed by 0 min (40% B), 5 min (70% B), 8 min (95% B), 10 min (95% B), 10.01 min (40% B), and 15 min (40% B). The column effluent was monitored at 280 nm and mass spectra data were acquired by electrospray ionization (ESI) in the positive ion mode with a Tandem Mass Spectrometry(API 3200). The source temperature was 500 °C. LC-MS data were collected and processed by Analyst 1.6.2. (SCIEX) (Concord, Ontario, Canada).

4.5. Cell Culture

The 3T3-L1 cells (American Type Culture Collection (ATCC), Manassas, VA, USA) were grown to confluence in a basal medium—Dulbecco's modified Eagle's medium (Sigma) with 50 IU/mL penicillin (Sigma), 50 μg/mL streptomycin (Sigma) and 2 mM l-glutamine (Sigma)—supplemented with 10% newborn calf serum (Linus, Madrid, Spain). Two days after the cells reached confluence (referred as Day 0), they were induced to differentiate in a basal medium containing 10% fetal bovine serum (FBS; Invitrogen, Carlsbad, CA, USA), 1 μM dexamethasone (DEX; Sigma), 0.5 mM methylisobutylxanthine (MIX; Sigma) and 1 μg/mL insulin (Sigma) for 48 h. This was followed by 48 h in a basal medium containing 10% FBS and 1 μg/mL insulin. The cells were subsequently refed fresh basal medium supplemented with 10% FBS (without insulin) every other day. For browning induction of white adipocytes, cells were treated with dibutyryl cyclic adenosine monophosphate (cAMP) (Bt2-cAMP; 0.5 mM) for 6 h at the end of adipocyte differentiation.

C3H10T1/2 cells were obtained from the ATCC and maintained in DMEM supplemented with 2 mM L-glutamine, 100 units/mL penicillin, 100 g/mL streptomycin and 10% (v/v) heat-inactivated fetal bovine serum in a humidified 5% CO_2 atmosphere at 37 °C. To induce adipocyte differentiation, cells (1 × 10^6 cells/mL) were grown to 70–80% confluency. Differentiation was induced 2 days later by adding 10 μM of rosiglitazone in an adipogenesis-inducing medium containing 1 μM dexamethasone, 0.5 mM isobutyl-methylxantine, 0.01 mg/mL insulin and 10% FBS in DMEM. After 72 h, the medium was changed every other day to an adipogenesis-inducing medium containing insulin (0.01 mg/mL) and 1 μM of rosiglitazone.

To prepare the primary cultures of rodent adipocytes, ear mesenchymal stem cells (EMSC) were obtained from the ears of Balb/c mice. Briefly, EMSC were isolated from the pools of 6–8 ears from adult mice (n = 3–4) by collagenase digestion (2 mg/mL). The confluent cultures of EMSC were stimulated with an adipogenic differentiation mixture according to standard adipocyte differentiation protocols [39].

HepG2 cells were kindly provided by Dr. Shin and cells were originally obtained from the Korean Cell Line Bank (KCLB, Seoul, South Korea). HepG2 cells were maintained in Dulbecco's modification of Eagle's medium (DMEM)/Ham's F12 containing 1 mM glucose, 1% L-glutamine, 10% fetal bovine serum, 100 units/mL penicillin, and 100 g/mL streptomycin in 5% CO_2 at 37 °C and used for OCR.

4.6. Cell Viability Assay

The cytotoxic effect of PSE was determined using the XTT cell viability kit (Cell Signaling Technology, Beverly, MA, USA) according to the manufacturer's protocol. Briefly, 3T3-L1 and C3H10T1/2 cells were cultured in 96-well plates with a seeding density of approximately 20,000 cells/well. Cells were incubated with either DMSO or increasing concentrations of PSE for 24 h (Figure 1). After this, the medium was replaced with a fresh medium containing XTT solution for 3 h at 37 °C before measurement of OD 450 nm using a microplate reader.

4.7. Lipid Accumulation

To measure the lipid accumulation in adipocytes, cells were fixed with 10% formalin and stained with oil red O (ORO). Bright-field images were taken by CKX41 Inverted Microscope (Olympus, Melville, NY, USA) and ORO dye was extracted by isopropanol to quantify the relative triglyceride (TG) accumulation (at OD 500 nm).

4.8. Total RNA Extraction and qPCR

Gene-specific primers for qPCR were obtained from Cosmo Genetech (Seoul, Korea). Total RNA was isolated with Trizol reagent (Invitrogen). To remove potential genomic DNA contamination, mRNA was treated with DNase (Mediatech). RNA concentrations were measured by the NanoDrop (Nano-200 Micro-Spectrophotometer, Hangzhou City, China). A total of 1 μg of mRNA was converted into cDNA in a total volume of 20 μL (High-capacity cDNA reverse transcription kits, Applied Biosystems, Foster City, CA, USA). Gene expression was determined by real-time qPCR (CFX96™ Real-Time PCR Detection System, Bio-Rad, CA, USA) and relative gene expression was normalized by hypoxanthine guanine phosphoribosyltransferase (HPRT) and/or glyceraldehyde 3-phosphate dehydrogenase (GAPDH) (primer sequences are available in Table S1).

4.9. Western Blot Analysis

To prepare total cell lysates, monolayers of 3T3-L1 adipocytes were scraped with ice-cold radioimmune precipitation assay (RIPA) buffer (Thermo Scientific, Waltham, MA, USA) containing protease inhibitors (Sigma). Proteins were fractionated using 8% or 10% SDS-PAGE, transferred to PVDF membranes and incubated with the relevant antibodies. Chemiluminescence from the ECL (Western Lightning) solution was detected with ChemiDoc (Bio-Rad, Hercules, CA, USA). Polyclonal or monoclonal antibodies targeting phospho-AMPK (Thr172, #2535), total AMPK (#5831), PPARγ (#2435), and β-actin (#4967) were purchased from Cell Signaling Technology. The mouse monoclonal antibodies for aP2 (sc-271529) were purchased from Santa Cruz Biotechnology (Santa Cruz, CA, USA).

4.10. Fatty Oxidation Rate Using [^{3}H]-OA

To measure the FA oxidation rate, we followed the previously published methods by Olpin et al. and Kang et al. [17,40], who utilized cultures of mature 3T3-L1 adipocytes. Briefly, cells were incubated with serum-free low glucose (1000 mg/L d-(+)-glucose) overnight before the experiment. [^{3}H]-OA (Perkin Elmer, Norwalk, CT, USA; final concentration of 0.5 μCi/mL) were mixed with a sodium oleate–bovine serum albumin (BSA) complex (400 μM), before this was added to cells and the mixture incubated for 2 h. The [^{3}H] radioactive containing medium was harvested and precipitated using 100% trichloroacetic acid (TCA) solution. After precipitation, we added 6N NaOH to reach a final concentration of 0.8–1.0 N to obtain an alkaline supernatant. The supernatant was run through columns filled with Dowex ion-exchange resin (Acros Organics, AC202971000, Geel, Belgium) to capture [^{3}H]-H$_2$O. Radioactivity was measured by MicroBeta Microplate counters (PerkinElmer, (Norwalk, CT, USA).

4.11. Oxygen Consumption Rate (OCR) by Seahorse

To determine the mitochondrial respiration activities, the O$_2$ concentration in the 3T3-L1 adipocytes and human hepatoma HepG2 cells were measured using a XF24 extracellular flux analyzer (Agilent Technologies, Santa Clara, CA, USA). Briefly, 3T3-L1 cells were seeded in a gelatin-coated seahorse microplate (24-well) until they reached confluence, which was followed by adipogenic differentiation as described above. HepG2 cells were pre-incubated with PSE (50 µg/mL) or DMSO for 48 h and 0.8 mM BSA-palmitic acid (PA) complex was loaded for 3 h. The mitochondrial basal respiration was assessed in untreated cells. The cells were then treated with oligomycin (oligo, 2 µM) to measure the ATP turnover. The maximum respiratory capacity was assessed by the addition of carbonyl cyanide 4-trifluoromethoxy phenylhydrazone (FCCP, 0.5 µM), which is a chemical uncoupler of electron transport and oxidative phosphorylation. The mitochondrial respiration was blocked by a combination of antimycin A (1 µM) and rotenone (1 µM) (A + R). The OCR was calculated by plotting the O$_2$ tension of the medium in the microenvironment above the cells as a function of time, which was normalized by protein concentrations and expressed in pmol O$_2$/min/µg protein.

4.12. Statistical Analysis

All the data were expressed as means ± standard error (SE), and statistical calculations were performed using the *t*-test and ANOVA (one-way analysis of variance) with Tukey's and Bonferroni's multiple comparison tests. Results were considered significant if $p < 0.05$ (GraphPad Prism Version 7.0, La Jolla, CA, USA).

Supplementary Materials: Supplementary materials can be found at http://www.mdpi.com/1422-0067/20/5/1216/s1.

Author Contributions: Conceptualization, I.K.; methodology, J.K.; investigation, S.S., S.J., and I.K.; data curation, S.S., and S.J.; writing—original draft preparation, I.K.; writing—review and editing, M.L., J.K., and Y.L.; project administration, I.K.; funding acquisition, I.K.

Funding: The following are the results of a study on the "Leaders INdustry-university Cooperation+" Project, supported by the Ministry of Education, (MOE, No. 20180402) and a National Research Foundation of Korea (NRF) grant funded by the Korean government (MSIT) (No. 2018R1C1B5042821).

Acknowledgments: We are grateful to WooYoung E&T (Jeju, South Korea) for the kind gift of the peanut sprouts used in our experiments and Koptri (Seoul, South Korea) for analyzing the resveratrol contents.

Conflicts of Interest: The authors declare no conflict of interest.

Abbreviations

AMPK	AMP-activated protein kinase
aP2, FABP4	fatty acid binding protein 4
AUC	area under the curve
cAMP	Cyclic adenosine monophosphate
CPT1	Carnitine palmitoyltransferase I
EMSCs	Ear mesenchymal stem cells
FA	fatty acid
MDI	methylisobutylxanthine, dexamethasone, and insulin
OA	oleic acid
OCR	oxygen consumption rate
ORO	oil red O
PA	Palmitic acid
PGC1α	Peroxisome proliferator-activated receptor gamma coactivator 1-alpha

PPARγ	Peroxisome proliferator-activated receptor gamma
PS	Peanut sprouts
PSE	Peanut sprout extracts
SIRT1	sirtuin 1
TFC	total flavonoid contents
TPC	total polyphenol contents
UCP1	uncoupling protein 1

References

1. World Health Organization. *Obesity: Preventing and Managing the Global Epidemic: Report of a WHO Consultation*; World Health Organization: Geneva, Switzerland, 2000; p. 253.
2. Imes, C.C.; Burke, L.E. The Obesity Epidemic: The United States as a Cautionary Tale for the Rest of the World. *Curr. Epidemiol. Rep.* **2014**, *1*, 82–88. [CrossRef] [PubMed]
3. Jo, J.; Gavrilova, O.; Pack, S.; Jou, W.; Mullen, S.; Sumner, A.E.; Cushman, S.W.; Periwal, V. Hypertrophy and/or Hyperplasia: Dynamics of Adipose Tissue Growth. *PLoS Comput. Biol.* **2009**, *5*, e1000324. [CrossRef] [PubMed]
4. Sun, K.; Kusminski, C.M.; Scherer, P.E. Adipose tissue remodeling and obesity. *J. Clin. Investig.* **2011**, *121*, 2094–2101. [CrossRef] [PubMed]
5. Despres, J.P.; Lemieux, I. Abdominal obesity and metabolic syndrome. *Nature* **2006**, *444*, 881–887. [CrossRef] [PubMed]
6. Griel, A.E.; Eissenstat, B.; Juturu, V.; Hsieh, G.; Kris-Etherton, P.M. Improved diet quality with peanut consumption. *J. Am. Coll. Nutr.* **2004**, *23*, 660–668. [CrossRef] [PubMed]
7. Wang, K.H.; Lai, Y.H.; Chang, J.C.; Ko, T.F.; Shyu, S.L.; Chiou, R.Y. Germination of peanut kernels to enhance resveratrol biosynthesis and prepare sprouts as a functional vegetable. *J. Agric. Food Chem.* **2005**, *53*, 242–246. [CrossRef] [PubMed]
8. Lagouge, M.; Argmann, C.; Gerhart-Hines, Z.; Meziane, H.; Lerin, C.; Daussin, F.; Messadeq, N.; Milne, J.; Lambert, P.; Elliott, P.; et al. Resveratrol improves mitochondrial function and protects against metabolic disease by activating SIRT1 and PGC-1alpha. *Cell* **2006**, *127*, 1109–1122. [CrossRef] [PubMed]
9. Price, N.L.; Gomes, A.P.; Ling, A.J.; Duarte, F.V.; Martin-Montalvo, A.; North, B.J.; Agarwal, B.; Ye, L.; Ramadori, G.; Teodoro, J.S.; et al. SIRT1 is required for AMPK activation and the beneficial effects of resveratrol on mitochondrial function. *Cell Metab.* **2012**, *15*, 675–690. [CrossRef] [PubMed]
10. Choi, J.Y.; Choi, D.I.; Lee, J.B.; Yun, S.J.; Lee, D.H.; Eun, J.B.; Lee, S.C. Ethanol extract of peanut sprout induces Nrf2 activation and expression of antioxidant and detoxifying enzymes in human dermal fibroblasts: Implication for its protection against UVB-irradiated oxidative stress. *Photochem. Photobiol.* **2013**, *89*, 453–460. [CrossRef] [PubMed]
11. Limmongkon, A.; Nopprang, P.; Chaikeandee, P.; Somboon, T.; Wongshaya, P.; Pilaisangsuree, V. LC-MS/MS profiles and interrelationships between the anti-inflammatory activity, total phenolic content and antioxidant potential of Kalasin 2 cultivar peanut sprout crude extract. *Food Chem.* **2018**, *239*, 569–578. [CrossRef] [PubMed]
12. Kang, N.E.; Ha, A.W.; Woo, H.W.; Kim, W.K. Peanut sprouts extract (*Arachis hypogaea* L.) has anti-obesity effects by controlling the protein expressions of PPARgamma and adiponectin of adipose tissue in rats fed high-fat diet. *Nutr. Res. Pract.* **2014**, *8*, 158–164. [CrossRef] [PubMed]
13. Ha, A.W.; Kang, N.E.; Kim, W.K. Ethanol Extract of Peanut Sprout Lowers Blood Triglyceride Levels, Possibly Through a Pathway Involving SREBP-1c in Rats Fed a High-Fat Diet. *J. Med. Food* **2015**, *18*, 850–855. [CrossRef] [PubMed]
14. Kim, W.K.; Kang, N.E.; Kim, M.H.; Ha, A.W. Peanut sprout ethanol extract inhibits the adipocyte proliferation, differentiation, and matrix metalloproteinases activities in mouse fibroblast 3T3-L1 preadipocytes. *Nutr. Res. Pract.* **2013**, *7*, 160–165. [CrossRef] [PubMed]
15. Jefcoate, C.R.; Wang, S.; Liu, X. Methods that resolve different contributions of clonal expansion to adipogenesis in 3T3-L1 and C3H10T1/2 cells. *Methods Mol. Biol.* **2008**, *456*, 173–193. [CrossRef] [PubMed]

16. Choi, H.S.; Jeon, H.J.; Lee, O.H.; Lee, B.Y. Dieckol, a major phlorotannin in Ecklonia cava, suppresses lipid accumulation in the adipocytes of high-fat diet-fed zebrafish and mice: Inhibition of early adipogenesis via cell-cycle arrest and AMPKalpha activation. *Mol. Nutr. Food Res.* **2015**, *59*, 1458–1471. [CrossRef] [PubMed]

17. Kang, I.; Kim, Y.; Tomas-Barberan, F.A.; Espin, J.C.; Chung, S. Urolithin A, C, and D, but not iso-urolithin A and urolithin B, attenuate triglyceride accumulation in human cultures of adipocytes and hepatocytes. *Mol. Nutr. Food Res.* **2016**, *60*, 1129–1138. [CrossRef] [PubMed]

18. Hwang, J.T.; Park, I.J.; Shin, J.I.; Lee, Y.K.; Lee, S.K.; Baik, H.W.; Ha, J.; Park, O.J. Genistein, EGCG, and capsaicin inhibit adipocyte differentiation process via activating AMP-activated protein kinase. *Biochem. Biophys. Res. Commun.* **2005**, *338*, 694–699. [CrossRef] [PubMed]

19. Chen, S.; Xiao, X.; Feng, X.; Li, W.; Zhou, N.; Zheng, L.; Sun, Y.; Zhang, Z.; Zhu, W. Resveratrol induces Sirt1-dependent apoptosis in 3T3-L1 preadipocytes by activating AMPK and suppressing AKT activity and survivin expression. *J. Nutr. Biochem.* **2012**, *23*, 1100–1112. [CrossRef] [PubMed]

20. Vassiliou, E.K.; Gonzalez, A.; Garcia, C.; Tadros, J.H.; Chakraborty, G.; Toney, J.H. Oleic acid and peanut oil high in oleic acid reverse the inhibitory effect of insulin production of the inflammatory cytokine TNF-alpha both in vitro and in vivo systems. *Lipids Health Dis.* **2009**, *8*, 25. [CrossRef] [PubMed]

21. Jiang, R.; Manson, J.E.; Stampfer, M.J.; Liu, S.; Willett, W.C.; Hu, F.B. Nut and peanut butter consumption and risk of type 2 diabetes in women. *JAMA* **2002**, *288*, 2554–2560. [CrossRef] [PubMed]

22. Alper, C.M.; Mattes, R.D. Peanut consumption improves indices of cardiovascular disease risk in healthy adults. *J. Am. Coll. Nutr.* **2003**, *22*, 133–141. [CrossRef] [PubMed]

23. Langcake, P.; Pryce, R.J. The production of resveratrol and the viniferins by grapevines in response to ultraviolet irradiation. *Phytochemistry* **1977**, *16*, 1193–1196. [CrossRef]

24. Jeandet, P.; Bessis, R.; Gautheron, B. The Production of Resveratrol (3,5,4′-trihydroxystilbene) by Grape Berries in Different Developmental Stages. *Am. J. Enol. Vitic.* **1991**, *42*, 41–46.

25. Poulsen, M.M.; Vestergaard, P.F.; Clasen, B.F.; Radko, Y.; Christensen, L.P.; Stodkilde-Jorgensen, H.; Moller, N.; Jessen, N.; Pedersen, S.B.; Jorgensen, J.O. High-dose resveratrol supplementation in obese men: An investigator-initiated, randomized, placebo-controlled clinical trial of substrate metabolism, insulin sensitivity, and body composition. *Diabetes* **2013**, *62*, 1186–1195. [CrossRef] [PubMed]

26. Timmers, S.; Konings, E.; Bilet, L.; Houtkooper, R.H.; van de Weijer, T.; Goossens, G.H.; Hoeks, J.; van der Krieken, S.; Ryu, D.; Kersten, S.; et al. Calorie restriction-like effects of 30 days of resveratrol supplementation on energy metabolism and metabolic profile in obese humans. *Cell Metab.* **2011**, *14*, 612–622. [CrossRef] [PubMed]

27. Youn, C.K.; Jo, E.R.; Sim, J.H.; Cho, S.I. Peanut sprout extract attenuates cisplatin-induced ototoxicity by induction of the Akt/Nrf2-mediated redox pathway. *Int. J. Pediatr. Otorhinolaryngol.* **2017**, *92*, 61–66. [CrossRef] [PubMed]

28. Choi, D.I.; Choi, J.Y.; Kim, Y.J.; Lee, J.B.; Kim, S.O.; Shin, H.T.; Lee, S.C. Ethanol Extract of Peanut Sprout Exhibits a Potent Anti-Inflammatory Activity in Both an Oxazolone-Induced Contact Dermatitis Mouse Model and Compound 48/80-Treated HaCaT Cells. *Ann. Dermatol.* **2015**, *27*, 142–151. [CrossRef] [PubMed]

29. Chavey, C.; Mari, B.; Monthouel, M.N.; Bonnafous, S.; Anglard, P.; Van Obberghen, E.; Tartare-Deckert, S. Matrix metalloproteinases are differentially expressed in adipose tissue during obesity and modulate adipocyte differentiation. *J. Biol. Chem.* **2003**, *278*, 11888–11896. [CrossRef] [PubMed]

30. Oh, J.; Lee, H.; Park, D.; Ahn, J.; Shin, S.S.; Yoon, M. Ginseng and Its Active Components Ginsenosides Inhibit Adipogenesis in 3T3-L1 Cells by Regulating MMP-2 and MMP-9. *Evid. Complement. Alternat. Med.* **2012**, *2012*, 265023. [CrossRef] [PubMed]

31. Wang, S.; Liang, X.; Yang, Q.; Fu, X.; Zhu, M.; Rodgers, B.D.; Jiang, Q.; Dodson, M.V.; Du, M. Resveratrol enhances brown adipocyte formation and function by activating AMP-activated protein kinase (AMPK) alpha1 in mice fed high-fat diet. *Mol. Nutr. Food Res.* **2017**, *61*. [CrossRef] [PubMed]

32. Pan, Y.; Zhao, W.; Zhao, D.; Wang, C.; Yu, N.; An, T.; Mo, F.; Liu, J.; Miao, J.; Lv, B.; et al. Salvianolic Acid B Improves Mitochondrial Function in 3T3-L1 Adipocytes Through a Pathway Involving PPARgamma Coactivator-1alpha (PGC-1alpha). *Front. Pharmacol.* **2018**, *9*, 671. [CrossRef] [PubMed]

33. Skates, E.; Overall, J.; DeZego, K.; Wilson, M.; Esposito, D.; Lila, M.A.; Komarnytsky, S. Berries containing anthocyanins with enhanced methylation profiles are more effective at ameliorating high fat diet-induced metabolic damage. *Food Chem. Toxicol.* **2018**, *111*, 445–453. [CrossRef] [PubMed]

34. Wilkening, S.; Stahl, F.; Bader, A. Comparison of primary human hepatocytes and hepatoma cell line Hepg2 with regard to their biotransformation properties. *Drug Metab. Dispos.* **2003**, *31*, 1035–1042. [CrossRef] [PubMed]

35. Gomez-Zorita, S.; Fernandez-Quintela, A.; Macarulla, M.T.; Aguirre, L.; Hijona, E.; Bujanda, L.; Milagro, F.; Martinez, J.A.; Portillo, M.P. Resveratrol attenuates steatosis in obese Zucker rats by decreasing fatty acid availability and reducing oxidative stress. *Br. J. Nutr.* **2012**, *107*, 202–210. [CrossRef] [PubMed]

36. Kim, D.S.; Kim, M.B.; Lim, S.B. Enhancement of Phenolic Production and Antioxidant Activity from Buckwheat Leaves by Subcritical Water Extraction. *Prev. Nutr. Food Sci.* **2017**, *22*, 345–352. [CrossRef] [PubMed]

37. Singleton, V.L.; Rossi, J.A. Colorimetry of Total Phenolics with Phosphomolybdic-Phosphotungstic Acid Reagents. *Am. J. Enol. Vitic.* **1965**, *16*, 144–158.

38. Moreno, M.I.; Isla, M.I.; Sampietro, A.R.; Vattuone, M.A. Comparison of the free radical-scavenging activity of propolis from several regions of Argentina. *J. Ethnopharmacol.* **2000**, *71*, 109–114. [CrossRef]

39. Gawronska-Kozak, B. Preparation and differentiation of mesenchymal stem cells from ears of adult mice. *Methods Enzymol.* **2014**, *538*, 1–13. [CrossRef] [PubMed]

40. Olpin, S.E.; Manning, N.J.; Pollitt, R.J.; Clarke, S. Improved detection of long-chain fatty acid oxidation defects in intact cells using [9,10-3H] oleic acid. *J. Inherit. Metab. Dis.* **1997**, *20*, 415–419. [CrossRef] [PubMed]

International Journal of
Molecular Sciences

MDPI

Article

Isodon rugosus (Wall. ex Benth.) Codd In Vitro Cultures: Establishment, Phytochemical Characterization and In Vitro Antioxidant and Anti-Aging Activities

Bilal Haider Abbasi [1,2,3,4,*,†], Aisha Siddiquah [1,†], Duangjai Tungmunnithum [2,3,5,†],
Shankhamala Bose [4], Muhammad Younas [1], Laurine Garros [2,3,6], Samantha Drouet [2,3],
Nathalie Giglioli-Guivarc'h [4] and Christophe Hano [2,3,*,†]

[1]   Department of Biotechnology, Quaid-i-Azam University, Islamabad 45320, Pakistan;
      aisha_siddiquah@yahoo.com (A.S.); pk.younas@gmail.com (M.Y.)
[2]   Laboratoire de Biologie des Ligneux et des Grandes Cultures (LBLGC EA1207), INRA USC1328,
      Plant Lignans Team, Université d'Orléans, 45067 Orléans CÉDEX 2, France;
      duangjai.tun@mahidol.ac.th (D.T.); laurine.garros@etu.univ-orleans.fr (L.G.);
      samantha.drouet@univ-orleans.fr (S.D.)
[3]   Bioactifs et Cosmétiques, GDR 3711 COSM'ACTIFS, CNRS, 45067 Orléans CÉDEX 2, France
[4]   EA2106 Biomolécules et Biotechnologies Végétales, Université de Tours, 37200 Tours, France;
      shankhamala85@gmail.com (S.B.); nathalie.guivarch@univ-tours.fr (N.G.-G.)
[5]   Department of Pharmaceutical Botany, Faculty of Pharmacy, Mahidol University, 447 Sri-Ayuthaya Road,
      Rajathevi, Bangkok 10400, Thailand
[6]   Institut de Chimie Organique et Analytique, ICOA UMR7311, Université d'Orléans-CNRS,
      45067 Orléans CÉDEX 2, France
*   Correspondence: bhabbasi@qau.edu.pk (B.H.A.); hano@univ-orleans.fr (C.H.)
†   These authors contributed equally to this work.

Received: 14 December 2018; Accepted: 19 January 2019; Published: 21 January 2019

Abstract: *Isodon rugosus* (Wall. ex Benth.) Codd accumulates large amounts of phenolics and pentacyclic triterpenes. The present study deals with the in vitro callus induction from stem and leaf explants of *I. rugosus* under various plant growth regulators (PGRs) for the production of antioxidant and anti-ageing compounds. Among all the tested PGRs, thidiazuron (TDZ) used alone or in conjunction with α-napthalene acetic acid (NAA) induced highest callogenesis in stem-derived explants, as compared to leaf-derived explants. Stem-derived callus culture displayed maximum total phenolic content and antioxidant activity under optimum hormonal combination (3.0 mg/L TDZ + 1.0 mg/L NAA). HPLC analysis revealed the presence of plectranthoic acid (373.92 µg/g DW), oleanolic acid (287.58 µg/g DW), betulinic acid (90.51 µg/g DW), caffeic acid (91.71 µg/g DW), and rosmarinic acid (1732.61 µg/g DW). Complete antioxidant and anti-aging potential of extracts with very contrasting phytochemical profiles were investigated. Correlation analyses revealed rosmarinic acid as the main contributor for antioxidant activity and anti-aging hyaluronidase, advance glycation end-products inhibitions and SIRT1 activation, whereas, pentacyclic triterpenoids were correlated with elastase, collagenase, and tyrosinase inhibitions. Altogether, these results clearly evidenced the great valorization potential of *I. rugosus* calli for the production of antioxidant and anti-aging bioactive extracts for cosmetic applications.

Keywords: *Isodon rugosus*; pentacyclic triterpenoid; phenolic acid; plant growth regulators; anti-aging; antioxidant

1. Introduction

According to a survey, around 70,000 important plant species are consumed for health and wellness purposes. Besides this, industry is continuously producing a large amount of bioactive compounds, but herbal medicines and phytotherapy are still in practice in many areas of the globe [1]. Species members of the genus *Isodon* (Schrader ex Bentham) Spach from *Lamiaceae* family are known universally for their economical and medicinal worth [2]. Plants from family *Lamiaceae* have been explored well for their health and wellness properties such as the treatment of hypertension, fever, rheumatism, dementia, toothache, cancer, antimicrobial, hypoglycemic, phytotoxic, antidiarrheal, anticholinesterase, lipoxygenase inhibitory, bronchodilator, and anthelmintic [3–8]. Among them, *I. rugosus* (Wall. ex Benth.) Codd is one of the most challenging and attractive choices to characterize its potential compounds and screen the biological activities that are applicable for cosmetic aspects. This medicinal plant is present in Pakistan, and widely distributed in the Northern areas of the country, especially in Gilgit; this *I. rugosus* is also recognized by various vernacular names such as sperkai, boi, and phaypush [5,6]. This medicinal plant is an aromatic shrub, its stems erect with the quadrangular branches, its leaves are opposite, broadly ovate shape with green color; leaf blade consist of small stellate dendroid hairs. Its inflorescence is Cymose, each flower is white or spotted pink or violet, bilabiate form, Nutlets fruit is an oblong shape with dark brown color. From a pharmacological point of view, this *Isodon* species is rich in bioactive compounds, with potential applications in cosmetics, as well as, traditional and modern medicine industries. *I. rugosus* is an aromatic medicinal plant containing essential oils. Previous works examined its essential oil composition by GC and GC-MS analysis [5,9–11]. Most of the analyzed extracts came from wild fresh living plants or dry plant materials harvested from the forest or the field [5,9,10,12]. The presence of pentacyclic triterpenes and caffeic acid phenolic derivatives have been also reported in this plant [13]. Important phytochemicals including pentacyclic triterpenoids (plectranthoic acid (PA), oleanolic acid (OA), betunilic acid (BA)), and other phenolic compounds were detected in *I. rugosus* [13]. As in many *Lamiaceae* species, both rosmarinic acid (RA) and caffeic acid (CA) are the predominant phenolic compounds that could be the reason behind the antioxidant properties of this plant [14–16].

Human beings have extensively exploited medicinal plants as bioactive ingredients for therapeutic and cosmetic applications since ancient times [17]. The anti-aging activities of plants have been credited to their intrinsic ability to reduce free radical damages to the skin, along with their ability to modulate the activity of many enzymes involved in aging process. For example, their capacity to inhibit elastase, hyaluronidase, or collagenase involve the cleavage of extracellular matrix components, while tyrosinase inhibition involves hyperpigmentation related to skin aging, or more recently to their capacity to activate SIRT1, which is a key regulator involved in the control of both oxidative stress response and regulation of aging processes. Pentacyclic triterpenoids have been regarded as effective enzymes inhibitors that are involved in the cleavage of the extracellular matrix components [18,19], whereas phenolic acids are described as strong antioxidants and possible potent SIRT1 activators [19–21]. As a potential rich source of these compounds, *I. rugosus* could be an attractive plant for cosmetic applications, however, currently this possibility has never been explored and the rationale of the possible biological activities of this plant are therefore unknown. Moreover, slow growth, slow germination, and a conventional way of harvesting large amount of wild plants have threatened this species. Therefore, alternative strategies are now required for both its conservation and usage. In this regard, in vitro culturing techniques ensure preservation of the uncommon and scarce medicinal plant species [22]. Production of secondary metabolites *via* tissue culture techniques provides imperative benefits to amplify the assembly of appropriate substances. Consequently, biotechnological strategies magnify the estimation of these bioactive phytochemicals [23,24]. Herbal products have gained attention worldwide because of their production of specialized metabolites such as phenolics and pentacyclic triterpenoids [25,26].

Some in vitro cultivation has been reported on some *Isodon* species including *I. serra* [27], *I. wiggthii* [28] or *I. amethystoides* [29]. However, until now, small number of studies are available on the establishment of in vitro cultivation, phytochemical analysis, and biological activities of the resulting cultures of *I. rugosus* with the exception of biogenic synthesis of ZnONPs from in vitro callus culture of this plant [30]. In the present study, we report the in vitro callus establishment through optimization of hormonal combination, significant accumulation of pentacyclic triterpenoids (BA, OA and PA) and phenolic compounds (CA and RA), as well as antioxidant and anti-ageing properties of the resulting extracts for future potential cosmetic applications. As per our knowledge, the current optimization report is the first to address *I. rugosus* in vitro cultures as a feasible large-scale production system of bioactive phenolics and pentacyclic triterpenoids.

2. Results and Discussion

2.1. Optimization of Callogenesis from Different Initial Explants

For the determination of callus induction frequency, stem and leaf explants from *I. rugosus* were cultured on MS medium encompassing different concentrations (1.0–5.0 mg/L) of several PGRs (TDZ, NAA and BAP) used either alone or in conjunction with TDZ, as shown in Table 1.

Table 1. Callogenesis, initiation (day) and morphology of stem and leaf callus under different PGRs after 5 weeks of culture.

S. No.	Treatment (mg/L)	Callus Initiation (day)		Callus Induction Frequency (%)	Callus Color	Callus Texture		Degree of Callus Formation
		Stem	Leaf			Stem	Leaf	
0	Control (MS0)	-	-	-	-	-	-	-
1	MS + TDZ 1.0	10	14	80–100	DG	F	C	+++
2	MS + TDZ 2.0	10	14	80–100	DG	F	C	+++
3	MS + TDZ 3.0	10	14	80–100	DG	F	C	+++
4	MS + TDZ 4.0	10	14	80–100	DG	F	C	+++
5	MS + TDZ 5.0	10	14	80–100	DG	F	C	+++
6	MS + NAA 1.0	12	12	40–60	SG	F	C	++
7	MS + NAA 2.0	12	12	40–60	SG	F	C	++
8	MS + NAA 3.0	12	12	40–60	SG	F	C	++
9	MS + NAA 4.0	12	12	40–60	SG	F	C	++
10	MS + NAA 5.0	12	12	40–60	SG	F	C	++
11	MS + BAP 1.0	20	20	20–30	SLG	F	C	+
12	MS + BAP 2.0	20	20	20–30	SLG	F	C	+
13	MS + BAP 3.0	20	20	20–30	SLG	F	C	+
14	MS + BAP 4.0	20	20	20–30	SLG	F	C	+
15	MS + BAP 5.0	20	20	20–30	SLG	F	C	+
16	MS + TDZ 1.0 + NAA 1.0	8	8	90–100	FG	F	C	+++
17	MS + TDZ 1.0 + NAA 2.0	8	8	90–100	FG	F	C	+++
18	MS + TDZ 1.0 + NAA 3.0	8	8	90–100	FG	F	C	+++
19	MS + TDZ 1.0 + NAA 4.0	8	8	90–100	FG	F	C	+++
20	MS + TDZ 1.0 + NAA 5.0	8	8	90–100	FG	F	C	+++
21	MS + TDZ 1.0 + BAP 1.0	-	-	-	-	-	-	-
22	MS + TDZ 1.0 + BAP 2.0	-	-	-	-	-	-	-
23	MS + TDZ 1.0 + BAP 3.0	-	-	-	-	-	-	-
24	MS + TDZ 1.0 + BAP 4.0	-	-	-	-	-	-	-
25	MS + TDZ 1.0 + BAP 5.0	-	-	-	-	-	-	-

Values are means $\pm$ SD from three replicates. Note: No callus ($-$), Scanty callus (+), Moderate callus (++), profuse callus (+++). F friable, DG dark green, FG fresh green, SG snowy green, SLG snowy light green.

Callus was initiated after 10–12 days of culturing explants. In case of leaf explants, TDZ (1.0 mg/L, 2.0 mg/L and 3.0 mg/L) and 1.0 mg/L TDZ + NAA (1.0 mg/L, 2.0 mg/L and 3.0 mg/L) led to highest callus induction (95–100%) as compared to BAP alone or combination of BAP with TDZ. NAA alone resulted in around 80% callus induction, but the value greatly increased (up to 90%) when TDZ was used in combination. Similarly, the induction frequency for stem explants was close to 100% when TDZ was employed either alone or combined with NAA (Figure 1A). However, higher concentration of all the tested PGRs restricted callus induction in both stem and leaf explants, possibly due to repression of some endogenic PGRs retarding callus formation. Indeed, changes in callus response formation have already been ascribed to diverse endogenous hormonal responses pointing to the variable sensitivity of tissues toward these PGRs. Sreedevi et al. (2013) and Anjum et al. (2017) reported similar observations [31,32]. No callogenesis was observed on MS medium lacking these PGRs, which has already been observed for various other plant species such as *Stevia rebaudiana* [33].

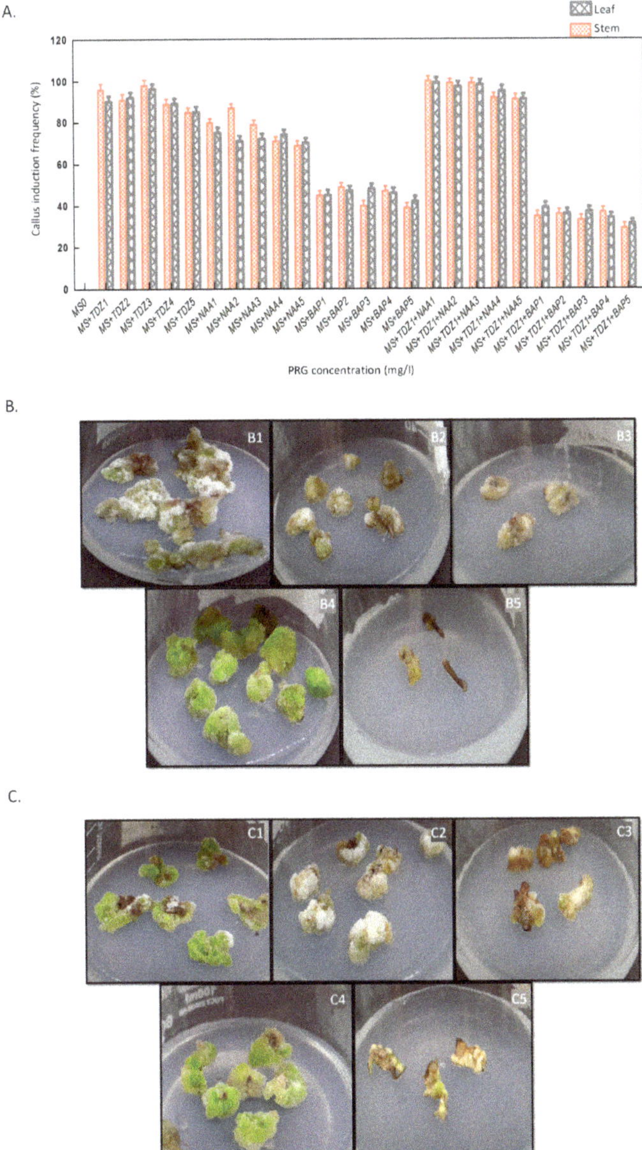

Figure 1. (**A**) Callus induction frequency (%) for stem and leaf explants under different PGRs concentrations. Values are means ± SE from three replicates; (**B**) Effect of 5 weeks old culture media containing various PGRs on callus induction and growth of stem explants; (**B1**) TDZ 1.0 mg/L; (**B2**) NAA 4.0 mg/L; (**B3**) BAP 2.0 mg/L; (**B4**) 1:1 TDZ 1.0 + NAA 1.0 mg/L; (**B5**) TDZ 1.0 + BAP 4.0 mg/L; (**C**) Effect of 5 weeks old culture media containing various PGRs on callus induction and growth of leaf explants; (**C1**) TDZ 3.0 mg/L; (**C2**) NAA 3.0 mg/L; (**C3**) BAP 4.0 mg/L; (**C4**) TDZ 1.0 + NAA 1.0 mg/L; and (**C5**) TDZ 1.0 + BAP 5.0 mg/L.

Visual morphological variations were also detected in calli (Figure 1B,C). Generally, stem-derived calli were more friable, while leaf-derived calli were compact in texture. Similar results have previously been reported for several other medicinal plant species [34,35]. We also observed that in *I. rugosus*, the callogenic response and morphological changes were markedly influenced by the exogenously applied PGRs. Physiological response of calli also radically varied in accordance to the type of initial explant. The potential growth rate was higher in stem-derived calli, as compared to the calli derived from leaf as starting explants.

Murthy et al. (1998) estimated that TDZ is a potent PGR for in vitro culture successive growth [36]. TDZ at a concentration of 3.0 mg/L produced highest biomass (FW 250.65 g/L and DW 18.65 g/L) in stem-induced callus cultures (Figure 2A,B). Similarly, 1.0 mg/L TDZ + 3.0 mg/L NAA resulted in FW of 140.22 g/L and DW of 14.98 g/L. Conversely, application of BAP alone or in combination with TDZ showed least response for biomass accumulation (Figure 2A,B). As for NAA, maximum biomass (FW 80.79 g/L and DW 5.98 g/L) was observed at 1.0 mg/L concentration, which then gradually decreased with increase in the concentration of NAA, as shown in Figure 2A,B. In case of leaf-derived callus cultures, optimum biomass accumulation (FW 115.2 g/L and DW 10.81 g/L) was observed for 1.0 mg/L TDZ + 3.0 mg/L NAA (Figure 2C,D). However, BAP + TDZ resulted in minimum biomass accumulation (Figure 2C,D).

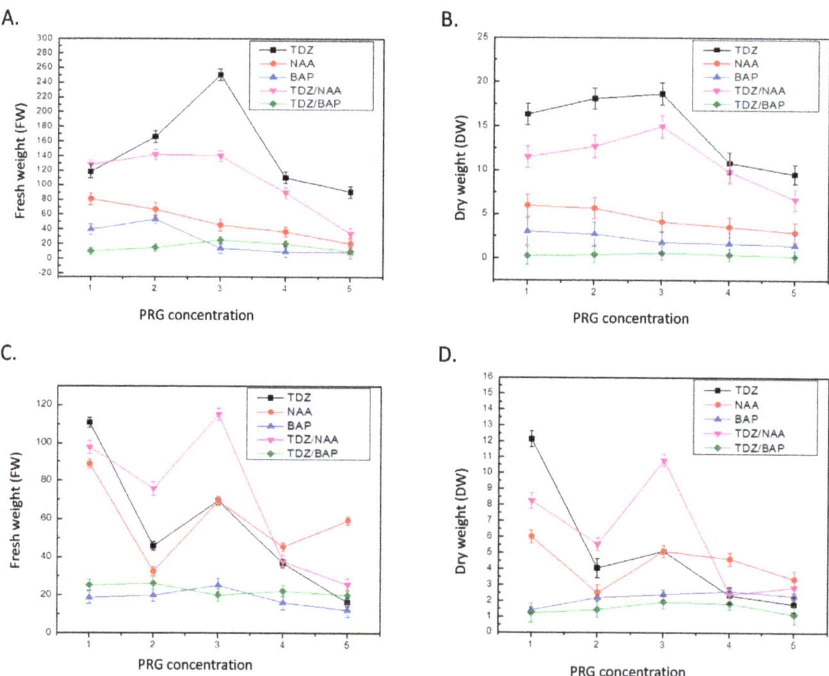

Figure 2. Time-course fresh and dry weight of callus cultures at different PGRs (in mg/L). (**A**) Fresh weight for stem-derived callus culture (in g/L); (**B**) dry weight for stem-derived callus culture (in g/L); (**C**) fresh weight for leaf-derived callus culture (in g/L); and (**D**) dry weight for leaf-derived callus (in g/L) cultured on MS medium fortified with TDZ, NAA, BAP (1.0–5.0 mg/L), TDZ (1.0 mg/L) + NAA (1.0–5.0 mg/L), TDZ (1.0 mg/L) + BAP (1.0–5.0 mg/L). Values are means of three replicates with standard deviation.

With respect to anatomical structure of callus, previous reports proved that callus induction frequency and proliferation increased considerably with the precise ratio of two exogenous hormones. Hypothetically, this was due to the effect and synthesis of endogenously grown regulators upon their exogenous presentation. Sahraoo et al. (2014) and Lee (1971) reported similar findings [37,38]. TDZ-treated tissues in combination with auxin maintain and enhance their accumulation and transport.

Our data is supported by Guo et al. (2011), who confirmed that the use of TDZ alone or in collaboration with other PGRs provoke high rate of callogenesis and cell proliferation due to high intrinsic activity and low absorbance in callus [39]. The current investigation explores that despite explants used for callogenesis, maximum growth is observed with lower concentration of all PGRs, used either alone or in combination, but biomass is gradually inhibited at higher concentrations. Our data also suggests that the combined treatment of TDZ and NAA is the best for callogenesis and biomass accumulation in *I. rugosus* callus cultures.

2.2. Evaluation of Secondary Metabolites Production

Total polyphenols accumulation in stem-derived calli of *I. rugosus* on all the tested PGRs ranged from 49.99 to 90.06 mg/g DW (Figure 3).

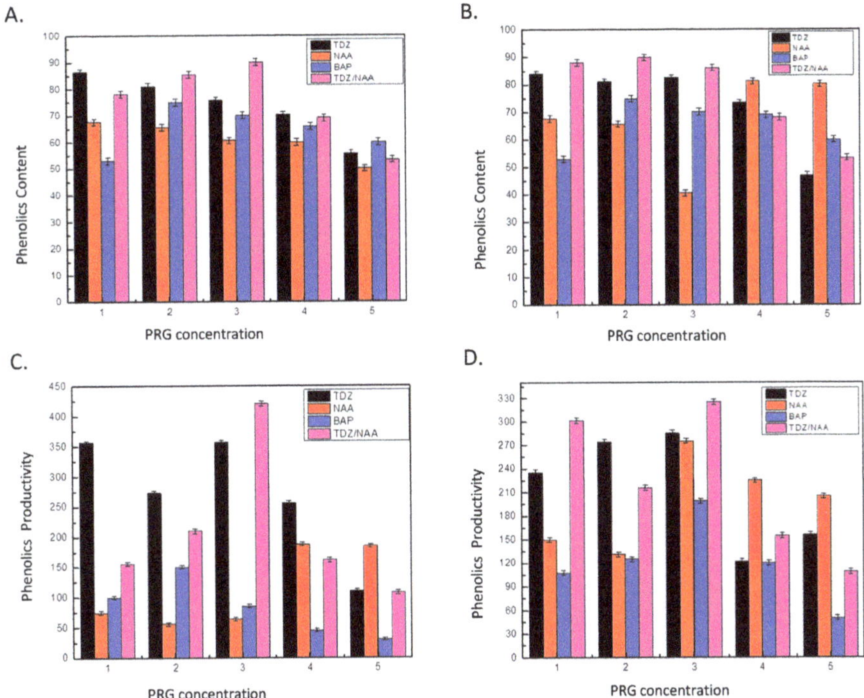

Figure 3. Comparison of total phenolic content (TPC) and total phenolic productivity (TPP) of extracts at different PGRs. (**A**) TPC (in mg/g DW) for stem-derived callus culture; (**B**) TPC (in mg/g DW) for leaf-derived callus culture; (**C**) TPP (in mg/L) for stem-derived callus culture; and (**D**) TPP (in mg/L) for leaf-derived callus cultured on MS medium fortified with PRG (TDZ, NAA, BAP (1.0–5.0 mg/L), TDZ (1.0 mg/L) + NAA (1.0–5.0 mg/L), TDZ (1.0 mg/L) + BAP (1.0–5.0 mg/L)). Values are means of three replicates with standard deviation.

Calli cultured on media supplemented with TDZ (1.0 mg/L) and NAA (3.0 mg/L) biosynthesized optimum levels (90.06 mg/g DW) of phenolic compounds (Figure 3A), while lowest accumulation (49.9 mg/g DW) was observed in media supplemented with high concentration (5.0 mg/L) of NAA. Phenolics accumulation in response to NAA and TDZ gradually declined with increases in hormonal concentration. However, Szopa, and Ekiert (2014) observed that PGRs directly influence the production of phenolic compounds in plants in vitro cultures [40]. Among all the PGRs, combined treatment of TDZ + NAA at low concentration exhibited maximum accumulation of TPC in stem-derived calli. Similar trend was observed for TPC in leaf-derived callus culture (Figure 3B) for which TDZ combined with NAA gave highest accumulation as compared to TDZ or NAA used alone. Faizal et al. (2017) reported that the best treatment for phenolic compounds production in red pitaya callus was 2.0 mg/L NAA + 4 mg/L TDZ, which is consistent with the results of our study [41]. Similarly, Tariq et al. (2014) also highlighted that growth regulators such as NAA and TDZ greatly influence the production of phenolic compounds, flavonoids, and antioxidants in *A. absinthium* cultures grown in vitro [42].

Antioxidant capacity generally correlated with TPC, thus collinear connection exists between these two variables, as evident from the literature [43–45]. Likewise, Khandaker et al. (2012) also indicated that the improved antioxidant activities in apple treated with different PGRs were linked with the increase in TPC [46]. A similar trend was also observed here with the quenching free radical activity (Figure S1). Our data suggests that *I. rugosus* extract could serve as a safe antioxidant agent.

The presence of pentacyclic triterpenes and caffeic acid phenolic derivatives have already been reported in this plant family [13–16]. Therefore, in order to make a step forward, the presence of these compounds were investigated in 12 callus cultures (hereafter called Ir#1 to Ir#12) grown on various culture media (precise PGRs composition of these media is shown in Supplementary Table S1) selected on the basis of DW accumulation, TPC and radical scavenging activity. A magnification showing and the quantification results are depicted in Table 2.

Table 2. Quantification of the main phytochemicals accumulated in twelve *I. rugosus* callus sample extracts (culture conditions are presented in Table S1).

Sample	CA (µg/g DW)	RA (µg/g DW)	BA (µg/g DW)	OA (µg/g DW)	PA (µg/g DW)
Ir#1	614.8 ± 20.2	1074.7 ± 18.9	98.0 ± 14.3	536.2 ± 18.8	454.8 ± 39.1
Ir#2	488.4 ± 24.6	751.6 ± 24.1	91.8 ± 19.5	201.7 ± 5.3	113.7 ± 33.9
Ir#3	784.0 ± 14.8	1519.5 ± 17.8	104.3 ± 18.8	348.4 ± 16.3	304.6 ± 33.0
Ir#4	735.6 ± 26.9	1158.3 ± 27.1	141.8 ± 16.3	317.1 ± 19.5	207.6 ± 28.2
Ir#5	728.2 ± 7.6	1685.2 ± 44.7	132.5 ± 24.8	631.0 ± 24.8	379.7 ± 14.2
Ir#6	979.5 ± 12.1	1259.0 ± 27.5	66.7 ± 9.4	198.2 ± 9.4	116.8 ± 23.6
Ir#7	575.6 ± 20.7	1279.0 ± 9.0	54.2 ± 5.4	389.0 ± 19.6	132.5 ± 18.8
Ir#8	901.6 ± 10.3	1708.9 ± 57.1	85.7 ± 9.1	386.0 ± 24.8	207.6 ± 8.8
Ir#9	647.2 ± 19.8	936.7 ± 13.1	22.9 ± 5.8	204.4 ± 23.6	69.9 ± 14.3
Ir#10	779.3 ± 18.0	797.1 ± 37.0	91.8 ± 10.8	248.3 ± 14.3	145.0 ± 23.6
Ir#11	886.8 ± 24.2	2013.5 ± 18.7	171.2 ± 9.2	331.2 ± 16.4	313.8 ± 14.1
Ir#12	835.8 ± 9.9	1335.9 ± 67.2	145.4 ± 5.1	429.8 ± 23.9	429.8 ± 14.3

Values are means ± standard deviations (n = 3).

Following HPLC analysis, phenolic acids contents ranged from 488.4 (Ir#2) to 979.5 (Ir#6) µg/g DW for CA and 751.6 (Ir#2) to 2013.5 (Ir#11) µg/g DW for RA, whereas, pentacyclic triterpenoids contents ranged from 54.2 (Ir#7) to 171.2 (Ir#11) µg/g DW for BA, 198.2 (Ir#6) to 631.0 (Ir#5) µg/g DW for OA and 69.9 (Ir#9) to 454.8 µg/g (Ir#1) for PA (Figure 4). Stem-derived calli were found to be the most suitable in accumulating highest levels of different phytochemicals, except CA, while lowest amount was observed for leaf-derived calli, except BA. These results clearly evidenced that the initial explant could be an important parameter to take into account for an optimal accumulation of secondary metabolites in callus cultures of *I. rugosus*. Similarly, TDZ supplementation resulted in a higher secondary metabolites production in stem-derived callus as compared to leaf-derived callus grown on the same medium. On the contrary, TDZ combined with NAA favored the secondary metabolites production in callus initiated from leaf explants (Table 2). TDZ at lower concentration (1.0 mg/L) resulted in optimal production of PA (Ir#1), while higher concentration (3.0 mg/L) of TDZ favored maximum accumulation of CA (Ir#6) and OA (Ir#5), but interestingly the nature of the initial explant differed between these two conditions: leaf-derived callus for optimal CA production vs. stem-derived callus for optimal OA accumulation. CA and RA are regarded as major phenolic metabolites of plants from *Lamiaceae* family [47,48]. In this study, we endeavor an effort to report in vitro culture condition for accumulation and production of these phenolics and pentacyclic triterpenes metabolites in *I. rugosus* callus culture. Until now, no report is available on phytochemical composition of *I. rugosus* in vitro cultures, except GC analysis of the wild extract and characterization and composition of essential oil [11,49–51]. However, in vitro production and accumulation of pentacyclic triterpenes have been documented in the literature [47,52–54]. Fedoreyev et al. (2005) and Hagina et al. (2008) also established callus culture of *Eritrichium sericeum* that produced a higher amount of both CA and RA [55,56]. Some other studies dealing with biotechnological approaches to produce RA using plant in vitro cultures reported a higher amount of this phenolic [57]. However, our *I. rugosus* callus culture system has the advantage of producing both types of secondary metabolites. We also anticipate that elicitation strategies could further stimulate the production of these compounds in future studies.

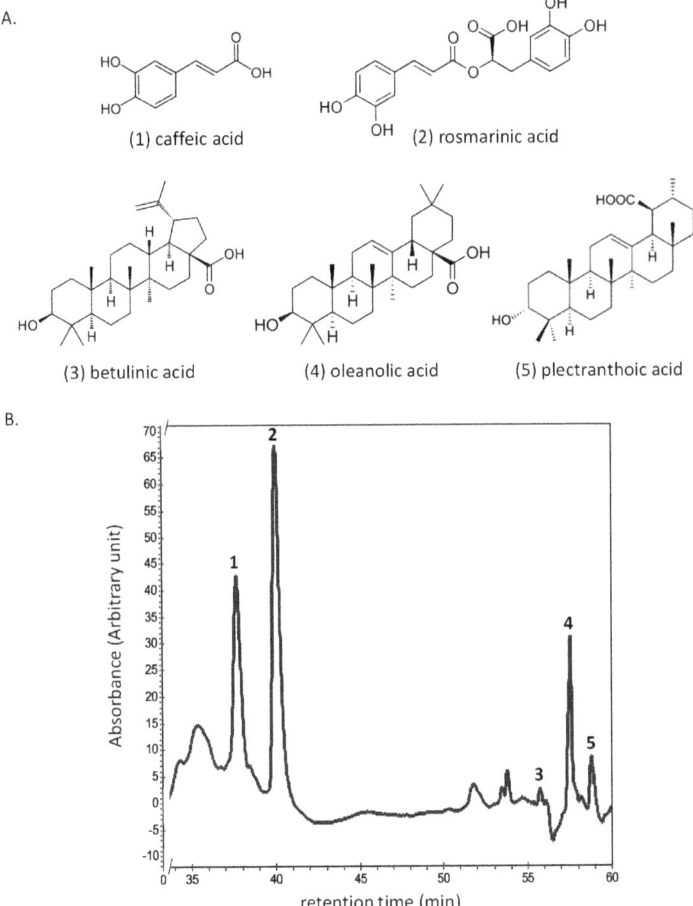

Figure 4. (**A**) Chemical structures of main phytochemicals accumulated in callus cultures of *I. rugosus* caffeic acid (CA, **1**) rosmarinic acid (RA, **2**) betulinic acid (BA, **3**) oleanolic acid and (OA, **4**) plectranthoic acid (PA, **5**); (**B**) Magnification of typical HPLC chromatograms showing the correct separation of the main phytochemicals accumulated in callus cultures of *I. rugosus*.

2.3. Evaluation of Antioxidant and Anti-Aging Potential of I. rugosus Callus Extracts

A complete screen of antioxidant and anti-ageing capacities of these 12 extracts with contrasting phytochemical profiles was also evaluated in the current study (Figure 4A). For in vitro antioxidant screening, antioxidant mechanisms were based on both electron transfer (FRAP and CUPRAC assays) and hydrogen atom transfer (ABTS and ORAC assays) [58]. Besides these two antioxidant mechanisms involved in the scavenging of reactive oxygen species, transient metal ion chelation was also considered as an antioxidant mechanism, since the Fenton reaction, responsible for the hydroxyl radical formation, and subsequently, radical chain reaction propagation, could be inhibited through this chelating mechanism. The chelation potential of these extracts was evaluated by both FRAP and metal chelating assays using ferrozine. The results of this antioxidant screening are reported in Table 3. All of the callus extracts of *I. rugosus* exhibited marked antioxidant and chelation activities. Extract from sample Ir#11 (stem-derived calli grown on 1.0 mg/L TDZ + 3.0 mg/L NAA) displayed highest antioxidant activities for all of the assays with values of 1203.7 TEAC for DPPH, 945.8 for ABTS, 733.3 TEAC for ORAC, 535.8 for FRAP, 460.2 for CUPRAC and 54.8 μmol of fixed Fe^{3+}. On the other hand, extract of sample Ir#2 (leaf-derived calli grown on 1.0 mg/L TDZ) presented the lowest antioxidant activities with values of 474.4 TEAC for DPPH, 434.5 TEAC for ABTS, 306.7 TEAC for ORAC, 211.9 TEAC for FRAP, 193.3 TEAC for CUPRAC, and 23.0 μmol of fixed Fe^{3+}. Whatever the test used, ET-based assays gave higher antioxidant capacities than HAT-based assays. The prominence of this action mode suggested the occurrence of at least one phytochemical involved in this type of antioxidant mechanism in *I. rugosus* callus extracts. Here, stem-derived callus extracts displayed higher antioxidant activities than the callus initiated from leaf explants. Combination of NAA and TDZ appeared to potentially further this biological activity.

Table 3. Antioxidant activities of 12 *I. rugosus* callus sample extracts. (Culture conditions are presented in Table S1).

Sample	DPPH (TEAC)	ABTS (TEAC)	ORAC (TEAC)	FRAP (AEAC)	CUPRAC (AEAC)	Chelation (μmol Fe^{2+})
Ir#1	674.4 ± 4.7	585.5 ± 9.0	421.8 ± 23.1	285.4 ± 6.3	260.3 ± 4.7	31.6 ± 0.9
Ir#2	474.4 ± 13.5	434.5 ± 15.4	306.7 ± 18.5	211.9 ± 3.2	193.3 ± 8.4	23.0 ± 1.4
Ir#3	911.7 ± 14.2	798.8 ± 15.1	529.4 ± 18.3	393.4 ± 6.9	354.2 ± 8.7	41.6 ± 1.0
Ir#4	721.2 ± 5.9	659.6 ± 25.1	453.2 ± 22.3	318.1 ± 6.4	273.5 ± 7.1	33.8 ± 1.7
Ir#5	1005.5 ± 13.8	879.5 ± 60.1	624.8 ± 9.0	456.8 ± 2.6	401.3 ± 2.7	45.2 ± 2.7
Ir#6	779.4 ± 5.4	708.1 ± 9.3	470.9 ± 14.5	354.8 ± 13.6	312.5 ± 8.8	34.3 ± 2.5
Ir#7	780.8 ± 6.2	688.2 ± 8.2	466.8 ± 9.6	349.3 ± 12.3	277.8 ± 8.4	34.9 ± 3.3
Ir#8	1043.2 ± 15.9	945.8 ± 6.4	641.0 ± 11.5	475.3 ± 10.0	410.8 ± 7.8	47.3 ± 1.5
Ir#9	563.1 ± 15.3	522.4 ± 5.2	350.1 ± 5.5	251.3 ± 9.0	235.7 ± 10.7	26.6 ± 2.8
Ir#10	516.3 ± 9.2	444.7 ± 18.2	318.1 ± 13.0	231.0 ± 8.8	186.1 ± 13.4	27.0 ± 2.8
Ir#11	1203.7 ± 53.2	944.7 ± 37.1	733.53 ± 7.3	535.8 ± 9.9	460.2 ± 5.5	54.8 ± 2.2
Ir#12	823.1 ± 25.6	727.1 ± 13.4	581.3 ± 173.5	353.8 ± 8.9	317.0 ± 4.8	35.9 ± 4.1

TEAC: Trolox C Equivalent Antioxidant Capacity (μM); AEAC: Ascorbic acid Equivalent Antioxidant Capacity (μM); Values are means ± standard deviations ($n = 3$).

The next step involved the evaluation of anti-aging action of *I. rugosus* callus extracts (at a fixed concentration of 50 μg/mL) determined as their in vitro capacities: (1) To inhibit elastase, hyaluronidase, collagenase (Matrix Metalloproteinase type 1 (MMP1)), tyrosinase and AGEs, and (2) to activate SIRT-1 activity. Elastase, hyaluronidase and collagenase have been found to degrade extracellular matrix components in the dermis, thus leading to skin alterations including skin tonus, deep wrinkles and resilience losses [19,59,60]. Tyrosinase dysfunctions advance with aging and can lead to malignant melanoma, as well as pigmentary disorders such as freckles or melisma [61]. Oxidative stress has been found to be associated with aging and age-related diseases [62] that could lead to the buildup of advanced glycation end products (AGEs) [63]. Therefore, compounds with the ability to inhibit these enzymatic activities or processes have attracted increasing attention in cosmetics. Several studies have challenged the classical radical theory of aging [64], and SIRT-1 (a class III deacetylase) have emerged as a new key factor of longevity controlling oxidative stress effects through the stimulation of antioxidant response via FOXOs and p53 pathways [65]. A stimulation of SIRT-1 activity has been reported to be crucial in the control of oxidative stress and in the regulation of aging process [65,66]. Interestingly, phytochemicals have been reported to activate SIRT-1 homologs and to prolong life span in yeast, drosophila, and *Caenorhabditis elegans* models [20,67–69]. The identification of SIRT-1 activators is also of great interest for cosmetic applications.

The results are presented in Table 4, expressed as a percentage of relative activities compared to control (consisting in extraction solvent addition to the assay). Here, we have considered an inhibition percentage of 30% as a marked inhibitory effect, we have detected strong inhibitory actions of our extracts toward tyrosinase (up to 72.2% inhibition observed with the stem-derived callus extract, Ir#3), collagenase (up to 36.3% inhibition with the stem-derived callus extract, Ir#5), and a strong inhibition of AGEs formation (up to 34.1% inhibition obtained with the leaf-derived callus extract, Ir#12). Inhibitory effects observed for elastase and collagenase were less marked (up to 25.3% and 22.2% respective inhibition observed with the same leaf-derived callus extract Ir#12). Interestingly, we observed that TDZ alone was more efficient in inducing the anti-aging action in stem-derived callus through the inhibition of these enzymes; whereas, addition of NAA appeared to limit the effect of initial explant origin and even seemed to reverse it (compare Ir#11 and Ir#12, Table 4). With the exception of tyrosinase inhibition, Ir#12 extract (leaf-derived callus grown on TDZ (1.0 mg/L) and NAA (3.0 mg/L)) appeared to be the most promising extract for these anti-aging activities.

Table 4. Anti-aging activities of 12 *I. rugosus* callus sample extracts expressed as percentage activities of control (DMSO) (culture conditions of the callus are presented in Table S1).

Sample	Elastase	Collagenase	Hyaluronidase	Tyrosinase	AGEs	SIRT1
Ir#1	77.8 ± 2.9	64.3 ± 3.0	85.8 ± 1.7	62.9 ± 2.3	78.5 ± 0.7	140.8 ± 5.0
Ir#2	90.7 ± 0.8	86.2 ± 2.2	88.4 ± 1.4	85.4 ± 1.7	79.8 ± 1.4	88.9 ± 3.3
Ir#3	79.8 ± 1.2	77.7 ± 3.0	88.6 ± 0.9	27.8 ± 12.1	84.4 ± 0.8	162.0 ± 7.0
Ir#4	83.0 ± 2.2	75.0 ± 3.3	87.7 ± 2.0	75.6 ± 2.6	77.6 ± 2.7	134.3 ± 10.3
Ir#5	76.8 ± 4.6	63.5 ± 4.3	80.1 ± 1.4	52.1 ± 4.4	73.8 ± 1.9	194.1 ± 6.4
Ir#6	87.9 ± 2.2	84.8 ± 0.3	85.6 ± 1.5	87.6 ± 1.9	76.1 ± 2.2	124.3 ± 11.0
Ir#7	85.8 ± 1.7	77.1 ± 1.9	86.6 ± 2.8	79.3 ± 1.4	76.9 ± 2.6	137.8 ± 5.8
Ir#8	85.9 ± 2.7	78.5 ± 0.9	89.2 ± 0.9	82.8 ± 1.7	83.2 ± 1.9	185.4 ± 11.0
Ir#9	90.2 ± 1.2	85.8 ± 1.6	88.7 ± 1.2	87.2 ± 1.5	83.7 ± 1.7	94.8 ± 4.2
Ir#10	86.2 ± 1.3	79.7 ± 1.3	91.5 ± 0.9	85.2 ± 1.1	85.1 ± 1.3	92.6 ± 3.2
Ir#11	79.2 ± 1.9	68.8 ± 2.1	78.7 ± 0.9	74.5 ± 2.7	70.8 ± 1.8	203.3 ± 6.2
Ir#12	74.7 ± 1.1	65.8 ± 2.2	77.8 ± 0.9	63.9 ± 2.5	65.9 ± 2.7	154.4 ± 7.9

Values are means ± standard deviations ($n = 3$).

Conversely, stem-derived callus was more prompted to stimulate SIRT1 activity with a maximum 2-fold increase measured with Ir#11 extract (stem-derived callus grown on TDZ (1.0 mg/L) and NAA (3.0 mg/L)). An activation level that is very similar to the stimulatory effect measured with resveratrol (the reference activator of SIRT1) [20]. However, here we have to note that a simple extract was used that could be very interesting since no purification steps are needed to obtain this activation. Moreover, without minimizing the potential synergistic effects, we assumed that all these anti-aging actions could be further reinforced following the purification steps.

2.4. Correlations Analysis

Hierarchical clustering analysis (HCA) was applied first to discriminate the different sample extracts based on their qualitative and quantitative phytochemical profiles (Figure 5). Regarding this HCA, decomposition into two main groups was observed. The first cluster (i.e., cluster A on Figure 5) grouped together sample extracts with the highest phytochemical accumulation capacities, with the sub-cluster A1 rich in pentacyclic triterpenes (BA, PA, and OA) and the sub-cluster A2 accumulating highest amount of phenolic acids (RA and CA). In contrast, cluster B shows sample extracts with the lowest accumulation capacities of these compounds. While we previously observed that stem-derived callus was generally more attractive than leaf-derived callus when considering each hormonal treatment individually, here the HCA pointed that the hormonal treatment is a prominent parameter over explant origin for an optimal accumulation since both stem and leaf-derived callus sample extracts were distributed equally in both clusters. Combination of NAA with higher concentrations of TDZ (Ir#11 and Ir#12) appeared to be the most favorable hormonal balances for accumulation of both pentacyclic triterpene and phenolic acid compounds in *I. rugosus* in vitro cultures.

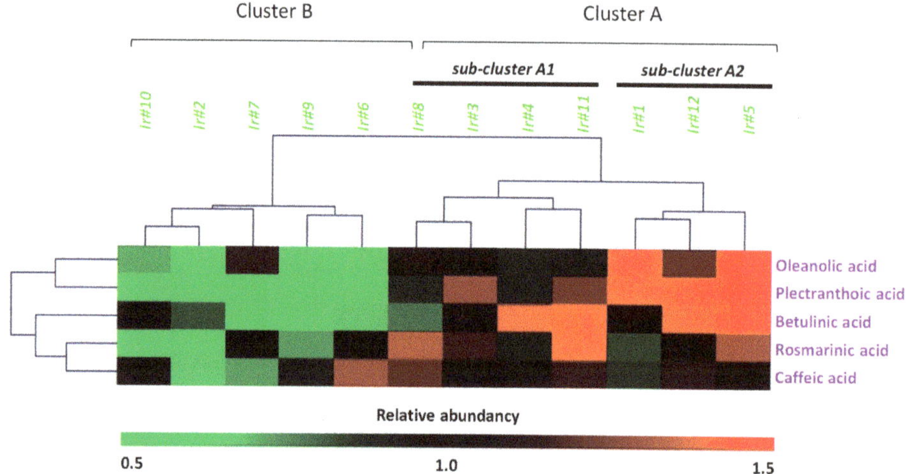

Figure 5. Hierarchical clustering analysis (HCA) of twelve *I. rugosus* callus extracts related to their phytochemical profile.

In order to rationale the apparent complex linkage between phytochemicals and biological activities, a principal component analysis (PCA) was then conducted (Figure 6). The obtained separation was satisfactory and allowed explaining 88.28% of the apparent complexity (F1×F2, Figure 6). The discrimination mainly occurred through the first dimension (F1 axis), explaining 70.13% of the apparent complexity by itself and allowing the separation of the sample extracts according to their phytochemical composition and biological activities. This point is of particular interest since it clearly evidences that it is possible to predict the antioxidant and anti-aging potential of a sample extract based on its phytochemical profile and that a direct linkage could exist between these two parameters. The second dimension axis (F2) accounted for only 18.14% of the initial variability, but intriguingly, it allowed the discrimination between: 1) Sample extracts rich in pentacyclic triterpenes and showing high inhibition capacities of tyrosinase, elastase and collagenase, and 2) sample extracts rich in phenolic acids showing the highest antioxidant activities, anti-AGEs, hyaluronidase inhibition, and SIRT1 activation. From this analysis, it appeared that sample extracts Ir#8 and Ir#11 were the most attractive for cosmetic applications looking for natural antioxidants, anti-hyaluronidase, anti-AGEs, and/or activator of SIRT1, whereas sample extracts Ir#1, Ir#2, and Ir#5 were the most promising for a natural anti-tyrosinase, anti-elastase, and/or anti-collagenase applications. Note that sample extracts Ir#3 appeared as the most potent extract for these cosmetic applications targeting the whole set of these activities.

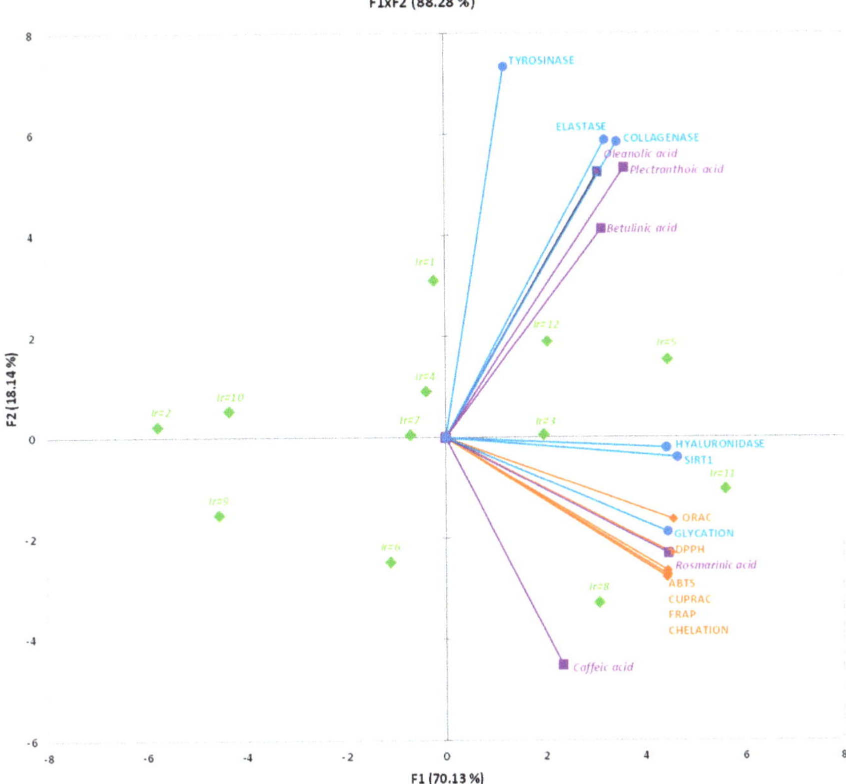

Figure 6. Principal component analysis (PCA) of the different phytochemicals and biological activities of *I. rugosus* callus extracts. Variance of factor 1 (F1) = 70.13% and of factor 2 (F2) = 18.14%.

To better assess the linkage between individual phytochemical and biological activities, Pearson coefficient correlations (PCCs) between these parameters were also calculated (Table 5). From this analysis, it appeared that the phenolic acid RA is the main contributor towards the antioxidant activities of *I. rugosus* in vitro cultures, with high (ranging from 0.982 for ABTS assay to 0.997 for DPPH and FRAP assays) and highly significant ($p < 0.001$) PCCs (Table 5). The anti-AGEs activity correlated with the presence of phenolic acids RA (PCC = 0.943, $p < 0.001$), and to a lesser extent, CA (PPC = 0.608, $p = 0.036$) could be linked to their well-described antioxidant activity [55]. Furthermore, it was observed that the pentacylic triterpene PA also significantly contributed towards the antioxidant ORAC assay (PCC = 0.604, $p = 0.038$). Concerning anti-aging activities, the analysis revealed a more complex linkage. The marked activation of SIRT1 activity and the anti-hyaluronidase activity appeared to be relied on RA and pentacylic triterpenes. Phenolic compounds are known as potent activator of SIRT1 activity [20], whereas both phenolic compounds and triterpenes are described as possible hyaluronidase inhibitors [70]. The marked anti-tyrosinase activity of our sample extracts was significantly linked with PA (PCC = 0.622, $p = 0.031$) and OA (PCC = 0.603, $p = 0.038$), but not with the third pentacylic terpene BA. From a structural point of view, PA and OA originate from the same olenyl cation precursor [71], which could be one explanation of this observation. The marked anti-collagenase, as well as less pronounced anti-elastase activities of *I. rugosus* sample extracts were correlated significantly with the pentacylic triterpenes.

Table 5. Pearson coefficient correlation linking the mains phytochemicals accumulated in *I. rugosus* callus extracts to their antioxidant and anti-aging activities.

	CA	RA	BA	OA	PA
DPPH	0.546	0.997 ***	0.471	0.477	0.537
ABTS	0.575	0.982 ***	0.484	0.447	0.466
ORAC	0.562	0.975 ***	0.511	0.550	0.604 *
FRAP	0.555	0.997 ***	0.447	0.423	0.510
CUPRAC	0.566	0.992 ***	0.454	0.466	0.513
Chelation	0.534	0.992 ***	0.456	0.465	0.559
Elastase	0.126	0.525	0.827 **	0.902 ***	0.748 *
Collagenase	0.097	0.571	0.900 ***	0.936 ***	0.720 **
Hyaluronidase	0.467	0.897 ***	0.572	0.602 *	0.538
Tyrosinase	-0.221	0.072	0.440	0.603 *	0.622 *
AGEs	0.608 *	0.943 ***	0.527	0.522	0.447
SIRT1	0.435	0.970 ***	0.665 *	0.646 *	0.625 *

$* p < 0.05, ** p < 0.01, *** p < 0.001.$

3. Materials and Methods

3.1. Chemicals and Reagents

All the extraction solvents employed in this study were of analytical grade, provided by Thermo Scientific (Illkirch, France). Standards and reagents were obtained from Sigma-Aldrich (Saint-Quentin Fallavier, France).

3.2. Plant Material

The seeds of *I. rugosus* were obtained from wild plant species located in Khyber Pakhtunkhwa province of Pakistan in December 2015. Seeds were surface sterilized prior to culturing in order to escape contamination. The air-dried seeds were then germinated on Murashige and Skoog (MS) (1962) medium comprising 0.8% agar and 30 g/L sucrose [72]. The pH of all the media was maintained at 5.8 before autoclaving at 121 °C for 20 min. For the establishment of callogenesis, stem and leaf explants were cut out from four weeks old in vitro grown plantlets of *I. rugosus* and cultured at various concentrations of PGRs, either alone or in combination.

3.3. Callogenic Frequency

Three different PGRs (NAA, TDZ and BAP) at varied concentrations (1.0–5.0 mg/L), used either alone or in combination with 1.0 mg/L TDZ, were employed for callus induction in the present study. The explants were maintained in a growth room for 16/8 h (light/dark cycle) at 25 ± 1 °C. Observation of callogenic frequency and callus morphology was done on weekly basis with visual eye. Respective calli were then sub-cultured on fresh medium supplemented with the same PGRs concentrations after every four weeks of the culture. Fresh weight and dry weight were also determined for subsequent examinations.

3.4. Determination of Total Phenolic Compounds Content

For phytochemical screening, extraction from calli was done according to the method presented by Ali et al. (2013) [73]. Briefly, 100 mg of dried callus was added to 10 mL of methanol (80%) to prepare the extract. The samples were then sonicated for 10–15 min, followed by vortexing for five–seven min and the procedure was repeated twice. After centrifugation at 6000 rpm for 15 min, the supernatants were collected and kept for further analysis.

Procedure described by Velioglu et al. (1998) was adopted for the estimation of total phenolic content (TPC) using FC reagent [74]. In brief, 20 µL of sample and 180 µL of FC reagent were mixed and incubated for five min. Then, 90 µL of sodium carbonate was added to the mixture and the OD was recorded at 630 nm with the help of a microplate reader. The calibration curve (0–50 µg/mL, R2 = 0.968) was generated by using gallic acid as a standard and the TPC was expressed as gallic acid equivalents (GAE) per gram of dry weight (DW) sample. Total phenolic production (TPP) was examined by using the following equation: TPP (mg/L) = TPC (mg/g) × Dry Weight (g/L).

3.5. Quantification and Identification by HPLC

The contents of caffeic acid, rosmarinic acid, oleanolic acid, plectranthoic acid and butelinic acid were determined by HPLC. Following a published protocol, a reversed-phase HPLC equipped with autosampler, Varian (Les Ulis, France) Prostar 230 pump and a Varian Prostar 335 photodiode array detector was used and controlled with Galaxie software (Varian v1.9.3.2) [19]. Briefly, the separation was achieved on an RP-18 column (5 µm, 250 × 4.0 mm id (Purospher Merck, Fontenay sous Bois, France)) at 35 °C. The mobile phase was comprised of acetonitrile (as solvent A) and 0.1% (*v/v*) formic acid acidified ultrapure water (as solvent B). The composition of the mobile phase varied during a 60 min run according to a linear gradient ranging from a 5:95 (*v/v*) to 100:0 (*v/v*) mixture of solvents A and B, respectively, at a flow rate of 0.6 mL/min. Detection was accomplished at 210 nm and 254 nm. Quantification and identification of each compound was done by coupling with retention times, UV spectra to those of authentic reference standards and by standard additions. Calibration curve (six points) was used to quantify each standard with the range of 0.05–1 mg/mL and correlation coefficient of at least 0.9994.

3.6. Antioxidant DPPH Assay

For this assay, 20 µL of each sample extract was combined with 180 µL of DPPH reagent and absorbance was recorded at 517 nm with the help of a microplate reader. The following equation was then used to calculate DPPH activity: % scavenging = 100 × (Abc-Abs/Abc).

Where, Abc denotes absorbance of the control, while Abs is absorbance of the sample or expressed as TAEC (Trolox C equivalent antioxidant capacity).

3.7. Antioxidant ORAC Assay

Oxygen radical absorbance capacity (ORAC) assay was carried out as suggested by Prior et al. (1998) [58]. In brief, 10 µL of the extracted sample was mixed with 190 µL of 0.96 µM fluorescein in 75 mM phosphate buffer (pH 7.4) and incubated for about 20 min at 37 °C. Then, 20 µL of 119.4 mM 2,2′-azobis-amidinopropane (ABAP) was added and the fluorescence intensity was recorded every five min during 2.5 h at 37 °C with the help of a fluorescence spectrophotometer (BioRad, Marnes-la-Coquette, France) set with an excitation at 485 nm and emission at 535 nm. Assays were made in triplicate and antioxidant capacity determined using ORAC assay was expressed as TAEC.

3.8. Antioxidant ABTS Assay

ABTS assay was accomplished with the method of Velioglu et al. (1998) [74]. Briefly, 2,2-azinobis (3-ethylbenzthiazoline-6-sulphonic acid (ABTS) solution was made by mixing equal proportion of ABTS salt (7 mM) with potassium persulphate (2.45 mM) and the mixture was kept in the dark for 16 h. The absorbance of the solution was measured at 734 nm and adjusted to 0.7 and then mixed with the extracts. The mixture was again kept in the dark for 15 min at 25 ± 1 °C and the absorbance was recorded at 734 nm using a BioTek ELX800 Absorbance Microplate Reader (BioTek Instruments, Colmar, France). Assays were made in triplicate and antioxidant capacity was expressed as TAEC.

3.9. Antioxidant FRAP Assay

Modified method of Benzie and Strain (1996) was used for the determination of ferric reducing antioxidant power (FRAP) assay [75]. In brief, 10 µL of the extracted samples were mixed with 190 µL of FRAP solution (composed of 20 mM FeCl3, 10 mM TPTZ, 6H$_2$O and 300 mM acetate buffer (pH 3.6) in a ratio 1:1:10 ($v/v/v$)). Reaction mixtures were then incubated at 25 ± 1 °C for 15 min. Absorbance of the reaction mixture was noted at 630 nm using a BioTek ELX800 Absorbance Microplate Reader (BioTek Instruments). Assays were made in triplicate and the antioxidant capacity, determined using this assay, was expressed as TAEC.

3.10. Antioxidant CUPRAC Assay

Cupric ion reducing antioxidant capacity (CUPRAC) assay was evaluated by some modifications in the method of Apak et al. (2004) [76]. Briefly, 10 µL of samples and 190 µL of CUPRAC reaction solution (containing 7.5 mM neocuproine, 10 mM Cu(II) and 1 M acetate buffer (pH 7) in a ratio 1:1:1 ($v/v/v$)) were mixed. Reaction mixtures were then incubated at 25 ± 1 °C for 15 min and absorbance was recorded at 450 nm using a BioTek ELX800 Absorbance Microplate Reader (BioTek Instruments). Assays were made in triplicate and antioxidant capacity determined using this assay was expressed as TAEC.

3.11. Metal Chelating Activity Assay

The ferrous ion chelating activity of *I. rugosus* extracts was evaluated following a slightly modified method of Srivastava et al. (2012) [77]. In brief, 10 µL of extracts were mixed with ferrous iron at a final concentration of 50 µM in HEPES (pH 6.8) buffers and 50 µL ferrozine (5 mM aqueous solution). All experiments were performed in a 96-well microplates at room temperature (25 ± 1 °C). Each sample was measured with and without (blank) the addition of ferrozine. Absorbance was noted at 550 nm instantaneously after addition of ferrozine and five min later with a BioTek ELX800 Absorbance Microplate Reader (BioTek Instruments). Chelating activity values were expressed in µM of fixed Fe^{3+}.

3.12. Collagenase Assay

Collagenase clostridium histolyticum (Sigma Aldrich) was employed for this assay and its activity was determined with the aid of a spectrophotometer by making use of *N*-[3-(2-furyl)acryloyl]-Leu-Gly-Pro-Ala (FALGPA; Sigma Aldrich) as a substrate in accordance to the protocol of Wittenauer et al. (2015) [78]. The absorbance decrease of FALGPA was followed at 335 nm for 20 min using a microplate reader (BioTek ELX800; BioTek Instruments, Colmar, France). Triplicated measurements were used and the anti-collagenase activity was revealed as a % of inhibition relative to corresponding control (adding same volume of extraction solvent) for every extract.

3.13. Elastase Assay

Elastase assay was performed by using porcine pancreatic elastase (Sigma Aldrich) and its activity was determined with spectrophotometer by making use of *N*-Succ-Ala-Ala-Ala-*p*-nitroanilide (AAAVPN; Sigma Aldrich) as a substrate and following p-nitroaniline release at 410 nm using a microplate reader (BioTek ELX800; BioTek Instruments) by adopting the method of Wittenauer et al. (2015) [78]. Triplicated measurements were used and the anti-elastase activity was expressed as a % of inhibition relative to the corresponding control (adding same volume of extraction solvent) for every extract.

3.14. Hyaluronidase Assay

Hyaluronidase inhibitory assay was carried out as suggested by Kolakul et al. (2017) using 1.5 units of hyaluronidase (Sigma Aldrich) and hyaluronic acid solution (0.03% (w/v)) as substrate [79]. The precipitation of undigested form of hyaluronic acid occurred with acid albumin solution (0.1% (w/v) BSA). The absorbance was measured at 600 nm using a microplate reader (BioTek ELX800; BioTek Instruments, Colmar, France). The hyaluronidase inhibitory effect was expressed as a % of inhibition relative to the corresponding control (adding same volume of extraction solvent) for every extract.

3.15. Tyrosinase Assay

Method of Chai et al. (2018) was used for the determination of tyrosinase assay [80]. In brief, L-DOPA (5 mM; Sigma Aldrich) was used as diphenolase substrate and mixed in sodium phosphate buffer (50 mM, pH 6.8) with 10 µL of *I. rugosus* extract. Finally, 0.2 mg/mL of mushroom tyrosinase solution (Sigma Aldrich) was added to this mixture to make a final volume of 200 µL. Control, with an equal amount of extraction solvent replacing the extract, was routinely carried out. The reaction processes were traced by using a microplate reader (BioTek ELX800; BioTek Instruments) at a wavelength of 475 nm. The tyrosinase inhibitory effect was expressed as a % of inhibition relative to the corresponding control for each extract.

3.16. Anti-AGE Formation Activity

The inhibitory capacity of AGE formation was determined as described by Kaewseejan and Siriamornpun (2015) [81]. *I. rugosus* extracts were mixed with 20 mg/mL BSA (Sigma Aldrich) solution prepared in 0.1 M phosphate buffer (pH 7.4), 0.5 M glucose (Sigma Aldrich) solution prepared in phosphate buffer and 1 mL of 0.1 M phosphate buffer at pH 7.4 containing 0.02% (w/v) sodium azide. This mixture was incubated at 37 °C for five days in the dark and then the amount of fluorescent AGE formed was determined using a fluorescence (VersaFluor fluorometer; Bio-Rad, Marnes-la-Coquette, France) set with 330 nm excitation wavelength and 410 nm emission wavelength. The percentage of anti-AGEs formation was revealed as a % of inhibition relative to the corresponding control (adding same volume of extraction solvent) for every extract.

3.17. SIRT-1 Assay

SIRT-1 activity was determined using the SIRT1 Assay Kit (Sigma Aldrich) following manufacturer instructions and fluorescent spectrometer (Biorad VersaFluor, Marnes-la-Coquette, France) set with 340 nm excitation and 430 nm emission wavelengths. The relative SIRT1 activity was revealed as a percentage relative to the corresponding control (adding same volume of extraction solvent) for every extract.

3.18. Statistical Analysis

Each experiment was carried out in triplicate and XL-stat_2018 (Addinsoft, Paris, France) was used for statistical analysis.

4. Conclusions

Our results hypothesized that cell culture protocols provide an excellent reproducible opportunity to optimize and obtain a uniform and high quality yield of the target compounds. HPLC analyses confirmed the presence of pentacyclic triterpenes namely plectranthoic acid (PA), betulinic acid (BA) and oleanolic acid (OA) and phenolic acids like caffeic acid (CA) and rosmarinic acid (RA) in all in vitro callus culture conditions. The impact of TDZ and NAA, as well as, the origin of initial explant phytochemical accumulation of the resulting *I. rugosus* was elucidated and correlated with relevant biological activities. Little is known about the in vitro biosynthesis, regulation, and accumulation of triterpenes and phenolic compound of *Isodon* genera. Hence, present research emphasizes a possible connection with respect to morphology, growth behavior and metabolic activity to produce fast-growing friable calli that is constantly able to generate the bulk of the target substances. Results showed the possibility to produce very contracting sample extracts in term of both phytochemical profiles and biological activities relying on simple and reproductive initial conditions. Taking advantage of these contrasting accumulation profiles, we have shown that *I. rugosus* in vitro cultures could represent a very promising and sustainable system for the production of anti-aging and antioxidant extracts for cosmetic applications. Correlation analysis further helped us to elucidate the complex link connecting phytochemicals accumulated in the callus to the biological activities of the resulting sample extracts. The antioxidant, anti-glycation, and SIRT1 activation actions relied on the presence of RA, whereas anti-tyrosinase, anti-elastase, and anti-collagenase activities were found to be linked with the occurrence of pentacylic triterpene derivatives. We anticipate that the methodology employed here could be applied to other health promoting activities of these extracts from *I. rugosus* in vitro cultures and to other plant production systems. Our research will facilitate, in the future, to enhance and examine the production of these bioactive metabolites on large-scale cultivation in bioreactors involving several biotechnological strategies like plant cell, tissue, and organ cultures.

Supplementary Materials: Supplementary materials can be found at http://www.mdpi.com/1422-0067/20/2/452/s1.

Author Contributions: Conceptualization, B.H.A. and C.H.; Methodology, A.S., M.Y., S.B., L.G., S.D. and D.T.; Validation, B.H.A., D.T. and C.H.; Formal Analysis, B.H.A., C.H. and N.G.-G.; Investigation, A.S., M.Y., S.B., L.G., S.D. and D.T.; Resources, B.H.A. and C.H.; Writing—Original Draft Preparation, M.Y. and C.H.; Writing—Review & Editing, B.H.A., D.T. and C.H.; Visualization, B.H.A., A.S., M.Y., S.B., L.G., S.D., D.T. and C.H.; Supervision, B.H.A. and C.H.; Project Administration, B.H.A. and C.H.; Funding Acquisition, B.H.A., N.G.-G. and C.H.

Funding: This research was partly supported by Cosmetosciences, a global training and research program dedicated to the cosmetic industry. Located in the heart of the Cosmetic Valley, this program led by University of Orléans is funded by the Région Centre-Val de Loire (VALBIOCOSM 17019UNI).

Acknowledgments: B.H.A. acknowledges research fellowship of Le Studium-Institute for Advanced Studies, Loire Valley, Orléans, France. D.T. gratefully acknowledges the support of French government via the French Embassy in Thailand in the form of Junior Research Fellowship Program 2018. S.B., L.G. and S.D. acknowledge research fellowships of Loire Valley Region.

Conflicts of Interest: The authors declare no conflict of interest.

Abbreviations

NAA	α-Naphthalene acetic acid
BAP	6-Benzyl adenine
TDZ	Thidiazuron
DPPH	2,2-Diphenyl-1-picrylhydrazyl
HPLC	High-performance liquid chromatography
CE	Callus Extract
WPE	Whole plant extract
MS	Murashige and Skoog
TPC	Total Phenolic Content
TFC	Total Flavonoid Content
DW	Dry Weight
FW	Fresh Weight
PGRs	Plant Growth Regulators
ROS	Reactive Oxygen Species
PAL	Phenylalanine ammonia-lyase
FRAP	Ferric reducing antioxidant power
CUPRAC	Cupric ion reducing antioxidant capacity
ORAC	Oxygen radical absorbance capacity
ABTS	2,2-Azinobis (3-ethylbenzthiazoline-6-sulphonic acid)
AGE	Advanced glycation end products

References

1. Abbasi, A.M.; Dastagir, G.; Hussain, F.; Sanaullah, P. Ethnobotany and marketing of crude drug plants in district Haripur. *Pak. J. Plant Sci.* **2005**, *11*, 103–114.

2. Sun, D.H.; Huang, S.X.; Han, Q.B. Diterpenoids from *Isodon* species and their biological activities. *Nat. Prod. Rep.* **2006**, *23*, 673–698. [CrossRef] [PubMed]

3. Akhtar, N.; Rashid, A.; Murad, W.; Bergmeier, E. Diversity and use of ethnomedicinal plants in the region of Swat. *North Pak. J. Ethnobiol. Ethnomed.* **2013**, *9*, 25. [CrossRef] [PubMed]

4. Rauf, A.; Muhammad, N.; Khan, A.; Uddin, N.; Atif, M. Antibacterial and phytotoxic profile of selected Pakistani medicinal plants. *World Appl. Sci. J.* **2012**, *20*, 540–544.

5. Janbaz, K.H.; Arif, J.; Saqib, F.; Imran, I.; Ashraf, M.; Zia-Ul-Haq, M.; Jaafar, H.Z.E.; De Feo, V. In-vitro and in-vivo validation of ethnopharmacological uses of methanol extract of *Isodon rugosus* Wall. ex Benth. (*Lamiaceae*). *BMC Complement. Altern. Med.* **2014**, *14*, 71. [CrossRef] [PubMed]

6. Khan, S.W.; Khatoon, S. Ethnobotanical studies on useful trees and shrubs of Haramosh and Bugrote valleys, in Gilgit northern areas of Pakistan. *Pak. J. Bot.* **2007**, *39*, 699–710.

7. Habtemariam, S. Molecular pharmacology of rosmarinic and salvianolic acids: Potential seeds for Alzheimer's and vascular dementia drugs. *Int. J. Mol. Sci.* **2018**, *19*, 458. [CrossRef]

8. Mothana, R.A.; Al-Said, M.S.; Al-Yahya, M.A.; Al-Rehaily, A.J.; Khaled, J.M. GC and GC/MS analysis of essential oil composition of the endemic *Soqotraen Leucas virgata* Balf. f. and its antimicrobial and antioxidant activities. *Int. J. Mol. Sci.* **2013**, *14*, 23129–23139. [CrossRef]

9. Zeb, A.; Sadiq, A.; Ullah, F.; Ahmad, S.; Ayaz, M. Phytochemical and toxicological investigations of crude methanolic extracts, subsequent fractions and crude saponins of *Isodon rugosus*. *Biol. Res.* **2014**, *47*, 57. [CrossRef]

10. Zeb, A.; Sadiq, A.; Ullah, F.; Ahmad, S.; Ayaz, M. Investigations of anticholinestrase and antioxidant potentials of methanolic extract, subsequent fractions, crude saponins and flavonoids isolated from *Isodon rugosus*. *Biol. Res.* **2014**, *47*, 76. [CrossRef]

11. Zeb, A.; Ullah, F.; Ayaz, M.; Ahmad, S.; Sadiq, A. Demonstration of biological activities of extracts from *Isodon rugosus* Wall. Ex Benth: Separation and identification of bioactive phytoconstituents by GC-MS analysis in the ethyl acetate extract. *BMC Complement. Altern. Med.* **2017**, *17*, 284. [CrossRef] [PubMed]

12. Sadiq, A.; Zeb, A.; Ullah, F.; Ahmad, S.; Ayaz, M.; Rashid, U.; Muhammad, N. Chemical characterization, analgesic, antioxidant, and anticholinesterase potentials of essential oils from *Isodon rugosus* Wall. ex. Benth. *Front. Pharmacol.* **2018**, *9*, 623. [CrossRef] [PubMed]

13. Duan, J.L.; Wu, Y.L.; Xu, J.G. Assessment of the bioactive compounds, antioxidant and antibacterial activities of *Isodon rubescens* as affected by drying methods. *Nat. Prod. Res.* **2017**, *26*, 1–4. [CrossRef] [PubMed]

14. Abdel-Mogib, M.; Albar, H.A.; Batterjee, S.M. Chemistry of genus *Plectranthus*. *Molecules* **2002**, *7*, 271–301. [CrossRef]

15. Batista, O.; Simoes, M.F.; Duarte, A.; Valderia, M.L.; Torre, M.C.; Rodriguez, B. An antimicrobial abietane from the roots of *Plectranthus hereroensis*. *Phytochemistry* **1994**, *38*, 167–169. [CrossRef]

16. Shetty, K.; Wahlqvist, M.L. A model for the role of the proline-linked pentose-phosphate pathway in phenolic phytochemical bio-synthesis and mechanism of action for human health and environmental applications. *Asia Pac. J. Clin. Nutr.* **2014**, *13*, 1–24.

17. Tiwari, S. Plants: A rich source of herbal medicine. *J. Nat. Prod.* **2008**, *1*, 27–35.

18. Ying, Q.L.; Rinehart, A.R.; Simon, S.R.; Cheronis, J.C. Inhibition of human leucocyte elastase by ursolic acid. Evidence for a binding site for pentacyclic triterpenes. *Biochem. J.* **1991**, *277*, 521–526. [CrossRef]

19. Coricovac, D.O.; Soica, C.O.; Muntean, D.A.; Popovici, R.A.; Dehelean, C.A.; Hogea, E.L. Assessment of the effects induced by two triterpenoids on liver mitochondria respiratory function isolated from aged rats. *Rev. Chim.* **2015**, *66*, 1707–1710.

20. Howitz, K.T.; Bitterman, K.J.; Cohen, H.Y.; Lamming, D.W.; Lavu, S.; Wood, J.G.; Zipkin, R.E.; Chung, P.; Kisielewski, A.; Zhang, L.L.; et al. Small molecule activators of sirtuins extend *Saccharomyces cerevisiae* lifespan. *Nature* **2003**, *425*, 191. [CrossRef]

21. Boots, A.W.; Haenen, G.R.; Bast, A. Health effects of quercetin: From antioxidant to nutraceutical. *Eur. J. Pharmacol.* **2008**, *585*, 325–337. [CrossRef] [PubMed]

22. Tiwari, K.N.; Sharma, N.C.; Tiwari, V.; Singh, B.D. Micropropagation of *Centella asiatica* (L.), a valuable medicinal herb. *Plant Cell Tissue Organ Cult.* **2000**, *63*, 179–185. [CrossRef]

23. Renouard, S.; Corbin, C.; Drouet, S.; Medvedec, B.; Doussot, J.; Colas, C.; Maunit, B.; Bhambra, A.S.; Gontier, E.; Jullian, N.; et al. Investigation of *Linum flavum* (L.) hairy root cultures for the production of anticancer aryltetralin lignans. *Int. J. Mol. Sci.* **2018**, *19*, 990. [CrossRef] [PubMed]

24. Thaniarasu, R.; Kumar, T.S.; Rao, M.V. Mass propagation of *Plectranthus bourneae* Gamble through indirect organogenesis from leaf and internode explants. *Physiol. Mol. Biol. Plants* **2016**, *22*, 143–151. [CrossRef]

25. Rakotoarison, D.A.; Greesier, B.; Trotin, F.; Brunet, C.; Luyckx, M.; Vasseur, J.; Cazin, M.; Cazin, J.C.; Pinkas, M. Antioxidant activities of polyphenolic extracts from flowers, in vitro cultures and cell suspension cultures of *Crataegus monogyna*. *Pharmazie* **1997**, *52*, 60–64. [PubMed]
26. Pietta, P.G. Flavonoids as antioxidants. *J. Nat. Prod.* **2000**, *63*, 1035–1042. [CrossRef] [PubMed]
27. He, H.; Xian, J.; Xiao, S.; Xu, H. In vitro culture and rapid propagation of *Isodon serra*. *Chin. Tradit. Herbal Drugs* **2001**, *32*, 255–256.
28. Thirugnanasampandan, R.; Mahendran, G.; Bai, V.N. High frequency in vitro propagation of *Isodon wightii* (Bentham) H. Hara. *Acta Physiol. Plant.* **2010**, *32*, 405–409. [CrossRef]
29. Duan, Y.; Su, Y.; Chao, E.; Zhang, G.; Zhao, F.; Xue, T.; Sheng, W.; Teng, J.; Xue, J. Callus-mediated plant regeneration in *Isodon amethystoides* using young seedling leaves as starting materials. *Plant Cell Tissue Organ Cult. (PCTOC)* **2018**, 1–7. [CrossRef]
30. Siddiquah, A.; Hashmi, S.S.; Mushtaq, S.; Renouard, S.; Blondeau, J.P.; Abbasi, R.; Hano, C.; Abbasi, B.H. Exploiting in vitro potential and characterization of surface modified Zinc oxide nanoparticles of *Isodon rugosus* extract: Their clinical potential towards HepG2 cell line and human pathogenic bacteria. *EXCLI J.* **2018**, *17*, 671. [PubMed]
31. Sreedevi, E.; Anuradha, M.; Pullaiah, T. Plant regeneration from leaf-derived callus in *Plectranthus barbatus* Andr. [Syn.: *Coleus forskohlii* (Wild.) Briq.]. *Afr. J. Biotechnol.* **2013**, *12*, 2441–2448.
32. Anjum, S.; Abbasi, B.H.; Hano, C. Trends in accumulation of pharmacologically important antioxidant-secondary metabolites in callus cultures of *Linum usitatissimum* L. *Plant Cell Tissue Organ Cult.* **2017**, *129*, 73–87. [CrossRef]
33. Mathur, S.; Shekhawat, G.S. Establishment and characterization of *Stevia rebaudiana* (Bertoni) cell suspension culture: An in-vitro approach for production of stevioside. *Acta Physiol. Plant* **2013**, *35*, 931–939. [CrossRef]
34. Cheng, H.; Yu, L.J.; Hu, Q.Y.; Chen, S.C.; Sun, Y.P. Establishment of callus and cell suspension cultures of *Corydalis saxicola* Bunting, a rare medicinal plant. *Z. Naturforsch. C* **2006**, *61*, 251–256. [CrossRef] [PubMed]
35. Vijaya, T.; Reddy, N.V.; Ghosh, S.B.; Chandramouli, K.; Pushpalatha, B.; Anitha, D.; Pragathi, D. Optimization of biomass yield and asiaticoside accumulation in the callus cultures from the leaves of *Centella asiatica*. URBAN. *World J. Pharm. Sci.* **2013**, *2*, 5966–5976.
36. Murthy, B.N.S.; Victor, J.; Singh, R.P.S.; Fletcher, R.A.; Saxena, P.K. In vitro regeneration of chickpea (*Cicer arietinum* L.): Stimulation of direct organogenesis and somatic embryogenesis by thidiazuron. *Plant Growth Regul.* **1996**, *19*, 233–240. [CrossRef]
37. Sahraoo, A.; Babalar, M.; Mirjalili, M.H.; Fattahi Moghaddam, M.R.; Nejad Ebrahimi, S.S.N. In-vitro callus induction and rosmarinic acid quantification in callus culture of *Satureja khuzistanica* Jamzad (*Lamiaceae*). *Iran. J. Pharm. Res.* **2014**, *13*, 1447–1456.
38. Lee, T.T. Cytokinin-controlled indoleacetic acid oxidase isoenzymes in tobacco callus cultures. *Plant Physiol.* **1971**, *47*, 181–185. [CrossRef]
39. Guo, B.; Abbasi, B.H.; Zeb, A.; Xu, L.L.; Wei, Y.H. Thidiazuron: A multi-dimensional plant growth regulator. *Afr. J. Biotechnol.* **2011**, *10*, 8984–9000.
40. Szopa, A.; Ekiert, H. Production of biologically active phenolic acids in *Aronia melanocarpa* (Michx.) Elliott in vitro cultures cultivated on different variants of the Murashige and Skoog medium. *Plant Growth Regul.* **2014**, *72*, 51–58. [CrossRef]
41. Faizal, A.; Fadzliana, N.; Sekeli Rogayah, S.; Shaharuddin Azmi, N.; Abdulla Janna, O.A. Addition of l-tyrosine to improve betalain production in red pitaya callus. *Pertanika J. Trop. Agric. Sci.* **2017**, *40*, 521–532.
42. Tariq, U.; Ali, M.; Abbasi, B.H. Morphogenic and biochemical variations under different spectral lights in callus cultures of *Artemisia absinthium* L. *J. Photochem. Photobiol.* **2014**, *130*, 264–271. [CrossRef] [PubMed]
43. Younas, M.; Drouet, S.; Nadeem, M.; Giglioli-Guivarc'h, N.; Hano, C.; Abbasi, B.H. Differential accumulation of silymarin induced by exposure of *Silybum marianum* L. callus cultures to several spectres of monochromatic lights. *J. Photochem. Photobiol. B* **2018**, *184*, 61–70. [CrossRef] [PubMed]
44. Kim, D.O.; Chun, O.K.; Kim, Y.J.; Moon, H.-Y.; Lee, C.Y. Quantification of polyphenolics and their antioxidant capacity in fresh plums. *J. Agric. Food Chem.* **2003**, *51*, 6509–6515. [CrossRef] [PubMed]
45. Djeridane, A.; Yousfi, M.; Nadjemi, B.; Boutassouna, D.; Stocker, P.; Vidal, N. Antioxidant activity of some Algerian medicinal plants extracts containing phenolic compounds. *Food Chem.* **2006**, *97*, 654–660. [CrossRef]

46. Khandaker, M.M.; Boyce, A.N.; Osman, N.; Hossain, A.S. Physiochemical and phytochemical properties of wax apple (*Syzygium samarangense* [Blume] Merrill & L. M. Perry var. Jambu Madu) as affected by growth regulator application. *Sci. World J.* **2012**, *13*, 728613.

47. Ikuta, A.; Kamiya, K.; Satake, T.; Saiki, Y. Triterpenoids from callus tissue cultures of *Paeonia* Species. *Phytochemistry* **1995**, *38*, 1203–1207. [CrossRef]

48. Raudone, L.; Zymone, K.; Raudonis, R.; Vainoriene, R.; Motiekaityte, V.; Janulis, V. Phenological changes in triterpenic and phenolic composition of *Thymus* L. species. *Ind. Crop Prod.* **2017**, *15*, 445–451. [CrossRef]

49. Zeb, A.; Ahmad, S.; Ullah, F.; Ayaz, M.; Sadiq, A. Anti-nociceptive Activity of Ethnomedicinally Important Analgesic Plant Isodon rugosus Wall. ex Benth: Mechanistic Study and Identifications of Bioactive Compounds. *Front Pharmacol.* **2016**, *7*, 200. [CrossRef]

50. Verma, R.S.; Pandey, V.; Chauhan, A.; Tiwari, R. Characterization of the essential oil composition of *Isodon rugosus* (Wall. ex Benth.) Codd. from garhwal region of western himalaya. *Med. Aromat. Plants* **2015**, *S3*, 002.

51. Hussain, I.; Khan, A.U.; Ullah, R.; SAlsaid, M.; Salman, S.; Iftikhar, S.; Marwat, G.A.; Sadique, M.; Jan, S.; Adnan, M.; et al. Chemical composition, antioxidant and anti-bacterialpotential of essential oil of medicinal plant *Isodon rugosus*. *J. Essent Oil Bear. Plants* **2017**, *20*, 1607–1613. [CrossRef]

52. Pandey, H.; Pandey, P.; Singh, S.; Gupta, R.; Suchitra Banerjee, S. Production of anti-cancer triterpene (betulinic acid) from callus cultures of different *Ocimum* species and its elicitation. *Protoplasma* **2015**, *252*, 64–655. [CrossRef] [PubMed]

53. Srivastava, P.; Chaturvedi, R. Simultaneous determination and quantification of three pentacyclic triterpenoids—Betulinic acid, oleanolic acid, and ursolic acid in cell cultures of *Lantana camara* L. *In Vitro Cell. Dev. Biol. Plant* **2010**, *46*, 549–557. [CrossRef]

54. Bakhtiar, Z.; Mirjalili, M.H.; Sonboli, A.; Farimani, M.M.; Ayyar, M. In vitro propagation, genetic and phytochemical assessment of *Thymus persicus* a medicinally important source of pentacyclic triterpenoid. *Biologia* **2014**, *69*, 594–603. [CrossRef]

55. Fedoreyev, S.A.; Inyushkina, Y.V.; Bulgakov, V.P.; Veselova, M.V.; Tchernoded, G.K.; Gerasimenko, A.V.; Zhuravlev, Y.N. Production of allantoin, rabdosiin and rosmarinic acid in callus cultures of the seacoastal plant *Mertensia maritima* (Boraginaceae). *Plant Cell Tissue Organ Cult. (PCTOC)* **2012**, *110*, 183–188. [CrossRef]

56. Hagina, Y.V.; Kiselev, K.V.; Bulgakov, V. Biotechnological analysis of caffeic acid metabolites possessing potent antinephritic activity. *Chin. J. Biotechnol.* **2008**, *24*, 2140–2141.

57. Khojasteh, A.; Mirjalili, M.H.; Hidalgo, D.; Corchete, P.; Palazon, J. New trends in biotechnological production of rosmarinic acid. *Biotechnol. Lett.* **2014**, *36*, 2393–2406. [CrossRef]

58. Prior, R.L.; Cao, G.; Matin, A.; Sofic, E.; McEwen, J.; O'Brien, C.; Lischner, N.; Ehlenfeldt, M.; Kalt, W.; Krewer, G.; et al. Antioxidant capacity as influenced by total phenolic and anthocyanin content, maturity, and variety of *Vaccinium* species. *J. Agric. Food Chem.* **1998**, *46*, 2686–2693. [CrossRef]

59. Liyanaarachchi, G.D.; Samarasekera, J.K.; Mahanama, K.R.; Hemalal, K.D. Tyrosinase, elastase, hyaluronidase, inhibitory and antioxidant activity of Sri Lankan medicinal plants for novel cosmeceuticals. *Ind. Crop. Prod.* **2018**, *31*, 597–605. [CrossRef]

60. Boran, R. Investigations of anti-aging potential of *Hypericum origanifolium* Willd. for skincare formulations. *Ind. Crop. Prod.* **2018**, *31*, 290–295. [CrossRef]

61. Briganti, S.; Camera, E.; Picardo, M. Chemical and instrumental approaches to treat hyperpigmentation. *Pigment Cell Res.* **2003**, *16*, 101–110. [CrossRef]

62. Finkel, T.; Holbrook, N.J. Oxidants, oxidative stress and the biology of ageing. *Nature* **2000**, *408*, 239. [CrossRef] [PubMed]

63. Gkogkolou, P.; Böhm, M. Advanced glycation end products: Key players in skin aging? *Dermato-Endocrinology* **2012**, *4*, 259–270. [CrossRef] [PubMed]

64. Harman, D. Aging: A theory based on free radical and radiation chemistry. *J. Gerontol.* **1956**, *11*, 298–300. [CrossRef] [PubMed]

65. Hori, Y.S.; Kuno, A.; Hosoda, R.; Horio, Y. Regulation of FOXOs and p53 by SIRT1 modulators under oxidative stress. *PLoS ONE* **2013**, *8*, e73875. [CrossRef]

66. Ghosh, H.S. The anti-aging, metabolism potential of SIRT1. *Curr. Opin. Investig. Drugs* **2008**, *9*, 1095–1102.

67. Viswanathan, M.; Kim, S.K.; Berdichevsky, A.; Guarente, L. A role for SIR-2.1 regulation of ER stress response genes in determining *C. elegans* life span. *Dev. Cell* **2005**, *9*, 605–615. [CrossRef]

68. Bauer, J.H.; Poon, P.C.; Glatt-Deeley, H.; Abrams, J.M.; Helfand, S.L. Neuronal expression of p53 dominant-negative proteins in adult *Drosophila melanogaster* extends lifespan. *Curr. Biol.* **2005**, *15*, 2063–2068. [CrossRef]

69. Viswanathan, M.; Guarente, L. Regulation *of Caenorhabditis elegans* lifespan by sir-2.1 transgenes. *Nature* **2011**, *477*, E1. [CrossRef]

70. Lee, K.K.; Cho, J.J.; Park, E.J.; Choi, J.D. Anti-elastase and anti-hyaluronidase of phenolic substance from *Areca catechu* as a new anti-ageing agent. *Int. J. Cosmet. Sci.* **2001**, *23*, 341–346. [CrossRef]

71. Moses, T.; Papadopoulou, K.K.; Osbourn, A. Metabolic and functional diversity of saponins, biosynthetic intermediates and semi-synthetic derivatives. *Crit. Rev. Biochem. Mol. Biol.* **2014**, *49*, 439–462. [CrossRef] [PubMed]

72. Murashige, T.; Skoog, F. A revised medium for rapid growth and bio assays with tobacco tissue cultures. *Physiol. Plant.* **1962**, *15*, 473–497. [CrossRef]

73. Ali, M.; Abbasi, B.H. Production of commercially important secondary metabolites and antioxidant activity in cell suspension cultures of *Artemisia absinthium* L. *Ind. Crop. Prod.* **2013**, *49*, 400–406. [CrossRef]

74. Velioglu, Y.S.; Mazza, G.; Gao, L.; Oomah, B.D. Antioxidant activity and total phenolics in selected fruits, vegetables, and grain products. *J. Agric. Food Chem.* **1998**, *46*, 4113–4117. [CrossRef]

75. Benzie, I.F.; Strain, J.J. The ferric reducing ability of plasma (FRAP) as a measure of "antioxidant power": The FRAP assay. *Anal. Biochem.* **1996**, *239*, 70–76. [CrossRef] [PubMed]

76. Apak, R.; Güçlü, K.; Özyürek, M.; Karademir, S.E. Novel total antioxidant capacity index for dietary polyphenols and vitamins C and E, using their cupric ion reducing capability in the presence of neocuproine: CUPRAC method. *J. Agric. Food Chem.* **2004**, *52*, 7970–7981. [CrossRef] [PubMed]

77. Srivastava, N.; Chauhan, A.S.; Sharma, B. Isolation and characterization of some phytochemicals from Indian traditional plants. *Biotechnol. Res. Int.* **2012**, *2012*, 549850. [CrossRef]

78. Wittenauer, J.; Mäckle, S.; Sußmann, D.; Schweiggert-Weisz, U.; Carle, R. Inhibitory effects of polyphenols from grape pomace extract on collagenase and elastase activity. *Fitoterapia* **2015**, *101*, 179–187. [CrossRef]

79. Kolakul, P.; Sripanidkulchai, B. Phytochemicals and anti-aging potentials of the extracts from *Lagerstroemia speciosa* and *Lagerstroemia floribunda*. *Ind. Crop. Prod.* **2017**, *109*, 707–716. [CrossRef]

80. Chai, W.M.; Huang, Q.; Lin, M.Z.; Ou-Yang, C.; Huang, W.Y.; Wang, Y.X.; Xu, K.L.; Feng, H.L. Condensed tannins from longan bark as inhibitor of tyrosinase: Structure, activity, and mechanism. *J. Agric. Food Chem.* **2018**, *66*, 908–917. [CrossRef]

81. Kaewseejan, N.; Siriamornpun, S. Bioactive components and properties of ethanolic extract and its fractions from *Gynura procumbens* leaves. *Ind. Crop. Prod.* **2015**, *74*, 271–278. [CrossRef]

International Journal of
Molecular Sciences

Review

DA-9701 (Motilitone): A Multi-Targeting Botanical Drug for the Treatment of Functional Dyspepsia

Mirim Jin [1,2,*] and Miwon Son [3]

1   Department of Microbiology, College of Medicine, Gachon University, Incheon 21999, Korea
2   Department of Health Science and Technology, GAIHST, Gachon University, Incheon 21936, Korea
3   Research Center & Phytotherapeutics Group, Viromed, Co. Ltd., Seoul 08826, Korea; mwson@viromed.co.kr
*   Correspondence: mirimj@gachon.ac.kr; Tel.: +82-032-899-6080

Received: 9 November 2018; Accepted: 7 December 2018; Published: 13 December 2018

Abstract: Functional dyspepsia (FD) is the most common functional gastrointestinal disorder (FGID). FD is characterized by bothersome symptoms such as postprandial fullness, early satiety, and epigastric pain or burning sensations in the upper abdomen. The complexity and heterogeneity of FD pathophysiology, which involves multiple mechanisms, make both treatment and new drug development for FD difficult. Current medicines for FD targeting a single pathway have failed to show satisfactory efficacy and safety. On the other hand, multicomponent herbal medicines that act on multiple targets may be a promising alternative treatment for FD. DA-9701 (Motilitone), a botanical drug consisting of Corydalis Tuber and Pharbitidis Semen, has been prescribed for FD since it was launched in Korea in 2011. It has multiple mechanisms of action such as prokinetic effects, fundus relaxation, and visceral analgesia, which are mediated by dopamine D_2 and several serotonin receptors involved in gastrointestinal (GI) functions. In clinical studies, DA-9701 has been found to be beneficial for improvement of FD symptoms and GI functions in FD patients, while showing better safety compared to that associated with conventional medicines. In this review, we provide updated information on the pharmacological effects, safety, and clinical results of DA-9701 for the treatment of FGIDs.

Keywords: functional dyspepsia (FD); DA-9701; botanical drug; multi-targeting

1. Introduction

1.1. Functional Dyspepsia: Evolution of Its Definition and Criteria

Functional gastrointestinal disorders (FGIDs) are the most common functional gastroenterologic abnormalities associated with physiological and morphological disturbances, and are often accompanied by conditions including motility dysfunction, visceral hypersensitivity, altered mucosal and immune functions, altered gut microbiota, and altered central nervous system processing [1–3]. In particular, functional dyspepsia (FD) has been recognized as an unexplained discomfort in the upper abdomen for over 100 years, and its symptoms include excessive fullness after eating or the inability to finish a normal-sized meal and recurrent epigastric pain [3,4]. FD is associated with not only a markedly impaired quality of life and negative impact on the work place, but also a significant economic burden [5]. The global prevalence of FD has been reported as 21%, with the range being 10% to 30% among various studies and geographies [6]. The definition and diagnostic criteria of FD have evolved within the Rome process [7]. The Rome Foundation was established in the late 1980s, at a time when there was little understanding of the pathophysiology of FGIDs. In Rome II criteria [8], FD was defined as recurrent upper abdominal pain and discomfort for at least 12 weeks during the preceding year. The Rome II classification divided patients having a wide range of dyspepsia symptoms into four groups on the basis of major symptomatic patterns: Reflux-like, ulcer-like, dysmotility-like, or nonspecific FD. However, the Rome II categorization lacked reliability, which was attributable to overlapping symptoms of various FGIDs such as irritable bowel syndrome (IBS), gastroesophageal reflux disease (GERD), postprandial distress syndrome (PDS), and epigastric pain syndrome (EPS). Therefore, the Rome III committee subdivided FD into two distinct syndromes: PDS characterized by postprandial fullness and early satiety, and EPS accompanied by epigastric pain or burning [9,10]. The Rome IV consensus published in 2016 emphasized that FD should not be considered a single disorder. It retained the subclassification of FD into PDS and EPS but strengthened the notion that these symptoms are separate entities that could overlap; it also emphasized that the symptoms should be severe enough to be bothersome (i.e., a discomfort score of at least 2 on a scale of 1 to 5 in daily life) and occur more frequently than that in the normal population. PDS and EPS were defined as "bothersome early satiety or postprandial fullness for three or more days per week in 3 months with at least a 6-month history" and "bothersome epigastric pain or epigastric burning for 1 or more days per week in the past 3 months with at least a 6-month history", respectively [3,11]. In addition, the Rome IV consensus accepted evidence that the symptoms of GERD and IBS are part of the spectrum of FD, and the brain-gut axis is an important factor in the etiology of functional gastrointestinal (GI) disorders. It further acknowledged the possibility that pathologic lesions are involved in FD development, even if FD is a functional disorder, based on the findings that duodenal inflammation and eosinophilia are associated with FD [12–14]. *Helicobacter pylori*-associated dyspepsia is classified separately from true FD in the Rome IV criteria; affected patients are defined as a subset of those with FD-like symptoms and *H. pylori* infection, with symptom resolution 1 year after successful eradication of *H. pylori* [11,15].

1.2. Pathophysiology of FD and Current Treatments

Undoubtedly, FD is a complex and heterogeneous disorder. Accumulating data indicate that multiple factors including environment factors (food and *H. pylori* infection), biological factors (duodenal inflammation, eosinophilia, and cytokines), physiological factors (acid, gastric emptying, and gastric accommodation), and psychological factors (visceral hypersensitivity, brain pain modulating circuits, and anxiety/depression) are involved in the pathophysiology of FD [7] (Figure 1). The multiple mechanisms involved in FD make successful treatment and the development of new drugs with satisfactory efficacy and safety difficult.

Historically, FD has been considered a motility disorder dominated by disturbances in gastric physiology including gastric emptying and relaxation. Although gastric emptying was not correlated with symptoms [9,10], delayed gastric emptying was reported, with an incidence of 20 to 50% among FD patients, and the overall gastric emptying of solids was 1.5 times slower in FD patients than in healthy subjects [16]. Usually, a prokinetic is administered as a first-line treatment for PDS patients [6]. Approximately 90% of the human body's total serotonin (5-hydroxytryptamine, 5-HT) is located in enterochromaffin cells in the gut. Intestinal movement is regulated via 5-HT receptors, including 5-HT_1, 5-HT_2, 5-HT_3, 5-HT_4, and 5-HT_7, which are activated by serotonin as their natural ligand. Serotonin receptors are drug targets for FGIDs [17–20]. After the withdrawal of cisapride, a 5-HT_4 agonist and 5-HT_3 antagonist, because of its cardiovascular side effects, several serotonergic drugs including tegaserod and mosapride were launched into the market. However, the efficacy and safety of these drugs are limited [21,22]. Mosapride failed to provide any benefit [23] and tegaserod was withdrawn due to cardiovascular side effects [24].

Itopride, a prokinetic agent that works as a dual dopamine D_2 receptor antagonist and acetylcholinesterase inhibitor, was reported to have good efficacy in global patient assessments and to improve FD symptom scores in a meta-analysis [25], but it failed to demonstrate any efficacy in a clinical trial [26]. Domperidone (a dopamine D_2 and D_3 receptor antagonist) and metoclopramide, (a dopamine D_1 and D_2 receptor antagonist) were also launched; however, their side effects, which included ventricular arrhythmia and extrapyramidal movement disorder, respectively, inhibited their long-term use [27,28].

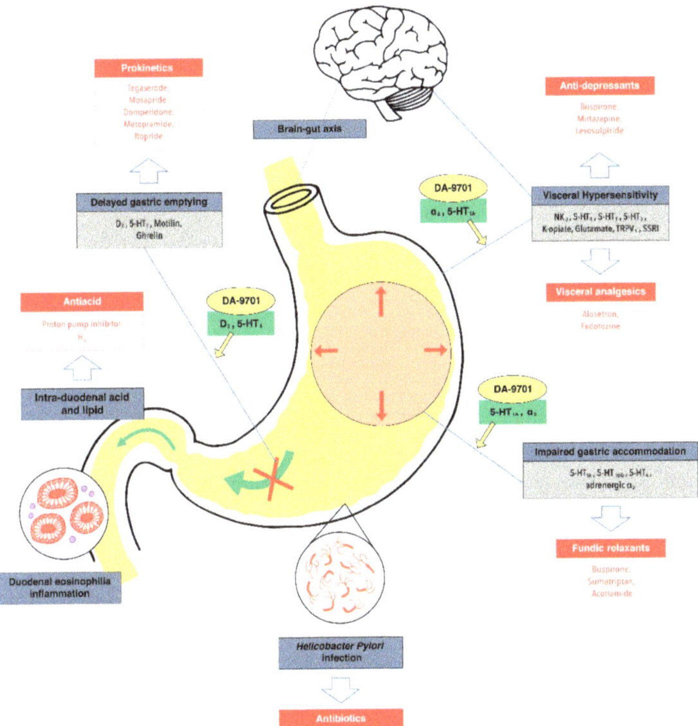

Figure 1. Major drug targets associated with pathophysiology of functional dyspepsia and current drugs. Targets of DA-9701 are presented.

Gastric accommodation is relaxation of the fundus, the upper part of the stomach, in response to food ingestion, which is mediated by a vasovagal reflex generated via nitrergic nerve activation. Tack et al. reported impaired gastric accommodation in 40% of FD patients [10], which was associated with early satiety [10]. Drugs that relax the gastric fundus appear to improve FD symptoms, particularly PDS. Buspirone, an antidepressant having a partial agonistic effect on 5-HT$_{1A}$, may enhance relaxation of the gastric fundus and body, and reduce the severity of symptoms including postprandial fullness, early satiety, and upper abdominal bloating [29]. Sumatriptan, a 5-HT$_{1B/D}$ agonist, was reported to alleviate FD symptoms by restoration of gastric accommodation with fundus-relaxing effects [30]. Recently, acotiamide, an M1/M2 muscarinic receptor blocker, was approved as a fundic relaxant for FD in Japan. This drug increases acetylcholine availability in gut nerve synapses [31], and has been reported to relax the gastric fundus and accelerate gastric emptying in humans [32,33].

Gastric hypersensitivity and/or brain-gut dysfunction has been associated with an impaired connection between gastric physiology and psychology in FD [34]. Compared to that of normal subjects, patients with FGIDs have been reported to be more sensitive to balloon distension. Many studies have confirmed intolerance to gastric distention in patients with FD [35,36]. Gastric hypersensitivity seems to be associated with postprandial epigastric pain, belching, and weight loss [37]. Approximately 34% of patients with FD had a lower threshold for the first perception of stimulus, discomfort, and pain during distention of the proximal stomach by barostat [37,38]. Visceral analgesics including alosetron, a 5-HT$_3$ antagonist [39], were reported to reduce visceral hypersensitivity and to be beneficial in symptom relief compared to that of placebo in clinical studies [40]. However, side effects, including constipation and ischemic colitis, were a concern [41]. Fedotozine, an opioid κ-agonist [42,43], seemed to be a visceral analgesic; however, in clinical trials, the results of an overall physician assessment of this drug were not satisfactory [44]. For EPS patients, acid suppression is beneficial, and proton pump inhibitors (PPIs) are primarily used. H_2 receptor antagonist therapy was reported to be superior to placebo control to reduce FD symptoms [45]. However, if first-line therapies fail, centrally acting drugs may be beneficial. Tricyclic antidepressants, including amitriptyline, appeared to provide a modest benefit for FD patients, especially for those with pain [46]. Interestingly, the tetracyclic antidepressant mirtazapine, which is an H_1 histamine receptor blocker, 5-HT_{2C} and 5-HT_3 antagonist, and $\alpha2$ adrenergic receptor blocker, seemed to be more beneficial than placebo in a preliminary randomized controlled trial of FD patients with weight loss [47,48], although it did not alter gastric function in healthy individuals [49]. Levosulpiride, an atypical antipsychotic drug and dopamine antagonist, may be efficacious in FD treatment [50]. However, more data including those from large placebo-controlled trials are needed [51].

Given the lack of medications with satisfactory efficacy and safety and the requirement of multiple drugs to achieve relief from various symptoms, there remains an unmet need for new treatments. Researchers have proposed that an agent capable of modulating multiple mechanisms will be more promising than an agent highly selective for a single mechanism and that a new treatment for FD should target many, if not all, pathophysiologies [52].

For centuries, herbal preparations have been traditionally used for a variety of GI disorders based on historical experiences and oriental medicine practices, but there has been a lack of scientific evidence and qualified clinical data. Herbal medicines, which generally contain multiple herbs, may be good options for FD treatment. The pathophysiology of FD is heterogeneous and multiple components of herbal medicines may enable simultaneous targeting of the multiple pathways involved in FD, enhancing efficacy compared to that of current chemical drugs targeting a single pathway [52]. DA-9701 (Motilitone) was launched as a new drug for FD in December 2011 in Korea; it is an herbal medicine with multi-acting mechanisms for FD treatment [53]. In this review, we provide updated information related to the multi-acting pharmacological effects, safety, and clinical data of DA-9701 and propose further research on it.

2. DA-9701

2.1. Herbal Composition

DA-9701 is a botanical drug for FD treatment in Korea, which is formulated with Pharbitidis Semen and Corydalis Tuber [53]; both constituent plants have been traditionally used for treating GI disorders [54–56]. Corydalis Tuber, the root of *Corydalis yanhusuo* W.T. Wang (Papaveraceae), is known to control gastric juice secretion and prevent gastric [56] and duodenal ulcers [57]. Extracts from Corydalis Tuber have been used as antispasmodic agents and analgesics [54] for abdominal pain because of its soothing and tranquilizing properties. Anti-inflammatory effects of the extract or its isolated compounds have been reported [58]. Pharbitidis Semen is the seed of *Pharbitis nil* Choisy (Convolvulaceae) and has been used as a traditional medicine for the treatment of abdominal pain [55] and as a strong purgative in Chinese medicine. The extract of Pharbitidis Semen is also used to stimulate and enhance intestinal peristalsis. Compared to that of a drug comprising a single synthetized chemical compound, a botanical drug requires more complicated quality control because of the properties of natural products. DA-9701 contains various components including corydaline (CD), dehydrocorydalin, chlorogenic acid, caffeic acid, coptisine, berberine, tetrahydroberberine (THB), palmatine, and tetrahydropalmatine (THP). For batch to batch control, CD ((13S, 13aR)-2,3,9,10-tetramethoxy-13-methyl-6,8,13,13a-tetrahydro-5H-isoquinolino[2,1-b]), the main constituent of Corydalis Tuber and CA (1S, 3R, 4R, 5R0-3-[(E)-3-(3,4-)prop-2-enoyl] oxy-1,4,5-trihydroxycyclohexane-1-carboxylic acid) from Pharbitidis Semen were selected as marker compounds for DA-9701 [59].

2.2. Pharmacology of DA-9701

The in vitro and in vivo pharmacological study results of DA-9701 are summarized in Table 1.

Table 1. Pharmacological profile summary of DA 9701.

Class	Study Type	Experimental	Results	Ref.
		Prokinetics		
D_2 antagonistic activity	In vivo	Rat, gastric emptying delayed by apomorphine and cisplatin treatment	Acceleration of gastric emptying in normal rats, improvement in delayed gastric emptying	[60]
Gastroprokinetic activity	In vitro	Whole cell patch clamp	Modulation of pacemaker activity	[61]
Gastroprokinetic activity	In vivo	Mouse, ^{13}C-octanoic acid breath test	Acceleration of gastric emptying	[62]
Gastroprokinetic activity	In vivo	Rat, clonidine-induced hypomotility of the gastric antrum	DA-9701 improved the clonidine-induced hypomotility of the gastric antrum	[63]
Gastroprokinetic activity	In vivo	Mouse, laparotomy or atropine injection	Enhancement of gastrointestinal transit	[60]
Gastroprokinetic activity	In vivo	Guinea pig, opioid-induced bowel dysfunction	Increase in amplitude of ileal muscle contraction, restoration of delayed GI transit	[64]
Gastroprokinetic activity	In vivo	Rat, postoperative ileus (POI)	Amelioration of POI by reduction in delayed GIT, improvement in defecation in rat model of POI	[65]
Prokinetic activity	In vivo	Guinea pig, postoperative ileus (POI)	Improvement in GI transit, inhibition of plasma ACTH levels via central CRF pathways	[66]
Gastroprokinetic activity	In vivo	Rat, stress-induced delayed gastric emptying	Improvement in delayed gastric emptying and inhibition of the hormonal changes induced by stress	[67]
		Fundic relaxants		
Fundus-relaxing activity	In vivo	Beagle dog, canine gastric compliance with barostat	Induction of gastric relaxation and increased gastric compliance	[60]
Fundus-relaxing activity	In vivo	Beagle dog, canine gastric compliance with barostat	CD from Corydalis Tuber induced gastric relaxation and facilitated gastric accommodation	[68]
Fundus-relaxing activity	In vivo	Beagle dog, meal-induced gastric accommodation with barostat	Improvement in gastric accommodation via increase in postprandial gastric volume	[69]
5-HT$_{1A}$ agonist activity	In vivo	Rat, restraint stress-induced feeding inhibition	DA-9701 was blocked by the 5-HT$_{1A}$ antagonist, WAY 100635	[70]
D_2 antagonistic activity, 5-HT$_{1A}$ agonist activity	In vivo	Rat, restraint stress-induced impaired gastric compliance	Tetrahydroberberine, an isoquinoline alkaloid isolated from Corydalis Tuber, enhanced gastric accommodation	[71]
D_2 antagonistic activity, 5-HT$_{1A}$ agonist activity	In vivo	Beagle dog, canine gastric compliance with barostat	Tetrahydroberberine, an isoquinoline alkaloid isolated from Corydalis Tuber, enhanced gastric accommodation	[71]
Down-regulation of 5-HT-induced contraction	In vitro	Feline, esophageal smooth muscle cells	Inhibition of 5-HT-induced contraction via inhibition of MLC20 phosphorylation	[72]
Gastric fundus relaxation	In vivo	Rat, electrical field stimulation (EFS)-induced contractile responses	Nitrergic pathway is major mechanism involved in relaxation of rat gastric fundus by DA-9701	[73]
		Visceral hypersensitivity		
Antinociceptive effect	In vivo	Neonatal rat colorectal distension (CRD)-induced visceral hypersensitivity	Significant decrease in mean arterial pressure (MAP) changes after CRD	[74]
Visceral pain modulation	In vivo	Rat, bee venom (BV)-induced persistent spontaneous nociception and pain hypersensitivity	Tetrahydropalmatine, an alkaloid constituent of plants from the genera *Stephania* and *Corydalis* effectively inhibited visceral nociception as well as thermal and mechanical inflammatory pain hypersensitivity	[75]
Visceral pain modulation	In vivo	Rat, CRD-induced visceral pain	Decrease in visceral pain via reduction in p-ERK in the dorsal root ganglion and spinal cord	[76]

2.3. Effects of DA-9701 on GI Motility

The effects of DA-9701 on gastric emptying have been evaluated in not only normal animals but also models of delayed gastric emptying induced by apomorphine, cisplatin, opioids, and clonidine. [77–79]. The prokinetic effects of DA-9701 were comparable to those of conventional drugs such as cisapride (10 mg/kg) in normal animals and a delayed gastric emptying model at doses of 0.3–3 mg/kg [60] and in an in vitro study of interstitial cells of Cajal, indicating pacemaker activity of DA-9701 [61]. Further, DA-9701 administration accelerated gastric emptying, as shown by a ^{13}C-octanoic acid breath test with repeated measurement in normal mice [62]. In rats sutured with a strain gauge force transducer, DA-9701 administration significantly restored clonidine-induced hypomotility of the gastric antrum in pre- and postprandial periods [63]. When laparotomy, atropine, or opioids were administered to animals to examine the effects of DA-9701 on delayed GI transit, DA-9701 enhanced GI transit in mice who had been subjected to laparotomy or atropine injection [60]. Further, in guinea pigs with opioid-induced bowel dysfunction, delayed GI transit was restored and fecal output was increased. In organ bath experiments using morphine-treated ileal muscle with reduced contractility, DA-9701 significantly increased the amplitude of contraction [64]. Additionally, DA-9701 administration to animals with abdominal surgery was able to ameliorate postoperative ileus (POI) by reducing delayed GI transit and improving defecation [65]. These effects might be mediated by decreased expression of corticotrophin releasing factor in the hypothalamus of DA-9701-pretreated animals with POI [66]. Diverse stresses usually induce dyspeptic symptoms [80], probably by alteration of gastric sensory and motor functions, and acute stressors are known to be associated with delayed gastric emptying in animal models and humans [81,82]. The administration of DA-9701 improved delayed gastric emptying induced by stress, including immobilization and subsequent immersion in a water bath, which was associated with inhibition of stress-induced increases in plasma levels of adrenocorticotropic hormone (ACTH) and ghrelin [67]. DA-9701 has a high affinity for multiple receptors related to GI function. It enhances gastric emptying and GI transit via dopamine D_2 antagonism and 5-HT$_4$ agonism [83]. The affinities of DA-9701 for the D_2 and 5-HT$_4$ receptors were 0.381 and 13.2 μg/mL, respectively [53]. THB and THP, two components of DA-9701, showed dopamine D_2 receptor antagonistic effects, and their IC$_{50}$ values against GTPγS binding to recombinant human dopamine D_2S receptor in Chinese hamster ovary cells were 0.622 μM (211 ng/mL) and 1.32 μM (469 ng/mL), respectively (Dong-A Pharmaceutical Co., Ltd., Seoul, South Korea). There is some controversy about the coexistence of relaxation and contraction effects on the stomach. However, this can be explained by the fact that contractile receptors are dominant in the antrum region and relaxing receptors are dominant in the fundus region of the stomach [84].

2.4. Effects of DA-9701 on Fundic Relaxation

Currently, the barostat is the most commonly available method for measuring fundus relaxation in human patients and canine models [37,85]. To evaluate the fundus relaxation effect, two end points can be measured: Accommodation and compliance [86,87]. The gastric accommodation response enables proximal stomach relaxation to provide space for receiving foods without an increase in gastric pressure [88]. Meal-induced gastric accommodation is thought to be the most important motor index that can be studied currently. Gastric compliance, which is tested in the fasting state, is a measure of gastric tone in the resting state and seems to be related to the pain and discomfort perception threshold [87,89]. DA-9701 significantly increased gastric accommodation in Beagle dogs, shifting the pressure-volume curve toward higher volumes that were comparable to those associated with the control dogs [60]. This finding was reproduced in the same experimental system by CD, a component of DA-9701 [68]. In DA-9701-treated dogs, meal-induced gastric volumes significantly increased and persisted, a finding comparable to that achieved with sumatriptan [69]. DA-9701 may be effective in restoring restraint stress-induced food intake inhibition in rats, and this effect appears to be related to enhanced gastric accommodation by 5-HT$_{1A}$ agonism [70]. Further, THB isolated from DA-9701 alleviated impaired gastric compliance in the rat after stress induction and relaxed the proximal stomach via 5-TH$_{1A}$ agonism [71]. Agonism of 5-HT$_{1A}$ and 5-HT$_{1B/D}$ is known to mediate relaxation of the gastric fundus through activation of a nitric oxide pathway [87,90] and to decrease the visceromotor response to noxious colorectal distention via smooth muscle relaxation [89]. The affinity of DA-9701 to 5-HT$_{1A}$ was shown to be 6.87 μg/mL. In in vitro studies, DA-9701 inhibited 5-HT-induced contraction in feline esophageal smooth muscle cells by reducing the phosphorylation of myosin light chain kinase (MLC20) [72]. Using gastric fundus muscle strips triggered with electrical field stimulation, it was demonstrated that the effect of DA-9701 on rat gastric fundus relaxation is mainly mediated by nitrergic rather than purinergic pathways [73].

2.5. Effects of DA-9701 on Visceral Hypersensitivity

Visceral hypersensitivity is one of the leading targets of FD drug development. In a neonatal rat model of colon irritation with colorectal distention (CRD), administration of DA-9701 significantly decreased mean arterial pressure in a dose-dependent manner, indicating an increase in the pain threshold with CRD-induced visceral hypersensitivity [74]. Pain signals are transmitted from the peripheral regions to the spinal cord (dorsal root ganglion) where the information is processed for transfer to the central nervous system (CNS). In rats treated with DA-9701, phosphorylation of extracellular signal-regulated kinase (p-ERK), a pain-related factor of the pain transmission pathway, significantly decreased in response to CRD [76], suggesting modulation of visceral pain. Further, THP and CD, two components of DA-9701, were reported to have antinociceptive effects on visceral and somatic nociception in animals [75]. It is known that the adrenergic nervous system plays a certain role in modulating visceral nociceptive processing. Adrenergic α_2 agonists, such as clonidine, have been shown to reduce pain perception during gastric and colonic distention [89] and to produce post-operative analgesia in humans [91]. This is probably mediated by modulation of spinal neurotransmitters at the levels of the dorsal horn, activation of descending inhibitory pathways, and/or emotional responses to visceral stimuli [92,93]. The affinity of DA-9701 to adrenergic α_2 receptors was shown to be 4.81 μg/mL.

3. Safety of DA-9701

In a single-dose toxicity study, the LD$_{50}$ of DA-9701 was over 2,000 mg/kg, and in a repeated-dose toxicity study for 26 weeks, the no-observed-adverse-effect level (NOAEL) was 150 mg/kg in rats. The NOAEL was 100 mg/kg in both 1-week and 13-week repeated-treatment studies in dogs. DA-9701 exhibited no genotoxicity. Due to the D$_2$ antagonism of DA-9701, hyperprolactinemia was the major safety concern; however, the prolactin ED$_{200}$ of DA-9701 was approximately 70-fold lower than that of itopride (3.78 vs. 270.1 mg/kg) in rats [53]. The pharmacokinetics and CNS distribution of THB and THP from DA-9701, both having dopamine D$_2$ receptor binding affinities, were examined because D$_2$ antagonists such as metoclopramide have a direct effect on the CNS by crossing the blood-brain barrier [59]. A tissue distribution study revealed that THB and THP were present at high concentrations in the stomach and small intestine compared to those in the plasma following administration of various oral doses of DA-9701, indicating considerable GI distribution. Increased concentrations of THB and THP, which cross the blood-brain barrier, were measured in the brain after repeated-dose administration of DA-9701. However, the brain concentration of DA-9701 was not expected to be sufficient to exert central dopamine D$_2$ receptor antagonism following oral administration of effective doses in humans [59].

4. Clinical Studies of DA-9701

The therapeutic efficacy of DA-9701 has been investigated in several clinical trials according to modern guidelines and these clinical studies are summarized in Table 2.

Table 2. Clinical trials using DA-9701.

Disease	Treatment	Study Design	Sample Size	End Point	Sign	(Ref.)
FD, Rome II	DA 9701 vs. itopride	Randomized, double-blind	464 patients	Change from baseline in the composite score of the 8 dyspeptic symptoms. Overall treatment effect (OTE) as rated on a 7-grade scale.	Significant improvement in symptoms and non-inferior efficacy and comparable safety to that of itopride	[94]
FD, Rome III	DA 9701 vs. pantoprazole	Randomized, double-blind	389 patients	Global symptom assessment, response rates, difference in each score and total score of FD symptoms, difference in dyspepsia-specific quality of life (QOL) outcomes, symptomatic relief according to the subtypes of FD, i.e., epigastric pain syndrome (EPS) and postprandial distress syndrome (PDS).	Improvement in global and individual symptoms and increase in dyspepsia-specific QOL among patients. Efficacy of the monotherapy was comparable to that of pantoprazole. There was no additive effect of the combination of DA-9701 and pantoprazole.	[95]
Minimal change esophagitis	DA 9701 vs. Placebo	Double blind, placebo-controlled	81 patients	Changes in Nepean dyspepsia index questionnaire-Korean version (NDI-K) symptom score	Although NDI-K symptom scores and QOL scores improved after 4 weeks of treatment compared to baseline values in patients with minimal change esophagitis, DA-9701 did not improve the symptom scores or QOL scores compared to those of placebo.	[96]
Parkinson's disease (PD)	DA 9701 vs. domperidone	Randomized, double-blind	40 patients	Gastric MRI, laboratory testing	Used for patients with PD to enhance gastric motility without aggravating PD symptoms	[97]
Healthy volunteers	DA 9701 vs. placebo	Randomized, double-blind, placebo-controlled	40 healthy volunteers	Gastric MRI	DA-9701 enhanced gastric emptying and did not significantly affect gastric accommodation in healthy volunteers.	[98]
Functional constipation, Rome III	DA 9701	Prospective study	37 patients	Colonic transit time (CTT) measurement	DA-9701 accelerated colonic transit and safely improved symptoms	[99]

The first clinical trial using DA-9701 was a multicenter, double-blind, randomized, and controlled trial with concealed allocation comparing the safety and efficacy of DA-9701 and itopride hydrochloride in Korea. Four hundred and sixty-four FD patients aged >20 years who were diagnosed with FD according to the Rome II criteria were randomized to the DA-9701 (30 mg t.i.d.) or itopride (50 mg t.i.d.) group. There was a 2-week medication-free run-in phase and a 4-week treatment period. Two primary efficacy end points, including the change in composite score from baseline of the eight dyspeptic symptoms and the overall treatment effect, were analyzed. Using the Nepean Dyspepsia Index (NDI) questionnaire, the impact on patient quality of life was assessed. DA-9701 significantly improved both FD symptoms and quality of life in patients with FD. The efficacy of DA-9701 was not inferior to that of itopride. Both drugs increased the NDI score of five domains without any differences between the groups. The safety profiles of both drugs were comparable. DA-9701 was well tolerated and did not show drug-related serious adverse effects [94].

To investigate the efficacy of DA-9701 for improving FD symptoms in comparison with that of pantoprazole, a PPI, and to evaluate the additive effect of DA-9701 over PPI treatment only, a multicenter, double-blind, randomized, parallel-comparative phase IV study was conducted. Three hundred and eighty-nine patients diagnosed with FD by using the Rome III criteria were allocated to three groups: 30 mg DA-9701 t.i.d., 40 mg pantoprazole q.d., and 30 mg DA-9701 t.i.d + 40 mg pantoprazole q.d. Patients in all treatment groups reported significant improvements in FD symptoms and dyspepsia-related quality of life ($p < 0.001$). The global symptomatic improvement was 60.5% in the DA-9701 group, 65.6% in the pantoprazole group, and 63.5% in the DA-9701 + pantoprazole group, as determined using a 5-point Likert scale at week 4; there were no significant intergroup differences [95].

In another study, 81 patients with minimal change esophagitis presenting with reflux or dyspeptic symptoms were enrolled and 42 and 39 patients were randomly assigned to receive either DA-9701 30 mg or placebo t.i.d., respectively. After 4 weeks, the NDI questionnaire-Korean version (NDI-K) symptom scores [96] were significantly reduced in the treatment ($p < 0.001$) and control groups ($p < 0.001$). However, changes in the symptom scores and quality of life scores did not differ between the two groups. The reflux symptom score significantly improved in the treatment group compared to that in the placebo group among patients aged 65 years or older ($p = 0.035$). Although the NDI-K symptom scores and quality of life scores improved after 4 weeks of treatment compared to baseline values in patients with minimal change esophagitis, DA-9701 did not improve the symptom scores or quality of life scores compared with those of the placebo [96].

While PD patients have impaired gastric function, which can cause altered responses to oral dopaminergic drugs, prokinetics with dopaminergic antagonism are commonly prescribed to prevent nausea and vomiting induced by anti-parkinsonian drugs. Therefore, to evaluate the clinical utility of DA-9701 in PD patients, the effects of DA-9701 on gastric motility were evaluated in PD patients by using magnetic resonance imaging (MRI). In a non-inferiority study with domperidone, 38 participants completed the 4-week treatment protocol. DA-9701 was not inferior to domperidone in terms of changes in the 2-h gastric emptying rate (GER) before and after treatment. However, a significant increase in the 2-h GER was observed only in the DA-9701 group ($p < 0.05$) without aggravation of PD symptoms [97].

A clinical study was conducted to evaluate the effects of DA-9701 on gastric accommodation and emptying after a meal in healthy volunteers by using 3-D gastric volume measurement with MRI. Forty healthy subjects randomly received DA-9701 or placebo. After 5 days of treatment, the subjects underwent gastric MRI (60 min before and 15, 30, 45, 60, 90, and 120 min after a liquid test meal). Whereas DA-9701 did not significantly affect gastric accommodation in healthy volunteers, pretreatment with DA-9701 increased postprandial proximal-to-distal total gastric volume ratio compared to that with placebo. Pretreatment with DA-9701 significantly increased gastric emptying compared to pretreatment with placebo [98].

In a recent prospective study conducted in 27 patients with functional constipation diagnosed based on the Rome III criteria, DA-9701 30 mg t.i.d. for 24 days was associated with a significantly reduced colonic transit time (CTT) ($p = 0.001$) and decreased segmental CTT ($p < 0.001$). In addition, all constipation-related subjective symptoms, including spontaneous bowel movement frequency, significantly improved compared to those before treatment, without any serious adverse effects [99].

5. Conclusion and Future Prospects

DA-9701, a natural product for FD treatment, is composed of multiple components that act on multiple targets involved in the pathophysiology of FD. DA-9701 has affinity for dopamine, serotonin, and adrenergic receptors. Therefore, it has multiple pharmacological actions through antagonistic effects on D_2 and agonistic effects on 5-HT$_4$, 5-HT$_{1A}$, and 5-HT$_{1B}$, which are associated with gastric emptying, relaxation, and hypersensitivity. Although clinical studies for DA-9701 have had some limitations, beneficial effects of DA-9701 on FD symptoms and GI functions have been demonstrated in FD patients. Further, the promising results indicate that extended use of DA-9701 for PD patients and functional constipation may be possible. The effective dose of DA-9701, which is well tolerated and safe, seems to be smaller than that of conventional drugs targeting a single pathway. Currently, DA-9701 is a prescription drug for FD in Korea. To be used as a medicine in the global market, further studies involving active compounds are definitely needed for a deeper understanding of the activities of DA-9701 in FD management. For example, efficacy and detailed molecular mechanism studies for each active compound targeting different receptors, alone or in combination for additive, synergistic, or even inhibitory effects, using various FD animal models are needed. Undoubtedly, larger and sophisticatedly designed clinical trials are warranted.

Author Contributions: M.J., M.S. wrote the paper. M.J. had primary responsibility for the final content of the manuscript. All authors read and approved the final manuscript.

Funding: This research was funded by the Bio-Synergy Research Project (NRF-2018M3A9C4076473) of the Ministry of Science and ICT through the National Research Foundation of Korea.

Acknowledgments: We thank Soo Jung Yoon for supporting manuscript preparation.

Conflicts of Interest: The authors declare no conflict of interest.

References

1. Brun, R.; Kuo, B. Functional dyspepsia. *Ther. Adv. Gastroenterol.* **2010**, *3*, 145–164. [CrossRef] [PubMed]
2. Pasricha, P.J. Neurogastroenterology: A great career choice for aspiring gastroenterologists thinking about the future. *Gastroenterology* **2011**, *140*, 1126–1128. [CrossRef] [PubMed]
3. Drossman, D.A. Functional Gastrointestinal Disorders: History, Pathophysiology, Clinical Features, and Rome IV. *Gastroenterology* **2016**, *150*, 1262–1279. [CrossRef] [PubMed]
4. Talley, N.J.; Ford, A.C. Functional Dyspepsia. *N. Engl. J. Med.* **2015**, *373*, 1853–1863. [CrossRef] [PubMed]
5. Lacy, B.E.; Weiser, K.T.; Kennedy, A.T.; Crowell, M.D.; Talley, N.J. Functional dyspepsia: The economic impact to patients. *Aliment. Pharmacol. Ther.* **2013**, *38*, 170–177. [CrossRef] [PubMed]
6. Talley, N.J.; Ford, A.C. Functional Dyspepsia. *N. Engl. J. Med.* **2016**, *374*, 896. [CrossRef] [PubMed]
7. Koduru, P.; Irani, M.; Quigley, E.M.M. Definition, Pathogenesis, and Management of That Cursed Dyspepsia. *Clin. Gastroenterol. Hepatol.* **2018**, *16*, 467–479. [CrossRef] [PubMed]
8. Drossman, D.A. The functional gastrointestinal disorders and the Rome II process. *Gut* **1999**, *45* (Suppl. 2), II1–II5. [CrossRef]
9. Tack, J.; Talley, N.J.; Camilleri, M.; Holtmann, G.; Hu, P.; Malagelada, J.R.; Stanghellini, V. Functional gastroduodenal disorders. *Gastroenterology* **2006**, *130*, 1466–1479. [CrossRef]
10. Tack, J.; Talley, N.J. Functional dyspepsia–symptoms, definitions and validity of the Rome III criteria. *Nat. Rev. Gastroenterol. Hepatol.* **2013**, *10*, 134–141. [CrossRef]
11. Stanghellini, V.; Chan, F.K.; Hasler, W.L.; Malagelada, J.R.; Suzuki, H.; Tack, J.; Talley, N.J. Gastroduodenal Disorders. *Gastroenterology* **2016**, *150*, 1380–1392. [CrossRef] [PubMed]

12. Powell, N.; Walker, M.M.; Talley, N.J. Gastrointestinal eosinophils in health, disease and functional disorders. *Nat. Rev. Gastroenterol. Hepatol.* **2010**, *7*, 146–156. [CrossRef]

13. Lee, K.J.; Tack, J. Duodenal implications in the pathophysiology of functional dyspepsia. *J. Neurogastroenterol. Motil.* **2010**, *16*, 251–257. [CrossRef]

14. Walker, M.M.; Salehian, S.S.; Murray, C.E.; Rajendran, A.; Hoare, J.M.; Negus, R.; Powell, N.; Talley, N.J. Implications of eosinophilia in the normal duodenal biopsy—An association with allergy and functional dyspepsia. *Aliment. Pharmacol. Ther.* **2010**, *31*, 1229–1236. [CrossRef] [PubMed]

15. Sugano, K.; Tack, J.; Kuipers, E.J.; Graham, D.Y.; El-Omar, E.M.; Miura, S.; Haruma, K.; Asaka, M.; Uemura, N.; Malfertheiner, P.; et al. Kyoto global consensus report on Helicobacter pylori gastritis. *Gut* **2015**, *64*, 1353–1367. [CrossRef]

16. Sarnelli, G.; Caenepeel, P.; Geypens, B.; Janssens, J.; Tack, J. Symptoms associated with impaired gastric emptying of solids and liquids in functional dyspepsia. *Am. J. Gastroenterol.* **2003**, *98*, 783–788. [CrossRef] [PubMed]

17. Lee, M.Y. Clinical Relevance of Serotonin Receptor Splice Variant Distribution in Human Colon. *J. Neurogastroenterol. Motil.* **2015**, *21*, 303–306. [CrossRef] [PubMed]

18. De Maeyer, J.H.; Lefebvre, R.A.; Schuurkes, J.A. 5-HT4 receptor agonists: Similar but not the same. *Neurogastroenterol. Motil.* **2008**, *20*, 99–112. [CrossRef] [PubMed]

19. Fayyaz, M.; Lackner, J.M. Serotonin receptor modulators in the treatment of irritable bowel syndrome. *Ther. Clin. Risk Manag.* **2008**, *4*, 41–48. [PubMed]

20. Chetty, N.; Coupar, I.M.; Tan, Y.Y.; Desmond, P.V.; Irving, H.R. Distribution of serotonin receptors and interacting proteins in the human sigmoid colon. *Neurogastroenterol. Motil.* **2009**, *21*, 551–e15. [CrossRef] [PubMed]

21. Acosta, A.; Camilleri, M. Prokinetics in gastroparesis. *Gastroenterol. Clin. N. Am.* **2015**, *44*, 97–111. [CrossRef] [PubMed]

22. Zala, A.V.; Walker, M.M.; Talley, N.J. Emerging drugs for functional dyspepsia. *Expert Opin. Emerg. Drugs* **2015**, *20*, 221–233. [CrossRef] [PubMed]

23. Bang, C.S.; Kim, J.H.; Baik, G.H.; Kim, H.S.; Park, S.H.; Kim, E.J.; Kim, J.B.; Suk, K.T.; Yoon, J.H.; Kim, Y.S.; et al. Mosapride treatment for functional dyspepsia: A meta-analysis. *J. Gastroenterol. Hepatol.* **2015**, *30*, 28–42. [CrossRef] [PubMed]

24. Pasricha, P.J. Desperately seeking serotonin. A commentary on the withdrawal of tegaserod and the state of drug development for functional and motility disorders. *Gastroenterology* **2007**, *132*, 2287–2290. [CrossRef] [PubMed]

25. Huang, X.; Lv, B.; Zhang, S.; Fan, Y.H.; Meng, L.N. Itopride therapy for functional dyspepsia: A meta-analysis. *World J. Gastroenterol.* **2012**, *18*, 7371–7377. [CrossRef] [PubMed]

26. Talley, N.J.; Tack, J.; Ptak, T.; Gupta, R.; Giguere, M. Itopride in functional dyspepsia: Results of two phase III multicentre, randomised, double-blind, placebo-controlled trials. *Gut* **2008**, *57*, 740–746. [CrossRef] [PubMed]

27. Ehrenpreis, E.D.; Roginsky, G.; Alexoff, A.; Smith, D.G. Domperidone is Commonly Prescribed With QT-Interacting Drugs: Review of a Community-based Practice and a Postmarketing Adverse Drug Event Reporting Database. *J. Clin. Gastroenterol.* **2017**, *51*, 56–62. [CrossRef]

28. Ganzini, L.; Casey, D.E.; Hoffman, W.F.; McCall, A.L. The prevalence of metoclopramide-induced tardive dyskinesia and acute extrapyramidal movement disorders. *Arch. Intern. Med.* **1993**, *153*, 1469–1475. [CrossRef]

29. Tack, J.; Janssen, P.; Masaoka, T.; Farre, R.; Van Oudenhove, L. Efficacy of buspirone, a fundus-relaxing drug, in patients with functional dyspepsia. *Clin. Gastroenterol. Hepatol.* **2012**, *10*, 1239–1245. [CrossRef]

30. Sekino, Y.; Yamada, E.; Sakai, E.; Ohkubo, H.; Higurashi, T.; Iida, H.; Endo, H.; Takahashi, H.; Koide, T.; Sakamoto, Y.; et al. Influence of sumatriptan on gastric accommodation and on antral contraction in healthy subjects assessed by ultrasonography. *Neurogastroenterol. Motil.* **2012**, *24*, 1083–e564. [CrossRef]

31. Tack, J.; Janssen, P. Acotiamide (Z-338, YM443), a new drug for the treatment of functional dyspepsia. *Expert Opin. Investig. Drugs* **2011**, *20*, 701–712. [CrossRef]

32. Kusunoki, H.; Haruma, K.; Manabe, N.; Imamura, H.; Kamada, T.; Shiotani, A.; Hata, J.; Sugioka, H.; Saito, Y.; Kato, H.; et al. Therapeutic efficacy of acotiamide in patients with functional dyspepsia based on enhanced postprandial gastric accommodation and emptying: Randomized controlled study evaluation by real-time ultrasonography. *Neurogastroenterol. Motil.* **2012**, *24*, 540–e251. [CrossRef]

33. Matsueda, K.; Hongo, M.; Tack, J.; Saito, Y.; Kato, H. A placebo-controlled trial of acotiamide for meal-related symptoms of functional dyspepsia. *Gut* **2012**, *61*, 821–828. [CrossRef] [PubMed]

34. Coffin, B.; Azpiroz, F.; Guarner, F.; Malagelada, J.R. Selective gastric hypersensitivity and reflex hyporeactivity in functional dyspepsia. *Gastroenterology* **1994**, *107*, 1345–1351. [CrossRef]

35. Bradette, M.; Pare, P.; Douville, P.; Morin, A. Visceral perception in health and functional dyspepsia. Crossover study of gastric distension with placebo and domperidone. *Dig. Dis. Sci.* **1991**, *36*, 52–58. [CrossRef] [PubMed]

36. Mearin, F.; Cucala, M.; Azpiroz, F.; Malagelada, J.R. The origin of symptoms on the brain-gut axis in functional dyspepsia. *Gastroenterology* **1991**, *101*, 999–1006. [CrossRef]

37. Tack, J.; Caenepeel, P.; Fischler, B.; Piessevaux, H.; Janssens, J. Symptoms associated with hypersensitivity to gastric distention in functional dyspepsia. *Gastroenterology* **2001**, *121*, 526–535. [CrossRef] [PubMed]

38. Farre, R.; Vanheel, H.; Vanuytsel, T.; Masaoka, T.; Tornblom, H.; Simren, M.; Van Oudenhove, L.; Tack, J.F. In functional dyspepsia, hypersensitivity to postprandial distention correlates with meal-related symptom severity. *Gastroenterology* **2013**, *145*, 566–573. [CrossRef] [PubMed]

39. Thompson, A.J.; Lummis, S.C. The 5-HT3 receptor as a therapeutic target. *Expert Opin. Ther. Targets* **2007**, *11*, 527–540. [CrossRef] [PubMed]

40. Delvaux, M.; Louvel, D.; Mamet, J.P.; Campos-Oriola, R.; Frexinos, J. Effect of alosetron on responses to colonic distension in patients with irritable bowel syndrome. *Aliment. Pharmacol. Ther.* **1998**, *12*, 849–855. [CrossRef] [PubMed]

41. Harris, L.A.; Chang, L. Alosetron: An effective treatment for diarrhea-predominant irritable bowel syndrome. *Womens Health* **2007**, *3*, 15–27. [CrossRef] [PubMed]

42. Pascaud, X.; Honde, C.; Le Gallou, B.; Chanoine, F.; Roman, F.; Bueno, L.; Junien, J.L. Effects of fedotozine on gastrointestinal motility in dogs: Mechanism of action and related pharmacokinetics. *J. Pharm. Pharmacol.* **1990**, *42*, 546–552. [CrossRef] [PubMed]

43. Diop, L.; Riviere, P.J.; Pascaud, X.; Junien, J.L. Peripheral kappa-opioid receptors mediate the antinociceptive effect of fedotozine (correction of fetodozine) on the duodenal pain reflex inrat. *Eur J. Pharmacol.* **1994**, *271*, 65–71. [CrossRef]

44. De Ponti, F. Drug development for the irritable bowel syndrome: Current challenges and future perspectives. *Front. Pharmacol.* **2013**, *4*, 7. [CrossRef] [PubMed]

45. Talley, N.J.; American Gastroenterological, A. American Gastroenterological Association medical position statement: Evaluation of dyspepsia. *Gastroenterology* **2005**, *129*, 1753–1755. [CrossRef] [PubMed]

46. Talley, N.J.; Locke, G.R.; Saito, Y.A.; Almazar, A.E.; Bouras, E.P.; Howden, C.W.; Lacy, B.E.; DiBaise, J.K.; Prather, C.M.; Abraham, B.P.; et al. Effect of Amitriptyline and Escitalopram on Functional Dyspepsia: A Multicenter, Randomized Controlled Study. *Gastroenterology* **2015**, *149*, 340–349. [CrossRef]

47. Vanheel, H.; Tack, J. Therapeutic options for functional dyspepsia. *Dig. Dis.* **2014**, *32*, 230–234. [CrossRef]

48. Tack, J.; Ly, H.G.; Carbone, F.; Vanheel, H.; Vanuytsel, T.; Holvoet, L.; Boeckxstaens, G.; Caenepeel, P.; Arts, J.; Van Oudenhove, L. Efficacy of Mirtazapine in Patients With Functional Dyspepsia and Weight Loss. *Clin. Gastroenterol. Hepatol.* **2016**, *14*, 385–392. [CrossRef]

49. Choung, R.S.; Cremonini, F.; Thapa, P.; Zinsmeister, A.R.; Talley, N.J. The effect of short-term, low-dose tricyclic and tetracyclic antidepressant treatment on satiation, postnutrient load gastrointestinal symptoms and gastric emptying: A double-blind, randomized, placebo-controlled trial. *Neurogastroenterol. Motil.* **2008**, *20*, 220–227. [CrossRef]

50. Mansi, C.; Borro, P.; Giacomini, M.; Biagini, R.; Mele, M.R.; Pandolfo, N.; Savarino, V. Comparative effects of levosulpiride and cisapride on gastric emptying and symptoms in patients with functional dyspepsia and gastroparesis. *Aliment. Pharmacol. Ther.* **2000**, *14*, 561–569. [CrossRef]

51. Mearin, F.; Rodrigo, L.; Perez-Mota, A.; Balboa, A.; Jimenez, I.; Sebastian, J.J.; Paton, C. Levosulpiride and cisapride in the treatment of dysmotility-like functional dyspepsia: A randomized, double-masked trial. *Clin. Gastroenterol. Hepatol.* **2004**, *2*, 301–308. [CrossRef]

52. Camilleri, M.; Bueno, L.; de Ponti, F.; Fioramonti, J.; Lydiard, R.B.; Tack, J. Pharmacological and pharmacokinetic aspects of functional gastrointestinal disorders. *Gastroenterology* **2006**, *130*, 1421–1434. [CrossRef] [PubMed]

53. Kwon, Y.S.; Son, M. DA-9701: A New Multi-Acting Drug for the Treatment of Functional Dyspepsia. *Biomol. Ther.* **2013**, *21*, 181–189. [CrossRef] [PubMed]

54. Ding, B.; Zhou, T.; Fan, G.; Hong, Z.; Wu, Y. Qualitative and quantitative determination of ten alkaloids in traditional Chinese medicine Corydalis yanhusuo W.T. Wang by LC-MS/MS and LC-DAD. *J. Pharm. Biomed. Anal.* **2007**, *45*, 219–226. [CrossRef]

55. Kumar, S.; Kumar, D.; Singh, J.; Narender, R.; Kaushik, D. Anti-inflammatory, Analgesic and Antioxidant Activities of Ipomoea hederaceae Linn. *Planta Med.* **2009**, *75*, p-100. [CrossRef]

56. Soji, Y.; Kadokawa, T.; Masuda, Y.; Kawashima, K.; Nakamura, K. Effects of Corydalis alkaloid upon inhibition of gastric juice secretion and prevention of gastric ulcer in experimental animals. *Nihon Yakurigaku Zasshi* **1969**, *65*, 196–209. [CrossRef]

57. Kamigauchi, M.; Iwasa, K. *Corydalis* spp.: In Vitro Culture and the Biotransformation of Protoberberines. *Biotechnol. Agric. For.* **1994**, *26*, 93–105. [CrossRef]

58. Yun, K.J.; Shin, J.S.; Choi, J.H.; Back, N.I.; Chung, H.G.; Lee, K.T. Quaternary alkaloid, pseudocoptisine isolated from tubers of Corydalis turtschaninovi inhibits LPS-induced nitric oxide, PGE_2, and pro-inflammatory cytokines production via the down-regulation of NF-kappaB in RAW 264.7 murine macrophage cells. *Int. Immunopharmacol.* **2009**, *9*, 1323–1331. [CrossRef]

59. Jung, J.W.; Kim, J.M.; Jeong, J.S.; Son, M.; Lee, H.S.; Lee, M.G.; Kang, H.E. Pharmacokinetics of chlorogenic acid and corydaline in DA-9701, a new botanical gastroprokinetic agent, in rats. *Xenobiotica* **2014**, *44*, 635–643. [CrossRef]

60. Lee, T.H.; Choi, J.J.; Kim, D.H.; Choi, S.; Lee, K.R.; Son, M.; Jin, M. Gastroprokinetic effects of DA-9701, a new prokinetic agent formulated with Pharbitis Semen and Corydalis Tuber. *Phytomedicine* **2008**, *15*, 836–843. [CrossRef]

61. Choi, S.; Choi, J.J.; Jun, J.Y.; Koh, J.W.; Kim, S.H.; Kim, D.H.; Pyo, M.Y.; Choi, S.; Son, J.P.; Lee, I.; et al. Induction of pacemaker currents by DA-9701, a prokinetic agent, in interstitial cells of Cajal from murine small intestine. *Mol. Cells* **2009**, *27*, 307–312. [CrossRef] [PubMed]

62. Lim, C.H.; Choi, M.G.; Park, H.; Baeg, M.K.; Park, J.M. Effect of DA-9701 on gastric emptying in a mouse model: Assessment by ^{13}C-octanoic acid breath test. *World J. Gastroenterol.* **2013**, *19*, 4380–4385. [CrossRef] [PubMed]

63. Kang, J.W.; Han, D.K.; Kim, O.N.; Lee, K.J. Effect of DA-9701 on the Normal Motility and Clonidine-induced Hypomotility of the Gastric Antrum in Rats. *J. Neurogastroenterol. Motil.* **2016**, *22*, 304–309. [CrossRef]

64. Hussain, Z.; Rhee, K.W.; Lee, Y.J.; Park, H. The Effect of DA-9701 in Opioid-induced Bowel Dysfunction of Guinea Pig. *J. Neurogastroenterol. Motil.* **2016**, *22*, 529–538. [CrossRef] [PubMed]

65. Lee, S.P.; Lee, O.Y.; Lee, K.N.; Lee, H.L.; Choi, H.S.; Yoon, B.C.; Jun, D.W. Effect of DA-9701, a Novel Prokinetic Agent, on Post-operative Ileus in Rats. *J. Neurogastroenterol. Motil.* **2017**, *23*, 109–116. [CrossRef]

66. Jo, S.Y.; Hussain, Z.; Lee, Y.J.; Park, H. Corticotrophin-releasing factor-mediated effects of DA-9701 in Postoperative Ileus Guinea Pig Model. *Neurogastroenterol. Motil.* **2018**, *30*, e13385. [CrossRef]

67. Jung, Y.S.; Kim, M.Y.; Lee, H.S.; Park, S.L.; Lee, K.J. Effect of DA-9701, a novel prokinetic agent, on stress-induced delayed gastric emptying and hormonal changes in rats. *Neurogastroenterol. Motil.* **2013**, *25*, 254–e166. [CrossRef]

68. Lee, T.H.; Son, M.; Kim, S.Y. Effects of corydaline from Corydalis tuber on gastric motor function in an animal model. *Biol. Pharm. Bull.* **2010**, *33*, 958–962. [CrossRef]

69. Kim, E.R.; Min, B.H.; Lee, S.O.; Lee, T.H.; Son, M.; Rhee, P.L. Effects of DA-9701, a novel prokinetic agent, on gastric accommodation in conscious dogs. *J. Gastroenterol. Hepatol.* **2012**, *27*, 766–772. [CrossRef]

70. Kim, Y.S.; Lee, M.Y.; Park, J.S.; Choi, E.S.; Kim, M.S.; Park, S.H.; Ryu, H.S.; Choi, S.C. Effect of DA-9701 on Feeding Inhibition Induced by Acute Restraint Stress in Rats. *Korean J. Helicobacter Up. Gastrointest. Res.* **2018**, *18*, 50–55. [CrossRef]

71. Lee, T.H.; Kim, K.H.; Lee, S.O.; Lee, K.R.; Son, M.; Jin, M. Tetrahydroberberine, an isoquinoline alkaloid isolated from corydalis tuber, enhances gastrointestinal motor function. *J. Pharmacol. Exp. Ther.* **2011**, *338*, 917–924. [CrossRef] [PubMed]

72. Oh, K.H.; Nam, Y.; Jeong, J.H.; Kim, I.K.; Sohn, U.D. The effect of DA-9701 on 5-hydroxytryptamine-induced contraction of feline esophageal smooth muscle cells. *Molecules* **2014**, *19*, 5135–5149. [CrossRef] [PubMed]

73. Min, Y.W.; Ko, E.J.; Lee, J.Y.; Min, B.H.; Lee, J.H.; Kim, J.J.; Rhee, P.L. Nitrergic Pathway Is the Major Mechanism for the Effect of DA-9701 on the Rat Gastric Fundus Relaxation. *J. Neurogastroenterol. Motil.* **2014**, *20*, 318–325. [CrossRef]

74. Kim, E.R.; Min, B.H.; Lee, T.H.; Son, M.; Rhee, P.L. Effect of DA-9701 on colorectal distension-induced visceral hypersensitivity in a rat model. *Gut Liver* **2014**, *8*, 388–393. [CrossRef] [PubMed]

75. Cao, F.L.; Shang, G.W.; Wang, Y.; Yang, F.; Li, C.L.; Chen, J. Antinociceptive effects of intragastric DL-tetrahydropalmatine on visceral and somatic persistent nociception and pain hypersensitivity in rats. *Pharmacol. Biochem. Behav.* **2011**, *100*, 199–204. [CrossRef] [PubMed]

76. Lee, S.P.; Lee, K.N.; Lee, O.Y.; Lee, H.L.; Jun, D.W.; Yoon, B.C.; Choi, H.S.; Hwang, S.J.; Lee, S.E. Effects of DA-9701, a novel prokinetic agent, on phosphorylated extracellular signal-regulated kinase expression in the dorsal root ganglion and spinal cord induced by colorectal distension in rats. *Gut Liver* **2014**, *8*, 140–147. [CrossRef] [PubMed]

77. Ramsbottom, N.; Hunt, J.N. Studies of the effect of metoclopramide and apomorphine on gastric emptying and secretion in man. *Gut* **1970**, *11*, 989–993. [CrossRef]

78. Tanila, H.; Kauppila, T.; Taira, T. Inhibition of intestinal motility and reversal of postlaparotomy ileus by selective alpha 2-adrenergic drugs in the rat. *Gastroenterology* **1993**, *104*, 819–824. [CrossRef]

79. Hirokawa, Y.; Yamazaki, H.; Yoshida, N.; Kato, S. A novel series of 6-methoxy-1H-benzotriazole-5-carboxamide derivatives with dual antiemetic and gastroprokinetic activities. *Bioorg. Med. Chem. Lett.* **1998**, *8*, 1973–1978. [CrossRef]

80. Haug, T.T.; Svebak, S.; Wilhelmsen, I.; Berstad, A.; Ursin, H. Psychological factors and somatic symptoms in functional dyspepsia. A comparison with duodenal ulcer and healthy controls. *J. Psychosom. Res.* **1994**, *38*, 281–291. [CrossRef]

81. Nakade, Y.; Tsuchida, D.; Fukuda, H.; Iwa, M.; Pappas, T.N.; Takahashi, T. Restraint stress delays solid gastric emptying via a central CRF and peripheral sympathetic neuron in rats. *Am. J. Physiol. Regul. Integr. Comp. Physiol.* **2005**, *288*, R427–R432. [CrossRef] [PubMed]

82. Tache, Y.; Martinez, V.; Million, M.; Wang, L. Stress and the gastrointestinal tract III. Stress-related alterations of gut motor function: Role of brain corticotropin-releasing factor receptors. *Am. J. Physiol. Gastrointest. Liver Physiol.* **2001**, *280*, G173–G177. [CrossRef] [PubMed]

83. Lim, H.C.; Park, H.; Lee, S.I.; Jahng, J.; Lee, Y.J. Tu1985 effects of DA-9701, a novel prokinetic agent on gastric motor function in guinea pig. *Gastroenterology* **2013**, *142*, S-893. [CrossRef]

84. Hansen, M.B. Neurohumoral control of gastrointestinal motility. *Physiol. Res.* **2003**, *52*, 1–30. [PubMed]

85. Azpiroz, F.; Malagelada, J.R. Abdominal bloating. *Gastroenterology* **2005**, *129*, 1060–1078. [CrossRef]

86. Azpiroz, F.; Malagelada, J.R. Physiological variations in canine gastric tone measured by an electronic barostat. *Am. J. Physiol.* **1985**, *248*, G229–G237. [CrossRef] [PubMed]

87. De Ponti, F.; Crema, F.; Moro, E.; Nardelli, G.; Frigo, G.; Crema, A. Role of 5-HT1B/D receptors in canine gastric accommodation: Effect of sumatriptan and 5-HT1B/D receptor antagonists. *Am. J. Physiol. Gastrointest. Liver Physiol.* **2003**, *285*, G96–G104. [CrossRef] [PubMed]

88. Azpiroz, F.; Malagelada, J.R. Vagally mediated gastric relaxation induced by intestinal nutrients in the dog. *Am. J. Physiol.* **1986**, *251*, G727–G735. [CrossRef] [PubMed]

89. Kuiken, S.D.; Tytgat, G.N.; Boeckxstaens, G.E. Review article: Drugs interfering with visceral sensitivity for the treatment of functional gastrointestinal disorders–the clinical evidence. *Aliment. Pharmacol. Ther.* **2005**, *21*, 633–651. [CrossRef]

90. Coulie, B.; Tack, J.; Sifrim, D.; Andrioli, A.; Janssens, J. Role of nitric oxide in fasting gastric fundus tone and in 5-HT1 receptor-mediated relaxation of gastric fundus. *Am. J. Physiol.* **1999**, *276*, G373–G377. [CrossRef]

91. Bonnet, F.; Boico, O.; Rostaing, S.; Loriferne, J.F.; Saada, M. Clonidine-induced analgesia in postoperative patients: Epidural versus intramuscular administration. *Anesthesiology* **1990**, *72*, 423–427. [CrossRef] [PubMed]

92. Mayer, E.A.; Gebhart, G.F. Basic and clinical aspects of visceral hyperalgesia. *Gastroenterology* **1994**, *107*, 271–293. [CrossRef]

93. Camilleri, M.; Coulie, B.; Tack, J.F. Visceral hypersensitivity: Facts, speculations, and challenges. *Gut* **2001**, *48*, 125–131. [CrossRef] [PubMed]

94. Choi, M.G.; Rhee, P.L.; Park, H.; Lee, O.Y.; Lee, K.J.; Choi, S.C.; Seol, S.Y.; Chun, H.J.; Rew, J.S.; Lee, D.H.; et al. Randomized, Controlled, Multi-center Trial: Comparing the Safety and Efficacy of DA-9701 and Itopride Hydrochloride in Patients With Functional Dyspepsia. *J. Neurogastroenterol. Motil.* **2015**, *21*, 414–422. [CrossRef] [PubMed]

95. Jung, H.K.; Lee, K.J.; Choi, M.G.; Park, H.; Lee, J.S.; Rhee, P.L.; Kim, N.; Park, K.; Choi, S.C.; Lee, O.Y.; et al. Efficacy of DA-9701 (Motilitone) in Functional Dyspepsia Compared to Pantoprazole: A Multicenter, Randomized, Double-blind, Non-inferiority Study. *J. Neurogastroenterol. Motil.* **2016**, *22*, 254–263. [CrossRef] [PubMed]

96. Park, C.H.; Kim, H.S.; Lee, S.K. Effects of the New Prokinetic Agent DA-9701 Formulated With Corydalis Tuber and Pharbitis Seed in Patients With Minimal Change Esophagitis: A Bicenter, Randomized, Double Blind, Placebo-controlled Study. *J. Neurogastroenterol. Motil.* **2014**, *20*, 338–346. [CrossRef] [PubMed]

97. Shin, C.M.; Lee, Y.J.; Kim, J.M.; Lee, J.Y.; Kim, K.J.; Choi, Y.J.; Kim, N.; Lee, D.H. DA-9701 on gastric motility in patients with Parkinson's disease: A randomized controlled trial. *Parkinsonism Relat. Disord.* **2018**, *54*, 84–89. [CrossRef] [PubMed]

98. Min, Y.W.; Min, B.H.; Kim, S.; Choi, D.; Rhee, P.L. Effect of DA-9701 on Gastric Motor Function Assessed by Magnetic Resonance Imaging in Healthy Volunteers: A Randomized, Double-Blind, Placebo-Controlled Trial. *PLoS ONE* **2015**, *10*, e0138927. [CrossRef]

99. Kim, S.Y.; Woo, H.S.; Kim, K.O.; Choi, S.H.; Kwon, K.A.; Chung, J.W.; Kim, Y.J.; Kim, J.H.; Kim, S.J.; Park, D.K. DA-9701 improves colonic transit time and symptoms in patients with functional constipation: A prospective study. *J. Gastroenterol. Hepatol.* **2017**, *32*, 1943–1948. [CrossRef]

International Journal of
Molecular Sciences

Article

Melanogenesis Inhibitors from the Rhizoma of *Ligusticum Sinense* in B16-F10 Melanoma Cells In Vitro and Zebrafish In Vivo

Min-Chi Cheng [1,†], Tzong-Huei Lee [2,†], Yi-Tzu Chu [3], Li-Ling Syu [3], Su-Jung Hsu [4], Chia-Hsiung Cheng [5], Jender Wu [1,*] and Ching-Kuo Lee [1,3,6,*]

[1] School of Pharmacy, Taipei Medical University, Taipei 11031, Taiwan; d301100008@tmu.edu.tw
[2] Institute of Fisheries Science, National Taiwan University, Taipei 10617, Taiwan; thlee1@ntu.edu.tw
[3] Graduate Institute of Pharmacognosy, Taipei Medical University, Taipei 11031, Taiwan;
    m303098006@tmu.edu.tw (Y.-T.C.); goat1201@hotmail.com (L.-L.S.)
[4] Department of Food Science, National Chiayi University, Chiayi 60004, Taiwan; r1101815@yahoo.com.tw
[5] Department of Biochemistry and Molecular Cell Biology, School of Medicine, College of Medicine,
    Taipei Medical University, Taipei 11031, Taiwan; chcheng@tmu.edu.tw
[6] Ph.D. Program in Biotechnology Research and Development, Taipei Medical University,
    Taipei 11031, Taiwan
*   Correspondence: jd0332@tmu.edu.tw (J.W.); cklee@tmu.edu.tw (C.-K.L.);
    Tel.: +886-2-27361661 (ext. 6150) (C.-K.L.); +886-2-27361661 (ext. 6135) (J.W.); Fax: +886-2-23772265 (C.-K.L.)
†   These authors contributed equally to this work.

Received: 3 November 2018; Accepted: 9 December 2018; Published: 11 December 2018

Abstract: The rhizoma of *Ligusticum sinense*, a Chinese medicinal plant, has long been used as a cosmetic for the whitening and hydrating of the skin in ancient China. In order to investigate the antimelanogenic components of the rhizoma of *L. sinense*, we performed an antimelanogenesis assay-guided purification using semi-preparative HPLC accompanied with spectroscopic analysis to determine the active components. Based on the bioassay-guided method, 24 compounds were isolated and identified from the ethyl acetate layer of methanolic extracts of *L. sinense*, and among these, 5-[3-(4-hydroxy-3-methoxyphenyl)allyl]ferulic acid (**1**) and *cis*-4-pentylcyclohex-3-ene-1,2-diol (**2**) were new compounds. All the pure isolates were subjected to antimelanogenesis assay using murine melanoma B16-F10 cells. Compound **1** and (3*S*,3a*R*)-neocnidilide (**8**) exhibited antimelanogenesis activities with IC$_{50}$ values of 78.9 and 31.1 μM, respectively, without obvious cytotoxicity. Further investigation showed that compound **8** demonstrated significant anti-pigmentation activity on zebrafish embryos (10-20 μM) compared to arbutin (20 μM), and without any cytotoxicity against normal human epidermal keratinocytes. These findings suggest that (3*S*,3a*R*)-neocnidilide (**8**) is a potent antimelanogenic and non-cytotoxic natural compound and may be developed potentially as a skin-whitening agent for cosmetic uses.

Keywords: *Ligusticum sinense*; rhizoma; antimelanogenesis; B16-F10 melanoma cell; zebrafish; pigmentation

---

1. Introduction

Melanin is a brown black pigment that is principally responsible for the color of skin, hair, and eyes [1]. Melanin is a biopolymer and includes two major classes of pigments in human skin, brownish black eumelanin and reddish yellow pheomelanin [1]. Melanin biosynthesis is a complex multistep process called melanogenesis, which is a physiological response of human skin to prevent deleterious effects of ultraviolet (UV) radiation and environmental pollutants. However, over-melanogenesis can lead to the darkening of the skin, and abnormal hyperpigmentation causes various dermatological problems, such as freckles, melasma, senile lentigines, and even skin cancer [2].

Melanin is produced in membrane-bound organelles referred to as melanosomes, which are present in specialized cells called melanocytes. Melanin synthesis starts from the hydroxylation of L-tyrosine to L-dihydroxyphenylalanin (DOPA) and is followed by oxidation to dopaquinone. The two reactions are catalyzed by a rate-limiting enzyme tyrosinase. In the absence of thiol substances, dopaquinone cyclizes to leukodopachrome, followed by a series of oxidoreduction reactions which involve tyrosinase-related protein-2 (Tyrp-2) to produce the intermediate dopachrome and 5,6-dihydroxyindole-2-carboxylic acid (DHICA). DHICA undergoes subsequent oxidation catalyzed by Tyrp-1 and polymerization to form eumelanin. Dopaquinone also conjugates with cysteine and glutathione to yield cysteinyldopa and glutathionyldopa, which are progressively transformed into pheomelanin [1].

The cells surrounding melanocytes such as keratinocytes and fibroblasts are involved in the regulation of melanogenesis [3]. Continuous exposure to UV irradiation induces DNA damages in keratinocytes and leads to p53-mediated up-regulation of proopiomelanocortin (POMC). POMC undergoes posttranslational cleavage to produce the melanocyte stimulating hormone (α-MSH) and β-endorphin. In turn, α-MSH binds to the melanocortin 1 receptor (MC1R) on adjacent melanocytes, resulting in the upregulation of cAMP. Elevated cAMP stimulates expression of microphthalmia-associated transcription factor (MITF). MITF then regulates the transcription of pigmentation enzymes, including tyrosinase, Tyrp-1 and Tyrp-2. The UV-triggered pathway eventually leads to melanin synthesis and transfer of melanosomes to keratinocytes [4–6]. In addition to extrinsic stimuli, melanin production is also influenced by intrinsic factors, such as hormone change, genetic disorders, inflammation, and age [3].

In East Asia, most women expect to avoid uneven skin pigmentation and pursue skin lightening. For instance, arbutin, kojic acid, azelaic acid, ascorbic acid, and white mulberry and licorice root extract have been used as whitening ingredients in cosmetic preparations [7]. The exploitation of effective and preventive skin whitening agents from natural sources is of great interest in the cosmetic field, primarily due to relative nontoxicity and fewer side effects [7,8]. The rhizoma of *Ligusticum sinense* Oliv. (Umbelliferae) have long been used as traditional Chinese medicine for 2000 years and till today its roots are a highly recommended herbal tea [9]. *L. sinense*, namely "Gaoben" in Chinese, also known as Chinese lovage, was used for expelling wind-cold, relieving pain and rheumatic arthralgia, and alleviating anemofrigid headache. The main external use of *L. sinense* is for skin whitening and hydrating [10]. To date, the reported chemical constituents present in *L. sinense* include terpenoids, phthalide analogues, and phenylpropanoid glycosides [11–16]. Such constituents have been reported to exert numerous pharmacological effects. For example, the essential oils from the roots and rhizomes of *L. sinense* were reported to possess analgesic, sedative, and antimicrobial effects [15,17]. Ligustilide, a main phthalide widely found in Umbelliferae plant, demonstrated dilatory effect on myometrium, and reduced inflammatory and neurogenic pain [18]. Cnidilide, another phthalide abundant in Umbelliferae plant, was proven to possess antispasmodic and inflammatory effects [19,20]. Previous studies have indicated that the extract of *L. sinense* inhibited melanogenesis on B16-F10 murine melanoma cells [21]. However, the active components for melanogenesis inhibitory activity were still unreported.

In our preliminary biological screening, it was found that the methanolic extracts of *L. sinense* exhibited antimelanogenesis activity in B16-F10 cells with an IC_{50} value of 50 μg/mL [22], and the antimealnogenesis principles are still undisclosed thus far. We thus set out to investigate the active principle of the rhizoma of *L. sinense* by a bioassay-directed method, and that has led to the isolation and identification of two new compounds **1** and **2** along with 22 known compounds **3–24**. This article also aimed to investigate the effects of compounds **1** and **8** on B16-F10 melanoma cells in vitro and zebrafish in vivo, assess the safety by normal human epidermal keratinocyte MTT assay, and quantify **1** and **8** in the rhizoma of *L. sinense*.

2. Results and Discussions

2.1. Isolation and Structural Elucidation

In an attempt to isolate and identify the melanogenesis inhibitors efficiently from the active fractions, we employed a bioassay-guided fractionation strategy. A methanolic extract of the rhizoma of *L. sinense* was partitioned to give ethyl acetate, *n*-butanol and water soluble layers. The obtained three layers were then tested for antimelanogenesis activity in murine melanoma B16-F10 cells. The mouse B16 melanoma cell is a sensitive, reliable, and feasible platform for screening large number of small molecular melanogenesis regulators [23–25]. At the concentrations of 25–100 µg/mL, the ethyl acetate-soluble layer demonstrated the most potent inhibitory activity, while slight inhibition was observed for either the *n*-butanol or water-soluble layers (Figure 1A) as evidenced by melanin contents in lyzed B16-F10 melanoma cells (Figure 1B). Open column separation of ethyl acetate layer over silica gel followed by HPLC purification afforded two previously unreported chemical entities **1** and **2** (Figure 2A) together with 22 known compounds. The known compounds were characterized to be eugenol (**3**) [26], 2-hydroxy-4-methylacetophenone (**4**) [27], 3*S**,3a*R**,7a*S**-3-butylhexahydrophthalide (**5**) [28], carvacrol (**6**) [29], squalene (**7**) [30], (3*S*,3a*R*)-neocnidilide (**8**) [31], coniferyl alcohol 9-methylester (**9**) [32], bergapten (**10**) [33], methoxsalen (**11**) [34], methyl vanillate (**12**) [35], 2,5-dihydroxy-4-methylacetophenone (**13**) [36], 2-methoxy-4-nitrophenol (**14**) [37], 2,6-dimethoxyphenol (**15**) [38], falcarindiol (**16**) [39], (9*Z*)-heptadecene-4,6-diyne-1,8-diol (**17**) [40], 3-*O*-(*p*-coumaroyl)ursolic acid (**18**) [41], pregnenolone (**19**) [42], (9*Z*,11*E*,13*R*)-13-hydroxyoctadeca-9,11-dienoic acid (**20**) [43], ferulic acid (**21**) [44], coniferyl ferulate (**22**) [45], *p*-hydroxyphenethyl ferulate (**23**) [46], and vanillic acid (**24**) [47].

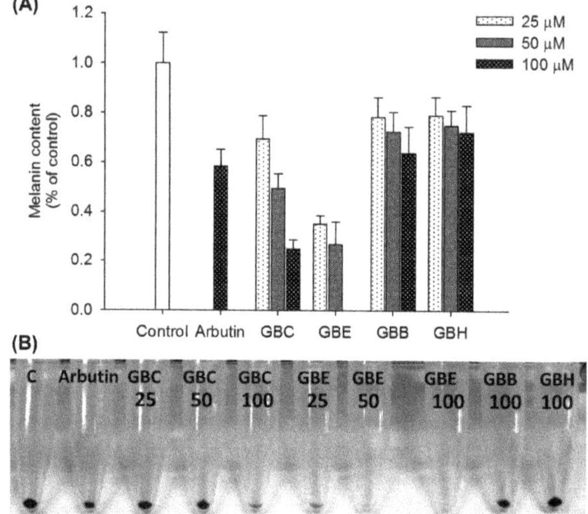

Figure 1. (**A**) The antimelanogenesis effect of different extraction layers with particular concentration on B16-F10 melanoma cells. The B16-F10 melanoma cells were seeded and incubated overnight to allow cells to adhere. The cells were exposed to various concentrations (25, 50 and 100 µM) of the different extraction layers or arbutin for 72 h in the presence of 100 nM α-MSH. At the end of the treatment, the cells were washed with PBS and lyzed with 150 µL of 1 N NaOH containing 10% DMSO for 1 h at 00 °C. The absorbance at 405 nm was measured using a microplate reader. (**B**) Melanin contents in lyzed B16-F10 melanoma cells of vehicle control (C), positive control (arbutin 100 µM), treatments of crude extract (GBC, 25, 50 and 100 µg/mL), ethyl acetate layer (GBE, 25, 50 and 100 µg/mL), *n*-BuOH layer (GBB, 100 µg/mL), and H$_2$O layer (GBH, 100 µg/mL) from rhizoma of *L. sinense*.

Compound **1**, obtained as colorless oil, had a formula of $C_{20}H_{20}O_6$ as deduced from ^{13}C NMR and positive-ion HRESI-MS, which showed a fragment ion at positive HRESI-MS m/z 357.1331 $[M + H]^+$ (calcd for $C_{20}H_{21}O_6$, 357.1333). Its IR absorptions at 3444, 1633, and 1509 cm^{-1} indicated the presence of hydroxy, olefinic, and aromatic functionalities, respectively. In the ^1H NMR of 1, a 1,3,4,5-tetrasubstituted aromatic moiety [δ_H 7.07 (d, J = 1.8 Hz, H-6) and 7.22 (d, J = 1.8 Hz, H-2)], an ABX-type aromatic functionality [δ_H 6.69 (dd, J = 7.9, 1.8 Hz, H-6'), 6.74 (d, J = 7.9 Hz, H-5') and 6.87 (d, J = 1.8 Hz, H-2')], two trans-mutual coupled olefinic protons [δ_H 6.33 (d, J = 15.9 Hz, H-8) and 7.56 (d, J = 15.9 Hz, H-7)], a terminal allylic group [δ_H 4.99 (ddd, J = 17.1, 1.8, 1.8 Hz, H-9'a), 5.10, (br d, J = 7.6 Hz, H-7'), 5.15 (ddd, J = 10.1, 1.8, 1.8 Hz, H-9'b) and 6.40 (ddd, J = 17.1, 10.1, 7.6 Hz, H-8')] and two methoxyl resonances [δ_H 3.77 (s, 3'-OCH$_3$) and 3.92 (s, 3-OCH$_3$)] were observed. Twenty carbon resonances, attributable to seven non-protonated aromatic carbons [δ_C 126.7 (C-1), 131.2 (C-5), 135.3 (C-1'), 146.1 (C-4'), 148.6 (C-3), 147.2 (C-4) and 148.2 (C-3')], one acid carbonyl (δ_C 168.4, C-9), one methine (δ_C 48.2, C-7'), eight olefinic methines [δ_C 108.9 (C-2), 113.2 (C-2'), 115.6 (C-5'), 116.1 (C-8), 121.8 (C-6'), 123.8 (C-6), 141.5 (C-8') and 146.2 (C-7)], one exomethylene (δ_C 115.9, C-9') and two methoxyls [δ_C 56.4 (3'-OCH$_3$) and 56.6 (3-OCH$_3$)], were observed in the ^{13}C NMR spectrum coupled with the DEPT spectrum of **1** (Table 1). The connectivity of **1** was further deduced by cross-peaks of δ_H 5.10 (H-7')/δ_C 113.2 (H-2'), 115.9 (H-9'), 121.8 (H-6'), 123.8 (C-6), 131.2 (C-5), 135.3 (C-1'), 141.5 (C-8') and 147.2 (C-4), δ_H 7.56 (H-7)/δ_C 108.9 (C-2), 116.1 (C-8), 123.8 (C-6), 126.7 (C-1) and 168.4 (C-9), δ_H 3.77 (3'-OCH$_3$)/δ_C 148.2 (C-3') and δ_H 3.92 (3-OCH$_3$)/δ_C 148.6 (C-3) in the HMBC spectrum (Figure 2B), which were further corroborated by the mutually-correlated signals of δ_H 3.77 (3'-OCH$_3$)/δ_H 6.87 (H-2') and δ_H 3.92 (3-OCH$_3$)/δ_H 7.22 (H-2) in the NOESY spectrum (Figure 2B). Accordingly, **1** was characterized as shown, and was named as 5-[3-(4-hydroxy-3-methoxyphenyl)allyl]ferulic acid. To our knowledge, **1** with two sets of C_6–C_3 unit connected at C-7' was a new skeletal type of lignan.

Compound **2** was isolated as colorless oil with molecular formula $C_{11}H_{20}O_2$ as deduced by positive-ion HR-ESIMS, showing an $[M + H]^+$ ion at m/z 185.1501 (calcd for $C_{11}H_{21}O_2$, 185.1541). Conspicuous absorptions at 3445 and 1660 cm^{-1} in the IR spectrum of **2** indicated the presence of hydroxy and olefinic functionalities, respectively. The ^1H NMR (Table 2) coupled with COSY spectrum of **2** showed two aliphatic chains at δ_H 0.85–1.97 (–H$_2$-7–H$_3$-11) and δ_H 1.66–5.44 (–H-3–H-2–H-1–H$_2$-6–H$_2$-5–). The above assignments also reflected in the ^{13}C NMR of **2** supported by DEPT spectra, in which one methyl (δ_C 14.2, C-11), six methylenes [δ_C 22.7 (C-10), 26.3 (C-6), 27.2 (C-5), 27.4 (C-8), 31.8 (C-9) and 37.4 (C-7)], three methines [δ_C 67.1(C-2), 69.2 (C-1) and 121.0 (C-3)] and one quaternary carbon (δ_C 144.1, C-4) were observed. In the HMBC spectrum of 2 (Figure 2C), cross-peaks of δ_H 5.44 (H-3)/δ_C 27.2 (C-5) and 37.4 (C-7), δ_H 1.92–2.01 and 2.04–2.10 (H2-5)/δ_C 144.1 (C-4) and δ_H 1.97 (H2-7)/δ_C 144.1 (C-4) indicated C-1–C-6 was a cyclohexene moiety with a double bond at Δ^3, and C-7–C-11, a saturated linear aliphatic chain, was attached at C-4. The relative configurations of hydroxy-bearing chiral C-1 and C-2 were approached by the J values of carbinoyl protons H-1 (9.1, 4.2 Hz) and H-2 (4.2 Hz), which indicated that H-1 and H-2 were pseudo-axial- and pseudo-equatorial-oriented, respectively (Figure 2C). The relative configuration of 1,2-dihydroxy in **2** was thus deduced to *cis*. Conclusively, the structure of **2** was elucidated as shown, and was named as *cis*-4-pentylcyclohex-3-ene-1,2-diol.

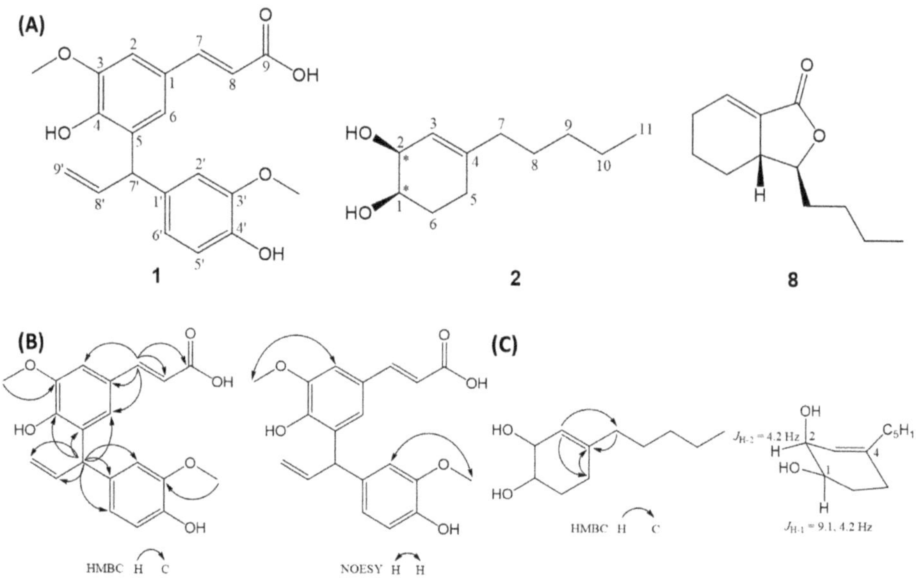

Figure 2. (**A**) Chemical structures of compounds **1**, **2**, and **8** isolated from the rhizoma of *L. sinense*. (**B**) Selected HMBC and NOESY of **1**. (**C**) Selected HMBC and half-chair conformation of **2**.

Table 1. ^{13}C (125 MHz), ^{1}H NMR (500 MHz), and HMBC data for compound **1** (in acetone-d_6, δ in ppm).

Position	^{13}C NMR [a]	^{1}H NMR	HMBC
	δ_C (multi.)	δ_H (multi., J in Hz)	H→C
1	126.7 (s)		
2	108.9 (d)	7.22 (d, 1.8)	C-1, C-3, C-4, C-6
3	148.6 (s)		
4	147.2 (s)		
5	131.2 (s)		
6	123.8 (d)	7.07 (d, 1.8)	C-2, C-4, C-7′
7	146.2 (d)	7.56 (d, 15.9)	C-1, C-2, C-6, C-8, C-9
8	116.1 (d)	6.33 (d, 15.9)	C-1, C-7, C-9
9	168.4 (s)		
1′	135.3 (s)		
2′	113.2 (d)	6.87 (d, 1.8)	C-1′, C-3′, C-4′, C-6′, C-7′
3′	148.2 (s)		
4′	146.1 (s)		
5′	115.6 (d)	6.74 (d, 7.9)	C-1′, C-3′, C-4′
6′	121.8 (d)	6.69 (dd, 7.9, 1.8)	C-2′, C-4′, C-7′
7′	48.2 (d)	5.10 (br d, 7.6)	C-4, C-5, C-6, C-1′, C-2′, C-6′, C-8′, C-9′
8′	141.5 (d)	6.40 (ddd, 17.1, 10.1, 7.6)	C-5, C-1′, C-7′
9′	115.9 (t)	4.99 (ddd, 17.1, 1.8, 1.8)	C-7′, C-8′
		5.15 (ddd, 10.1, 1.8, 1.8)	C-7′
3-OCH$_3$	56.6 (q)	3.92 (3H, s)	C-3
3′-OCH$_3$	56.4 (q)	3.77 (3H, s)	C-3′

[a] Multiplicities were obtained from DEPT experiments.

Table 2. ^{13}C (125 MHz), ^{1}H NMR (500 MHz) and HMBC data for compound **2** (in chloroform-*d*, δ in ppm).

Position	^{13}C NMR ^{a}	^{1}H NMR	HMBC
	δ$_C$ (multi.)	δ$_H$ (multi., *J* in Hz)	H→C
1	69.2 (d)	3.73 (dt, 9.1, 4.2)	C-3, C-5
2	67.1 (d)	4.08 (br t, 4.2)	C-1, C-3, C-4, C-6
3	121.0 (d)	5.44 (m)	C-1, C-2, C-5, C-7
4	144.1 (s)		
5	27.2 (t)	1.92–2.01 (m)	C-1
		2.04–2.10 (m)	C-1, C-3, C-4, C-6, C-7
6	26.3 (t)	1.66–1.78 (m)	C-1, C-2, C-5
7	37.4 (t)	1.97 (br t, 7.3)	C-4, C-9
8	27.4 (t)	1.38 (m)	C-4, C-7, C-9, C-10
9	31.8 (t)	1.20 (m)	C-7, C-10
10	22.7 (t)	1.27 (m)	C-8, C-9, C-11
11	14.2 (q)	0.85 (t, 7.0)	C-9, C-10

^{a} Multiplicities were obtained from DEPT experiments.

2.2. Effects of Compounds **1** and **8** on Melanin Content in α-MSH-Stimulated B16-F10 Cells

To ascertain the depigmentation constituents in rhizoma of *L. sinense*, all the pure isolates were subjected to antimelanogenesis assay in B16-F10 melanoma cells. Murine melanoma B16-F10 cells are a well-established model for antimelanogenic principles discovery, and have been widely adopted in previous studies [48]. α-Melanocyte-stimulating hormone (α-MSH) is a peptide hormone and responsible for the production of melanin by melanocytes through activating melanocortin **1** receptor [49]. In this research, B16-F10 cells were stimulated with α-MSH (100 nM) and simultaneously treated with each compound at concentrations of 25, 50, or 100 μM. The melanin content of B16-F10 melanoma cells without compound treatment was assigned as 100%. Arbutin, a common skin whitening agent in cosmetic products, was used as positive control. Of these compounds, **1** and **8** inhibit α-MSH-induced melanin production in a dose-dependent manner. The IC$_{50}$ values of compounds **1** and **8** were 78.9 and 31.1 μM, respectively (Figure 3B). Compound **1** and **8** at the effective concentrations did not show obvious effects on MTT assay (Figure 3A). The MTT results may arise from the combined effects of cell proliferation reduction and cell viability inhibition according to the experimental conditions. These results suggested that the anti-melanogenic effects of compound **1** and **8** did not attribute to the cell death or growth inhibition.

Using the HPLC-DAD method, the main effective constituents (**1** and **8**) were characterized from the crude extract (Figure 4) by comparing with the retention time of pure **1** (from laboratory synthesis, data not published yet) and **8** (from Sigma) as standards. Stock solutions of 1, 10, 20, 40, 60, 80 and 100 μg/mL were utilized. Each concentration was injected in triplicate. The content ratios of **1** and **8** in the extract of dried material were quantified to be 0.009% and 0.15% (*w*/*w*) by linear regression of the respective peak areas.

2.3. In Vivo Zebrafish Pigmentation Assay

In addition to evaluate the effect of **8** on in vitro antimelanogenesis, its in vivo anti-pigmentation ability through zebrafish pigmentation assay was further investigated. Zebrafish has been considered as an advantageous vertebrate model organism due to its small size, high fecundity, and similar gene sequences and organ systems to human beings. Additionally, the melanin pigmentation process on its surface allowing easy observation makes zebrafish a particular useful model for investigating in vivo melanogenic inhibitors or stimulators [50]. In this study, arbutin at 20 mM was used as positive control, and arbutin of 20 μM was included to compare with compound **8** at same concentration level. After the incubation of zebrafish embryos from 7 hpf to 72 hpf, compound **8** exhibited higher pigmentation inhibitory activity at concentrations of 10 and 20 μM compared to arbutin (20 μM) (Figure 5). Upon the treatment of 20 µM compound **8**, the pigmentation level of zebrafish markedly decreases about 31%; while compound **8** at 10 µM decreases 26.2% pigmentation level.

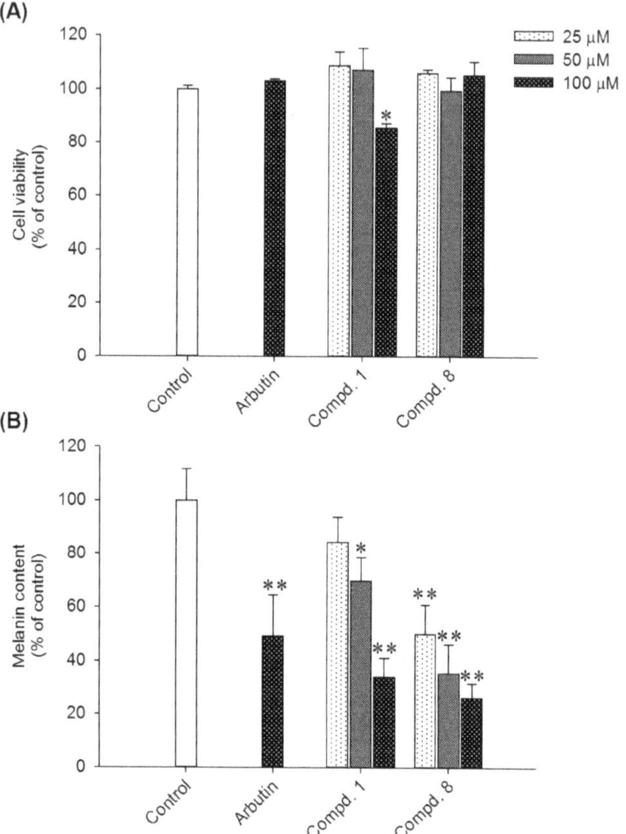

Figure 3. (**A**) The effects of compounds **1** and **8** on cell viability determined by MTT assay. Melanoma cells pretreated with 100 nM α-MSH were seeded at a density of 1×10^4 cells/well in a 12-well plate. Then, the melanoma cells were left to adhere overnight. Pure isolates (25, 50 and 100 µM) or arbutin (100 µM) were added to each well and incubated for another 72 h. Subsequently, the treated cells were labelled with MTT dye reagent in PBS (2 mg/mL) for 3 h. The formazan precipitates were dissolved by DMSO, and the concentrations were measured at 570 nm in a microplate reader. (**B**) The effects of compounds **1** and **8** on melanin contents in B16-F10 cells. Melanoma cells were seeded at a density of 1×10^4 cells/well in a 6-well plate and incubated overnight. The cells were exposed to various concentrations (25, 50 and 100 µM) of the pure isolates or arbutin for 72 h in the presence of 100 nM α-MSH. The cells were washed with PBS and lyzed with 150 µL of 1 N NaOH containing 10% DMSO for 1 h at 80 °C. The absorbance at 405 nm was measured using a microplate reader. Results were expressed as % control and data mean ± S.D. $n = 3$ in each group. * $p < 0.05$, ** $p < 0.01$ compared to the control group.

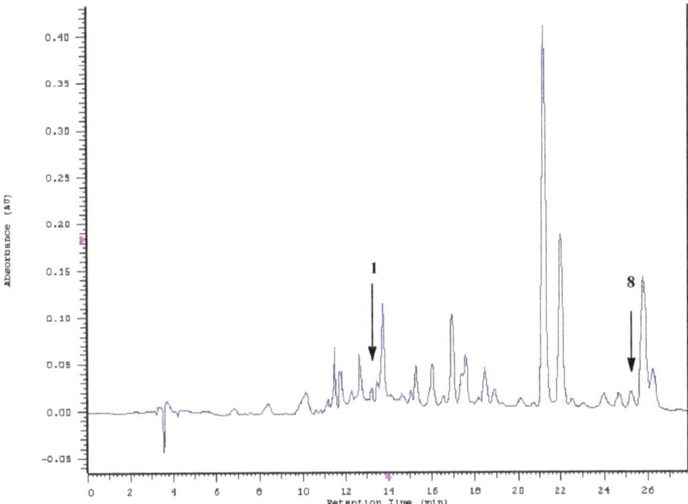

Figure 4. The HPLC-DAD chromatogram of compounds **1** and **8** in the crude extract of rhizoma of *L. sinense*.

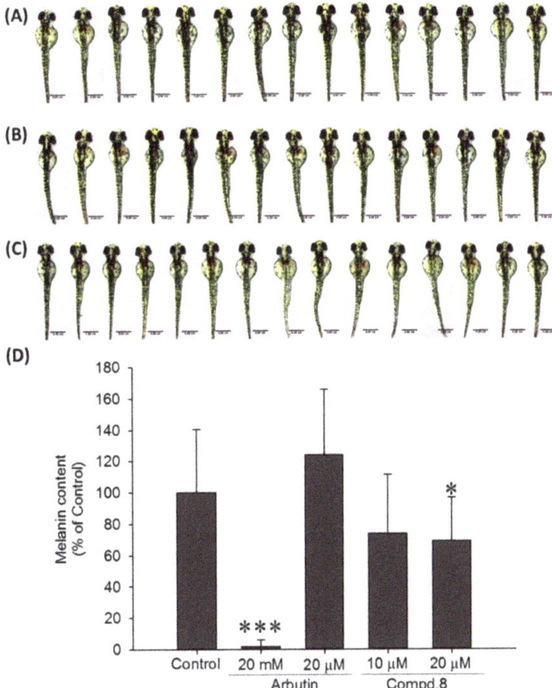

Figure 5. Depigmenting effect of compound **8** and melanogenic regulators on melanogenesis of zebrafish in an in vivo phenotype-based system. Zebrafish embryos were exposed to (**A**) E3 buffer (containing 1% alcohol), (**B**) arbutin (20 μM), or (**C**) compound **8** (20 μM) from 7 hpf (post-fertilization) to 72 hpf. (**D**) Pigmentation levels of zebrafish treated with arbutin (20 mM and 20 μM) and compound **8** (10 μM and 20 μM). Results were expressed as % control and data mean ± S.D. n = 15 in each group. * $p < 0.05$ and *** $p < 0.001$ compared to the control group.

2.4. Viability Assay of Human Epidermal Skin Equivalents

The safety of cosmetic products is a serious concern. For instance, hydroquinone, an effective skin lightening agent, has been banned from the market because of numerous adverse reactions and controversy over the potential carcinogenic risk [7]. Kojic acid, a potent tyrosinase inhibitor, may induce contact dermatitis and potential genotoxicity [51,52]. In order to assess the safety of compounds **1** and **8** on human skin, we performed a cell viability assay on Human skin equivalents (HSEs) model. HSEs are three-dimensional culture systems that are generated by seeding human keratinocytes onto an appropriate dermal substrate pre-seeded with human fibroblasts. HSEs are physiologically comparable to the natural skin and provide suitable alternatives for animal testing. Under controlled culture conditions, the HSEs demonstrate high similarity with the native tissue from which it was derived [53]. In this study, viability assays on the normal human epidermal keratinocytes (NHEKs) in Leiden epidermal models (LEMs) were performed. Compounds **1** and **8** did not affect the cell viability at the concentrations of 10–100 μM. The cell viabilities of compound **1** are 98% and 106 % at the concentrations of 10 and 100 μM, respectively. The cell viabilities of compound **8** are 97% and 92% at the concentrations of 10 and 100 μM, respectively. Accordingly, compounds **1** and **8** did not exert cytotoxicity against NHEKs in the Leiden epidermal models at concentrations below 100 μM (Figure 6). Since the effective concentration of **8** is below 100 μM, which suggests that **8** might be safe for skin whitening below the tested concentrations. However, in practice, cosmetic products may be applied on human skin for a long time, thus an extended experimental period of tested compounds on HSEs has to be further conducted.

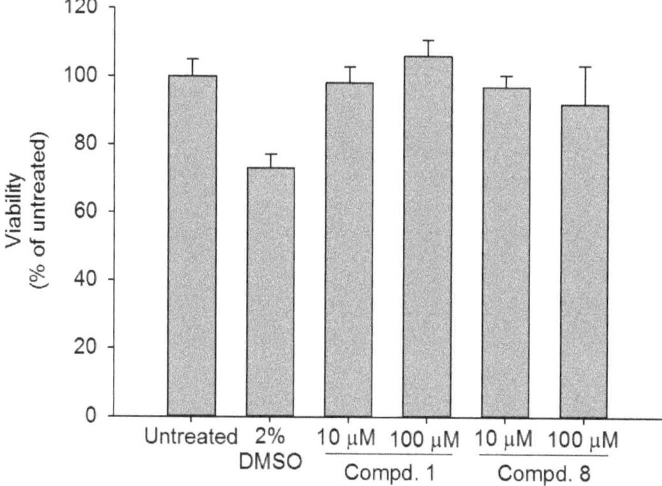

Figure 6. The effect of compounds **1** and **8** on cell viability in normal human epidermal keratinocytes (NHEKs). NHEKs were exposed to **1**, **8**, and vehicle at the indicated concentrations for 24 h. Cell viability was determined by MTT assay.

2.5. Molecular Docking Study of B16-Mus Musculus Tyrosinase

Tyrosinase catalysis is the rate-limiting step of melanin biosynthesis. Thus, inhibition of tyrosinase is the most common approach to achieve skin whiteness [54,55]. In order to determine whether *L. sinense* suppressed melanogenesis in B16-F10 cells through tyrosinase inhibition, 3D stick models (Figure 7A) and a 2D diagram (Figure 7B) of molecular docking using DS software were performed, which revealed the possible inhibitory mechanism of compound **8** on mouse (Mus musculus) tyrosinase. It was shown that the active site of mouse tyrosinase was located at a domain surrounded by amino acids His377, Asn378, His381, Gly389, Thr391, Ser394 together with two copper ions. The CDOCKER interaction energy between the enzyme and inhibitor was −46.0067 kcal/mol. The simulation results showed that hydrophobic amino acids including Gly389, Asn 378, Thr391 Ser394, His 404, and His192 around the catalytic site formed van der Waals forces with compound **8**. Additionally, the oxygen atom of γ-lactone moiety of **8** forms hydrogen bonds with His377 and His215, while its carbonyl group forms coordination bonds with two copper ions. It was also observed that the cyclohexene ring and butane exerted weak π-alkyl interaction with His381 and His215, respectively. Previous studies have demonstrated that *L. sinense* extract displayed mushroom tyrosinase inhibition and down-regulation of tyrosinase mRNA expression in B16-F10 cells [21,56]. Since the molecular docking illustrated a strong interaction between the active domain of mouse tyrosinase and compound **8**, it was thus proposed that (3*S*,3a*R*)-neocnidilide (**8**) exhibited antimelanogenesis activity due to tyrosinase activity attenuation and further decreasing melanin production. However, in addition to tyrosinase inhibition, the other underlying mechanisms of antimelanogenesis activity of (3*S*,3a*R*)-neocnidilide (**8**) remains to be further investigated.

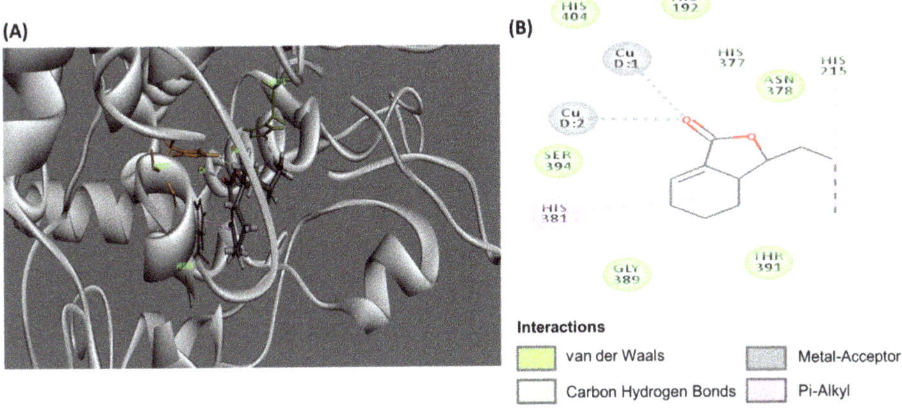

Figure 7. Molecular docking of compound 8 to Mus musculus tyrosinase. Binding conformations of (**A**) 3D stick model and (**B**) 2D diagram of molecular interactions to the active site.

3. Materials and Methods

3.1. General

HPLC-grade solvents, *n*-hexane, ethyl acetate, methanol and acetonitrile, were purchased from J. T. Baker (Phillipsburg, NJ, USA). Ethanol was purchased from Merck (Darmstadt, Germany). α-Melanocyte-stimulating hormone (α-MSH), dimethyl sulfoxide (DMSO), phosphate-buffered saline (PBS), 3-(4,5-dimethylthiazol-2-yl)-2,5-diphenyl tetrazolium bromide (MTT) were purchased were purchased from Sigma Aldrich (St. Louis, MO, USA). Open column chromatography was performed on silica gel (70–230 mesh, Merck, Darmstadt, Germany). Pre-coated silica gel plates 60 F254 and Aluminum Sheets RP-18 F254S for TLC were purchased from Merck (Darmstadt, Germany). Optical rotations were measured on a JASCO P-1020 polarimeter (Jasco, Tokyo, Japan). ^{1}H and ^{13}C NMR were acquired with a Bruker Avance DRX-500 spectrometer (Bruker, Rheinstetten, Germany). Low resolution and high resolution mass spectra were obtained using an ABI API 4000 Q-TRAP ESI-MS (Applied Biosystem, Foster City, CA, USA) and Q-Exactive Plus HR-ESI-MS (Thermo Fisher Scientific, MA, USA), respectively. IR spectra were recorded on a JASCO FT/IR 4100 spectrometer (Jasco, Tokyo, Japan).

3.2. Plant Materials

Dried rhizoma of *L. sinense* Oliv. was purchased from Sheng Chang Pharmaceutical Co., Ltd., Taoyuan, Taiwan.

3.3. Isolation and Structural Elucidation

Dried rhizoma (9.9 kg) of *L. sinense* was smashed and extracted with methanol (40 L × three times), which was filtered and evaporated to give a black residue (1758 g). This residue was then suspended in H_2O (3.0 L) and partitioned with equal volume of ethyl acetate and *n*-BuOH for three times, successively. Each layer was concentrated under reduced pressure to obtain EtOAc (415 g), *n*-BuOH (147 g), and H_2O (896 g) layers. Subsequently, the dried ethyl acetate layer (250 g) was mixed with 375 g silica gel, and was loaded onto a conditioned open column packed with 3550 g silica gel and eluted in a step-wise gradient method by mixtures of *n*-hexane, ethyl acetate and methanol. Each 500 mL was collected for one fraction and analyzed by TLC. Then, all the fractions were combined into eight portions I–VIII according to the results of TLC analyses, which were re-dissolved in a minimum volume of *n*-hexane—ethyl acetate mixtures used in the subsequent HPLC system. Portion II eluted by *n*-hexane—ethyl acetate (95:5) was purified by a semi-preparative HPLC (Hibar®Fertigäute, 10 × 250 mm) using *n*-hexane—ethyl acetate (96:4) as eluent at a flow rate of 3 mL/min to afford **6** (4.2 mg, t_R = 19.8 min), **3** (6.1 mg, t_R = 21.2 min), **5** (32 mg, t_R = 23.5 min) and **7** (79 mg, t_R = 28.0 min). The same portion was purified by a semi-preparative HPLC (Phenomenex®Luna, 10 × 250 mm) using *n*-hexane—ethyl acetate (99:1) as eluent at a flow rate of 3 mL/min to afford **4** (36 mg, t_R = 32.5 min). Portion III eluted by *n*-hexane—ethyl acetate (90:10) was purified by a semi-preparative HPLC (Phenomenex®Luna, 10 × 250 mm) using *n*-hexane—acetone (95:5) as eluent at a flow rate of 3 mL/min to afford **8** (1.7 g, t_R = 19.3 min). Portion IV eluted by *n*-hexane—ethyl acetate (80:20) was purified by a semi-preparative HPLC (Phenomenex®Luna, 10 × 250 mm) using *n*-hexane—ethyl acetate (85:15) as eluent at a flow rate of 3 mL/min to afford **15** (19 mg, t_R = 24.4 min), **12** (11 mg, t_R = 26.9 min), **9** (25 mg, t_R = 33.8 min), **10** (31 mg, t_R = 37.0 min), **11** (24 mg, t_R = 42.5 min). The same portion was purified by the same column using *n*-hexane—ethyl acetate (78:22) as eluent at a flow rate of 3 mL/min to obtain **14** (10 mg, t_R = 20.1 min) and **13** (22 mg, t_R = 24.6 min). The same portion was purified by the same column using *n*-hexane—ethyl acetate—acetone (80:10:10) as eluent at a flow rate of 3 mL/min to obtain **20** (25 mg, t_R = 13.2 min), **17** (15 mg, t_R = 16.3 min), **18** (28 mg, t_R = 18.5 min), **16** (132 mg, t_R = 21.8 min) and **19** (39 mg, t_R = 25.6 min). Portion V eluted by *n*-hexane—ethyl acetate (60:40) was purified by a semi-preparative HPLC (Hibar®Fertigäute, 10 × 250 mm) using *n*-hexane—ethyl acetate—acetone (68:27:5) as eluent at a flow rate of 3 mL/min to afford **21** (5.3 g, t_R = 10.2 min), **23** (63 mg, t_R = 20.0 min), **22** (84 mg, t_R = 22.0 min) and **24** (33 mg, t_R = 26.5 min). The same portion was purified by a semi-preparative HPLC (Hibar®Fertigäute, 10 × 250 mm) using *n*-hexane—ethyl acetate (72:28) as eluent at a flow rate of 3 mL/min to afford **2** (24 mg, t_R = 24.3 min). Portion VI eluted by *n*-hexane—ethyl acetate (40:60) was purified by a semi-preparative HPLC (Hibar®Fertigäute, 10 × 250 mm) using *n*-hexane—ethyl acetate (53:47) as eluent at a flow rate of 3 mL/min to afford **1** (47 mg, t_R = 13.2 min).

3.4. Spectroscopic Data

5-[3-(4-Hydroxy-3-methoxyphenyl)allyl]ferulic acid (**1**): colorless oil; $[\alpha]_D^{25}$ +5.6° (*c* 0.12, CH_3OH); IR (neat) ν_{max} 3444, 2935, 1633, 1509, 1434, 1376, 1267, 1153; positive ESI-MS *m/z* 357.2 $[M + H]^+$; positive HRESI-MS *m/z* 357.1331 $[M + H]^+$ (calcd for $C_{20}H_{21}O_6$, 357.1333); ^1H and ^{13}C NMR data see Table 1.

Cis-4-pentylcyclohex-3-ene-1,2-diol (**2**): colorless oil; $[\alpha]_D^{22}$ -20.3° (*c* 0.20, CH_3OH); IR (neat) ν_{max} 3445, 2919, 1660, 1455, 1371, 1225, 1084; ESIMS *m/z* 185.2 $[M + H]^+$; HRESI-MS *m/z* 185.1501 $[M + H]^+$ (calcd for $C_{11}H_{21}O_2$, 185.1541); ^1H and ^{13}C NMR data see Table 2.

3.5. HPLC DAD Analysis

Chromatographic analyses were performed on a Hitachi HPLC system consisting of L-7100 pump, L-7200 autosampler, L-7455 detector and D-7000 system manager data acquisition software, on an XBridge™ C18 column (250 mm length, 4.6 mm internal diameter, 5 um particle size; Waters).

The mobile phase consisted of H_2O (solvent A) and CH_3CN (solvent B). The flow rate was 1.0 mL/min. The elution program was as follows: isocratic with 2% B (0–5 min), 2–20% B (5–25 min), 20–90% B (25–30 min), and isocratic with 90% B (30–35 min). The injection volume was 10 L. UV–visible spectra were recorded at 240 nm.

3.6. Cell Culture

The B16-F10 murine melanoma cells (CRL6475) were purchased from the Food Industry Research and Development Institute (FIRDI, Hsinchu, Taiwan). The cells were cultured in 90% Dulbecco's Modified Eagle's Medium (DMEM, Sigma-Aldrich, St. Louis, MO, USA) supplemented with 10% fetal bovine serum (FBS, Sigma, St. Louis, MO, USA) and 1% penicillin–streptomycin solution in culture flasks in a CO_2 incubator with a humidified atmosphere containing 5% CO_2 in the air at 37 °C. The culture medium was changed every two days. The cells were harvested by trypsinization when they were about 85% confluent, counted with a haemocytometer (Neubauer Improved., Marienfeld, Germany) and seeded at the appropriate numbers into wells of cell culture plates for further experiments.

3.7. Cell Viability Assay

To determine the safety of the various extracts the viability of cells following treatment with extracts was determined by the MTT assay. This method is based on the reduction of 3-(4,5-dimethylthiazol-2-yl)-2,5-diphenyl tetrazolium bromide (MTT) to formazan by mitochondrial enzymes in viable cells [57]. The quantity of formazan formed is proportional to the number of viable cells present and can be measured spectrophotometrically. Briefly, 100 nM α-melanocyte stimulating hormone (α-MSH)-pretreated cells seeded at a density of 1×10^4 cells/well in a 12-well plate were left to adhere overnight. Pure isolates or arbutin were then added to each well and incubated for another 72 h. Then, the treated cells were labelled with MTT dye reagent (Applichem, Denmark) in PBS (2 mg/mL) for 3 h. The formazan precipitates were dissolved by DMSO and the concentrations were measured at 570 nm in a microplate reader. Cell viability was calculated using the following formula: cell viability (%) = (A sample/A control) $\times$ 100, where A sample and A control are the absorbances from the mixture with, or without the addition of test sample, respectively.

3.8. Melanin Content Assay

Melanin content was measured as described previously with slight modifications [58]. The B16-F10 melanoma cells were seeded with 1×10^4 cells/well in 3 mL of medium in 6-well culture plates and incubated overnight to allow cells to adhere. The cells were exposed to various concentrations (25, 50 and 100 µM) of the pure isolates or arbutin for 72 h in the presence of 100 nM α-MSH. At the end of the treatment, the cells were washed with PBS and lyzed with 150 µL of 1 N NaOH (Merck, Germany) containing 10% DMSO for 1 h at 80 °C. The absorbance at 405 nm was measured using a microplate reader. The melanin content of B16-F10 melanoma cells without compound treatment was assigned as 100%, and the melanin content of compound-treated cells was calculated relative to the control group.

3.9. In Vivo Zebrafish Pigmentation Assay

The animal use protocol has been reviewed and approved by the Institutional Animal Care and Use Committee or Panel (IACUC/IACUP) of Taipei Medical University (No. LAC-2017-0311). The methods were carried out in compliance with the relevant laws and the approved guidelines. Wild-type zebrafish embryos were collected from Zebrafish Core Facility of Taipei Medical University. Phenotype-based evaluation of zebrafish embryo was performed according to the previous study with slight modification [50]. The embryos were incubated at 28 °C with 1% ethanol as control and a different concentration of compounds from 7 hpf (post-fertilization) to 72 hpf. To evaluate the anti-melanogenesis effects of melanogenic modulators on zebrafish developmental process, the pigmentation of zebrafish was analyzed at 72 hpf. The embryos were mounted in 1% Low

Melting Agarose (Bioshop Canada, Burlington, ON, Canada) and captured images with a ZEISS Stemi 508 stereomicroscope (ZEISS, Oberkochen, Germany). Images pixel measurement analysis was carried out by Fiji package of ImageJ (http://rsb.info.nih.gov/ij/index.html, NIH, USA). The quantification of pigmentation data was analyzed (the area below the eyes of the zebrafish) and compared to control group.

3.10. Viability Assay on Normal Human Epidermal Keratinocytes

All primary human skin cells from healthy donors used by the Department of Dermatology of the Leiden University Medical Center are isolated from surplus tissue collected according to article 467 of the Dutch Law on Medical Treatment Agreement and the Code for proper Use of Human Tissue of the Dutch Federation of Biomedical Scientific Societies. According to article 467 surplus tissues can be used if no objection is made by the patient. This means that the patient who will undergo plastic surgery is well informed on the research. None of the authors were involved in the tissue sampling. The Declaration of Helsinki principles were followed when working with human tissues.

The fresh mamma reduction surplus skin of a single female individual was used for isolation of normal human epidermal keratinocytes (NHEKs) as described previously [59]. The NHEKs in Leiden epidermal models (LEMs) were incubated overnight under submerged conditions in keratinocyte medium. Within four days, fetal bovine serum was gradually omitted and the NHEKs in LEMs were cultured serum-free at the air-liquid interface for seven days, while culture medium was refreshed twice a week. Viability assays were performed by adding 0.5 mL of 1 mg/mL MTT to each of the NHEKs in LEMs for 3 h, after 24 h exposure to the test compounds **1**, **8**, and DMSO (negative control). The precipitated blue formazan product was extracted from the cells within 2 h with 0.5 mL isopropanol per well. The concentration of formazan was measured by determining the OD at 570 nm using a Tecan Infinite F50 microplate reader.

3.11. Statistical Analysis

All the data in our study were obtained as averages of experiments that were performed at least in triplicate and expressed as means $\pm$ SD (Standard deviation). Statistical analysis was performed by Student's t-test. The statistical significance of results was set at $p < 0.05$ (*) and $p < 0.01$ (**).

3.12. Molecular Docking Study of B16-Mus Musculus Tyrosinase

3.12.1. Homology Modeling

As no three-dimensional structures for Mus musculus tyrosinase are available now, Homology modeling is the most assured method for prediction of three-dimensional structures of unknown protein based on the assumption that the structure of the unknown protein is similar to the known structures of some homologous reference proteins. We acquired the Mus musculus tyrosinase amino acid sequence from the National Center for Biotechnology Information (NCBI, https://www.ncbi.nlm.nih.gov/) protein sequence database. A homolog protein template of the query protein Mus musculus tyrosinase was identified by Phyre2 (Protein Homology/analog Y Recognition Engine V 2.0) [60]. The 446 residues (81% of Mus musculus sequence) have been modelled with 100.0% confidence by the single highest scoring template as PDB code: c5m8pA was chosen as a receptor for the docking calculation studies.

3.12.2. Analysis of Ligand-Protein Interaction

The binding site of the Mus musculus tyrosinase was determined based on reference human tyrosinase (PDB 5M8M) six amino acid residues and the binding pocket have two copper. Therefore, the binding site sphere was defined. Subsequently, the 3D structure of the compound **8** was prepared and optimized by energy minimization docked into the binding pocket of the Mus musculus tyrosinase using the CDOCKER program and the number of docking runs was set to 50 for inhibitor. All other parameters were set as default to the analysis of Ligand-Protein Interaction through using Biovia Discovery Studio v.4.0 (Accelrys Software Inc., San Diego, CA, USA). Finally, from the 50 docking conformations, the top one with the highest CDOCKER energy score was chosen to explore the binding mode of docked compound in the Mus musculus tyrosinase active site.

4. Conclusions

In the present study, we adopted a bioassay-guided method using B16-F10 cells to isolate the antimelanogenesis constituents from *L. sinense* rhizoma extracts. The active constituents were determined to be 5-[3-(4-hydroxy-3-methoxyphenyl)allyl]ferulic acid (**1**) and (3*S*,3a*R*)-neocnidilide (**8**). According to the HPLC analysis, the content of compound **8** accounts for 0.15% of the crude extract. The antimelanogenesis activity of **8** was verified by both in vitro B16-F10 cells and in vivo zebrafish assays. All these findings suggested that **8** may, at least, provide a rationale for the potential antimelanogenesis effect of *L. sinense* for its high potency and quantity. The cell viability data on B16-F10 cells and NHEKs imply that **8** could be developed potentially as an antimelanogenesis agent. The mode of action of **8** on antimelanogenesis was speculated to be the inhibition of the tyrosinase activity based on the results of molecular docking; however, that remains to be further confirmed. Our finding revealed that compound **8** exhibited anti-melanogenesis effects and safety through in vitro, in vivo, and ex vivo studies, however, the depigmenting efficacy and biosafety on human skin need to be evaluated and validated by clinical trials in the future.

5. Patents

An earlier version of the manuscript has been published as a patent (patent number: EP 2832719, TW I507390).

Author Contributions: Conceptualization, M.-C.C., T.-H.L., Y.-T.C. and C.-K.L.; Formal analysis, S.-J.H.; Funding acquisition, C.-K.L.; Investigation, Y.-T.C. and L.-L.S.; Resources, C.-H.C.; Supervision, J.W. and C.-K.L.; Visualization, M.-C.C.; Writing—original draft, T.-H.L.; Writing—review & editing, M.-C.C., T.-H.L., J.W. and C.-K.L.

Funding: This research was funded by Ministry of Science and Technology of ROC for financial supports (MOST104-2320-B-038-020-MY3).

Acknowledgments: We are grateful to Shwu-Huey Wang, Shou-Ling Huang and Shu-Yun Sun of the Instrumentation Center of Taipei Medical University and the Instrumentation Center of the College of Science, National Taiwan University, respectively, for the NMR and MS data acquisition and Biomimiq Company provides technical support for ex vivo human skin equivalents (HSEs) assay.

Conflicts of Interest: The authors declare no conflict of interest.

References

1. Slominski, A.; Tobin, D.J.; Shibahara, S.; Wortsman, J. Melanin Pigmentation in Mammalian Skin and Its Hormonal Regulation. *Physiol. Rev.* **2004**, *84*, 1155–1228. [CrossRef] [PubMed]
2. Hwang, I.; Hong, S. Neural Stem Cells and Its Derivatives as a New Material for Melanin Inhibition. *Int. J. Mol. Sci.* **2018**, *19*, 36. [CrossRef] [PubMed]
3. D'Mello, S.A.; Finlay, G.J.; Baguley, B.C.; Askarian-Amiri, M.E. Signaling Pathways in Melanogenesis. *Int. J. Mol. Sci.* **2016**, *17*, 1144. [CrossRef] [PubMed]
4. Lo, J.A.; Fisher, D.E. The melanoma revolution: From UV carcinogenesis to a new era in therapeutics. *Science* **2014**, *346*, 945–949. [CrossRef] [PubMed]
5. Chen, H.; Weng, Q.Y.; Fisher, D.E. UV signaling pathways within the skin. *J. Investig. Dermatol.* **2014**, *134*, 2080–2085. [CrossRef] [PubMed]
6. Pillaiyar, T.; Manickam, M.; Jung, S.H. Recent development of signaling pathways inhibitors of melanogenesis. *Cell Signal.* **2017**, *40*, 99–115. [CrossRef] [PubMed]
7. Couteau, C.; Coiffard, L. Overview of Skin Whitening Agents: Drugs and Cosmetic Products. *Cosmetics* **2016**, *3*, 27. [CrossRef]
8. Zhu, W.; Gao, J. The use of botanical extracts as topical skin-lightening agents for the improvement of skin pigmentation disorders. *J. Investig. Dermatol. Symp. Proc.* **2008**, *13*, 20–24. [CrossRef] [PubMed]

9. Hu, S.-Y. *Food Plants of China*; Chinese University Press: Hong Kong, China, 2005.

10. Chen, D.W. *Shen Nong's Herbal Classic Collections*, 1st ed.; Tianjin Science and Technology Press: Tianjin, China, 2010.

11. Ma, J.-P.; Tan, C.-H.; Zhu, D.-Y. Chemical Constituents of *Ligusticum sinensis* Oliv. *Helv. Chim. Acta* **2007**, *90*, 158–163. [CrossRef]

12. Yu, D.Q.; Chen, R.Y.; Xie, F.Z. Structure elucidation of ligustilone from *Ligusticum sinensis* Oliv. *Chin. Chem. Lett.* **1995**, *6*, 391–394.

13. Yu, D.Q.; Xie, F.Z.; Chen, R.Y.; Huang, Y.H. Studies on the structure of ligustiphenol from *Ligusticum sinense* Oliv. *Chin. Chem. Lett.* **1996**, *7*, 721–722.

14. Zhang, Y.-C.; Chen, C.; Li, S.-J.; Xu, H.-Y.; Li, D.-F.; Wu, H.-W.; Yang, H.-J. Chemical analysis and observation on vascular activity of essential oil from *Ligusticum sinensis*, *Conioselinum tataricum* and *Ligusticum jeholense*. *Chin. J. Exp. Tradit. Med. Formulae* **2011**, *17*, 159–165. [CrossRef]

15. Wang, J.; Xu, L.; Yang, L.; Liu, Z.; Zhou, L. Composition, antibacterial and antioxidant activities of essential oils from *Ligusticum sinense* and *L. jeholense* (Umbelliferae) from China. *Rec. Nat. Prod.* **2011**, *5*, 314–318.

16. Wang, J.; Yang, J.-B.; Wang, A.-G.; Ji, T.-F.; Su, Y.-L. Studies on the chemical constituents of *Ligusticum sinense*. *Zhong Yao Cai* **2011**, *34*, 378–380. [PubMed]

17. Zhang, J.; Zhou, Z.; Chen, R.; Xie, F.; Cheng, G.; Yu, D.; Zhou, T. Study on chemistry and pharmacology of genus Ligusticum. *Chin. Pharm. J.* **2002**, *37*, 654–657. [CrossRef]

18. Du, J.; Yu, Y.; Ke, Y.; Wang, C.; Zhu, L.; Qian, Z.M. Ligustilide attenuates pain behavior induced by acetic acid or formalin. *J. Ethnopharmacol.* **2007**, *112*, 211–214. [CrossRef] [PubMed]

19. Lee, W.S.; Shin, J.S.; Jang, D.S.; Lee, K.T. Cnidilide, an alkylphthalide isolated from the roots of Cnidium officinale, suppresses LPS-induced NO, PGE2, IL-1beta, IL-6 and TNF-alpha production by AP-1 and NF-kappaB inactivation in RAW 264.7 macrophages. *Int. Immunopharmacol.* **2016**, *40*, 146–155. [CrossRef] [PubMed]

20. Ozaki, Y.; Sekita, S.; Harada, M. Centrally acting muscle relaxant effect of phthalides (ligustilide, cnidilide and senkyunolide) obtained from Cnidium officinale Makino. *Yakugaku Zasshi* **1989**, *109*, 402–406. [CrossRef]

21. Wu, C.-Y.; Pang, J.-H.S.; Huang, S.-T. Inhibition of Melanogenesis in Murine B16/F10 Melanoma Cells by *Ligusticum sinensis* Oliv. *Am. J. Chin. Med.* **2006**, *34*, 523–533. [CrossRef]

22. Chu, Y.T. *Studies on Constituents and Melanogenesis-Inhibitory Effects of Ligusticum sinense*; Taipei Medical University: Taipei, Taiwan, 2011.

23. Niu, C.; Yin, L.; Aisa, H.A. Novel Furocoumarin Derivatives Stimulate Melanogenesis in B16 Melanoma Cells by Up-Regulation of MITF and TYR Family via Akt/GSK3beta/beta-Catenin Signaling Pathways. *Int. J. Mol. Sci.* **2018**, *19*, 746. [CrossRef]

24. Chan, Y.Y.; Kim, K.H.; Cheah, S.H. Inhibitory effects of Sargassum polycystum on tyrosinase activity and melanin formation in B16F10 murine melanoma cells. *J. Ethnopharmacol.* **2011**, *137*, 1183–1188. [CrossRef] [PubMed]

25. Oh, J.; Kim, J.; Jang, J.H.; Lee, S.; Park, C.M.; Kim, W.K.; Kim, J.S. Novel (1E,3E,5E)-1,6-bis(Substituted phenyl)hexa-1,3,5-triene Analogs Inhibit Melanogenesis in B16F10 Cells and Zebrafish. *Int. J. Mol. Sci.* **2018**, *19*, 1067. [CrossRef] [PubMed]

26. Miyazawa, M.; Hisama, M. Suppression of Chemical Mutagen-Induced SOS Response by Alkylphenols from Clove (*Syzygium aromaticum*) in the *Salmonella typhimurium* TA1535/pSK1002 umu Test. *J. Agric. Food Chem.* **2001**, *49*, 4019–4025. [CrossRef] [PubMed]

27. Yoshida, Z.-I.; Haruta, M. Effects of substituents on intramolecular hydrogen bonds of 2-hydroxy-4-substituted acetophenones. *Tetrahedron Lett.* **1965**, *6*, 3745–3751. [CrossRef]

28. Bartschat, D.; Beck, T.; Mosandl, A. Stereoisomeric Flavor Compounds. 79. Simultaneous Enantio-selective Analysis of 3-Butylphthalide and 3-Butylhexahydro-phthalide Stereoisomers in Celery, Celeriac, and Fennel. *J. Agric. Food Chem.* **1997**, *45*, 4554–4557. [CrossRef]

29. Yamada, K.; Murata, T.; Kobayashi, K.; Miyase, T.; Yoshizaki, F. A lipase inhibitor monoterpene and monoterpene glycosides from Monarda punctata. *Phytochemistry* **2010**, *71*, 1884–1891. [CrossRef] [PubMed]

30. Nishiyama, Y.; Moriyasu, M.; Ichimaru, M.; Tachibana, Y.; Kato, A.; Mathenge, S.G.; Nganga, J.N.; Juma, F.D. Acyclic triterpenoids from Ekebergia capensis. *Phytochemistry* **1996**, *42*, 803–807. [CrossRef]

31. Oguro, D.; Watanabe, H. Synthesis and sensory evaluation of all stereoisomers of sedanolide. *Tetrahedron* **2011**, *67*, 777–781. [CrossRef]

32. Naito, T.; Niitsu, K.; Ikeya, Y.; Okada, M.; Mitsuhashi, H. A phthalide and 2-farnesyl-6-methyl benzoquinone from Ligusticum chuangxiong. *Phytochemistry* **1992**, *31*, 1787–1789. [CrossRef]

33. Kong, C.-S.; Um, Y.-R.; Lee, J.-I.; Kim, Y.-A.; Yea, S.-S.; Seo, Y. Constituents isolated from Glehnia littoralis suppress proliferations of human cancer cells and MMP expression in HT1080 cells. *Food Chem.* **2010**, *120*, 385–394. [CrossRef]

34. Miyazawa, M.; Tsukamoto, T.; Anzai, J.; Ishikawa, Y. Insecticidal Effect of Phthalides and Furanocoumarins from *Angelica acutiloba* against *Drosophila melanogaster. J. Agric. Food Chem.* **2004**, *52*, 4401–4405. [CrossRef] [PubMed]

35. Yin, X.J.; Xu, G.H.; Sun, X.; Peng, Y.; Ji, X.; Jiang, K.; Li, F. Synthesis of Bosutinib from 3-Methoxy-4-hydroxybenzoic Acid. *Molecules* **2010**, *15*, 4261–4266. [CrossRef] [PubMed]

36. Li, J.-T.; Fu, X.-L.; Tan, C.; Zeng, Y.; Wang, Q.; Zhao, P.-J. Two new chroman derivations from the endophytic *Penicillium* sp. DCS523. *Molecules* **2011**, *16*, 686–693. [CrossRef] [PubMed]

37. Ritter, T.; Stanek, K.; Larrosa, I.; Carreira, E.M. Mild Cleavage of Aryl Mesylates: Methanesulfonate as Potent Protecting Group for Phenols. *Org. Lett.* **2004**, *6*, 1513–1514. [CrossRef] [PubMed]

38. Faustino, H.; Gil, N.; Baptista, C.; Duarte, A.P. Antioxidant activity of lignin phenolic compounds extracted from kraft and sulphite black liquors. *Molecules* **2010**, *15*, 9308–9322. [CrossRef] [PubMed]

39. Tamura, S.; Ohno, T.; Hattori, Y.; Murakami, N. Establishment of absolute stereostructure of falcarindiol, algicidal principle against *Heterocapsa circularisquama* from Notopterygii Rhizoma. *Tetrahedron Lett.* **2010**, *51*, 1523–1525. [CrossRef]

40. Ohta, T.; Uwai, K.; Kikuchi, R.; Nozoe, S.; Oshima, Y.; Sasaki, K.; Yoshizaki, F. Absolute stereochemistry of cicutoxin and related toxic polyacetylenic alcohols from *Cicuta virosa. Tetrahedron* **1999**, *55*, 12087–12098. [CrossRef]

41. Bibi, N.; Tanoli, S.A.K.; Farheen, S.; Afza, N.; Siddiqi, S.; Zhang, Y.; Kazmi, S.U.; Malik, A. In vitro antituberculosis activities of the constituents isolated from Haloxylon salicornicum. *Bioorg. Med. Chem. Lett.* **2010**, *20*, 4173–4176. [CrossRef] [PubMed]

42. Shingate, B.B.; Hazra, B.G.; Pore, V.S.; Gonnade, R.G.; Bhadbhade, M.M. Stereoselective syntheses of 20-epi cholanic acid derivatives from 16-dehydropregnenolone acetate. *Tetrahedron* **2007**, *63*, 5622–5635. [CrossRef]

43. Cantrell, C.L.; Case, B.P.; Mena, E.E.; Kniffin, T.M.; Duke, S.O.; Wedge, D.E. Isolation and Identification of Antifungal Fatty Acids from the Basidiomycete Gomphus floccosus. *J. Agric. Food Chem.* **2008**, *56*, 5062–5068. [CrossRef] [PubMed]

44. Xiao, H.; Parkin, K. Isolation and Identification of Phase II Enzyme-Inducing Agents from Nonpolar Extracts of Green Onion (*Allium* spp.). *J. Agric. Food Chem.* **2006**, *54*, 8417–8424. [CrossRef] [PubMed]

45. Yu, Y.; Zhang, Q.-W.; Li, S.-P. Preparative purification of coniferyl ferulate from *Angelica sinensis* oil by high performance centrifugal partition chromatography. *J. Med. Plants Res.* **2011**, *5*, 104–108.

46. Gopalakrishnan, S.; Subbarao, G.V.; Nakahara, K.; Yoshihashi, T.; Ito, O.; Maeda, I.; Ono, H.; Yoshida, M. Nitrification Inhibitors from the Root Tissues of Brachiaria humidicola, a Tropical Grass. *J. Agric. Food Chem.* **2007**, *55*, 1385–1388. [CrossRef] [PubMed]

47. Lee, C.-K.; Lu, C.-K.; Kuo, Y.-H.; Chen, J.-Z.; Sun, G.-Z. New Prenylated Flavones from the Roots of Ficus Beecheyana. *J. Chin. Chem. Soc.* **2004**, *51*, 437–441. [CrossRef]

48. Yao, C.; Oh, J.H.; Oh, I.G.; Park, C.H.; Chung, J.H. [6]-Shogaol inhibits melanogenesis in B16 mouse melanoma cells through activation of the ERK pathway. *Acta Pharmacol. Sin.* **2013**, *34*, 289–294. [CrossRef] [PubMed]

49. Takahashi, A. Subchapter 16B—Melanocyte-Stimulating Hormone. In *Handbook of Hormones*; Takei, Y., Ando, H., Tsutsui, K., Eds.; Academic Press: San Diego, CA, USA, 2016; p. 120-e16B-7.

50. Choi, T.Y.; Kim, J.H.; Ko, D.H.; Kim, C.H.; Hwang, J.S.; Ahn, S.; Kim, S.Y.; Kim, C.D.; Lee, J.H.; Yoon, T.J. Zebrafish as a new model for phenotype-based screening of melanogenic regulatory compounds. *Pigment Cell Res.* **2007**, *20*, 120–127. [CrossRef]

51. Nakagawa, M.; Kawai, K.; Kawai, K. Contact allergy to kojic acid in skin care products. *Contact Dermatitis* **1995**, *32*, 9–13. [CrossRef]

52. Nohynek, G.J.; Kirkland, D.; Marzin, D.; Toutain, H.; Leclerc-Ribaud, C.; Jinnai, H. An assessment of the genotoxicity and human health risk of topical use of kojic acid [5-hydroxy-2-(hydroxymethyl)-4H-pyran-4-one]. *Food Chem. Toxicol.* **2004**, *42*, 93–105. [CrossRef]

53. El Ghalbzouri, A.; Commandeur, S.; Rietveld, M.H.; Mulder, A.A.; Willemze, R. Replacement of animal-derived collagen matrix by human fibroblast-derived dermal matrix for human skin equivalent products. *Biomaterials* **2009**, *30*, 71–78. [CrossRef]

54. Solano, F.; Briganti, S.; Picardo, M.; Ghanem, G. Hypopigmenting agents: An updated review on biological, chemical and clinical aspects. *Pigment Cell Res.* **2006**, *19*, 550–571. [CrossRef]

55. Smit, N.; Vicanova, J.; Pavel, S. The hunt for natural skin whitening agents. *Int. J. Mol. Sci.* **2009**, *10*, 5326–5349. [CrossRef] [PubMed]

56. Tiechi, L.; Wenyuan, Z.; Mingyu, X. Studies on the Effect of TCM on Melanin Biosynthesis I. Inhibitory Actions of Ethanolic Extracts of 82 Different Chinese Crude Drugs on Tyrosinase Activity. *Chin. Tradit. Herb. Drugs* **1999**, *30*, 336–339. [CrossRef]

57. Mosmann, T. Rapid colorimetric assay for cellular growth and survival: Application to proliferation and cytotoxicity assays. *J. Immunol. Methods* **1983**, *65*, 55–63. [CrossRef]

58. Siegrist, W.; Eberle, A.N. In situ melanin assay for MSH using mouse B16 melanoma cells in culture. *Anal. Biochem.* **1986**, *159*, 191–197. [CrossRef]

59. Commandeur, S.; De Gruijl, F.R.; Willemze, R.; Tensen, C.P.; El Ghalbzouri, A. An in vitro three-dimensional model of primary human cutaneous squamous cell carcinoma. *Exp. Dermatol.* **2009**, *18*, 849–856. [CrossRef] [PubMed]

60. Kelley, L.A.; Mezulis, S.; Yates, C.M.; Wass, M.N.; Sternberg, M.J. The Phyre2 web portal for protein modeling, prediction and analysis. *Nat. Protoc.* **2015**, *10*, 845–858. [CrossRef] [PubMed]

MDPI

St. Alban-Anlage 66

4052 Basel

Switzerland

www.mdpi.com

International Journal of Molecular Sciences Editorial Office

E-mail: ijms@mdpi.com

www.mdpi.com/journal/ijms